AF538613

SP
Pvt. Ltd.

PIG PRODUCTION

P.N. Bhat, N.H. Mohan and Sukh Deo

Centre for Integrated Animal Husbandry Dairy Development,
Flat No. 205, Block No. F – 64/C9,
Sector – 40, Noida – 201 301

2010

Studium Press (India) Pvt. Ltd.

PIG PRODUCTION

ISBN: 978-93-80012-26-1
SERIES ISBN: 978-93-80012-00-1

Published by:

Studium Press (India) Pvt. Ltd.
4735/22, 2nd Floor, Prakash Deep Building
(Near Delhi Medical Association),
Ansari Road, Darya Ganj, New Delhi-110 002
Tel.: 23240257, 65150447; Fax: 91-11-23240273;
jngovil@gmail.com; jngovil@hotmail.com

Printed at:

Salasar Imaging Systems
Delhi-110035 (India)

ABOUT THE SERIES

According to the 2003 Census data, the country had 485 million (M) livestock and 489 M poultry, having the second highest number of cattle 185 M, the highest number of buffaloes 97 M, the third highest number of sheep 61 M, the second highest number of goats 124 M, the sixth highest number of camels 632 M, the fifth highest number of chickens 489 M and the fourth highest number of ducks 33 M in the world. The number of pigs in India was 13.5 M.

Livestock Sector has been playing an important role in Indian economy and is an important sub-sector of Indian agriculture. The contribution of livestock to GDP was 4.36% in 2004–05 at current prices. According to CSO estimates, gross domestic product from livestock sector at current prices was about Rs 935 billion during 1990–2000, (about 22.51% of agriculture and allied GDP). This rose to Rs 1239 billion during 2004–05 with 24.72% share in agriculture and allied GDP. But the share of livestock sector in the plan allocation hovered at around 7% of the agricultural out lay.

This sector plays an important and vital role in providing nutritive food, rich in animal protein to the general public and in supplementing family incomes and generating gainful employment in the rural India, particularly among the small, marginal farmers, land less labourers and women. Distribution of livestock wealth in India is more egalitarian, compared to land. Hence, from the equity and livelihood perspectives, it is an important component in poverty alleviation programmes. This fact however has not been appreciated by Policy planners and implementers.

The development of animal husbandry has been envisaged as an integral part of system of diversified agriculture. With its large livestock population, India has vast potential for meeting the growing need of millions, in respect of livestock products such as milk, eggs, meat and wool. This sector has the greatest potential in creating new self sustaining jobs in villages, if the knowledge base in veterinary and animal husbandry technology is improved and is used in transforming India by creating entrepreneurships, small and big, poverty can be banished from India in five years.

Livestock production systems are based on low cost agro by-products as nutritional inputs, using current day technologies. The spectacular growth of livestock products especially milk, meat, eggs and poultry meat is attributable to the several initiatives taken by Government and the organized private sector, which has primarily been driven by horizontal increase in numbers. It has been observed that with increasing income, demand for cereals is decreasing, which is causing a demand driven livestock revolution.

With the livestock sector assuming an important role in the national economy, there is a requirement to improve the present state of knowledge gathering and information dissemination. Although considerable resources have been directed towards collecting and disseminating information on basic crops, little attention has been given to collecting, analyzing and disseminating information on livestock.

It is necessary that livestock units are made financially viable through generating a service provider industry which becomes a technology catalyser in a low educated farming community. This requires several initiatives, one of the major initiative is to upgrade the knowledge base and make the information available to students, teachers, planners and farmers.

The **Studium Press (India) Pvt. Ltd.** has decided in association with **Centre for Integrated Animal Husbandry and Dairy Development (CIAH&DD)** to bring out a series of books for under graduate and post graduate scholars in Animal and Veterinary Sciences in several volumes under the chief editorship of Professor (Dr) P.N. Bhat, Former Vice Chancellor and Director of Indian Veterinary Research Institute, Izatnagar–243122 (U.P) and former Animal Husbandry Commissioner of India and Deputy Director General (Animal Science) of Indian Council of Agricultural Research, Ministry of Agriculture, Krishi Bhawan, New Delhi. The basic idea of the series is to provide first rate text books to students and scholars in developing countries based on the experiences of developing countries themselves with special focus on Southern Asia in conformity with standards laid out by regulatory agencies in India (VCI, ICAR, UGC, AICTE) and similar agencies in other developing countries.

The titles to be brought out in present series are given below.

1. Goat Production
2. Dairy co-operatives in India
3. Sheep Production
4. Buffalo Production

5. Dairy cattle production
6. Pig Production
7. Poultry Production
8. Cross breeding of cattle for improved milk production in tropics
9. Camel Production
10. Yak Production
11. Mithun Production
12. Rabbit Production
13. Laboratory Animal Production
14. Dog Management, Breeding and Health
15. Animal Biodiversity
16. Livestock Statistics
17. Livestock Economics
18. Livestock Extension
19. Breeding and Health of Equine
20. Animal Nutrition
21. Animal Physiology

The following three books have already been published:

1. Sheep Production
2. Goat Production
3. Buffalo Production

It is hoped that in the next two years, all these books will be available for the benefit of the students, teachers and professionals in the area and fill the gap which is currently wide.

Prof (Dr) Pushkar Nath Bhat
Chairman- World Buffalo Trust
and
Centre for Integrated Animal Husbandry & Dairy Development
and
Chief Editor

ABOUT THE AUTHORS

Prof. (Dr.) P.N. Bhat got his Ph.D in population genetics from Institute of Population Genetics Purdue University, West Laffayette, Indiana, USA followed by several post doctoral assignments and visiting professorship in genetics, biotechnology, livestock production systems. He returned to India and joined Punjab Agricultural University at Hisar–Ludhiana followed by several professional assignments in India and abroad. He joined as Project co-ordinator (Animal Breeding) and subsequently as Head, Division of Animal Genetics at Indian Veterinary Research Institute in 1971. He was responsible for establishing coordinated projects on cattle, buffalo, sheep, goat, pigs and poultry during 1970–74. He was founder Director of Central Institute for Research on Goats. In 1984 he became Vice-Chancellor and Director of IVRI. He joined as Deputy Director General (Animal Science) in May, 1992 and Animal Husbandry Commissioner in December, 1992. He is fellow of several National and International Science Academies.

Dr. N.H. Mohan

Dr. N.H. Mohan, presently Senior Scientist, IVRI was the first regular staff to join National Research Centre (NRC) on Pig, Asom and was closely associated with the establishment of the NRC. Dr. Mohan, before joining ICAR, had served as Assistant Professor of Veterinary Physiology, N.D.University of Agriculture and Technology, Faizabad (UP). He has acted as an investigator for about 13 research projects, including two externally funded ones. From 2003-2009 he is also an associated scientist with AICRP on Pigs and Mega Seed Project on Pigs since its inception in 2007 in the coordinating unit at NRC on pig. Dr. Mohan has authored about 23 research papers in peer reviewed international and national journals and contributed chapters to published books and edited 4 books / monographs. He has organized two training programmes for skill upgradation of staff from line agencies. Dr. Mohan was closely associated with organization of various fora for discussion on development of pig husbandry in India.

Dr. Sukh Deo

Dr. Sukh Deo got his PhD in Genetic and Animal Breeding from IVRI. He has got 33 years of experience of research, teaching, farm management and administration in animal breeding, out of which for 23 years he has worked in Livestock Production Research (Pigs) in one of the research unit of All India Coordinated Research Project on Pigs at IVRI and was responsible for management, feeding and breeding of pigs. He worked as a member of "Board of Studies" at IVRI Deemed University. He worked as Officer-in-Charge for more than 10 years in Livestock Production Research (Pigs), IVRI, Izatnagar (1984 to 1994). He has authored 30 Research Papers. He has retired as Principal Scientist, IVRI.

Other than these three main authors, many scientists have contributed in the contents of the book namely Dr. Anubrata Das, Director and Project Coordinator, NRC on Pig, Guwahati (AICRP on Pigs and NRC on Pig, Section 7.6 and 7.7 and Chapter 25), Dr. C.N Dinesh, Asstt. Professor, Dept of Animal Genetics and Breeding, College of Veterinary and Animal Science, Kerala Agriculture University, Pookode (Chapter 5), Dr. M.K.Tamuli, Principal Scientist, NRC on Pigs, ICAR, Guwahati and Sanjeev Borah, Dept of Veterinary Physiology, College of Veterinary Science, AAU, (Chapter 10), Dr. P.K.Pankaj, Scientist, NRC on Pig, ICAR, Guwahati (Chapter 15 and 24), Dr. Chintu Ravishankar, Asst. Prof., Dept of Microbiology, College of Veterinary Science, Pookode, Kerela (Chapter 18). Dr. R. Thomas, Scientist, NRC on Pig, ICAR, Guwahati and A.S.R. Anjaneyulu, Emeritus Scientist, NRC on Meat, ICAR, Hyderabad (Chapter 20) and Dr. A. Kumaresan, Sr. Scientist, LPM Divn, NDRI, Karnal (Chapter 22). We are thankful to Dr. J. Suresh, Sr. Scientist and Head, AICRP on Pigs Tirupati for sending details of pig breed along with the photograph. Also Dr. A.P. Usha Professor, Dept of Animal Genetics and Breeding, College of Veterinary and Animal Science, Kerala Agriculture University, Pookode for sending details of Ankamali breed of pigs along with photograph. We gratefully acknowledge the valuable contribution of all these scientists without which it would have been very difficult to publish this book. The photographs received from NRC on pig, Guwahati through Dr. N.H. Mohan is acknowledged.

PREFACE

The total population of pigs in the world during 2005 was 944 million heads. Major concentration of pigs was in China (465 million), Vietnam (23 million), Brazil (30 million) and India (13.5 million). Amongst the developed countries USA had 60, Germany 26, Spain 24, Canada 15, Japan 9.6, UK 5.5, Australia 2.9 million heads of pigs. (FAOSTAT–Website year, 2006). During 2005, in pig meat production also, China topped the list by producing 48 million ton, followed by USA (9 million), Germany (4.5 million), Spain (3.1 million), Brazil (3.1 million), Canada (1.9 million). India produced only 0.5 million ton during the same year.

The primary purpose of pig farming all over the world is the production of meat. In the tropics fresh pork has always been and continues to be the most important type of pig meat, but elsewhere processed meat is produced in large quantities. The advantage of pig farming is that on account of the pig's high fecundity and growth rate, pig production can yield a relatively rapid rate of return on the capital invested and can provide employment round the year. However, in India and other developing countries pig raising and pork industry are in the hands of traditional pig keepers belonging to the lowest socio-economic stratum. They have no means to undertake intensive pig farming with good foundation stock, proper housing, feeding and management. Though pigs are maintained for the production of pork, their role in progressive agriculture is not fully recognized. Although, pig meat production went up from 0.12 million tones in 1982 to 0.42 million tones in 1995, 0.47 million tones in 2000 and 0.63 million tones in 2003, it constituted only around 10% of the total meat production in the country. Apparently, the species is not being fully exploited taking into consideration its larger growth and prolificacy potential.

Several project complexes were created by the animal Husbandry Department, Govt. of India in collaboration with the State Governments, particularly of Uttar Pradesh, Rajasthan, West Bengal and Andhra Pradesh.. This was consistent with the general policy framework that poultry and pigs being fast growers, could replace local populations much faster than other livestock and at a much lower cost, to improve the livestock sector in general and livelihood of small and marginal farmers in particular.

The Indian Council of Agricultural Research has been in the forefront of pig development. All India Coordinated Research Project on Pigs was launched as back as 1970 by revamping its research programme in pig production based on review of the bacon factory development programme of the Animal Husbandry Department which would provide improved breeding material of developed breeds through genetic improvement and adaptability under India's eco-climatic conditions. It would also focused on studies on nutrition of these breeds and developed economically sustainable low cost rations using conventional and non-conventional feed ingredients. The third focus in its objective was to study the disease portfolio and how to develop a system of disease control so that the small and marginal farmers would benefit from the technology.

As a consequence of various research and development efforts, pig husbandry and pork production has gained impetus during the recent past and the concept of pig farming is changing from a minimum input enterprise to that of a semi-commercial one. This is due to the realization of its positive qualities like short generation interval, higher growth rate, higher litter size at weaning, yield of around 2 crops per sow per year, ability to convert efficiently agro-industrial and grain by-products into meat, etc.

In this book we have tried to incorporate all the relevant topics of Pig Production which would be useful for the students, researchers and entrepreneurs interested for academic, research or establishment of pig enterprises.

The first draft of the manuscript prepared by the authors has been revised by Dr. A. Bandyopadhyay, who worked very studiously and carefully on the draft. It was edited by Mrs. Aruna T. Kumar, ICAR, New Delhi. I am grateful to her for carefully going through the manuscript and preparation of index and for making several suggestions which have improved the text. The advantage of having outstanding colleagues and friends like Dr R.M. Acharya, Dr N.K. Bhattacharyya, Dr. V.K. Taneja and Dr. M.C.Sharma for referral discussion is acknowledged.

Dr J.N. Govil, Publishing-Director and Managing Editor, Researchco Books & Periodicals Pvt. Ltd., Daryaganj, New Delhi, who is the brain behind this initiative deserves special thanks for making it possible to see that this volume is brought out in time and to the expected standards. Mr Anil Jain and Mr Shrey Jain, proprietors of Studium Press (India) Pvt. Ltd. need to be complemented for sustained support. The hard work put up by Dr. A. Bandyopadhyay for proof reading the manuscript is gratefully acknowledged. The coordination work of Mr G.P. Gangadharan Pillai, Executive Assistant to Chairman, in preparation of the draft manuscript and typing by Mr Pius Joseph and Lal Babu Singh is gratefully acknowledged.

New Delhi
2nd January, 2010

Prof (Dr) Pushkar Nath Bhat
Chief Editor

CONTENTS

List of Tables

List of Figures

List of Colour Figures

CHAPTER 1

INTRODUCTION

1.1 Scope of Swine Farming in the Country

Livestock production significantly contributes to agriculture production and national health of the country. It plays vital role in supplying essential nutrients of animal origin to the large human population besides providing gainful employment to large section of the people, majority of them being small, marginal farmers and agricultural labourer. The quality and productivity of livestock is generally taken as an index of industrial prosperity of a country. In some thickly populated countries of the world like China, piggery and poultry, which give quick and successful returns have made substantial contribution towards solving problem of food shortages.

The primary purpose of pig farming all over the world is the production of pork. Secondary considerations are the production of pig skin, bristles, manure and gainful employment round the year.

In the tropics fresh pork has always been and continues to be the most important type of meat, but elsewhere processed meat is produced in large quantities, probably because pig flesh can be more effectively preserved with salt than other types of meat. Processed pork is now finding a ready acceptance among many consumers in tropical countries and consumer preferences are slowly changing everywhere as industrialization advances.

Pig skin has generally been used only for the manufacture of light leather goods and its production has been localized, as has production of pig bristles. The introduction of synthetic leather fabric and bristles will ultimately reduce demand for this product. The bristles are widely used for preparation of brushes.

Pig manure can be used as a fertilizer, to enrich the soil or for fish feed by fertilizing the ponds; for the production of biogas for electricity generation and for the culture of algae such as chlorella that is also used as fish feed. Pig manure contains on an average 0.70, 0.68 and 0.70% of nitrogen, phosphorous and potassium, respectively.

Another advantage of pig farming is that on account of the pig's high fecundity and growth rate, pig production can yield a relatively rapid rate of return on the capital invested and can provide employment round the year for the entrepreneur.

The potential of pig farming can be summarized as follows:

- The pig has highest feed conversion efficiency *i.e.* they produce more live weight gain from a given weight of feed than any other class of meat producing animals except broilers.
- The pig can utilize wide variety of feed stuffs *viz.* grains, forages, damaged feeds and garbage and convert them into valuable nutritious meat.
- They are prolific breeders with short generation interval. A sow can be bred as early as 8–9 months of age and can farrow twice in a year. They produce 6–12 piglets in each farrowing.
- Pig farming requires small investment on buildings and equipments.
- Pigs are known for their meat yield, which in terms of dressing percentage ranges from 65–80% in comparison to other livestock species whose dressing yields may not exceed 65%.
- Pork is nutritious with high fat and low water content and has got better energy value than that of other meats. It is rich in vitamins like thiamin, niacin and riboflavin.
- Pig manure is widely used as fertilizer for crop farms and fish ponds.
- Pig stores fat rapidly for which there is an increasing demand from poultry feed industry, soap industry, paints and other chemical industries.
- They produce bristles which have many uses.
- Pig farming provides quick returns since the marketable weight of fatteners can be achieved within a period of 6–8 months.
- There is good demand from domestic as well as export market for pig products such as pork, bacon, ham, sausages etc.
- Pig farming provides an opportunity to integrate animals farming with poultry cum fish culture.

1.2 Contribution of Pigs

1.2.1 Contribute food/meat

(a) The food supplied by the pork is of the highest quality.

(b) Pork contains 15 to 20% rich quality protein, on a fresh basis. The pork protein provides all the essential amino acids, including lysine and methionine.

(c) Pork is a rich source of energy, the energy value depends largely upon the amount of fat it contains.

(d) Pork is a rich source of several minerals but it is especially good as a source of phosphorus and iron.

(e) Pork is the richest source of the important B group of vitamins, especially thiamin, riboflavin, niacin and vitamin B-12.

(f) Pork is highly digestible, about 97% of meat proteins and 96% fats are digested.

1.2.2 Convert inedible feeds into valuable products

Pigs are better adapted than any other class of livestock in utilizing many wastes and by-products that are not suited for human consumption.

1.2.3 Aid in maintaining soil fertility

Swine helps in maintaining fertility of the soil at the farms as is the case with other livestock, provided their manure is properly utilized at the farm.

1.2.4 Serve as an important companion of grain production

Swine provide a large and flexible outlet for the year-to-year changes in grain supplies. When there is a large production of grain, (i) more sows can be bred to farrow, and (ii) market pigs can be carried to heavier weights. On the other hand, when grain prices are high, (i) pregnant sows can be marketed without too great a sacrifice in price, (ii) market pigs can be slaughtered at lighter weights, and (iii) the breeding herd can be maintained by reducing the grain that is fed and increasing the pasture of ground hay. Thus swine give elasticity and stability to grain production system.

1.2.5 Supplement other enterprises like dairying and crop farming

Supplement dairying

Where cream or butter is marketed, rather than whole milk, the skim milk or buttermilk is available for feeding. Swine supplement the dairy enterprise admirably. Finer protein supplement in the form of dairy by-products for swine can be obtained which will bring handsome returns.

Supplement crop production

Pig also supplements crop production through hogging down certain crops. In addition to doing own harvesting, maximum fertility value of the manure is conserved. This contribution of pigs is valuable especially where crops have been damaged or lodged, where harvesting labour is not available or where crop prices are disastrous.

1.2.6 Slaughterhouse by-products

In western countries maximum utilization of slaughterhouse and meat factory waste and by-products are made, which has enabled them to improve their economic return from such units, as they are able to sell their finished products at a much cheaper rate. In India most of these materials are generally wasted and full benefits are not derived from them. Proper utilization of these products can substantially contribute towards improving the economy of these units provided care in collection, preservation and facilities for their proper utilization are made available. Wastes and by-products which can effectively be used are:

(a) Blood
(b) Bone
(c) Meat: condemned parts and organs
(d) Fat
(e) Viscera
(f) Lung, liver, kidney, ears, head
(g) Hooves

1.2.6.1 *Blood*

Dried blood is a good source for fertilizer, and it contains nitrogen which is required for growth of plants. It is also used as manure in tea gardens, coffee and rubber plantation and agriculture farms. Fresh blood, if properly collected, can be converted into blood meal by dry rendering or blood dryer.

1.2.6.2 *Bone*

Bone meal is made out of skeletal bones, head bones, feet, ribs etc., from which meat had been scraped and bone meal is produced in dry rendering mill or bone digester. While producing bone meal, some technical fat is also produced (about 10%). It is used in livestock/poultry feed.

1.2.6.3 *Meat cuttings and condemned meat*

Meat cuttings and condemned meat after steaming and drying, is converted into meat meal. It is mostly used as a supplement for the livestock feed. The drying rate is 4:1.

1.2.6.4 *Fat*

Fat available from slaughtered animal is rendered and converted into good quality edible lard and canned and sold at good price. The other inferior quality fat after rendering is utilized by soap manufacturers.

1.2.6.5 *Casings and gut*

After stripping of intestine of all food material and then washing and cleaning, they are processed in automatic gut making machine for making casing which is utilized for sausage making. Roughly 0.4 rings of grade A per animal can be produced.

1.2.6.6 *Viscera*

Viscera can be utilized for animal feed after cleaning and rendering.

1.2.6.7 *Glands*

Glands like pancreas, pituitary, and ovaries are collected and used for manufacture of pharmaceuticals. It requires proper collection and preservation in proper manner under hygienic conditions.

1.2.7 Manure

Pig manure may be sun dried and sold as a fertilizer. It can also be used for the production of methane gas or for the culture of chlorella. In many places pig farming is associated with fish pond culture. Effluent from the piggeries is run in to fish ponds as it is believed that it improves the growth of micro-organisms and plants on which the fish feeds. A mature pig produces about 14 kg of manure per day.

1.2.8 Bristles

Pig bristles are used for manufacture of brushes.

1.3 Pig Production in Developing Countries

In India and other developing countries pig raising and pork industry are in the hands of traditional pig keepers belonging to the lowest socio-economic stratum with no means to undertake intensive pig farming with good foundation stock, proper housing, feeding and management. They are compelled to follow old and primitive methods with common village hogs which could properly be designated as scrub animals. The small sized animals do not have any definite characteristics, grow slowly, produce small litters and the meat is of inferior quality. The poor farmers cannot afford to provide the minimum attention in their managerial affairs and as such most of the time the animals are left loose to pick up feed stuffs from the waste areas of neighboring localities. The most unhygienic and unimpressive life for the indigenous pigs creates an aversion to such animal products in the minds of the majority of Indians. But they are, nevertheless, raised as a very essiential part of their diet and has immence value for the owner.

Though pigs are maintained for the production of pork, their role in progressive agriculture is not fully recognized. Pig farming is adapted to both diversified and intensive agriculture. Pigs convert inedible feeds, forages, certain grain by-products obtained from mills, damaged feeds and garbage into valuable nutritious meat. Most of these feeds are either not edible or not very palatable to humans. The faeces of pigs are useful in maintaining soil fertility as about 80% of the fertilizing value of the feed is excreted in the faeces and urine.

During the Second and Third Five Year Plans, however, a coordinated programme for piggery development was taken up in some states in India. The scheme involved establishment of bacon factories, regional pig breeding stations and pig breeding farms/units and piggery development blocks. Some exotic breeds of pigs, *viz.* Landrace, Large White Yorkshire, Tamworth and Hampshire were introduced at different pig breeding farms. The major objective was to acclimatize and use them for upgrading the native pigs.

As a consequence of various research and development efforts, pig husbandry and pork production has gained impetus during the recent past and the concept of pig farming is changing from that of a zero input enterprise to that of a semi-commercial one. This is due to the realization of its positive qualities like short generation interval, higher growth rate, higher litter size at weaning, yield of around 2 crops per sow per year, ability to convert efficiently agro-industrial and grain by-products into meat, etc. Although pig meat production went up from 0.12 million tonnes in 1982 to 0.42 million tones in 1995 and 0.47 million tonnes in 2000, (Table 1.1 and Table 1.2) it constituted only around 10% of the total meat production in the country. Apparently, the species is not being fully exploited taking into consideration its larger growth and prolificacy potential.

Table 1.1 Swine Meat Production in India

Qty in 000 MT

	Year				
	1985	**1990**	**1995**	**2000**	**2003**
Quantity	85	360	420	578	630

Source: FAO production year book and FAOSTAT website.

Table 1.2 Export of Swine Meat from India 2005–06 to 2007–08

Qty in MT, value in Lakh

2005–06		2006–07		2007–08	
Quantity	**Value**	**Quantity**	**Value**	**Quantity**	**Value**
320.70	207.38	1523.47	865.30	1710.89	2463.69

Source: DGCIS annual data.

CHAPTER 2

CLASSIFICATION, ORIGIN AND DOMESTICATION

2.1 Origin and Domestication of Pigs

The wild boar is widespread in Eurasia and occurs in Northwest Africa; the existence of at least 16 different subspecies has been proposed (Ruvinsky and Rothschild 1998). Domestication of the pig is likely to have occurred first in the near east and may have occurred repeatedly from local populations of wild boars (Bokonyi 1974). However, it is not yet established whether modern domestic pigs showing marked morphological differences compared with their wild ancestor have a single or multiple origin. Darwin (1868) recognized two major forms of domestic pigs, a European (*Sus scrofa*) and an Asian form (*Sus indicus*). The former was assumed to originate from European wild boar, while the wild ancestor of the latter are unknown. Darwin considered the two forms as distinct species on the basis of profound phenotypic differences. It is well documented that Asian pigs were used to improve European pig breeds during the 18th and early 19th centuries (Darwin 1868; Jones 1998) but to what extent Asian pigs have contributed genetically to different European pig breeds is only now being investigated. In a recent study the divergence between major European breeds and the Chinese Meishan breed was estimated using micro satellite markers (Paszek *et al.* 1998). Limited studies on mitochondrial DNA (mtDNA) have indicated genetic differences between European and Asian pigs but no estimate of the time since divergence has been provided (Watanabe *et al.* 1986; Okumura *et al.* 1996).

Archeological evidence indicates that swine were first domesticated in the Eastern India and South-eastern Asia, in the Neolithic period or New Stone Age.

Beginning about 9000 BC, in the eastern part of New Guinea (now known as Papua New Guinea) island in the Pacific Ocean just North of Australia; and about 7000 BC, in Jerico, which lies in Jordan vally, north of the Dead Sea.

The domestication of the European wild boar came independently and later than the East Indian pig.

The East Indian pig was taken to China about, 5000 BC. Chinese pigs were taken to Europe in the last century, where they were crossed on the descendants of the European wild boar, thereby fusing the European and Asiatic strains of *Sus indicus* and forming the foundation of present day Euro American breeds.

2.2 Place of Pigs in Animal Kingdom

The wild pigs belong to **Class *Mammalian*** which is warm-blooded, hairy animals that produce their young alive and suckle them for a variable period on a secretion from the mammary glands. They belong to **Sub-class *Eutheria*, Order *Artiodactyla*** (even toed, hoofed animals) and **Family *Suidae***, the family of non-ruminant, artiodacty ungulates, consisting of wild and domestic swine. In modern classification, they exclude the peccaries, which belong to the family ***Tayassuidae*. Genus Sus *Linn*,** the typical genus of swine includes several wild species besides the domesticated pig. Some of them are: **Eurasian wild boar** (***Sus scrofa*,** distributed in Europe, North Africa and Asia; the Eurasian wild boar will cross freely with domestic swine and the offspring are fertile; **Sus scrofa cristatus;** the Indian wild boar Sus scrofa Andamanesis is native of the Andaman Island; **Sus scrofa salvanus** is found in the parts of Himalayas and **Sus scrofa vittatus,** found in south Indian mountain ranges. **Sus scrofa barbatus** is native to Malaysia.

2.3 Purpose of domestication

Domestication of the pig is likely to have occurred first in the near east and may have occurred repeatedly from local populations of wild boars. By seeing the characteristics of the pigs as a meat animal, it was felt necessary to domesticate the pigs to exploit its full potential. Pigs are raised solely for meat production. They are efficient converters of feed into meat, quick to multiply and can fit to diverse system of management. Tethering of animals in the field or close to the home is practised widely to collect dung for crop production. Pigs are mainly fed with kitchen wastes and rice bran and occasionally purchased concentrates are given. There is practically no investment in housing. All these factors influenced people to domesticate wild pigs particularly to benefit the poor community of tribals.

2.4 The Worldwide Distribution of Pigs

The pig is omnivorous and in some respects competitive with man for food, but is also very useful utilizer of the by-products and wastes from human feeding. Thus pigs are usually most numerous where human food is cheap and plentiful and where there are large quantities of by-products or offal available. The size of the pig population of any given region also, depends upon other factors, *e.g.* the climate, only a small number of pigs being found in the arid areas of the world and the social and religious beliefs of the indigenous people, there being few pigs in countries with a predominantly Muslim population.

Today there is a very wide distribution of wild and feral pigs and it is generally believed that all domesticated breeds have been derived in one way or another from two wild types: *Sus vittatus*, synonyms *S. scrofa cristatus*, the wild pig of east and southeast Asia, and *S. scrofa*, the present European wild pig, which may also have existed during the past in western Asia.

From 1770 to 1870 Chinese pigs were introduced into Britain and crossbred with the Old English pigs. It is believed that these imported pigs originated mainly from the Canton area. They were mostly white in colour, but a few were pied or black; possessing a wide head and a dished face, short, erect ears, short legs with light hams and a drooping back. Some Siamese pigs were also imported in Britain at about the same time. Later in 1830, pigs of the Neopolitan breed, black with no bristles, were also introduced into Britain and crossbred with local types. It was the crossbreeding of Chinese, Siamese and Neopolitan pigs with the Old English pig that produced the ancestors of the modern British breeds.

In early colonial days in America, pigs of the Old English type were imported, as were pigs from continental Europe. Later these pigs were crossed with improved British breeds and with pigs from southeast Asia and other parts of the world. These became the ancestors of American breeds of present days.

Domestic pigs are scare in the African countries inhabited by Hamintic and Semitic peoples, and in the Congo. There are, however, domestic pigs in the Cameron Republic and in other countries in the West African Coast.

The distribution on a continental basis of the world's pig population is shown in Table 2.1. It will be seen that approximately one-fifth of the world's pig population are in the tropics and that the pig population in the tropics is increasing more rapidly than that in the mid latitude regions.

In India, North Eastern Region (NER) is characterized by a high proportion of tribal people for whom pig keeping is integral to their way of life. Assam is the major state and it has the biggest pig herd (1.54 million). The increasing demand for animal-source foods in the NER and in India generally matching with current low productivity of the NER pig population, suggests that well targeted interventions to improve pig production could deliver significant livelihood benefits for the tribal and other marginalized groups in the region.

2.5 Importance of Pig Farming and its Contribution to National Economy

The pig population of the country is 13.52 million as per the 2003 livestock census and constitutes around 1.30% of the total world's population. The state-wise pig populations are given in Table 2.2. During 2001–02 the production of pork and pork products were estimated to be 630 thousand MT with 3.03% growth rate in last decade. Indian share in world pork production moderately increased from 0.53 in 1981 to 0.63 % in 2002. The contribution of pork products in terms of value, works out to 0.80% of total livestock products and 4.32% of the meat and meat products. The contribution of pigs to Indian exports is very small. About 1711 tonnes of pork and pork products were exported during 2007–08. The value of pork and pork products exported was Rs 2464 lakh.

Table 2.1 World Pig Population

Unit: 1000

Country	1992	1999	2000	2001	2002	Annual growth rate (%) 1992–02
Developing countries						
Southeast Asia						
1. Cambodia	2043.0	2189.3	1933.9	2114.5	2105.4	0.2
2. Indonesia	8135.0	7041.8	5356.8	5867.0	6000.0 F	-4.2
3. Lao PDR	1560.5	1320.0	1425.0	1425.9	1425.9 F	-2.0
4. Malaysia	2842.5	1954.9	1807.6	1972.5	1824.2	-5.5
5. Myanmar	2630.0	3715.0	3914.3	4138.9	4498.7	5.7
6. Philippines	8021.9	10397.0	10712.9	11063.1	11652.7	4.1
7. Thailand	4655.5	6369.7	6558.1	6688.9	6688.9 F	3.7
8. Viet Nam	13891.7	18885.8	20193.8	21740.7	23169.5	4.9
South Asia						
9. Bangladesh						
10. Bhutan	44.5	53.0 F	48.0 F	45.0 F	41.4	-0.1
11. India	12788.0	16500.0 F	17000.0 F	17500.0 F	18000.0 F	3.5
13. Nepal	599.0	825.1	877.7	912.5	934.5	5.2
15. Sri Lanka	90.8	73.6	70.8	68.3	67.0 F	-3.5
Central Asia						
16. Kazakhstan	2794.0	891.8	984.2	1076.0	1123.8	-11.2
17. Tajikistan	128.0	1.2	1.1	0.6	0.7	-43.6

Table 2.1 *(Contd...)*

Unit: 1000

Country	1992	1999	2000	2001	2002	Annual growth rate (%) 1992–02
18. Uzbekistan	653.6	80.0	80.0	89.0	90.0 F	-20.9
Other Asia						
19. China	379910.5	429201.6	437541.2	454410.0	464695.0	1.7
20. DPR Korea	5000.0 F	2970.0	3120.0	3137.0	3152.0	-3.2
21. Iran (Islamic Rep. of)	0.0	0.0	0.0	0.0	0.0	0.0
22. Mongolia	83.3	21.7	14.7	14.8	15.0 F	-13.2
23. Rep. of Korea	5462.7	7863.7	8214.4	8719.9	8811.0*	5.1
24 Pacific Islands	452946.0	512473.3	521971.4	543237.8	556552.1	
Developed countries						
36. Australia	2792.4	2626.0	2433.0	2763.0	2912.0*	0.1
37. Japan	10966.0	9879.0	9806.0	9788.0	9612.0	-1.2
38. New Zealand	411.1	368.9	368.8	354.5	358.1	-1.8
Sub-total	14169.5	12873.9	12607.8	12905.5	12882.1	-1.0
Asia and pacific*	467.0	525.0	534.0	556.0	569435.1	1.7
	116.4	348.3	580.1	144.3		
Rest of world	401.0	377.0	373.0	368.0	371586.6	-0.7
	259.2	544.7	700.5	694.1		
World	868.0	902.0	908.0	924.0	941021.7	0.7
	375.6	892.9	280.6	838.5		

F=FAO estimates, *=Unofficial figures

Source: FAO 2003

Table 2.2 State Wise Pig Population in India (2003)

Unit: 1000 heads

Sl No.	States/U.T.s	Total
1.	Andhra Pradesh	570
2.	Arunachal Pradesh	330
3.	Assam	1543
4.	Bihar	672
5.	Chhattisgarh	552
6.	Goa	87
7.	Gujarat	351
8.	Haryana	120
9.	Himachal Pradesh	3
10.	Jammu and Kashmir	2
11.	Jharkhand	1108
12.	Karnataka	312
13.	Kerala	76
14.	Madhya Pradesh	358
15.	Maharashtra	439
16.	Manipur	415
17.	Meghalaya	419

Table 2.2 *(Contd...)*

Unit: 1000 heads

Sl No.	**States/U.T.s**	**Total**
18.	Mizoram	218
19.	Nagaland	644
20.	Orissa	662
21.	Punjab	29
22.	Rajasthan	338
23.	Sikkim	38
24.	Tamil Nadu	321
25.	Tripura	209
26.	Uttar Pradesh	2284
27.	Uttarakhand	33
28.	West Bengal	1301
Union Territories		
29.	Andaman and Nicobar Islands	52
30.	Chandigarh	0
31.	Dadra and N Haveli	3
32.	Daman and Diu	0
33.	Delhi	28
34.	Lakshadweep	0
35.	Pondicherry	1
All India		**13519**

CHAPTER 3

PRODUCTION SYSTEMS AND POPULATION TREND

3.1 Pig Production System

3.1.1 Pig production in India and developing countries

In India and other developing countries pig raising and pork industry is in the hands of traditional pig keepers belonging to the low socio-economic stratum with no means to undertake intensive pig farming with good foundation stock, proper housing, feeding and management. The poor farmers cannot afford to provide the minimum attention to management and as such most of the time the animals are left loose to pick up feed from the waste areas of neighboring localities. This system can be described as free-range scavenging. This is a low-input/low-output extensive system whose main purpose is to guarantee subsistence and household's emergency funds as coping strategy, whilst also supplying the farmers food security with some meat from time to time. There is no major investment interms of money, and it is typical of small farmer mixed holdings. In traditional farming investment remains mostly restricted to time and physical labour.

The main constraints with scavenging pigs are the high rates of piglet loss, and slow growth rates. Pigs kept in a free-range system will not grow quickly, because they expend a lot of energy in their scavenging activities. Worm infestation is also an important problem resulting in slower growth rates. However, it is also important to recognize that with low levels of inputs this systems under certain situations is the only sustainable method of production for these marginal people.

It was observed that in hilly areas of North East India the pig farmers constructed their pigsty with locally available materials like bamboo and woods, located in road side slope area with a raised platform above 2–3 feet from the ground. The floor space per adult was inadequate (average 12 sq.ft) in majority (97%) of the farms. The farm equipments included mainly iron vessel (*Kerahi*) for boiling feeds, empty mustard oil tin (modified form) or cut piece of woods or bamboos, vehicle tyers as feeding trough. Further it was recorded that supply of water mostly dependent either on rain or nearby streams. Separate water storage facility for pigs and electricity were absent in most of the farms.

During the Second and Third Five Year Plans, however, a co-ordinated programme for piggery development was taken up in some states in India. The scheme involved establishment of bacon factories, regional pig breeding stations and pig breeding farms/units and piggery development blocks. Some exotic breeds of pigs, *viz.* Landrace, Large White Yorkshire, Tamworth and Hampshire were introduced at different pig breeding farms. The major objective was to acclimatize these breeds and use for upgrading the native pigs.

As a consequence of various research and development efforts, pig husbandry and pork production has gained impetus during the recent past and the concept of pig farming is changing from that of a scavenging to that of a semi-commercial one. This is due to the realization of its positive qualities like short generation interval, higher growth rate, higher litter size at weaning, yield of around 2 crops per sow per year, ability to convert efficiently agro-industrial and grain by-products into meat, etc. Although, pig meat production went up from 0.12 million tones in 1982 to 0.42 million tonnes in 1995 and 0.47 million tonnes in 2000, it constituted only around 10% of the total meat production in the country. Apparently, the species is not being fully exploited taking into consideration its larger growth and prolificacy potential.

3.1.2 Pig production in developed countries

Pig production in developed countries has become an increasingly specialized activity. Two main factors are involved, on the one hand, market segments require exact carcass specifications, and on the other, the economics of scale resulting from intensive production units.

As a result, production is increasingly being concentrated in the hands of specialist and large scale producers capable of controlling genetics and formulation of feed to produce carcasses that the markets demand.

Through intensive pig keeping, the type and scale of production aims at producing meat for the market efficiently and profitably, usually with a large numbers of pigs. The system requires significant inputs of both time and money, with careful calculation of the costs and the resulting benefits. An important component is the specialization of jobs, and the specialized knowledge required to operate such an enterprise successfully.

The pig also differ considerably, with intensive systems specializing in varieties that have been bred specifically for production. In practice, this also means that these breeds require significantly greater inputs in terms of health care, feeding and nutrition, as well as general livestock husbandry and management.

3.2 Population Growth

3.2.1 Trend in pig population (India)

During the year 1992 the total pig population of India was 12.79 million, which increased to 18 million in 2002 showing an increase in 3.5% growth as against world growth rate of 0.7 % during the same period. Table 3.1 indicates the world pig population vis-a-vis India.

3.2.2 Trend in pig population (World)

Over the last half century, the world pig population have been trebled from about 282 million in 1935–37 to 868 million in 1992. However, from 1979–81 to 1992 the rate of increase has showed down to 11% (FAO, 1992). Hence, the pattern of change has been far from uniform. An analysis of the data presented in Table 3.1 shows that from 1979–81 to 2002, pig numbers in countries designated by FAO as "developing" increased by 25% whereas there was 58% decrease in pig numbers in 'developed' nations. Mainly due to expansion of pig numbers in China and other Far Eastern countries, developing countries as classified by FAO, now account for more than 62% of the world pig population.

However, the efficiency gap between the developed and the developing nations is closing; according to data from FAO (1992), between 1979–81 and 1992, productivity of the developing nations increased from 53 to 66% of that of the developed nations. There was 33% improvement in the productivity measures (pig slaughtered/pig population) for the developing nations (from 0.69 in 1979–81 to 0.92 in 1992). *e.g.*, for each pig in China, 0.64 pigs were slaughtered in 1979–81, but in 1992, there were 0.95 pigs slaughtered/pig population. As another measure of productivity 8.86 pigs were marketed per sow in China in 1981, compared to 13.0 pigs per sow in 1991.

Table 3.1 Pig Population

Unit 1000 heads

Country	1992	1999	2000	2001	2002	Annual growth rate (%)
India	12788.0	16500.0	17000.0	17500.0	18000.0	3.5
Asia and pacific	467116.4	525348.3	534580.1	556144.3	569435.1	1.7
World	868375.6	902892.9	908280.6	924838.5	941021.7	0.7

Source: FAO, 2003.

Table 3.2 The Change in Pig Numbers in Developed and Developing Nations

	Pig numbers (in millions)		
	Total	Developed nations	Developing nations
1979–81	779	335	444
1990	856	341	415
1991	864	338	526
1992	864	316	539
2002	868	141	556

Source: FAO, 1992.

3.2.3 Factors affecting population

Growth of population depends up on number of factors. The optimum growth of population can be achieved if the following factors are taken care of:

1. Good animal husbandry practises
2. Controlling diseases
3. Proper nutrition
4. Good housing
5. Proper selection of breed conducive to the prevailing environment
6. Improved marketing facilities of the poor pig raiser
7. Improving the market demand of pork meat

All these factors have been discussed in detail in their respective chapters in the book.

3.2.4 Trend in pork production

The globalization of the swine industry has caused major changes in national and international swine production over the past decade and these changes are likely to continue. The easing of international trade barriers has meant that less competitive countries are under increasing pressure from imports by more efficient countries with lower cost of production.

3.2.4.1 *Consumption of pork*

More pork is consumed than any other meat in the world. In 1998 it represented 39% of the world's total meat consumption compared to 26.5% for beef and 28% for poultry. World pork consumption increased from 34 to 88 million tonnes per year between 1970 and 1999. World population expansion undoubtedly contributed to a substantial portion of this, but average per capita intake also increased from 10 to 14.3 kg/year (Black, 2000). Pork consumption varies widely among countries and regions with per capita intake in 1998 ranging from 2 kg/year in many African countries to 60 kg/year in Germany and Spain. During the same year consumption in the US was 30.7, in Brazil 9.3 and in Australia 18.8 kg/year. During this period worldwide consumption of beef remained fairly stable at 9 to 10 kg/year, but consumption of poultry increased form 4.4 to 10.4 kg/year.

3.2.4.2 *Changes in pig performance*

During the 1980s there was a major global emphasis in production of leaner pork and more efficient pigs that met the market demand for less fat and more 'healthful' meat. Under intense genetic selection for fast growing lean animals there were sizable increases in growth rate and feed efficiency.

In the UK feed conversion efficiency improved from 3.6 in 1960 to 2.69 in 1990 (Close, 1999). Between 1990 and 1999 there was only a small improvement in growth and feed efficiency. The growth rate in Australia from birth to slaughter increased from 500 g/day in 1960 to 700 g/day in 1990. Again, as in the UK, improvements in performance during the 1990s were small.

Even with these improvements, there is still a significant difference between the performance of pigs raised in commercial operations and those raised under ideal experimental conditions and environments. Swine raised in typical commercial environments grow 15 to 25% more slowly, are fatter, and are not as efficient as pigs of the same genotype grown in individual pens and in a controlled environment (Black and Carr, 1993; Morgan *et al.*, 1998).

There is a significant opportunity for continued improvement in commercial operations that would improve the competitiveness of the swine industry relative to other forms of meat protein.

During 2005, China has become the world's leading meat producer (48.27 %) followed by USA (9.42%), Germany (4.51%), Brazil (3.12%), Spain (3.11), Vietnam (2.30 %), France (2.26%), Poland (1.96), Canada (1.92%), Mexico (1.11%) and India (0.50%).

Table 3.3 Top 11 Pig Producer Countries Worldwide (FAO Pig Data, Year 2005)

Countries	Live pig heads (million)	%	Pig meat tonnes (million)	%	Slaughter pig heads (million)
China	488800.000	50.78	48117.790	48.27	630309.610
USA	60644.500	6.30	9392.000	9.42	103691.500
Brazil	33200.000	3.45	3110.000	3.12	38400.000
Vietnam	27434.895	2.85	2288.315	2.30	33000.000
Germany	26857.800	2.79	4499.991	4.51	48251.550
Spain	24884.000	2.58	3100.718	3.11	38029.666
Poland	18112.380	1.88	1955.500	1.96	22525.704
France	15020.198	1.56	2257.000	2.26	24885.000
Canada	14675.000	1.52	1913.520	1.92	22319.800
Mexico	14625.199	1.52	1102.940	1.11	14307.996
India	14300.000	1.49	497.000	0.50	14200.000

Table 3.4 Pig Meat Production in Different Regions of the World (in million tones)

World region	2003	2004	2005
Africa	07.8204	08.0714	08.0388
	(0.80)	(0.81)	(0.78)
America	17.1292	17.5830	17.6954
	(17.47)	(17.56)	(17.22)
Asia	54.5106	56.6757	59.7088
	(55.61)	(56.62)	(58.10)
Europe	25.6063	25.0386	24.5617
	(26.12)	(25.01)	(23.90)
World	98.0281	100.1046	102.7701

Figures in parenthesis indicate % of the world production.

Source: FAO, Stat 2006.

Table 3.5 World production of meat including pork, beef and poultry

(million tones)

Year	Total meat	Pig meat	Beef	Poultry
2003	253.48	9.858	5.830	6.580
2004	257.50	100.39	5.870	6.772

Source: FAO, December 20, 2004.

CHAPTER 4

BREEDS OF PIGS

Pig breeds useful in tropical environments may be classified in several ways: firstly according to their utility and the major products that they produce, *i.e.* pork meat, bacon, lard, pig skin, bristles or manure; secondly with regard to their skin colour that can be black, some other colour, or white, as this characteristic determines in some respects how they should be managed; and finally, whether they are developed breeds of worldwide importance that has waned but may still be useful in the tropics, developed breeds of local importance and undeveloped indigenous breeds that could become extinct.

Porter (1993) has published a comprehensive and useful guide to the pig breeds of the world while King (1991) has attempted to asses the relative importance of the breeds and their adaptability.

Upgraded indigenous stock developed by crossing them with imported exotic stock of different grades is available in the country at organized piggeries as well as with private farms in rural areas in different regions of the country and they are thriving well. Their characteristics vary depending upon the degree of exotic blood level and genotype composition, which is exhibited in physical characteristics as well as in economic returns of the upgraded pigs.

The early domesticated pigs descended from wild forest pigs of Europe and Africa, which were short, heavy shouldered, razor backed with relatively large head, neck and poorly developed loin and have been resembling wild boars of Medieval England. In middle ages, selection was based on length, depth and overall size providing an animal with better balance of hind quarter to fore and with shoulders and head remaining large. This situation continued in UK till later half of

18th and first half of 19th century when stock from China and Mediterranean was imported, but overall conformation was not greatly affected, although growth rate was improved. Later National Pig Breeders Associations were formed in most developed world when they played important role in improvement of type and conformation and selection of more prolific strains of pigs

4.1 Indian Subcontinent

4.1.1 Indian breeds

In India four kinds of pigs are found *viz.* wild pigs, domesticated or indigenous pigs, exotic breeds and crossbred (upgraded stock of pigs) pigs. In order to raise the productivity of indigenous pigs and thereby obtain better meat yield, exotic breeds *viz.* Large White Yorkshire, Middle White Yorkshire, Landrace, Large black, Hampshire, Berkshire, Wessex Saddleback, Duroc and Charmukha were imported for cross breeding work from developed countries such as UK., New Zealand, Australia, USA and Russia.

4.1.1.1 *Wild pigs*

Three strains of wild pigs are present in different agro-climatic conditions of India.

1. *Sus scrofa cristatus*, are commonly found in low jungles or forests of Himalayas up to an elevation of 4500 ft. The animal measures about 1.5 m in length from nose to vent, and 71–91 cm height at shoulder. It exceeds 136 kg in weight. The wild pig has a long snout, short ribs and long legs. Males are larger than females. Colour of the animal is rusty grey when young and as it advances in age, it becomes dark chestnut brown with its hairs tinged with grey at the extremities.
2. *Sus salranius* are distinctive in possessing a sparse coat and a mane of black bristles running from the neck down to back. It has no wooly under coat. The tusks are well developed in the males, both the upper and lower tusks curving outwards and projecting from the mouth. They are extremely active and when provoked may also attack human beings.
3. *Sus scrofa Andamanesis* and Sus scrofa nicobarians are the wild boar found in the forest of Andaman Nicobar Islands.

The wild pigs are poor producer of piggery products. The meat is however delicious.

4.1.1.2 *Domesticated or indigenous pigs*

They are a distinct group and formed due to domestication of wild pigs at different places through both natural and artificial selection and hence they have different names.

These pigs differ in their characteristics and colour from region to region within the country depending on the topography and climatic conditions. Different colour pattern are found *viz.* black, brown, rusty grey and even an admixture of any two colours and they differ considerably in size and appearance. They have long face tapering towards nostrils, head and shoulders are heavier in comparison to hind quarters, back is slightly arched and rump drooping, ears are small or medium sized. Tail reaches nearly to hocks which has a tuft of hair. Hairs on neck and back are thick and bristly. Females have 6–12 teats. Adult pigs weigh up to 150 kg.

4.1.1.3 *Pigs of Indo-Gangetic plain (Izatnagar strain)* (*Plate 1*)

The animals of this group are distributed in a wide area covering almost entire northern and north western India. The regions include the plains of Uttar Pradesh, Bihar, Madhya Pradesh, Punjab, Haryana and areas of Himachal Pradesh. The animals are of large size may be owing to the abundance of the feed/fodder available to the animals. The adult animals may weigh up to more than 160 kg. The body colour varies from rusty grey to brown to black. The hairs of the animal on the neck and part of the back are bristly thick and long and those on flank and sides are comparatively thinner and shorter. Average litter size is 7–8 and litter weight at birth is about 5–5.5 kg. The average birth wt of piglet is 0.79 kg. The pigs have good reproductive performance having number of service/conception only 1.25. The carcass characteristics like average slaughter weight is 48 kg having hot carcass weight about 35 kg. The average dressing % is 72. The carcass length is about 60 cm and backfat thickness 2.25 cm.

4.1.1.4 *Jabalpur strain*

The animals of this strain are black or brown in colour. They have tapering head, head and shoulder heavier then hind quarter, tail almost reaches nearly to hocks. Bristles are thick on the neck and back. Growth rate is slow. Average litter size is 6.74 and litter weight at birth is about 5–5.5 kg. The average birth weight of piglet is 0.70 kg. The pig's reproductive performance is not as good as Izatnagar strain (average number of service/conception 2.1). Average litter size at weaning is 5.52 and average litter weight at weaning is 38.62 kg. The carcass characteristics like average slaughter weight is 45 kg having hot carcass weight about 31 kg. The average dressing % is 68. The carcass length is about 54 cm and back fat thickness 2.06 cm.

41.1.5 *Khanapara strain*

The strain of pigs are availble mainly in Assam and other adjoining states. The animals are mainly black and may have an admixture of light brown. There is always a row of coarse and straight bristles starting from the neck to back like that of wild pigs. Average litter size at birth is 4.84. The average birth wt of piglet is 0.70 kg. (average number of service/conception is 2.65). The average birth wt of piglet is 0.90 kg. Average litter size at weaning is 3.10 and average litter weight at weaning is 21.32 kg. The carcass characteristics like average slaughter weight is 23 kg at 35 weeks of age having hot carcass weight about 18 kg; the average dressing % is 71; the carcass length is about 44 cm, and backfat thickness 2.25 cm.

4.1.1.6 *Gannavaram (Tirupati) strain (Plate 1)*

This strain of pigs popularly known as local pigs/country pigs are scavengers by nature. As far as the breed characters are concerned, the body colour is by and large black and occasionally presence of white patches on legs and snout are also seen. The body is entirely covered with thick and strong bristles which are more prominent on the mane region. Erect ears is also a common feature. The face is long and narrow with strong snout suitable for digging the soil. Tusks are more prominent in male adult animals compared to females. Animals are highly active and ferocious by nature. It is a common sight to see these pigs scavenging in the streets and taking shelter in middy soils including small drainage ponds to beat the summer heat. Average number of teats present in a sow is 10 to 12.

The animals are found in the districts of Andhra Pradesh and southern or peninsular region of Karnataka, Kerala, Tamil Nadu, and southern Maharashtra. They have a back coat with white patches on the body. However, the rusty grey specimens are also not uncommon. The adults may weigh from 40–70 kg and from 90 cm to one meter in their body length. The tail reaches the hock and has a tuft of hair. Average litter size at birth is 6.58. The average birth wt of piglet is 0.72 kg. The pig's reproductive performance is very good having no. of service/conception 1.04. The average birth wt of piglet is 0.72 kg. Average litter size at weaning is 5.31 and average litter weight at weaning is 45.28. The carcass characteristics like average slaughter weight is 48 kg having hot carcass weight about 37 kg; the average dressing % is 78; the carcass length is about 55 cm and backfat thickness 1.09 cm.

The information on all these four strains was taken from the All India Coordinated Research Project (AICRP) on Pigs for the research and development during 1971 to 1992.

4.1.1.7 *Ankamali (Plate 1)*

This breed is the domesticated native pigs of Kerala and are black with white patches, the overall appearance being rusty-grey. Weight 40–70 kg with a length of about 91 cm. Sow produce 12–15 piglets at a time but only 6–8 survive. The introduction and popularity of the exotic white pigs led the black pigs of Kerala to an endangered level. The scavenging practice was also a reason for the rejection of the local variety. But there have been some farmers who retain and breed the black pigs. The change towards more refined toilet system resulted in the cleaner feeding habits. The aversion towards the black pig has vanished. Not only that, the market trend for this pork has changed to the extent of becoming a high-priced delicacy on the dining table. But the scarcity is the problem requiring immediate attention.

4.1.1.8 *Ghoongroo (Plate 1)*

This breed of pig with distinctive productive and reproductive characteristics has been identified in the eastern Sub-Himalayan region of the state of West Bengal, India. The breed is also found in the eastern part of Nepal adjoining the Darjeeling district. Farmers manage the animals both under stall-feeding and stall-feeding-cum-grazing systems. Simple housing principally made up of bamboo and jute stick is used with an emphasis on giving protection from the rain. The population in the breeding tract varies depending on market demand. Generally, the population varies from 8000 to 100000. The pigs are black (>98%) to tan in colour with occasional white patches at front and hind feet with a compact body, long thick coarse hair, the tail extends up to hock joint. It has typical Bulldog type head with folded skin at face and neck. Face line slightly convex with loose skin at chin. Ears are large and drooping.The hindquarters are heavier and rumps are drooping. Body back line is straight in male and slightly concave in females. The scrotum loosely hangs from the body (Sahoo, 2009). Average litter size at birth is11.92 ± 0.06 and a litter size of up to eighteen are not uncommon on a low to medium plane of nutrition. Body weights at birth, five months and one year of age are 1.08 ± 0.22, 38.91 ± 1.49 and 106.3 ± 0.31 kg, respectively, irrespective of sex. This unique germ-plasm has the potential to replace exotic breeds from temperate zones currently used in improved pig production programmes. However the breed is under constant threat due to indiscriminate crossbreeding with other varieties. Thus the immediate implementation of conservation and improvement programmes is essential to salvage the breed.

4.1.1.9 *Gahuri (north-east Indian)*

Gahuri pigs are mainly black with an admixture of light brown colour. These pigs are kept by tribal people. It is a dwarf type and is found also in Nepal. All are hardy, scavenger pigs. There is a similar type of pig in Sri Lanka known as the *Sri Lanka* native.

4.1.1.10 *Pigmey pig -S. salvanius* (Hodgson)

This type of pigs are found to inhabit in the dense moist forests at the base of the Himalayas in Sikkim, Assam and other north east states and Nepal and Bhutan, It is nocturnal in habits and prefers to remain in high grasses and therefore is rarely seen. It lives in herds of 5–20. The animal measures about 32 cm over the shoulder and 66 cm from snout to rump. It weighes 7.7 kg, colour is brown/black. No distinct crest is present. There is no wooly cover under fur; the hairs on the hind part of the neck and middle of back are rather long whereas those of the ears are small. Its habits are those of the wild boars.

4.1.1.11 *Dom (Plate 2)*

Dom pigs are native of Assam. The colour is light black with or without white mark on the forehead, snout, lower abdomen and switch of the tail. Some pigs also could be observed with dark brownish black colour with fine hair sparsely distributed. Ears are small in size, erect and placed inwardly having a length ranging from 15–35 cms. There are 8–12 number of teats along the thorax and abdomen.

4.1.1.12 *Pigs of Andaman and Nicobar group of Islands*

The status of the pigs of Andaman and Nicobar Islands is a subject of conjecture and dispute. It is not certain whether they represent an endemic species or feral population. There are two quite distinct and apparently table pig morphotypes in the Andaman and Nicobar islands. The two distinct forms are the long snouted (*Suc scrofa nicobaricus*) and short snouted (*Sus scrofa andamanensis*) (Abdulali, 1962). However, both types remain poorly known and their origin is far from certain. The dwarf wild pigs *Sus scrofa andamanensis* and *Sus scrofa nicobaricus* was thought to be endemic. However now several experts are of the opinion that these populations are feral (Oliver, 1984).

The pigs of Andaman and Nicobar islands are associated with the most isolated tribal populations of the world, the Jarawa, Sentenilese, and the nearly extinct Andamanese and Ongesnegritos. These tribes are closely associated with the wild pigs which are primary source of food and also have ritual and religious significance. Despite being protected, wild pigs are under threat due to poaching by immigrant groups, high level of deforestation and logging, agricultural encroachment and other developments (Whitaker, 1988).

4.1.2 Bangladesh

The pigs of Bangladesh are mostly scavengers, are of Dom breed as in parts of Assam. The rest are non-descript and live by scavenging and therefore very prone to parasitic infection.

4.1.3 Nepal

There are four types of indigenous breeds available in Nepal which constitute 58% of total pig population. They are black coloured Chwanch in hills (adult weight 35 kg with litter size 6–8), the rusty brown Hurrah (adult wt 46 kg with litter size 5–8), rusty brown Bampudke (adult weight 25 kg with litter size 6–8) and the cross breed Pukhribas which have been produced by crossing Tamworth, Saddleback and Fayuen (adult weight 100–150 kg with litter size 10).

4.1.4 Bhutan

Bhutanise pigs have much similarity with Nepalese pigs of hilly region. This is because of constant live pig trade activity. Nepal exports piglets to North-East India via which piglets are smuggled into Bhutan through Bhutan-India border. Locally, the pigs are called phap (in Dzongkha) or phagpa (in Sharchop). Local pigs are preferred over exotic ones for meat quality.

4.2 Southeast Asia

Many of the indigenous pigs in the region are of Chinese type ancestry but there are exceptions. There has also been an extensive upgrading using developed, breeds, mainly British breeds.

4.2.1 Myanmar

Most native pigs are black in colour. The head is small and of moderately dished profile, concave back, and pendulous belly, characterized by slow growth, thick fat and hardiness. A well fed pig weighs about 60 kg at 12 months of age. In mountainous region, small miniature pigs, Chin Dwarf, characterized by long snout, small body size, early maturity with no excess fat, weighing 30 kg at maturity, are commonly raised by different tribes of that area.

4.2.2 Thailand

Native pigs still exist in the remote areas, especially in south Thailand, where the livestock industry is not well developed. Hill tribe people are still keeping pigs as scavengers around houses and farms. The distinct breeds of native pigs are Hailum, Raad, Puang and Kwai (Na Puket, S.R. 1980). Virtually all breeds are of the Chinese type.

First breed is the Hailum, also called Hainan are raised in the southern region. Hainan breed is morphologically characterized by black and white coat colour, short straight face (snout), hollowed back, large belly and small erect ear.

Second breed is called "Raad", similar to short ear breed of Taiwan and mainly raised in northern regions of Thailand. The type is characterized by black coat colour, long straight face (snout), straight or slightly hollowed back (chine) and small, erect ears.

The third breed is called 'Kwai' mainly raised in central region of Thailand. The characters are almost the same as Hainan breed except that the coat colour is black with white legs and big body size.

Virtually all breeds are of the Chinese type. All have been extensively upgraded using the developed European or American breeds.

4.2.3 Malaysia

Under the government encouragement, Malaysia has introduced a number of superior breeds from European countries, the United States and Australia. This introduction has led to a great progress in swine industry. 85% of the pig population of the country consists of various exotic crossbreds, and the purebred local varieties are just becoming a rarity. There are three kinds of native pigs in Malaysia:

1. South China breed-the upper parts of the body including the head is black, while the abdominal part including the legs is white. The forehead has a white patch. The texture of skin is fine and sparsely covered with hair, there is also a mane.
2. Cantonese is entirely black in colour.
3. Wild pigs are found in the jungles and are black in colour, with densely thick long hairs around the body and legs (Mukherjee, T.K. 1980).

4.2.4 Indonesia

There are several breeds of native pig amongst which Java, Bali and Sumatra pigs are important. Java pig originated from the crossing of European breeds with indigenous pigs. This pig is short and fat and displays a mild swayback position with a heavy mane of bristles on the neck and a long snout. Bali pigs are of the

Chinese type, with an extreme swayback position. In fact the belly almost touches the ground. There is also a great deal of skin folding in adult animals. It is a hardy and prolific scavenger pig that has been exported to other islands in the Indonesian archipelago. Sumatra pigs appear to be more nearly related to the feral pigs, of which there are still thousands in the jungle. They are small with a tight skin and has well developed tusk.

4.2.5 Philippines (*Plate 2*)

During the Second World War, swine industry in the Philippines was totally destroyed. After the war they introduced from European countries and USA a number of exotic breeds such as Berkshire, Poland China, Duroc Jersey, Hampshire and Landrace. These breeds were distributed to different government breeding station/centers, agricultural schools, private hog farms to improve their size and feed efficiency (Eusebio, A.N. 1980). The introduction of these standard breeds greatly influenced the development of the existing stock of pigs raised in the Philippines. At present, the Philippine swine raised in backyards consist of several strains, which are widely distributed in the country.

There are four common strains of swine in the Philippines, the 'Kaman' and 'Koronadal' hogs which are red and the 'Diani' and 'Ilocos' strains which are black. The Kaman is common in the province of Batangas and Koronadal in the province of Cotabato. The Kaman is an upgraded native pig with Duroc Jersey blood. The Koronadal pig is an amalgamation of Berkjala, Poland China and Duroc Jersey and is red with dark spots all over its body. The black strains of pigs in the Philippines have either the Berkshire or Poland China blood. They are swayback breeds that are usually black in colour. They are small and less prolific than the Cantonese and are almost extinct, being continuously upgraded by pigs of introduced American and European breeds.

4.2.6 Vietnam, Cambodia and Laos (*Plate 2*)

Apart from breeds such as the *Vietnamese Pot belly*, *Meo* and the *Huang Kong* raised in the mountains, the numerous local breeds are of the south China type.

4.2.7 Sarawak

The *iban* (syn. *Kayan)* breed is said to be a domesticated wild pigs, *Sus scrofa vittatus.* These pigs are rather small, black or black and white in colour, with a narrow head, a long snout, a short neck and small, erect ears. They are used as scavengers, partly of human faeces.

4.2.8 New Guinea

The main functions of the native pigs are aesthetic and cultural; their importance is as a measure of prestige, wealth and as an exchange medium. European pigs are used for pork production.

4.2.9 Taiwan

There are four types of native breeds of pigs in Taiwan: Taoyuan, Meinung, Ting-shuang-hsi and small ear black. The former three types were introduced from Canton Province of southern China, and the latter one which has been raised exclusively by the aborigines and is considered belonging to the same lineage as the native pigs in the Island areas of Malaysia, Indonesia and Ploynesia.

4.2.9.1 *Taoyuan*

The Taoyuan has a very short wide head and a dished and deeply wrinkled face, broad snout, large nostrils, small eyes, moderately thick and drooping ears, narrow chest wide, flat ribbed and hollow back, thin and flat ham, short and thick legs, and a long straight tail. The thin neck is badly set to the coarse shoulders and has several vertical skin folds. The skin is very thick, its deep folds extending over the major part of the animal except the shoulders and hams. The bristles are sparse, short and rather soft. Skin colour is black or grey, and the bristles are black (Koh, F.K. 1952).

4.2.9.2 *Meinung*

This variety is very similar to the Taoyuan, but smaller. It is found in the south west of Taiwan and is named after the town of Meinung in Kaohsiung country.

4.2.9.3 *Ting-shuang-hsi*

Ting-shuang-hsi named after the town in Taipei country, north eastern Taiwan, is now almost extinct.

4.2.9.4 *Small ear pig*

The small ear pig had a long narrow head, with a straight profile, long nose and strongly developed snout. The ears are very small and usually erect. This breed had a short muscular neck, very strong shoulders, narrow slightly hollow back, large often pendulous barrel, short stooping rump, moderately long straight legs, and straight tail. The skin is deeper black than that of the Taoyuan. The body is densely covered with black bristles and adult boars has a ridge of long thick black bristles from the poll to the mid back.

The fact that the native breeds of pigs have become nearly extinct is due to increase of human population and meat requirement. In addition, the native breeds could not compete with the fast growing and better quality cross breeds. Under the condition that the rural environment for pig raising has been improved in feed supply, especially increase of protein feed, the cross breeds could obtain their utmost efficiency and almost all the farmers have been willing to raise them.

4.3.1 Indigenous tropical breeds of Africa

Indigenous breeds of pigs exist mainly in West Africa. Although these are wild species of the Suidae family in Africa, there is no evidence that they have been domesticated. Present 'indigenous' breeds in West Africa are descended from imported pigs. The domestic pig was originally introduced into North and Northeast Africa but since the Arab invasions, only remnant populations remain in North Africa, Egypt and in isolated areas in the southern Sudan. In East, Central and South Africa, developed breeds have been introduced from Europe. However, in South Africa there is a breed known as the *Bantu,* (*Plate 3*) believed to be derived from introduced European and Asian pigs. There are no indigenous domestic pigs in the tropical areas of western Africa.

4.3.2 West Africa

Domestic pigs are found throughout the forest areas of west Africa. Well-known breeds are the *Bakosi* in Cameroun, the *Ashanti Dwarf* in Ghana and the Nigerian Native. They vary in colour from black to brown and are very hardy. The Ashanti Dwarf and possibly others are said to be trypano-tolerant (Jollans, 1959). They were considered by Epstein (1971) to be of Iberian ancestry, but it is likely that they are pigs of more ancient ancestry that have been crossbred with Iberian-type pigs introduced by the Portuguese. In Cote d'lvoire there is a breed known as Ikorhogo that has apparently evolved from crosses between Berkshire Large White and West African pigs.

4.4 Exotic Breeds of International Importance

King (1991) listed four breeds of international importance; Large White, Landrace, Duroc and Hampshire.

The primary objective of swine production is to get maximum lean meat in the form of bacon and ham. For this reason it is essential to know the different germ plasm available in the country and all over the world in relation to these traits. In

Table 4.1 and 4.2 the list of the most popular old established exotic breeds and of new breeds according to the predominant hair colour and type of ears is given.

4.4.1 Large White Yorkshire (*Plate 3*)

The Large White Yorkshire is native breed of United Kingdom and is reported to produce better bacon when crossed with other suitable types. This bred is imported into India from UK, New Zealand and Australia. It is large in size with a long and slightly dished face. Body is covered with fine hair, free from curves. Skin is pink colored and is free form wrinkles with long and moderate fine coat. Ears are thick, long and slightly inclined forward and fringed with fine hair. Neck is long and full to the shoulder with deep and wide chest. Shoulders are not too wide. Back is slightly arched. Loin is long and broad with a well developed wide rump. Hump is fleshy extending up to the hocks. Tail is set high. Mature boars and sows of this breed generally weigh 295–408 kg and 227–317 kg respectively. This breed is very popular for the bacon. The sows are prolific breeders and good milkers.

4.4.2 Landrace (*Plate 3*)

It is native of Denmark. It is a bacon breed. It is white in colour, large in size, ears are lopped, head and neck small, light shoulders, great length of side and heavy hams. Sows have good mothering quality. It is noted for its smoothness and length of body and for a carcass that contains a high proportion of lean.

4.4.3 Hampshire (*Plate 4*)

The Hampshire breed of pig originated from southern England. It is a black pig with a white belt encircling the body including the legs. Head and tail are black, and the ears are erect. The pigs are short legged. Sows are very prolific and good mothers. The weight of a mature boar and sow is about 400 and 250 kg respectivly.

4.4.4 Duroc (*Plate 4*)

Duroc has its origin in the USA. It is red in colour, with the shades varying from golden to very dark red. The ears are medium sized and tipped forward. It is a large breed with excellent feeding capacity and prolificacy. The sows are good mothers. The weight of a mature boar and sow is about 400 and 250 kg respectively.

4.5 Breeds of Limited and/or Regional Importance

There are number of developed breeds in Britain, Europe and North America that have been imported into the tropics, but have made no particular impact. These include the Craon, Edelschwein, Gloucester old spot and Pietrain. Other developed breeds have been used frequently in the tropics, either as purebreds or for crossbreeding purposes. These include the Berkshire, Large Black, Middle White, Tamworth and Poland China.

4.5.1 Large black (*Plate 4*)

The breed developed by crossbreeding indigenous pigs from the eastern countries of England and Neapolitan pigs. It is a long, black pig with lop ears and good hams and is considered a good grazer and mother. It can be utilized for the production of pork or bacon and has been used extensively for crossing with indigenous pigs in various regions of the tropics.

4.5.2 Chinese pigs

About half of the world's pigs are raised in China. There are many Chinese breeds, bred for different human requirements in several different climatic environments (Epstein, 1969; Cheng Peilieu, 1984; Porter, 1993). Breeds from the tropical and subtropical regions of China such as the Cantonese have been introduced into most Southeast Asian countries, probably by Chinese immigrants. Some were also introduced into Portugal and Spain in the 15th and 16th centuries. Subsequently, Chinese type pigs or Chinese crossbreds were exported to islands in the Caribbean, central and South America and west Africa. South Chinese pigs were also introduced into Britain in the 18th century. In Britain, they played an important role in the development of many British breeds whilst in America they were used in the formation of the Poland China and Chester white breeds.

4.5.2.1 *The Cantonese*

The Cantonese, synonym Pearl River delta, is the characteristic black and white sway back type of pig indigenous to south China. It is usually called the Chinese in Britain and the Macao in Portugal and Brazil. The head is small with a moderately dished profile; the back is hollow and the belly pendulous. It is very fecund. The average litter size is 12 and litters of up to 20 are not uncommon (Epstein, 1969). The number of teats possessed by the sows, range from 14 to 16. Fat pigs weigh approximately 75 kg at 12 months of age. Sows farrow twice a year and gilts are bred at 5 months of age (Phillips *et al.,* 1945). The sows are said to be excellent mothers and piglet mortality due to 'overlaying' is low as the sow always lies down very carefully. Pigs of this breed are said to exhibit some tolerance of kidney worm and liver fluke.

Other breeds found in the tropical and subtropical areas of southern China are the Wenchang (Hainan), a small breed from Hainan Island, lop eared pigs known as the northern Kwangtung that are also very fecund, the Luchuan of Kwangsi and the Ningsiang and Dawetze of Human (Epstein, 1969). The *Taoyuan* breed of Taiwan is similar to the Cantonese.

At the present time the world's pig breeders are very interested in the early maturing and highly prolific Chinese breeds of Taihu, a non-tropical region in the lower Changjiang river basin. There are at least three types: the *Meishan,* (*Plate 3*) *Fengjing* and *Jiaxing* black (Cheng Peilieu, 1984). These are pot-bellied pigs adapted to roughage feeding hardy and long lived, with lop-ears, a wrinkled skin and black or black/grey hair. Gilts come into first oestrus at 3 months when they weigh 15–25 kg, while boars can mount and fertilize females at 3 months of age. Sows possess 16–18 teats and by the third litter an average of 15 piglets are born while on an average 12.5 are weaned. Growth rate and efficiency of food conversion are low, the back fat is 20–50 mm thick and the carcass only yields 40% lean meat. The quality of the meat, however, is excellent.

The French, the British and the Americans (McLaren, 1990) have imported pigs of the Taihu-type in an attempt to incorporate their characteristic of high prolificacy in breeds of international importance. As the heritability of litter size is low at 0.10 in the pig breeds it will be some years before it is known whether these attempts have been successful.

4.5.3 Portuguese and Spanish pigs

Pigs of the Portuguese and Spanish Iberian type breeds such as the *Alentejana, black Iberian* and *Extremadura Red* and/or Celtic type breeds such as the Bisaro, together with crosses with imported Chinese pigs, were introduced in to Caribbean Islands, central and South America from the 15th century onwards. Some Chinese type pigs may have also been introduced to Mexico from the Philippines by the Spanish. Coloured Liberian type pigs were also introduced by the Portuguese to West Africa.

4.5.4 Middle White Yorkshire (*Plate 5*)

The Middle While Yorkshire was evolved as a result of crossing Large White Yorkshire and Small Yorkshire breeds of UK. It is a medium sized bacon pig and a good porker at light weights. It is white in colour with a short head unturned dished face wide between the ears. Neck is blended neatly from head to shoulder. Ears are nearly erect but somewhat inclined forwards. Hams are broad and fleshy up to the hocks. It is a prolific breeder, maturing early and the sows make good mother. Mature boars and sows generally weigh 249–340 kg and 181–272 kg respectively.

4.5.5 Berkshire (*Plate 5*)

The Berkshire is one of the oldest English breed of swine. This breed is valued as producer of quality meat, especially suitable for the pork market. This breed is used in upgrading programs. The pigs are black with white markings usually on the feet, head and tail. It has a short head with dished face. The snout is short. The body is long and ribs well sprung. Mature boars weigh about 280–360 kg or more.

4.5.6 Tamworth (*Plate 5*)

The Tamworth originated in Ireland. It is possibly the purest modern representative of the native English pig. The colour is reddish or chestnut, typically golden red hairs on a flesh coloured skin. The head is long and narrow with long snout and erect ears. It has a strong back and thin shoulders. The carcass produces bacon of best quality. Sows are prolific breeders. Mature boars weigh up to 300 kg.

4.5.7 Russian Chazmukha

They are of large in size, black in colour with white spots. Sows are prolific and possess good mothering quality.

4.5.8 Wessex saddleback (*Plate 6*)

Wessex Saddleback, an English breed is essentially a bacon breed, easily adaptable for pork production. It is known for its prolificacy and has a robust make up; head, neck, hind quarters, hind legs and tail of this breed are black. Head is fairly long with straight snout and ears having forward pith without being floppy.

4.5.9 Chester white

The Chester White had its origin in Chester and Delaware counties in Pennsylvania. The breed has white hair and skin. The ears are drooping. It is lard type, hardy and fairly good feeder. Chester White sows are very prolific and are exceptional mothers. The pigs adapt well to a variety of conditions; they mature early and the finished barrows are very popular on the market. They are intermediate in size and mature boars weigh 400 kg and over.

4.5.10 Poland China (*Plate 6*)

The breed originated from Warren and Butler counties in Ohio (USA) by a fusion of Polish pigs and Big China. The colour of the breed is black with six white points, the feet, face and tip of tale. The typical Poland China has thick, even flesh, and is free from wrinkles and flabbiness. The breed has good length and excellent hams. The head is trim, and the ears are drooping. The Poland China is efficient feed converters. The breed is less prolific and produce excellent carcass.

4.5.11 Hereford (*Plate 6*)

The breed originated about 1900 by R.V. workers of the Plata Missouri. They are crosses of white and red-blooded stock of Duroc, Chesters and OIC's (Ohio Improved Chesters). In the years 1920 to 1925 a group of breeders in Iowa and Nebraska, led by John Schulte, Norway, Iowa, established a breed that was also called Hereford. Modern specimens of the breed trace to this foundation.

4.6 New Breeds of Pigs

4.6.1 Beltsville No. 1

The breed carries approximately 75% Landrace and 25% Poland China blood and about 35% inbred. Animals of this breed are black and white spotted. They have long bodies, little arch back, moderate depth of body, smooth sides, and plump hams.

4.6.2 Beltsville No. 2

This breed was also developed at the Agricultural Research Center at Beltsville from crosses which begun in 1940. The pigs carry 58% Danish Yorkshire, 32% Duroc, 5% Landrace and 5% Hampshire blood. Beltsville no.2 pigs are usually solid red in colour and have white underlines. The ears are usually short and erect. The head is intermediate in length and has a moderately trim jowl. Pigs of this breed have the length of the Yorkshire. The back is of medium width and has little arch.

4.6.3 Lacombe (*Plate 6*)

Bloodlines are now stabilized at 55% Landrace, 23% Berkshire, and 22% Chester White. Lacombe pigs are white with drooping ears of medium length. Their general appearance resembles the landrace breed.

4.6.4 Maryland no. 1

The Maryland no. 1 line was established in 1941 and carries approximately 62% of landrace and 38% of Berkshire blood. Pigs of this breed are black and white spotted and are intermediate in conformation between the Landrace and Berkshire. The back is slightly arched and medium in width. The head is long and the jowl is somewhat heavier than that of the Landrace. The ears are medium in size and are usually erect.

4.6.5 Minnesota no. 1

This breed was developed by the Minnesota Agricultural Experiment Station and USDA. The breed is a cross between Canadian Tamworth and Danish Landrace. The breed contains about 55% landrace and 45% Canadian Tamworth. The colour is red with frequently a tinge of black and occasionally a few black spots. The body is long, about two inches longer than most American breeds. The back usually has no arch. The jowl is refined, and the neck is thin. The snout is usually long and trim. The ears are fine textured and vary from erect to drooping.

4.6.6 Minnesota no. 2

The Minnesota no. 2 pig was developed by the Minnesota Agricultural Experiment Station in cooperation with the Regional Swine Breeding Laboratory of the USDA. The breed contains 40% Yorkshire and 60% Poland China blood. Animals of this breed are black and white in colour, having long bodies, well muscled loins, full and deep hams, and have heads with shorter snouts than the Minnesota No. 1. The ears are medium in size and are erect.

4.6.7 Minnesota no. 3

The Minnesota no.3 breed is an inbred line developed from eight other breeds. The foundation for the line was established in 1950.

4.6.8 Palouse

The Palouse breed of swine was developed by Washington State University in 1945 by crossing three Landrace boars with 18 Chester White gilts and sows. The pigs are solid white in colour and resemble the Landrace in body type and conformation.

4.6.9 San Pierre

The San Pierre pig originated at the Inbred Swine Farm, San Pierre, Indiana, then owned by Gerald Johnson. The foundation stocks were Canadian Berkshire and Chester White. This is the only new breed which has been developed by a private producer. San Pierre pigs are black and white in colour, have length similar to the Berkshire, and the growth vigor of the Chester White. The breed is intermediate in type and is characterized by excellent stretch of body, neatly turned loin and plumped well muscled hams. Their ears are erect. They are a meat type pig and have been used in crossbreeding programs.

4.6.10 Montana no. 1 or Hamprace

Blood lines derived 52% from Landrace and 42% from Hampshire ancestry. The Montana no.1 is a pig of medium size, solid black with slightly arched back, medium length of legs, small narrow head of medium length and neat jowl. The ears tend to be large and may either stand erect or drop forward full length. The body is uniformly deep, the sides smooth, and the hams deep and full. The sows are gentle, have 12 to 14 well spaced teats, and are good milkers.

Table 4.1 Old Popular Established Breeds: Place of Origin, Physical Characteristics and Economic Importance

Breed	Place of origin	Colour of hair	Type of ears	Economic importance	Remarks
Large white	Northern England	White	Erect (dished face)	Good mothers raise large litter and they are great milkers. Growth is excellent under confinement	Animals on slaughter yield a high dressing % and produce good quality meat.
Middle white	Northern England	White	Erect at face (considerably dished face)	Large type pig, hardy and fairly good feeder. Sows are prolific, usually milk well, carcass quality is intermediate.	–
Berkshire	South and south east of England	Black with six white point on face and tail switch	Erect	Some what less prolific and grower in gaining ability, but excellent in milking ability. It is good for cross breeding programme	–
Landrace	Denmark	White	Large, slightly drooping	Noted for prolificacy and for efficiency of feed utilization, carcass contain a high proportion of lean	–
Hampshire	England	Black with white belt	Erect	Famous for prolificacy, hardiness vigour and outstanding killing qualities	Breed has been used in cross breeding because of its quality feeding and prolificness

Table 4.1 (*Contd...*)

Breed	Place of origin	Colour of hair	Type of ears	Economic importance	Remarks
Tamworth	Ireland and UK	Red (slightish)	Erect	Extreme bacon type confirmation.The sows are prolific and careful mother	Used frequently in cross breedign Tamworth × Berkshire crosses are popular
Larege black	East Anglia	Black or blue black	Ear are long and thin and inclined forward and slightly inwards over the face	Hardy and docile breed. Sows are good mothers and are reasonably prolific	Cross with large white either way is very successful
Duroc	USA	Red	Drooping	Breed is prolific and the sows are good mothers	Breed ranked first in both rate of gain and feed efficiency
Chester white	USA	White	Drooping (slightly dished)	Large type pig, hardy and fairly good feeder. Sows are prolific usually milk well, carcass quality is intermediate	–
Poland China	USA	Black with white on face feet legs and switch	Drooping	Less prolific, produce excellent carcass	–
Hereford	USA	Red with white head, feet and switch underling	Drooping	Individuals are generally quite maturing	–

Table 4.2 New Breeds Place and Year of Origin, Physical Characteristics, Economic Importance

Breed	Place of origin	Year of origin	Colour of hair	Type of ears	Economic importance	Remarks
Beltsville no.1	USA	1934	Red with white head, feet and switch underling	Drooping	–	75% Landrace and 25% Poland China
Beltsville no.2	USA	1952	Red	Erect	–	58% Yorkshire 32% Landrace, 5% Duroc and 5% Hampshire
Lacombe	Canada	1947	White	Drooping	Famous for rapid weight gain	55% landrace 23% Berkshire +22% Chester white

Table 4.2 (*Contd...*)

Breed	Place of origin	Year of origin	Colour of hair	Type of ears	Economic importance	Remarks
Maryland no. 1	USA	1941	White	Erect	–	62% Landrace and 38% Berkshire
Minnesota no. 1	USA	1936	Red	Slightly erect	It gains rapidly and economically, and carcass yield well in terms of high priced cuts of meat	48% Landrace and 52% Tamworth
Minnesota no. 2	USA	1941	Red	Slightly erect	It gain rapidly and economically and carcass yields within terms of high prices cuts of meat	40% Yorkshire and 60% Poland China
Minnesota no. 3	USA	1956	Light red with black spot	Slightly erect	–	Combination of gloucesterld spot Poland China Welsh, Large white Beltsville no. 2, Minnesota no. 1, Minnesota no. 2, and San Pierre
Montana no.1	USA	1936	Black	Slightly in breeding	Recommended Hampshire for the production of market pigs	55% Landrace Hamprace
Palouse	USA	1945	White	Slightly erect to dropping	Animals of this breed produces a carcass	65% landrace 35% chester white
San Pierre	USA	1953	Black and white	Erect	These are meat type pigs and have been used in crossbreeding programmes	Berkshire and Cheste white crosses

Plate - 1, Breeds of Pigs

♂ = Male, ♀ = Female

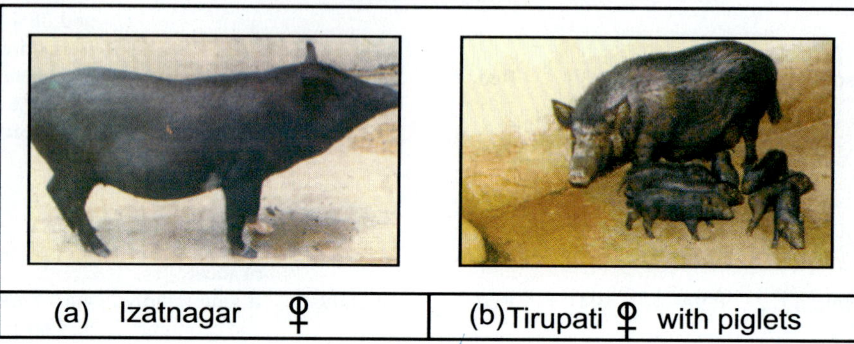

(a) Izatnagar ♀ | (b) Tirupati ♀ with piglets

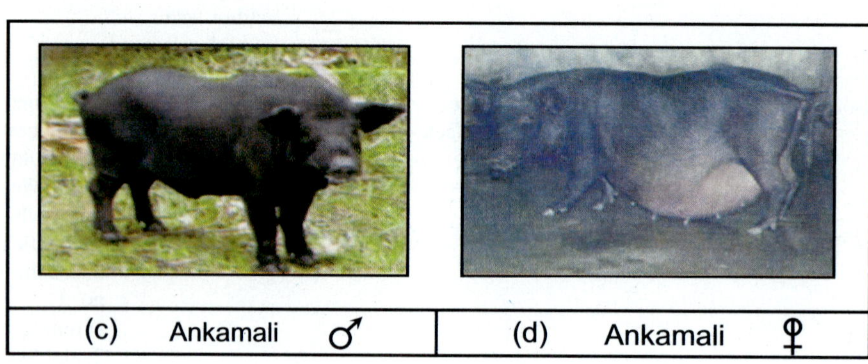

(c) Ankamali ♂ | (d) Ankamali ♀

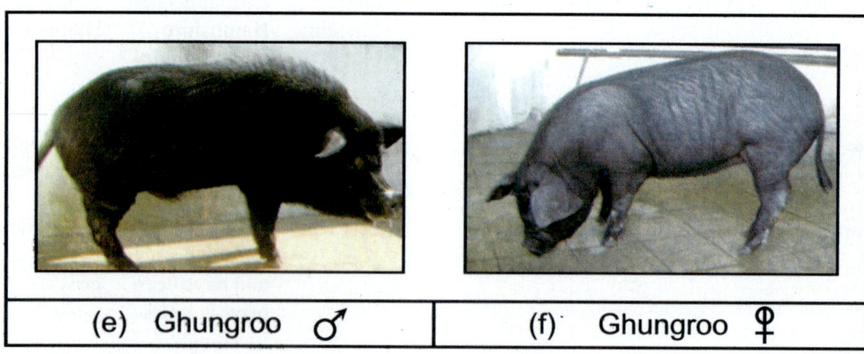

(e) Ghungroo ♂ | (f) Ghungroo ♀

Plate - 2, Breeds of Pigs

♂ = Male, ♀ = Female

(a) Dom ♂ | (b) Dom ♀

(c) Vietnamese Potbelly ♂ | (d) Vietnamese Potbelly ♀

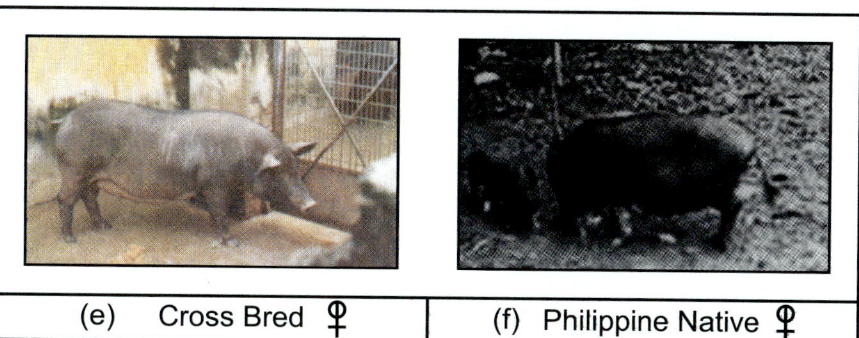

(e) Cross Bred ♀ | (f) Philippine Native ♀

Plate - 3, Breeds of Pigs

♂ = Male, ♀ = Female

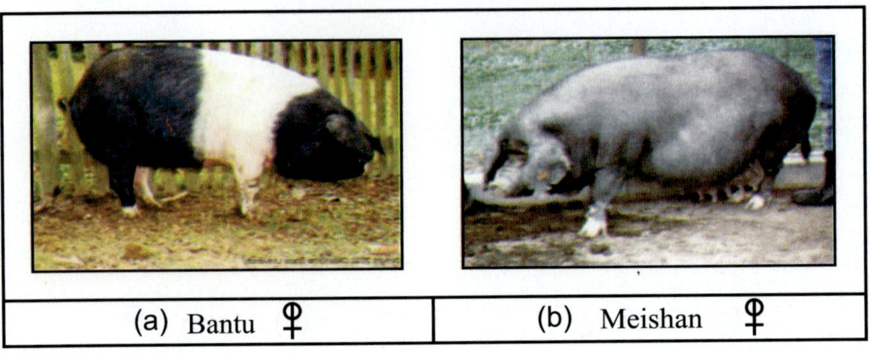

(a) Bantu ♀ | (b) Meishan ♀

(c) Large WhiteYorkshire ♂ | (d) Large WhiteYorkshire ♀

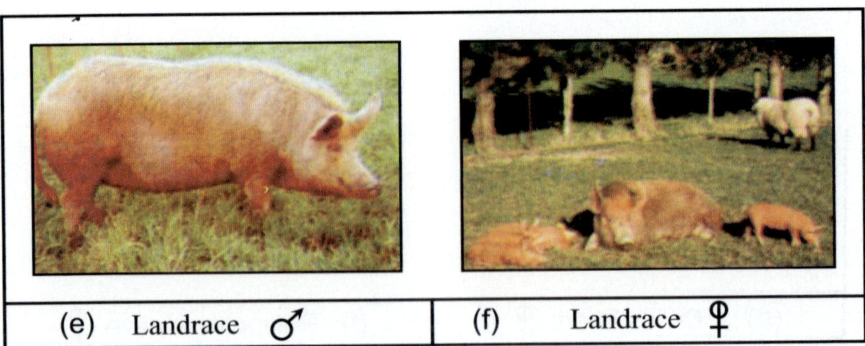

(e) Landrace ♂ | (f) Landrace ♀

Plate - 4, Breeds of Pigs

♂ = Male, ♀ = Female

(a) Hampshire ♂

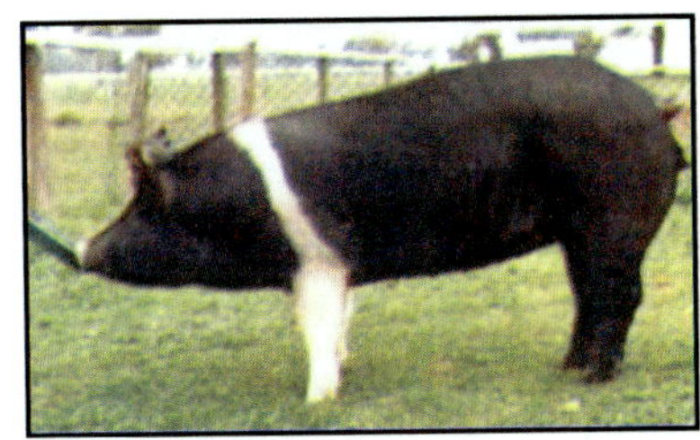

(b) Hampshire ♀

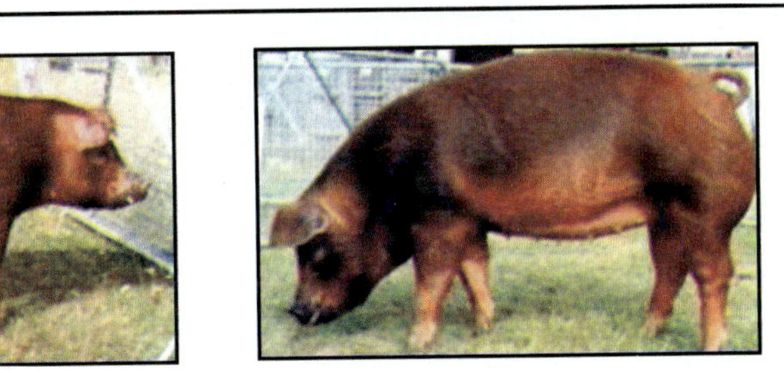

(c) Duroc ♂

(d) Duroc ♀

(e) Large Black ♂

(f) Large Black ♀

Plate - 5, Breeds of Pigs

♂ = Male, ♀ = Female

(a) Middle White Yorkshire ♂

(b) Middle White Yorkshire ♀

(c) Berkshire ♂

(d) Berkshire ♀

(e) Tamworth ♂

(f) Tamworth ♀

Plate - 6, Breeds of Pigs

♂ = Male, ♀ = Female

(a) Saddleback ♂

(b) Saddleback ♀

(c) Poland China ♂

(d) Hereford ♂

(e) Lacombe ♂

CHAPTER 5

GENETICS

5.1 Basic Genetics

5.1.1 Introduction

There is very wide distribution of wild and feral pigs in the world and it is generally believed that all domesticated breeds have been derived in one way or another from two wild types: *Sus vittatus*, synonyms *S. scrofa cristatus*, the wild pig of east and southeast Asia, and *S. scrofa*, the present European wild pig, which may also have existed during the past in western Asia. While considering the distribution on a continental basis, approximately one-fifth of the world's pig population are to be found in the tropics and that the pig population in the tropics is increasing more rapidly than that in other regions.

Domestication of the pig is likely to have occurred first in the near east and may have occurred repeatedly from local populations of wild boars. By seeing the characteristics of the pigs as a meat animal, it was felt necessary to domesticate the pigs to exploit its full potential.

The knowledge of genetics is important for improvement of the production of animals through breeding and selection.

5.1.2 Karyotypes and chromosomal polymorphism

It is generally agreed that domesticated pig breeds have a chromosomal complement of (2n = 38), the number and morphology was the same in both the sexes except for sex chromosomes. There are 7 pairs of subcentric, 5 pairs of metacentric, 6 pairs of acrocentric and one pair of sex chromosomes.

The chromosome number in pigs is polymorphic. The Japanese wild boar *Sus scrofa leucomystax* has a diploid count of 38 chromosomes (Muarmoto *et al.,* 1965). The same chromosome number has been reported in the wild boar of Israel. A chromosomal count of 38 has been described in all the domesticated species. The European wild pigs have a chromosomal profile of 36 or 37 chromosomes (McFee *et al.*, 1966). This may be owing to the intermixing of the domestic and wild populations. Presumably both the populations have reproductive compatibility. Animals with a chromosome profile of 37 are fertile leading to their coexistence.

One of the mammalian species showing chromosomal polymorphism is the *European wild boar*, *Sus scrofa*. It was stated that the chromosome number of *S. scrofa* in continental Europe, Central and Far East Asia, varied from 36 to 38, and in the Mediterranean islands was 38 (McFee *et al.*, 1966; Rary *et al.*, 1968; Gustavsson *et al.,* 1973; Tikhonov and Troshina, 1974; Bosma, 1976; Macchi *et al.*, 1995). In addition, it was reported that the diploid chromosome number of the domestic pig was 38 (Hansen-Melander and Melander, 1974; Gustavsson, 1988; Bosma *et al.* 1991). Hsu and Benirschke (1967) reported that the diploid chromosome number of the wild boar distributed in the USA, firstly imported from Germany in 1912, was 36. McFee *et al.* (1966) pointed out that the polymorphism in the diploid number was caused by the Robertsonian translocation. Some authors (Tuncoks 1935; Erencin, 1977; Kumerloeve, 1978; Turan, 1984) have stated that the whole of Turkey is within the distribution area of *Sus scrofa.* Some authors (Steiner and Vauk, 1966; Hus 1967; Kumerloeve, 1978; Mayer and Brisbin, 1991) also gave the distribution of *Sus scrofa* on a provincial basis in Turkey. Mohr (1960) and Mursaloglu (1964) pointed out that the wild boar in Anatolia was represented by *Sus scrofa libycus*.

In the karyotype of the domestic pig *Sus scrofa* the chromosomes of pair 10 have a marked secondary constriction in the short arm near the centromere (Reading Conference, 1980; Committee for the Standardized Karyotype of the Domestic pig, 1988). The karyoptype of *S. scrofa* was examined and observed that in addition to displaying the secondary constriction typical of pair 13 (equal to pair 10), one of the chromosomes of pair 8 also exhibited the same characteristic. The occurrence of such constrictions was not sex-linked (Haag and Nizza, 1969).

Indian domestic pig revealed a modal chromosome number 2n=38 Kanadkhedkar *et al.* (2006). The number and morphology were same in the male and female pigs except for that of the sex chromosomes. Among these there were

7 pairs of submetacentric (1-7), 5 pairs of metacentric (8–12), 6 pairs of acrocentric (13–18) and one pair of sex chromosomes. The X chromosome was large submetacentric. However, Y chromosome was metacentric and smallest one in the chromosome complement. Conventionally stained preparation permitted the identification of chromosome complement. It also permitted the identification of chromosome pair no. 1 and 13 based on length and centromere position. The chromosome pair no. 8 and 10 showed an unstained region near the centromere in all breeds. Unstained region in chromosome 8 was more distinct and noted easily than chromosome no. 10.

NOR-band polymorphisms in the pig are rare. Veijalainen and Rimaila-Parnanen (1978) described a chromosomal polymorphism in pair 10 found in a Yorkshire female showing progressive ataxia and incoordination syndrome. This female had one chromosome lacking the secondary constriction which was NOR-band negative. Vischnevskaya and Vsevodolov (1986), Czaker and Mayr (1982) and Mellink *et al.* (1991) detected variations in the number and/or size of NOR bands in pigs.

Jorge Luis Armada and Ana Clecia Vieira Santos (1993) showed that NOR-banded metaphase chromosomes characterized in the domestic pig (*Sus scrofa*). Only the number 10 pair showed NOR banding in the region adjacent to the centromere of the short arm, following silver staining. NOR region association at metaphases was not observed, though intraspecific variation of labeled NORs was seen. Polymorphism was evident in two animals in which the NOR band was duplicated. A possible association of this polymorphism with reproductive problems detected in a female were observed.

Albayrak (2007) analyzed the data from 6 countries (Table 5.1) which describes the present status of chromosomal polymorphsm in pigs which perhaps may be one of the main reasons for such a large number of translocation and genetic disorder and abnormality related to translocations.

Ducos *et al.* (2002) reported eight cases of reciprocal translocation in the domestic pig. All the rearrangements were highlighted using GTG banding techniques. Chromosome painting experiments were also carried out to confirm the proposed hypotheses and to accurately locate the break points. Three translocations, rep (4;6) (q21;p14), rep (2;6) (p17;q27) and rep (5;17) (p12; q13) were found in boars siring small litters (8.3 and 7.4 piglets born alive per litter, on average, for translocations $^{2}/_{6}$ and $^{5}/_{17}$, respectively). The remaining five, rep (5;8) (p12:q21), rep (15;17) (q24;q21), rep (7;8) (q24;p21), rep (5;8) (p11;p23) and rep (3;15) (q27;q13) were identified in young boars controlled before entering reproduction. A decrease in prolificacy of 22% was estimated for the $^{3}/_{15}$ translocation after reproduction of the boar carrier. A parental origin by inheritance of the translocation was established for the (5;8) (p11;p23) translocation. The overall incidence of reciprocal translocations in the French pig populations over the 2000/2001 period was estimated (0.34%).

Table 5.1 Karyotypic Characteristics of *Sus scrofa* from the USA, Holland, Yugoslavia, Poland, Italy, Europe and Turkey

Country	Species or subspecies	2n	NFa	M/S/M/ST	A	X	Y
USA (Hsu and	S. scrofa	37	–	–	–	–	–
Benirschke, 1967;	S. scrofa	36	60	26	8	SM	M
Rary *et al.*, 1968)	S. scrofa domestica	38	60	24	12	SM	M
Holland	S. scrofa scrofa	38	–	–	–	–	–
(Bosma, 1976)	S. scrofa scrofa	37	–	–	–	–	–
	S. scrofa scrofa	36	–	–	–	–	–
Yugoslavia (Zivkovic *et al.*, 1971)	S. scrofa scrofa	38	–	–	–	–	–
Italy (Macchi *et al.*, 1995)	S. scrofa scrofa (2 ♀♀, 1 ♂)	38	60	24	12	SM	M
	S. scrofa scrofa (2 ♀♀)	37	60	25	10	SM	M
	S. scrofa scrofa (6♂♂, 2 ♀♀)	36	60	26	8	SM	M
	S. scrofa domestica (3 ♀♀, 1 ♂)	38	60	24	12	SM	M
Europe	S. scrofa	38	–	–	–	–	–
(Bosma *et al.*, 1991; Groves, 1981)	S. scrofa	36	–	–	–	–	–
Poland (Rejduch *et al.*, 2003)	S. scrofa scrofa (1 ♂)	38	60	24	12	–	–
	S. scrofa scrofa (2 ♀♀, 1 ♂)	37	60	25	10	–	–
	S. scrofa scrofa (1 ♂)	36	60	26	8	–	–
Turkey[1]	S. scrofa (3 ♀♀, 1 ♂)	38	60	24	12	SM	M

2n Diploid chromosome number, NFa: Number of autosomal arms, M: Metacentric, SM: Submetacentric, ST: Subtelocentric, A: Acrocentric, X: X chromosome, Y: Y chromosome. (Adopted from Albayrak and Inci, 2007).

Uses of Karyotyping

1. Identification of species.
2. To detect numerical and structural chromosomal abnormalities.
3. Identification of the sex of the fetus.
4. Examining Y chromosome polymorphism.
5. Chromosome banding technique is used to establish evolutionary relationship.

Chromosomal abnormalities

1. Trisomy–presence of an extra chromosome resulting from duplication of a portion of a chromosome may attach to a chromosome or remain as a separate fragment.
2. Monosomy–missing of chromosome or portion of it.
3. Translocation–is the result of chromosomal breakage but the broken segment transfers itself to a broken segment of another chromosome. Translocation may be balanced or unbalanced. If the total genetic combination is retained, it will be a balanced one, otherwise it is unbalanced.
4. Deletion occurs when a chromosome breaks and a portion of the chromosome is lost.
5. Inversion–a section of the chromosome is inverted or reversed on the same chromosome.

The reported chromosome abnormalities in pig include trisomy of chromosome 14, aneuploidy of sex chromosomes (XO, XXY, XXXY), paracentric inversion of chromosome 8, sex reversal (XX male), translocation through centric fusion ($^{13}/_{15}$ and 15.17) and a wide range of reciprocal translocations ($^{1}/_{6}$, 1.7, $^{1}/_{11}$, $^{1}/_{14}$, $^{4}/_{14}$, 13, 14 etc). For further details on chromosomal aberrations in animals the readers may see the URL www.angis.org.au/ocoa.

Among the chromosomal aberrations, translocations are important and are extensively studied due of their severe effects on reproduction. More than 80 different reciprocal translocations are known in pig. The imbalanced translocations cause severe embryonic mortality as early as 6–18 days, that is, during implantation stage. Balanced translocations result in reduction of litter size by 30–50% resulting in fetal mortality. The first step in studying the chromosomal abnormalities is detection of animals producing litters reduced in size. A boar can be arbitrarily considered as hypoprolific when the mean number of offspring from six litters is 8/litter.

5.1.3 Blood groups in pigs

The characteristics of the blood of men and animals have long attracted the interest of the scientists and at the present time more is known about the genetic variations of blood components than of any other animal tissue or fluid. Differences between blood of animals from different species had already been reported by Landois at the end of the nineteenth century, who found that agglutination or haemolysis occurred when human blood was mixed with that of higher animals. That there were differences between the bloods of individuals from the same species was established by Landsteiner in 1990, when he made his fundamental discovery of the A, B and O groups of human blood.

Investigation into the blood groups of farm animals also began in 1900, when Ehrlich and Morgenroth (1900) demonstrated differences between the bloods of different goats. The genetic classification of the various constituents of blood is based mainly on the immunological and biochemical methods. The immunological approach is by far the oldest and consequently the term 'blood groups' has tended to be more or less synonymous with blood characteristics detectable by immunological techniques. However, the term 'blood group' is sometimes used more broadly to include other inherited blood characters.

5.1.3.1 *Natural blood group system*

The blood typing in animals is based on development of iso-immune sera. The information in porcine blood groups is based primarily on the studies from US, Germany, Poland, Chezkoslovakia, Russia and Denmark. Sixteen blood group and seven serum protein systems have been identified in the domestic pig. They are A, B, C, D, E, F, G, H, I, J, K, L, M, N, O and P (Table 5.2) The pig A system is similar to A in human, J in cattle and R in sheep. However, the significance of these polymorphisms in pig populations is less known. The relative viabilities of different genotypes or phenotypic classes can be studied using segregation data from known mating (Smith *et al.,* 1968). Seven of the blood group systems (A, C, F, H, J, K, and M) are termed "open" because some pigs did not react to any of the reagents available for these systems. Andersen (1966) has reported that the blood group system C and J are closely linked and a close linkage between the locus for hemoglobin binding proteins and the K blood-group system.

Humans have 3 major alleles (A, B, and O), whereas pigs are known to have only A and O alleles. The porcine A gene is homologous to the ABO genes in humans and other species. The immunodominant structures of A and B antigens are defined as *N*-acetyl-*D*-galactosamine (GalNAc) $\alpha\ 1\rightarrow 3$ (Fuc $\alpha\ 1\rightarrow 2$) Gal- and Gal 1 3 (Fuc $\alpha\ 1\rightarrow 2$) Gal-, respectively. The blood group A gene encodes A transferase, which transfers GalNAc to the galactose residue of the acceptor H substrates (Fuc $\alpha\ 1\rightarrow 2$ Gal-), whereas the B gene encodes B transferase, which transfers galactose to the same substrates (Yamamoto and Yamamoto, 2001). This A and O antigens are not part of intrinsic components of cell membrane. A is dominant than O allele and is suppressed by S allele. The soluble antigens of N system can be seen in serum and blood.

A blood group factor Kf in the K blood group system of pigs, controlled by alleles Kacf, Kacef and Kbf and a new allele Kae has been reported. The K system with 6 alleles, 11 phenotypes and 21 combinations of genotypes remains is recognized as an open system. The H system with alleles H^1=H^a, H^2= H^b, H^3= H^{ab}, H^4=H^{cd}, H^5= H^{bd}, H^6=H^{be} and H^7= H- continues to be a genetically open system.

Table 5.2 Blood Groups in Pig

System	Chromosome location	Blood factor	No of alleles
EAA	1	AO	2
EAB	unknown	a,b	2
EAC	7	A	2
EAD	12	a,b	2
EAE	9	a, b, c, d, e,e, f, g, h I, j, k, l, m, n, o, p, q, r, s, t	17
EAF	8	a, b, c, d	4
EAG	15	a, b	2
EAH	6	a, b, c, d, e	7
EAI	18	a, b	7
EAJ	7	a, b	3
EAK	9	a, b, c, d, e, f, g	6
EAL	4	a, b, c, d, f, g, h I, j, k, l, m	6
EAM	11	a, b, c, d, f, g h I, j, k, l, m	20
EAN	9	a, b, c	3
EAO	6	a, b	2
EAP	unknown	a	2

(From Feldman *et al.,* 2000. Schalm's Veterinary hematology, Lippinkott Williams and Wilkins, USA).

A-O system

The occurrence of A factor was detected through naturally occurring anti-A in the normal pig serum (Table 5.1). Goodwin and Coombs (1956) observed that A antigen was not present on the red cells of the newborn piglets of type A and developed only after 7 to 10 days. Some piglets did not show strongly positive reaction until about the 30th day. This variation in the rate of appearance was evident even among litter mates. The existence of soluble A substance was observed in saliva and gastric mucin of some newborn pigs which were later found to be A positive but it was not detected in A negative pigs. The A-O system in pigs has a striking similarity with R-O system of sheep.

B-system

The antigenic factor constituting the B-system was first detected by Andersen (1962). The inheritance pattern showed an independent system with 3 phenotypes, *viz* Ba, Bb and BaBb, corresponding to 3 genotypes B^aB^a, B^bB^b and Ba Bb with 2 alleles B^a and B^b.

Table 5.3 A-O Blood Group System in Pigs

Locus	Alleles	Genotypes
A	$A^A A^O$	$A^A A^A$, $A^A A^O$,
$A^0 A^0$		
S	S^S	$S^S S_{s\ ss}$
Possible combinations of genotypes and phenotypes		
Phenotype	Genotypes	
A	$A^A A^A SS$	$A^A S_S$, $A^A A^A{}_{ss}$
O	$A^A A^O S_s$	$A^A A^O S_s$

Hojny and Hala (1965) distinguished 2 types of A antigens in pigs by the differing capacities of inhibiting anti-A serum from rabbits immunized with human A_1. These subtypes were designated as A_p and $A_{c.}$

Seventy erythrocytic antigenic factors comprising 16 blood group systems have been established so far.

C-system

Andersen and Baker (1964) described the red cell antigen Ca constituting the C-system and also showed that it was determined through an independent locus. It was closely linked with locus determining the J-system.

E-system

It is the most complex blood group system in pigs. Andersen *et al.* (1959) described 5 antigenic factors constituting this system, *Eb, Ee, Ed, Ef* and *Eq* constituted closed sub-systems within the E-system (Andersen, 1962). Five alleles segregated among different populations. Additional alleles have been added afterwards (Rasmusen, 1965; Hojny *et al.,* 1966; Dinklage and Major, 1968; Dinklage *et al.,* 1969).

F-system

Andersen (1957) detected the Fa blood group factor. It was controlled by an independent locus (Andersen *et al.,* 1959).

G-system

Andersen (1957) described this as a closed system with 2 alleles and 3 phenotypes.

H-system

Andersen *et al.* (1959) originally reported it to include only 1 factor. Hb was soon detected and it became a 3 allelic open system (Andersen and Wroblewski, 1961).

Subsequently, more blood group factors were included in this system (Hojny and Hradecky, 1972). Presently this system is known to have at least 7 alleles and continues to be an open system.

I-system

The factor Ia was first reported by Andersen (1957). It is controlled by an independent locus.

J-system

The Ja antigen was observed by Andersen (1957) and reported as 2 allele open system (Andersen *et al.*,1959)

K-system

Andersen *et al.,* (1959) reported detection of 4 antigenic factors at the K locus.

L-system

This system was first reported to have 1 blood factor with 2 phenotypes and 3 genotypes (Andersen *et al.*, 1959).

N-system

Saison (1967) reported the existence of N blood group system in pigs with 2 antigens Na and Ng.

O-system

Hojny and Hala (1965) detected Oa antigenic factor. It is independent of other systems.

5.2 Biochemical Polymorphisms in Domestic Pigs

Biochemical polymorphisms can be used to characterize populations, to investigate the levels of genetic variability exhibited by breeds and to verify the relationships among them. Within this context, several studies have been performed on pig populations of various origins (Oishi and Tomita, 1976; Oishi *et al.*, 1980; Tanaka *et al.*, 1983; Van Zeveren *et al.*, 1990). Moreover, the gene frequencies of various polymorphic biochemical loci have been used for paternity control (Oishi and

Abe, 1970). The establishment of this procedure is important for the prevention of erroneous paternity on outstanding boars.

5.2.1 Electrophoretic variants of serum proteins

Electrophoretic variants of serum protein have been referred to in swine (Kristjansson, 1961) as heptaglobins. Imlah (1964) found that these same electrophoretic variants of swine could be demonstrated only if old haemoglobin or alkaline hematin rather than fresh haemoglobin was added to swine sera prior to electrophoresis. Consequently Imlah (1964) renamed this polymorphic protein as "haem-binding globulin".

5.2.2 Albumin (Alb)

Kristjansson (1966) discovered a triallelic polymorphism for the major albumin fraction in weakly acidic gels. The different phenotypes were distinguishable by the pattern of small sub fractions classifiable into 6 discrete phenotypes, *viz.* AA, AB, BB, AO, BO and OO. Mating data showed control through alleles A, B and O.

5.2.3 Ceruloplasmin (Cp)

Polymorphism for this copper binding protein was discovered by Imlah (1964). The existence of 2 alleles and 3 phenotypes at this locus was reported on the basis of segregation analysis among offsprings. These observations have since been confirmed by Hesselholt (1969).

5.2.4 Transferrin (Tf)

Intraspecific variation of 7 transferrin types was first described independently by Kristijansson (1960) and Kristjansson and Cipera (1963). Two alleles Tf^A and Tf^B were identified. Three bands represented homozygous expression, and 5 to 6 bands the heterozygous condition. Subsequently more alleles determining transferrin heterogeneity were identified.

5.2.5 Haemopexin (Hpx)

Kristijansson (1961), using electrophoresis, demonstrated 10 benzidine stainable components. Three components were assumed to be haemoglobin binding haptoglobins. Genetic investigations suggested a triallelic system (Hp^1, Hp^2 and Hp^3). Subsequently, these proteins were identified as haemopexin.

5.2.6 Acid phosphates (Acp)

Meyer and Verhorst (1973) described phenotypes A, AB and B in porcine haemolysates. A phenotype represented the most predominant type and consisted of 2 fast moving fractions. B showed 2 slow moving fractions. The heterozygote phenotype (AB) showed the presence of 3 electrophoretic fractions.

5.2.7 Carbonic anhydrase (Ca)

Porcine haemolysates during electrophoresis revealed 2 zones of activity, *viz.* Ca-I (low activity zone) and Ca-II (high activity zone). Kloster *et al.* (1970) described 3 phenotypes controlled by 2 autosomal alleles Ca-II^{A} and Ca-II^{B}. Ca-II^{B} was predominant in most of the populations.

5.2.8 Amylase (Am)

Gene controlled variation for porcine serum amylase was discovered by Graetzer *et al.* (1965). Six phenotypes consisting of three variants, *viz.* A^{m-1}, A^{m-2} and A^{m-3}, in the order of decreasing anodic mobility were observed. This polymorphic system was extensively investigated and confirmed. Hesselholt (1969) described an additional amylase variant Am-2F. Tanake and Masangkey (1978) reported the occurrence of Am-X and Am-Y variants in the Philippine native pigs.

5.3 Genetic Relationship

Fourteen protein systems encoded by 15 structural loci were used to investigate genetic variability in three swine breeds (Landrace, Large White and Duroc), reared in Southern Brazil. The degree of genetic variability was similar in the three breeds (Landrace, He-0.116; Large White, He-0.119; Duroc, He-0.095). These values are close to those computed for other populations of these breeds and higher than those obtained for wild pig populations. The gene frequencies at the polymorphic loci were employed to evaluate the usefulness of these systems for parent identification. The combined probabilities of paternity exclusion were estimated at 59% for Landrace, 54% for Large White and 50% for Duroc animals. Analysis of genetic relationships revealed that Landrace and Large White are the most similar breeds (D-0.044), while the Duroc breed presents lower levels of genetic similarity to the other two breeds (Landrace/Duroc: D-0.084; Large white/ Duroc: D-0.106). These findings are in agreement with the historical development of these breeds.

Table 5.4 shows allele frequencies estimated for the systems that were polymorphic in at least one of the samples under consideration.

Table 5.4 Frequencies of Various Blood Protein Alleles in Populations of Landrace, Large White and Duroc Breeds (Tagliaro *et al.* (1993))

Locus	Allele	Allele frequencies		
		Landrace (N-109)	Large white (N-116)	Duroc (N-57)
Pgd	Pgd *A	0.628	0.621	0.228
	Pgd *B	0.372	0.379	0.772
EsD	EsD*A	0.931	1.000	0.851
	EsD*B	0.069	0.000	0.149
Amy1	Amy1*A	0.133	0.090	0.000
	Amy1*B	0.862	0.910	1.000
	Amy1*C	0.005	0.000	0.000
Phi	Phi*A	0.156	0.444	0.184
	Phi*B	0.844	0.556	0.816
Cp	Cp*A	0.018	0.000	0.000
	Cp*B	0.982	1.000	1.000
Hpx	Hpx*0	0.064	0.004	0.000
	Hpx*1	0.624	0.746	0.070
	Hpx*2	0.046	0.000	0.140
	Hpx*3	0.266	0.250	0.790
Tf	Tf*A	0.037	0.168	0.096
	Tf*B	0.963	0.832	0.904

5.4 Physical Traits

5.4.1 Colour

In a number of experiments the inheritance of colour has been studied. The most detailed study of colour has been carried out by H.O. Hetzer (1945–1948). The Scandinavian Landrace, the English Yorkshire and Large White are examples of white breeds of pigs. The English Large Black is black whereas Tamworth and Duroc from England and the USA respectively are red.

Hetzer is of the opinion that the inheritance of black and red colour in pigs is genetically similar to that of rodents. In the Hampshire, for example, the black is determined by a dominant gene, E; the almost black Berkshire and Poland China are assumed to be homozygous for the gene e^p, but the spotted red and black colour has been obtained by an accumulation of modifying genes. The all white colour of Scandinavian Landrace and Yorkshire is controlled by a dominant gene usually denoted by *I*.

Adult wild pigs are recognized by a dark greyish-brown colour, but the piglets, up to 4–5 months of age, have a red colour with longitudinal creamy-white stripes on each side of the body. The difference between the wild pig type and black colour of the Berkshire is apparently due to a dominant gene in the wild pig.

5.4.2 Hair characteristics

In the majority of farm animal species there are individual differences in hair length, diameter and general appearance such as waviness, curliness etc. A part of this variation is clearly genetic.

5.5 Genetic Abnormalities

Abnormalities are deviations from normal development and can involve any part of the pig, internal or external. These defects can impair the pig's ability to function or even cause death. Anatomical abnormalities or defects occur in at least 1% of newborn pigs. These defects may be caused by genetic or environmental factors. However, the frequent enough occurrence in an individual herd causes substantial economic loss. That is why they are important.

5.5.1 Chromosomal aberrations

Chromosomes occur in pairs in body cells. A sperm or ovum contains only one of each pair of chromosomes. There are two types of chromosomes. One pair of chromosomes is known as the sex chromosomes because they are involved in determining the sex of an animal. In mammals the sex chromosomes are called X and Y, with the X chromosome being much larger than the Y chromosome. Females have two X chromosomes, and males have one X and one Y chromosome. All chromosomes other than sex chromosomes are called autosomes.

1. ***Number of chromosomes***: Number of chromosomes in pigs may be increas or decreas from normal (19 pairs). The effects of increased or decreased number of chromosomes are usually so severe that early embryonic death occurs. The exception is increased number of sex chromosomes, which usually results in infertility.
2. ***Structural Alterations***: Structural alterations usually are the result of pieces of chromosomes breaking off and recombining in a non-normal manner during the process of sperm or egg formation. Such defects also tend to result in major abnormalities that often cause fatal death or early death of the newborn pigs. However, data from Europe show that some translocations (movement of pieces of one chromosome to another) are not fatal to some pigs and in boars may cause lower fertility. A sharp reduction in litter size in a group of sows, mated to a specific boar, may indicate that the boar possesses a translocation.

Simple genetic inheritance

The gene is the smallest unit of inheritance and is a structural part of a chromosome.

If genes at only one location on the pair of chromosomes are responsible for the disorder, it is considered simply inherited.

Multigenic inheritance

Multigenic disorders are those controlled by genes at two or more locations on the chromosomes.

5.5.2 Important genetic abnormalities

In breeding research work on the identification of genes involved in genetic defects in swine is the focus.

Splayleg

Splayleg is a genetic defect seen in newborn piglets that are unable to hold the front and/or back legs together. Up to 2% of the born piglets can be affected. The mobility of the piglet is impaired which makes teat access difficult. One of the theories is that this defect is caused by immaturity of the muscle fibres in the hind leg (myofibrillar hypoplasia) (Thurley *et al.,* 1967). However not much and only very old literature is available on the physiology of muscle development and glycogen storage and release in the period around birth. Splayleg is more common in Landrace (Ward, 1978). It is described in literature that probably two recessive genes are responsible for the observed defect (Stigler *et al.,* 1991).

Maak *et al.* (2003) describes the selection of candidate genes based on differential display. Sixteen genes were selected for further analysis. One candidate gene (CDKN3) has been described in detail. Several SNPs and mutations that result in altered RNA are described. Association of the mutations, however, with splayleg, have not been found. In this project 16 candidate genes that were selected based on their possible involvement in the development of splayleg in piglets. Several hundred animals that were diagnosed with splayleg and at least one sibling that is unaffected were used in this study. The candidate genes were screened for mutations that might be associated with the occurrence of splayleg in the collected animal population. In total, more than 40 SNPs in these genes were observed in a panel of 4 affected animals with their 4 unaffected siblings. The most interesting SNPs were selected for high throughput typing on the complete animal data set. This will hopefully result in association of the mutations with the phenotype. Based on these results easy to use DNA tests will be developed that can be applied in the breeding program to eliminate this genetic defect from the population.

Scrotal hernia

Scrotal hernia and inguinal hernia are variants of a defect in which intestines or other abdominal organs pass into the inguinal canal. Scrotal hernia is the more exaggerated form of the defect in that the abdominal organs protrude into the scrotum. Scrotal hernia can occur in males that have very large inguinal canals. Without castration, most of the animals having scrotal hernia will grow without problems to slaughter weight. Castrated animals, however, that have scrotal hernia have higher risks of problems.

Frequencies vary from 1.68% to 6.7% described in literature for several breeds. Based on several studies, heritability for scrotal hernia ranges from 0.15 – 0.86 (Vogt and Ellersieck, 1990). There is agreement that development of this defect is genetically influenced, but no major genes or any clear pattern of inheritance has been identified. Different studies report on scrotal hernia being influenced by one incompletely dominant factor, 2 loci, 2 pairs of homozygous recessive genes, or multiple genes (Vogt and Ellersieck, 1990). Some groups have tried to identify the genes involved in scrotal hernia. Some candidate genes were screened for association with scrotal hernia (Beck *et al.*, 2002, Knorr *et al.*, 2002), in addition to a total genome scans (Bornemann *et al.*, 2002). No associations of SNPs in these genes with the trait are described in literature. Some genes were excluded as a common genetic basis of hernia inguinalis/scrotalis in pigs based on the absence of association.

From several different lines, animals that showed scrotal hernia and their unaffected siblings were collected. In total, several hundreds of tissue samples were collected. For scrotal hernia 7 relevant candidate genes could be selected from literature and biological databases. These genes were screened for polymorphisms in the introns and exons. Four interesting SNP were detected, which are being typed on all animals, affected and unaffected. As described for splayleg, DNA tests can be developed out of these results that can be used for lowering the incidence of scrotal hernia.

Gene defects increase susceptibility of pigs to infectious diseases

Gene defects that increase susceptibility of pigs to infectious diseases have been established by Lillie and coworkers. They have specifically shown that the normal pig mbl-1 and mbl-2 genes supply the MBL-A and MBL-C proteins that are produced in the liver and circulate in the blood. A defect in the mbl-1 gene was discovered and a genetic test for this was developed. The mbl-1 defect was more frequently found in pigs culled with various common infections. Low MBL-C producers were more frequently sick. Several defects were identified in the pig mbl-2 gene, and some of these explain why liver production of MBL-C is highly variable among pigs.

Other important defects due to gene abnormalities are stated here:

Procine stress syndrome (PSS)

This condition is characterized by a progressive increase in body temperature, muscle rigidity, and metabolic acidosis leading to sudden death of heavy muscled pigs. PSS can also lead to the production of pale, soft, and exudative (PSE) meat. PSS is inherited as an autosomal recessive.

Umbilical hernia

This defect may have a genetic liability that is magnified by adverse environmental conditions, such as crowding to conserve heat during cold weather.

Atresia Ani

This condition is characterized by a pig being born without a rectal opening. This condition has a genetic basis, but is definitely not due to a single gene.

Cryptorchidism

Cryptorchids or ridglings are male pigs with one or both testicles retained in the body cavity. Animals with both testicles retained are sterile. Sex limited inheritance with at least two gene pairs seems possible.

Hermaphrodites

Hermaphrodites are frequently observed among the Large White and Landrace breeds of Europe and with a frequency of 0.1 to 0.5% in Yorkshire and Landrace in the United States. Sex chromatin studies show most hermaphrodites to be genetic females (XX genotype), but to posses portions of the male sex organs.

Nipple abnormalities

Inverted nipples are the underlying abnormality of the greatest concern. This condition is characterized by failure of nipples to protrude from the udder surface. The teat canal is held inward, forming a small crater so that normal milk flow is prevented. This abnormality has a genetic cause, but the number of pairs of genes involved is unknown. The heritability is estimated to be approximately 20%.

5.6 DNA Polymorphism

5.6.1 Sequencing of the porcine genome

After the human genome sequence was completed some time ago, the intention was made to sequence genomes from important livestock species. Much international effort is required to make the tools available that are needed for the sequencing effort (*e.g.* ESTs, BAC libraries and comparative maps)

5.6.2 Dissection of complex traits–QTLs and candidate genes

Two major strategies are used to identify genes that are involved in complex traits. The candidate gene approach tries to identify genes based on their possible role in the physiology of the trait. The Quantitative Trait Locus (QTL) strategy relies on a scan of the entire genome using anonymous markers combined with phenotypic measurements.

The best-described examples of genes controlling variation in quality traits are the Halothane gene (RYR1) and the RN locus (PRKAG3). The Halothane gene is associated with stress sensitivity. Homozygote recessive animals are more sensitive to stress, have higher carcass lean meat and lower meat quality. The DNA test that was developed in the early 1990s made it possible to distinguish between all three genotypes and thus allowing breeders to change the frequencies of the alleles in their commercial populations. The RN phenotype is common in Hampshire pigs and is characterized by large effects on meat quality traits. Animals carrying the dominant allele designated RN (-) have lower meat quality but stronger taste and smell. The difference with the RN (+) animals is caused by higher glycogen content storage in the muscle. It is expected that in the coming years several more candidate genes will be identified for complex traits and will be used by the commercial pig industry.

A large number of QTLs have been reported on nearly all chromosomes for growth, carcass and meat quality traits. In addition, QTLs for disease resistance and reproduction have been reported for several chromosomes. In most studies crosses between exotic breeds (*e.g.* Meishan, Wild Boar) were used to detect QTLs. In only a few studies commercial populations were used. The results from the latter crosses are more relevant to pig breeders. It remains, however, still very difficult to find the gene or mutation in the QTL regions that are responsible for the observed phenotypic variation. One way to identify the underlying genes is called the positional candidate approach, where a directed search for candidate genes based on biological function in the QTL region is conducted. A large QTL effect for muscle mass and fat deposition that is only expressed (seen) in boars and not in sows, caused by one single basepair mutation near the IGF2 gene.

Quantitative traits are generally regulated by multiple genes and their interactions between these genes and their environment. The Quantitative Trait Loci (QTLs) are stretches of DNA that are closely linked to the genes that controlling the trait, which may not necessarily be the genes themselves. QTLs can be identified by many methods such as AFLP to map genomic regions that contain genes involved in specifying a quantitative trait. QTLs are often found on different chromosomes.

Use of QTL data

1. The number of QTLs explains variation in the phenotypic trait and gives an idea about genetic architecture of a trait. For example, QTL data can be used to identify genes affecting litter size in pig such that it is controlled by many genes of small effect, or by a few genes having large effect.
2. Identification of candidate genes controlling a trait. Once a region of DNA is identified as contributing to a phenotype, it can be sequenced. The DNA sequence of any genes in this region can then be compared to a database of DNA for genes whose function is already known.
3. QTL information along with gene expression profiling data from microarrays and transcriptome profiling can identify regulatory elements of gene expression (cis- and trans elements).

QTL mapping

QTL mapping in detail is beyond the scope of this book; hence a brief account is included. QTL mapping is the statistical analysis of the alleles in a locus and the phenotypes represented by these alleles. Since most traits are polygenic, analysis of entire locus of genes related to a trait gives an understanding of genotype. QTLs identify a particular region of the genome as containing a gene that is associated with the trait being assayed or measured. They are shown as intervals across a chromosome, where the probability of association is plotted for each marker used in the mapping experiment.

Steps in QTL mapping

Defining genetic marker for pig, which is an identifiable region of variable DNA is the first step in mapping QTL. This is done by identifying gene sequences likely to co-occur with traits of interest through statistical analysis. One may exclude genes of known function from the DNA sequences identified to finally arrive in QTL. If no genome data is available, one may sequence the DNA segments and determine possible functions to identify a QTL. Several on-line tools such as BLAST at NCBI site are available for this purpose. Several methods of QTL mapping has been identified such as analysis of variance (ANOVA), interval mapping, Composite interval mapping, Family-pedigree based mapping and analysis of single nucleotide polymorphisms (SNP).

QTL in pigs

The first QTL that was discovered was a major locus for fat deposition on chromosome 4 in 1994 (Andersson *et al.,* 1994) following which several QTLs have been mapped. Initially many QTL experiments were undertaken by using initial linkage maps to help determine regions underlying traits of importance to the pig industry. Recently researchers have used two commercial breeds for F2 families or large commercial synthetic lines or breeds for candidate gene studies and large scale SNP association analyses (Rothschild *et al.,* 2007). Excellent information on QTL in pigs can be found in http://www.animalgenome.org/cgi-bin/QTLdb/SS/index. The pig QTL database (Pig QTLdb) contains 4928 QTLs representing 499 different traits. For further information the readers may also consult Zhiliang *et al.* (2005, 2007) and Zhiliang and Reecy, (2007). The following figure shows QTL map of chromosome 1 where QTL for average daily gain has been mapped (Quintanilla *et al.* 2002).

The details of QTP mapping methods are presented in the Table below.

Table 5.5 Method of QTL mapping

Method of QTL mapping	Remarks
Analysis of variance (ANOVA)/ Marker regression	1. Separate estimates of QTL location and QTL effect cannot be found out. 2. Missing marker genotypes have to be discarded and cannot be included in breeding programme. 3. Efficiency for QTL detection will decrease when they are distantly placed from marker.
Interval mapping	Overcomes the three disadvantages of analysis of variance at marker loci. The method makes use of a genetic map of the typed markers, and, like analysis of variance, assumes the presence of a single QTL. Each location in the genome is posited, one at a time, as the location of the putative QTL.
Composite interval mapping (CIM)	Interval mapping using a subset of marker loci as covariates is done in CIM. These markers serve as proxies for other QTLs to increase the resolution of interval mapping, by accounting for linked QTLs and reducing the residual variation. In CIM the main concern is the choice of suitable marker loci to serve as covariates.
Pedigree based mapping	Plant geneticists are attempting to incorporate some of the methods pioneered in human genetics. There are some successful attempts to do so.

(Adapted from http://en.wikipedia.org/wiki/Quantitative_trait_locus)

Besides this, one may use pedigree mapping or new methods developed (Rosyara *et al.,* 2007).

Marker assisted selection

Recent development in molecular genomic analysis has revolutionized the evaluation of animals for breeding among the existing population. One of this is the application of markers in selection of animal. A marker is a DNA segment, gene which marks a section of chromosome affecting the performance. The gene for an economically important trait, the presence of which it detects, is known as a quantitative trait locus (QTL), with linkage between the marker and the QTL. The relation between marker and the QTL is used by the pig breeders and industry to improve swine production by marker-assisted selection. Selection with the aid of information at genetic markers is termed marker assisted selection (MAS). MAS are immensely supported by the tremendous progress made in mapping and characterizing the swine genome, which has been very recently completed.

Selection based on DNA markers is most useful for traits that are hard to measure and have low heritability. It allows earlier and more accurate selection, increasing short-and medium-term selection response, and may aid in targeting genotypes for specific production environments or markets. The use of genotypic information in breeding programmes for within-breed selection will generally have limited extra benefit, unless selection based on phenotype is difficult or advanced reproductive technologies are used (Werf and Marshall, 2005).

Association between a quantitative trait and genetic markers can be evaluated using single markers or multiple markers. When using one single marker, it is possible to make inference about the segregation of a QTL linked to that marker. However, with use of single markers it is not possible to distinguish between size of a QTL effect and its position relative to the marker. If multiple markers are used in an analysis, there is less confounding between the position and size of QTL effect, and subsequent increased possibility in detecting a QTL, even if the markers are far apart. Inference about the QTL effect as well as the recombination rate between QTL and markers is possible. The recombination rate between markers is usually assumed known. Therefore successful mapping of a QTL requires the use of multiple marker genotypes in the analysis (Werf, 2009).

The strategies for application of MAS in pig breeding programmes has been reviewed. The genetic markers could be codominant or DNA based. The markers may be applied for pig breeding programmes such as gene introgression, selection from synthetic populations and within line selection.

The MAS can apply in following conditions where index selection will be inefficient (Weller, 2001):

(i) Low heritability of trait,

(ii) Difficult or impossible to score the trait for *e.g.* traits cannot be measured in young animals, sex limited traits,

(iii) When there is negative correlation between traits,

(iv) Presence of non additive genetic variance and

(v) Cryptic genetic variations.

The application of MAS increases all the components of genetic gain, which is increasing accuracy of selection, increasing selection intensity and decreasing generation interval. In some cases, MAS is about 1.4 times more efficient than conventional methods. For detailed description on MAS, the readers may consult Weller (2001).

5.6.3 Genetic defect that causes infertility in pigs

The defective KPL2 gene in porcine chromosome 16 caused pig spermatozoa to be short tailed and immotile. The recessive genetic defect did not cause any other symptoms in the pigs. Sequence analysis of the candidate gene KPL2 reveled the presence of an inserted retrotransposon, a DNA sequence which moves around independently in the host genome. These transposable elements are found in all plants and animals.

Sironen also developed an accurate DNA test which can be used to identify animals carrying the defective gene with 100% certainty. The method, based on PCR technology, multiplies part of the KPL2 gene and detects the retrotransposon if it is present. The test has been used as a tool in Finnish pig breeding since 2006.

In a breakthrough study, a university of Missouri researcher is producing pigs born with cystic fibrosis (CF) that mimic the exact symptoms of human CF. This may help in further studying the deadly lung disease of humans. Table 5.6 indicates other anatomical defects and inherited disorder caused by genetic abnormality in swine.

Table 5.6 Other Anatomical Defects and Inherited Disorder of Swine

Disorder	Description	Probable cause
Blood warts (Melanotic tumors)	Moles or skin tumors. Increase in size with age. Tumors heavily pigmented and contain hair. Injury causes depigmentation. Common in Durocs and Hampshires.	Inheritance unknown but multigenic inheritance has been postulated.
Brain hernia	Skull fails to close and brain protrudes. Generally lethal.	Simple recessive inheritance suggested.

Table 5.6 (*Contd...*)

Disorder	Description	Probable cause
Cleft palate	Palate does not close. Harelip results. Generally lethal.	Recessive lethal has been theorized but may result from multigenic genetic liability influenced by an environmental effect.
Gastric ulcers	Erosion of the epithelial lining of the stomach. Generally in the esophageal region.	Heritability estimates ranging from low to high have been reported. Pelleted and finely ground diets, high unsaturated fats and low selenium in the diet, copper toxicity and psychosomatic factors have been found to cause that problem.
Hemophilia (bleeders)	Slow clotting time. Death results from slight wounds or from navel cord hemorrhage.	Known to be caused by mycotoxins in feed or vitamin K deficiency. One confirmed case of simple recessive inheritance.
Humpback	Crooked spine behind shoulder.	Likely to have genetic cause but inheritance is unknown.
Hydrocephalus	Fluid on the brain. Brain cavity much enlarged.	A lethal gene inherited as a simple recessive.
Lymphosarcoma (Leukemia, lymphoma)	Malignant tumors of the lymph nodes with increased lymphocyte count. Stunted growth and death before 15 months of age.	Convincing evidence of an autosomal recessive
Motor neuron disease	Distinctive locomotor disorder of nursery pigs, characterized by inability to coordinate muscle movements and slight paralysis.	Strong suggestion of autosomal dominant inheritance.
Oedema (myxoedema, dropsy, hydrops)	Abnormal accumulation of fluid in tissue and body cavities, suggested. Possibly associated with a thyroid defect.	Autosomal recessive disorder.
Pseudo-vitamin D deficiency (rickets)	Indistinguishable from non-genetic lack of vitamin D resulting from deficiency of calcium or insufficient exposure to sunlight. The most noticeable effect is bowing of the limbs.	Inherited as an autosomal recessive.

Table 5.6 (*Contd...*)

Disorder	Description	Probable cause
Rectal prolapse	Protrusion of the terminal part of the rectum and anus.	Many environmental influences including coughing, piling, feed constituents, antibiotics, diarrhea have been implicated though genetic liability may exist.
Persistent frenulum	A close attachment of the prepuce to the body by a mucous membrane resulting in inadequate protrusion of the penis and inability to breed.	Inheritance unknown
Screw tail (kinky tail)	Flexed, crooked, or screw tail caused by fusion of caudal vertebrae.	Multigenic recessive inheritance has been postulated.
Swirls hair (hair whorls)	Forms a cowlick or swirl on neck or back, are involved.	At least 2 pairs or recessive genes
Wattles fleshy, bells)	Cartilaginous appendages covered with normal skin and suspended from the jaw.	Single locus (tassles, autosomal recessive inheritance.

CHAPTER 6

SELECTION AND GENETIC IMPROVEMENT

6.1 Introduction

The process in which certain individuals in a population are preferred to others for the production of the next generation is known as selection. Selection in general is of two types: natural, due to natural forces, and artificial, due to the efforts of man.

No new genes are created by selection. Under selection pressure there is a tendency for the frequency of the undesirable genes to be reduced whereas the frequency of the more desirable ones is increased. Thus, the main genetic effect of selection is to change gene frequencies, although there may be a tendency for an increase in homozygosity of the desirable genes in the population as progress is made in selection.

One of the most important decisions which breeders make is choosing which traits to be improved in their herds. Breeders must decide among numerous traits of economic importance and determine whether to improve performance of a small amount in several traits or make larger amounts of improvement in fewer traits.

Selection is similar to developing a financial budget when one has a limited amount of money to spend each month. Just as monthly income is limited, selection intensity is also limited. The breeder must decide how many traits to attempt to improve and how much selection pressure to put to each trait. Similar to compounding interest, genetic improvements accumulate over generations and hence affect the performance of the herd in subsequent generations. And like investment opportunities, returns resulting from selection are not the same for all traits. Expected response to selection is proportional to the heritability and selection differential of the traits. Traits with higher heritability have a greater response with a given selection intensity than traits with lower heritability. However, not all traits have the same economic value. So, while progress may be more rapid in a trait with a high heritability, the value of the progress may be greater for a trait with a lower heritability. The challenge to breeders is to determine which traits to improve based on the heritability and the economic values among them.

Once the selection objective is chosen, breeders should apply the appropriate selection criteria over a period of years to achieve a positive change in herd performance. The selection criterion may include any number of traits and methods of selection. Developing the criterion to maximize the rate of genetic improvement in the selection objective, results in maximum economic gain.

It is important to keep in mind that the objective and criterion are not the same. The objective is the goal of the program, whereas the criterion is the traits measured on animals and/or their relatives and used as the basis for selection to achieve the objective. The objective and criterion may even include different traits *e.g.*, the objective might be to improve pork quality of the carcass by increasing % lean, colour, and flavour. The criterion used to select breeding animals might be ultrasonic back fat depth and loin area (as estimators of percent lean) measured directly on the selected candidates plus colour and marbling score (as an indicator of flavour) measured on sibs or progeny. The selection criterion is developed to maximize the genetic improvement of the selection objective, as constrained by the cost and or ability to gather data on selected candidates and their relatives to use for the selection criterion.

Table 6.1 Relative Response in one Trait from Selection for Multiple Traits

Number of traits	Relative response[1]
1	1.00
2	0.71
3	0.58
4	0.50
5	0.44
10	0.31
20	0.22

Relative response=$1/\sqrt{n}$ where, n= number of traits.

Improving the performance in multiple traits simultaneously is usually desired in genetic improvement programmes. It is important that only traits of economic importance to the breeder and customers are included in selection objectives. Expanding the number of traits in the objectives, reduces the rate of improvement in individual traits but may increase overall productivity.

6.1.1 Natural selection

The main force responsible in nature for selection is the survival of the fittest in a particular environment. Natural selection is of interest because of its apparent effectiveness and because of the principles involved. Natural selection can be illustrated by considering the ecology of some of our wild animal species.

Some of the most interesting cases of natural selection are those involving man himself. All races of man that now exist belong to the same species, because they are interfertile, or have been in all instances where mating have been made between them. All races of man now in existence had a common origin, and at one time probably all men had the same kind of skin pigmentation. As the number of generations of man increased, mutations occurred in the genes affecting pigmentation of the skin, causing genetic variations in this trait over a range from light to dark or black. Man began to migrate into the various parts of the world and lived under a wide variety of climatic conditions of temperature and sunshine. In Africa, it is supposed, the dark skinned individuals survived in larger numbers and reproduced their kind, because they were better able to cope with environmental conditions in that particular region than were individuals with a lighter skin. Likewise, in the northern regions of Europe, men with white skins survived in a greater proportion, because they were better adapted to that environment of less intense sunlight and lower temperatures. But the Eskimos who live in the polar regions of the North are dark skinned. This is because, Eskimos are more recent migrants from Asia to the polar region as compared to the Negros in Africa and the Whites in Europe, they have not lived so long in that region.

Further, evidence is available that there is a differential selection for survival among humans for the A, B and O blood groups. It has been found that members of blood group A have more gastric carcinoma than other types and that members of type O have more peptic ulcers. This would suggest that natural selection is going on at the present time among these different blood groups, and the frequency of the A and O genes might be gradually decreasing unless, of course, there are other factors that have opposite effects and have brought the gene frequencies into equilibrium.

Natural selection is a very complicated process and many factors determine the proportion of individuals that will reproduce. Among these factors, the differences in mortality of the individuals in the population, especially early in life; differences in the duration of the period of sexual activity; the degree of sexual activity itself and differences in degrees of fertility of individuals in the population.

It is interesting to note that in the wild state, and even in domesticated animals to a certain extent, there is a tendency toward an elimination of the defective or detrimental genes that have arisen through mutations, through the survival of the fittest.

6.1.2 Artificial selection

Artificial selection is that which is practiced by man. Under this, man determines to a great extent which animals to be used to produce the next generation of offspring. Even in this, selection seems to have a part. Some research workers have divided selection in farm animals into two types, one known as automatic and the other as deliberate selection.

Litter size in swine can be used as an illustration to define these two terms. Here, automatic selection would result from differences in litter size even if parents were chosen entirely at random from all individuals available at sexual maturity. Under these conditions, there would be twice as much chance of saving offsprings for breeding purposes from a litter of eight than from a litter of four. Automatic selection here differs from natural selection only to the extent that the size of the litter in which an individual is reared influences the natural selective advantage of the individual for other traits. In deliberate selection, this term is applied to selection in swine for litter size above and beyond that which was automatic. In one study by Dickerson (1973) involving selection in swine, most of the selection for litter size at birth was automatic and very little was deliberate; the opportunity for deliberate selection among pigs, however, was utilized more fully for growth rate.

Definite differences between breeds and types of farm animals within a species prove that artificial selection has been effective in many instances. This is true, not only from the standpoint of colour patterns which exist in the various breeds, but also from the standpoint of differences in performance that involve certain quantitative traits. For instance, in dairy cattle there are definite breed differences in the amount of milk produced and in butterfat percentage of the milk.

6.2 Basis of Selection

The changes in traits due to selection, affects directly the changes in the frequency of gene influencing the traits. Selection in practice can seldom be for genes at single locus. Most of the traits of economic importance for farm animals are quantitative in nature and posses the following characteristics. Estimates of genotypes can probably never be perfect in quantitative traits. Information on (i) individual (ii) on his ancestors and (iii) collateral relatives and (iv) on his progeny are useful in arriving at genotype estimates.

Characteristics of quantitative traits are influenced by many pair of hereditary factors most of which individually have minor effects. It is seldom or never possible to identify the individual gene effects. They have continuous distribution with no sharp demarcation between 'good' and 'bad'.

Although we are far from having complete knowledge of the type of hereditary factor action, it appears that additive gene action, dominance (probably including over dominance) and epistasis are all of importance, the relative importance varying from character to character. The expression is greatly affected by environmental influences.

6.2.1 Selection on the basis of individuality

Selection on the basis of individuality means that the animals are selected on the basis of their own phenotype.

6.2.2 Traits considered useful of individual selection

In the case where the character or characters being selected are expressed in both sexes, the use of individual selection has much to be recommended. In first place, information on the individual is the most readily available. Such traits as body type, growth rate, litter size etc. Evaluation on the basis of individuality of all animals can be made, as information is available. After a female comes into production her records represent its phenotypes.

6.2.2.1 *Traits consideration*

(i) Coat colour (ii) Type and conformation and (iii) Carcass quality

Type may be defined as the ideal of body construction that makes an individual body suited for particular purpose. Increased emphasis is now being placed on selection for performance and carcass quality, because breeders realize that type or conformation of an individual is not the best indicator of its potential performance or its carcass quality and culling can be done on the basis of records representing their phenotype.

When the heritability of the trait is high, (range approx. 0.1 to 0.25) indicating that the trait is greatly affected by additive gene action, selection based on individual trait is most effective. High h^2 also suggests that phenotype strongly reflects the genotype and that the individuals that are superior for a particular trait also possess the desirable gene for that trait and would transmit them to their offspring.

Chance combinations of genes may make an individual outstanding, but his offspring may turn out to be inferior, because he cannot transmit gene combination to his offspring. The breeder should avoid keeping superior individuals from very mediocre parents and ancestors. For breeding purpose it would be much more desirable to keep superior individuals from parents and ancestors that themselves were outstanding.

6.2.2.2 *Individuality*

Selection on the basis of individuality means that animals are kept for breeding purposes on the basis of their own phenotype. Selection may be made for several traits, such as coat colour, conformation, performance or carcass quality. In the past, the emphasis in selection probably was based on coat colour and conformation, although performance and carcass quality have received more attention in recent years.

Most of the breeds of livestock are characterized by a particular coat colour or colour pattern, and this is one of the requirements for entry into the registry associations. Selection for coat colour has been practiced because of its aesthetic value rather than its possible correlation with other important economic traits.

Attempts to relate variations of coat colour to performance within a breed have not met with success although many livestock men feel that there is a relationship. There is a strong belief of horse breeders that there is a strong relationship between colour and temperament which has no basis as per the evidence. There is however evidence that animals of some colours are better able to cope with certain environmental conditions, such as high temperatures and intense sunlight in some regions of the tropics or in the south and the south-western portions of the United States. Coat colour in some instances is closely related to lethal and undesirable genes in farm animals. Further, other species such as the mouse, dog, cat, mink, and fox, also show such relationships. Certain coat colours are the trademark of the some breeds of livestock. This is probably because this can be easily recognized. It is thus important that the breeder must conform to the breed requirements for this trait otherwise he will not be in the purebred business for long.

Type and conformation have been used as the basis of selection for many years throughout the world. Type may be defined as the ideal of body construction that makes an individual best suited for a particular purpose. This basis of selection has merit in some instances. The conformation of a draught horse is such that he is better suited to pulling heavy loads than he is to racing. On the other hand, the reverse is true of the thoroughbred.

The performance of individuals has also been given some attention in the development of some of our breeds of livestock. For many years thoroughbred horses have been selected for breeding purposes for their speed. Dairy cows have been selected for their ability to give large amounts of milk and butter fat. In beef cattle and swine, however, less attention has been paid to selection for performance and carcass quality until recently.

Increased emphasis is now being placed on selection for performance and carcass quality, because breeders realize that the type or conformation of an individual is not the best indicator of its potential performance or its carcass quality. Appropriate measures of these traits must be applied before progress can be made in selection for them.

The correlation between type and carcass quality is greater in some instances than is the correlation between type and performance. The meatiness of hogs by a visual inspection, can be assessed but this is not reliable. Better methods are backfat probes on live animals, actual weighing and measuring of lean meat in the carcass.

The fact that type and performance are not usually closely related, indicates the importance of selecting separately for the important traits in livestock production. If the correlation between type and other traits is low, it means that they are inherited independently and that they can be improved only if selection is practiced for each of them.

Individuality for certain traits should always be given some consideration in a selection programme. However, it is more important in some instances than in others. It is most important as the basis of selection when the heritability of a trait is high, showing that the trait is greatly affected by additive gene action. High heritability estimates also suggest that the phenotype strongly reflects the genotype and that the individuals that are superior for a particular trait should also possess the desirable genes for that trait and should transmit them to their offspring.

6.2.2.3 *Short comings of individual selection*

1. Several important characters including milk production in diary cattle, maternal abilities in cows, ewes and sows and egg production in poultry are expressed only by females. Thus selection of breeding males cannot be based on their own performance.
2. Performance of records of milk and egg production and other maternal qualities are available only after sexual maturity is reached and usually after such selection has taken place.
3. In cases where heritability is low, individuality is a poor indicator of breeding value.
4. The easy appraisal of appearance (or ‘type’) often tempts the breeder to over emphasize on this character in selection. For characters to which

individual selection is adopted certain procedure will tend to maximization of the selection differential and the accuracy of selection.

In spite of these short comings, individuality must be considered in selection. In general, for traits expressed by both sexes, which are above average, should be used for breeding, regardless of the merit of close relatives.

Selection should be directed only towards factors of real importance

Simultaneous selection for more than one character automatically reduces the amount of selection pressure for any one character, so that it can be only $^{1}/_{2}$n as intensive as if it were the only character selected for. Thus selection for more characters simultaneously reduces the intensity of selection for any one character to one half what it could be if it were the only character selected for.

Secondly for some characters repeated observations are possible. Use of all the available records increases the accuracy of selection for characters affected by temporary environmental conditions, by maximizing the effects of these conditions thus reducing the number of mistakes made in selection.

The greatest disadvantage of selection on the basis of individuality is that environmental and genetic effects are sometimes difficult to distinguish. Much of the confusion may be avoided by growing or fattening of the offspring being compared for possible selection purposes under a standard environment. Even then, it is still possible to mistake some genetic effects for environmental effects. This is less likely to happen, however, in the outstanding individuals than in those that have a mediocre record. For instance, a bull calf placed on a performance test may make a poor record because of an injury or because of sickness while on test. But if he makes an outstanding record, it is certain that he possessed the proper genes and in the right combination as well as the proper environment to make the good record. It cannot always be certain, however, whether an individual with a mediocre record would have done better even if adverse environmental factors had not interfered. We can be certain that his record is poor and by culling on this basis, elimination of the genetically poor individuals is possible. This chance is worth taking, even though we may discard some genetically superior individuals occasionally.

Studies of selection on the basis of individuality within inbred lines of swine have shown that selection favoured the less inbred litters. This is another way of saying that selection probably favoured the more heterozygous individuals, and this may be true also in many cases where inbreeding is not involved to a great extent. Chance combinations of genes may make an individual outstanding, but his offspring may be inferior, because he cannot transmit his heterozygosity to his offspring. The breeder should avoid keeping superior individuals from very mediocre parents and ancestors. For breeding purposes, it would be much more desirable to keep superior individuals from parents and ancestors that themselves were outstanding.

6.2.3 Pedigree information as an aid to selection

A pedigree is a record of an individual's ancestors that are related to him through his parents. Earlier, the information included in a pedigree has been simply the names and registration numbers of the ancestors, and little has been indicated as to the type and performance of the ancestors. Pedigrees now include information on the size of the litter at birth and weaning.

If full information is available on the ancestors as well as the collateral relatives, it may be of importance in detecting carriers of a recessive gene. Such information has been used to a great extent in combating dwarfism in beef cattle.

A disadvantage of the use of the pedigree information in selection against a recessive gene is that there are often unintentional and unknown mistakes in pedigrees that may result in the condemnation of an entire line of breeding when actually the family may be free of such a defect. On the other hand, the frequency of a recessive gene in a family may be very low, and records may be incomplete. Then later, it will be found that the gene is present.

Another disadvantage of pedigree selection is that the individuals in the pedigree, especially the males, may have been selected from a very large group, and the pedigree tells us nothing about the merit of their relatives.

Still another disadvantage of pedigree selection is that a pedigree may often become popular because of fashion or fad and not because of the merit of the individuals it contains. The popularity of the pedigree may change in a year or two, and the value of such a pedigree may decrease considerably or may even be discriminated against. If popularity is actually based on merit, there is less danger of a diminution of value in a short period of time.

In using pedigrees for selection purposes, weight should be given to the most recent ancestors. This is because the percentage of genes contributed by an individual's ancestors is halved in each new generation. Some breeders place much emphasis on some outstanding ancestor for which three or four generations has been removed in the pedigree, but such an ancestor contributes a very small percentage of the genes the individual possesses and has very little influence on type and performance, unless line breeding to that ancestor has been practised.

An individual's own performance is usually of more value in selection than its pedigree, but the pedigree may be used as an accessory to sway the balance when two animals are very similar in individuality but one has a more desirable pedigree than the other. Pedigree information is also quite useful when the animals are selected at a young age and their own type and conformation is not known. Pedigree is useful in identifying superior families if good records are kept and are available.

6.2.3.1 *General principles which limit the usefulness of pedigree information*

The accuracy of pedigree information as an aid to selection is limited because of the sampling nature of inheritance, wherever gene are in heterozygous state. This makes it impossible to be exactly sure of what an individual offspring will be, even if one were in the extreme position of knowing exactly what inheritance its sire and dam hard.

It is mostly for characters which are not highly heritable, for characteristics which only one sex manifest and in selection which must be made while the animals are yet too young to show clearly their own performance what their individual merit is.

The kind of errors in individual selection, which are most likely to be remedied by pedigree information are those arising from the immaturity of the individual and from mistaking difference caused by environment and epistasis interactions for differences in breeding value. It helps and rarely in errors are caused by dominance when fairly full information about collateral relatives is included, but is not of much help in this respect when only the ancestors are described.

Information of this kind is now being used in meat certifications purposes, where a barrow and a gilt from each litter may be slaughtered to obtain carcass data. This is done, because otherwise the animal himself has to be slaughtered and information on his own carcass quality is to be obtained. Information on collateral relatives is also used in selecting since prolificacy can be measured boars only in sows even though the boar transmits genes to his offspring for this trait.

The record of a close ancestor is more significant than that of a distant one since the proposition of genes expected to be common increases, as degree of relationship increases. Further more, when the genotype of a close ancestor is estimated with high accuracy, the records of the more remote ancestors in the same of pedigree lose importance. When an animal has its own performance record, accuracy is increased very little by considering the pedigree.

6.2.4 Information from collateral relatives

Collateral relatives are those that are not related directly to an individual, either as ancestors or as their progeny. Thus, they are the individual's brothers, sisters, cousins, uncles and aunts. The more closely they are related to the individual in question, the more valuable the information for selection purposes.

Complete information on collateral relatives, gives an idea of the kind of genes and combinations of genes that the individual is likely to possess. Information of this kind has been used in meat hog certification programs, where a barrow and gilt from each litter may be slaughtered to obtain carcass data. Information on collateral has been used in the All India Coordinated Research on Breeding wherein information on slaughter traits has been used from full brothers for selecting boars for future breeding.

6.2.5 Progeny test

Selection on this basis means that we estimate the breeding value of an individual through a study of the traits or characteristics of its offspring. In other words, the progeny of different individuals are studied to determine which group is superior, and on this basis the superior breeders are given preference for future breeding purposes. If information is complete, this is an excellent way of identifying superior breeding animals.

Progeny tests are very useful for determining characteristics that are expressed only in one sex, such as milk production in buffalo or egg production in hens. Even though the bull does not produce milk nor does the rooster lays eggs, they carry genes for these traits and supply one-half of the inheritance to each of their daughters for that particular trait.

Progeny tests are also useful in measuring traits which cannot be measured in the living individual. A good example of this is carcass quality in cattle, sheep and hogs.

Progeny tests are also being used at the present time by experiment stations in studies of reciprocal recurrent selection. This type of selection is used to test for the "nicking ability" of individuals and lines and is based on the performance of the line cross progeny. Selection of this type is for traits that are lowly heritable and in which non-additive gene action seems to be important.

In comparing individuals on the basis of their progeny, certain precautions should be taken to make the comparisons fair and accurate. In conducting a progeny test, it is very important to test a random sample of the progeny. It would be more desirable if all progeny could be tested, but where this cannot be done, as in litters of swine, those nearer the average of the litter should be tested. It is also important that the females to which a male is mated should be from a non-selected group. One would expect the offspring of a sire to be superior if he is mated to the outstanding females in the herd. Such a practice would be misleading in comparing males by a progeny test, since much of the superiority of the offspring of one male could come from the dams and not from the sire. Some breeders prefer using a rotation of different dams when testing males, but this is practical only in swine, where two litters may be produced each year.

Using a large number of offspring in testing a sire increases the accuracy of the test. Where the number of females in a herd is limited, the number of males that may be progeny tested will be less as the number of mating per sire is decreased. The point is, then that the breeder must make some decision as to how many sires to test and how many progeny must be produced to give a good test. The number of offspring required for an accurate progeny test will depend upon the heritability of a trait, with fewer offspring being required, when the trait is highly heritable, and more being required when it is lowly heritable.

To make accurate progeny tests, it is also important to keep the environment, as nearly as possible, the same for the offspring of the different sires. In progeny testing in swine, for instance, confusion would result when the progeny of one sire were fed in dry lot during the summer and the progeny of another were fed on pasture. This would be particularly true in progeny testing for rate of gain, where pigs fed with modern rations often grow considerably faster in dry lot than on pasture. When this environmental condition is not controlled, the inferior sire might actually be thought to be superior.

Progeny tests in most of our farm animals have certain definite limitations. In cattle especially, it takes so long to prove an animal on a progeny test that he may be dead before the test is completed and his merit actually known. Progeny tests may be now easily done in swine than in other farm animals, but even in this case the males are usually disposed off by the time they are thoroughly progeny tested.

The process of progeny testing may be speeded up by testing males at an earlier age than they would ordinarily be used for breeding purposes. By hand-mating them to a few females, or by using them on a larger number of females by artificial insemination, harmful effects that might occur from overuse at too early an age may be prevented.

Too often, farmers send their sires to market just as soon as their daughters are old enough to breed, in order to prevent inbreeding. This practice has resulted in much loss of good genetic material for livestock improvement. Actually a sire is not proved until his daughters come into production. Rather than being slaughtered, a sire that has proved himself to be of high genetic merit should be used more extensively. It is true that his usefulness in a particular herd may be finished when his daughters are of breeding age, but he should be sent to another herd to be used for additional breeding purposes. To be proved, a sire must have completed a satisfactory progeny test record of some kind. He may be considered proved if he has offspring who have completed one year's record, but this varies with the traits involved. This may be a lactation record, or one of litter size, egg production, or birth and weaning weights, fleece yield and quality. A sire so tested may be said to be proved whether his offspring are good or poor. Before buying a proven sire to use in a herd, a breeder should not neglect to find out if he has been proved good or a poor producer. Newer methods of progeny testing may be developed that are superior to those already available. For instance, the semen of a buffalo bull that has been proved highly superior could be collected at regular intervals, frozen, and stored for later use, even after his death. In swine, it might be possible to get quicker progeny tests on females by weaning their pigs at two or three weeks of age and breeding them again as soon as possible to produce three litters per year. Superovulation, by the injection of certain hormones, a female can be made to produce hundreds of eggs instead of the usual one or few. Embryo transfer technique has made possible using extra ova to other females, where the fertilized ova may develop to birth and possess the characteristics of the mother which ovulated the egg. The success of the embryo transplantation of ova has been limited, but future studies may make it more practical. If this could be done, it would be possible for an outstanding female to have many offspring in one year, rather than just a few.

6.2.5.1 *Basis of progeny testing*

It has been said, individuality tells us what an animal seems to be, his pedigree tells us what he ought to be but his performance as breeding animal tell us what he is? Progeny testing is an effort to evaluate the genotype of an animal on the basis of progeny performance. The progeny test is used in animal breeding to help to decide which animal, within a group all having progeny, to keep for the production of more offspring and which to cull.

Genetic differences among progeny groups arises not only from simple additive gene action, but also interaction among allelic as well as non allelic genes.

The principles of the progeny test come from the sampling nature of inheritance. Each offspring receives from the parent sample half of the parent's inheritance. Each additional offspring receives another independent sample from the same source. If one can find out what was in several such samples he will be fairly sure of what was in the parent.

6.2.5.2 *Boar testing*

One or two boars are selected from a litter consisting of at least 8 pigs weaned. Growth rate and feed efficiency are recorded having reached 90 kg weight. The thickness of the back fat is measured on live animal. If satisfactory results are obtained in all respect, the tested boars and also sibs are recommended for breeding. If not, the boars are castrated and sent for slaughter and rest of the litter discarded.

6.2.5.3 *Other methods of progeny testing*

According to the new system recently introduced by Pet Industry Distributors Association (PIDA) use is made of both performance and progeny testing. The unit of testing is a group of four litter mates consisting of one gilt and two boars. The castrated and gilt are penned and fed together and after slaughter at 90 kg the carcasses are examined in detail for carcass quality. The two boars are penned together but fed separately. At 90 kg they are assessed for rate of growth and feed conversion. In addition, their back fat thickness is measured by ultrasonics etc., and it supplements the carcass information of their litter mates. The intention is to increase the number of litter groups for a complete progeny test of boars from four to six.

In an efficient breeding programme the objectives should be simple and clearly defined. In the PIDA system selection is based on two characters: carcass quality and economy of performance. Lean percentage as estimated by progeny testing or ultrasonic measurements, is the principal method of assessing carcass quality. Other carcass characters will be recorded, so that it will be possible to detect any deterioration. Daily gain and feed conversion will be recorded separately to be later combined into a single figure representing economy of performance.

Extra care is taken to avoid the spreading of contagious diseases by boars which, after selection go back as breeding boars to elite or accredited herds. The use of pigs of both sexes for carcass traits eliminates the risk of selection bringing about under sizable sex differences in the carcass quality.

6.2.5.4 *Expectation on future trend*

In a pig breeding programme, the performance test selection system is of vital importance to control and maximize the genetic gain.

Indiana breed societies use a different system of testing. The participating breeder send to the testing station an in pig gilt. Feed consumed by the gilt during gestation and lactation is recorded. The litter size is tested in usual way. The entire litter is fattened and after it reaches a weight of 90 kg one barrow is slaughtered and carcass data obtained of the remaining litter back fat thickness is measured on the live animals. The breeder receives the results of the test in order to enable him to select his pig for breeding.

6.2.5.5 *The advantages of progeny test*

1. Testing of traits which cannot be measured in the potential breeding animal itself and have to be measured on the carcass (*e.g.* meat quality).
2. Accuracy of prediction, especially if traits with low heritability are involved, due to large number of animals tested.

6.2.5.6 Short coming of progeny testing

1. Slow progress due to increase in interval between generations. Thus the increased cost and generation interval must be balanced against the additional accuracy of the progeny test.
2. Only male can be adequately progeny tested.
3. Only a few males must be tested in order to find out one that is truly outstanding.
4. For traits which are weakly inherited.
5. A high percentage of sire breeding life will have been passed by the time he is proved.
6. Progeny test information will accumulate so slowly on animals that by the time an adequate sample of her progeny has been tested a female will have passed much of her useful life and high expenses.

6.2.5.7 *Performance testing*

Young boars, from good parents in breeding herds, are performance tested for feed conversion, growth rate and back-fat thickness; they are also scored for bacon type. Information about the boar's breeding value for other carcass traits is obtained from full and half sibs, which are tested at the progeny testing stations. These stations are still operating with the traditional two males and two females in each test litter. Finally, information about the fertility of the dams, and possibly the maternal and paternal grand parents, is available from sow recording in the breeding herds.

Advantages of performance testing

- Early availability of results thus reduced generation interval.
- In case of traits with high heritability to good source of information.

- Possibility to test physical fitness prior to use of a breeding animal, in particular leg weakness in pigs.

Disadvantages of performance testing

- Less reliable information in case of low heritability.
- Problem to objectively assess carcass quality.
- A special testing station where groups of pigs can be tested in standardized condition is built for this purpose. Every litter must be inspected before it can be tested. Out of the approved litters, two pig is (1 barrow and 1 gilt) are sent to the station, where
  - (a) Rate of growth
  - (b) Economy of gain are recorded from 63 day of age to 95 kg weight. Having reached this weight each pig is slaughtered. Record of (i) dressing percent (ii) weight of five primal cuts (iii) length of body (iv) back fat thickness (v) loin eye area are taken.

The results of test are sent to the breeder for selection

Advantages of testing stations

- Standardized environmental conditions and simultaneous group testing make connection superfluous.
- Testing can be done at a fixed age and stage of production.
- Both feed consumption and performance can be recorded.

Disadvantages of testing stations

- High expenses (building and personnel).
- Limited testing capacity.
- Possibility of bias due to selected material.

The things which may keep the progeny test from being perfectly accurate are: the first practical difficulty encountered in using the progeny test is that we do not know exactly what composion of genes the offsprings have. The second practical difficulty encountered in using the progeny test is that each offspring also has received half of its inheritance from its other parent. Since we usually do not know exactly what was in that parent and will be still farther from knowing just what it contributed to this particular offspring, we are often in doubt as to whether a certain good quality in one offspring came from its sire or from its dam.

One way of overcoming difficulty consists of progeny testing an animal by breeding it to a large number of different mates in the hope that the merits and defects of those other parents would just cancel each other. Any general difference, then between the average of the progeny and the average of the breed could be credited to the common parent. This method might of course lead to errors if the other parents were so selected that their average merit was distinctly different from the breed average.

The third practical difficulty in using the progeny test is that the offspring of a given individual aught to have been born on somewhere near the same date and to have been reared under much the same environmental conditions. If there was anything unusual about that environment and if proper allowances for that was not made, we will credit or blame the heredity of the parent for something which was really caused by the environment of the offspring. This is probably the most influencing general limitation on the accuracy of the progeny test and there seems to be no automatic way of overcoming it. One can merely study as closely as possible the environment under which these offspring were tested and make such allowance as he thinks fairest for any conditions which were not standard.

6.2.5.8 *Selection index procedure for sires*

1. $$Is = \frac{0.5nh^2}{1+(n-1)0.25h^2}(\bar{S}-\bar{P})$$

n = Weighted average number of full sibs in a family

h^2 = heritability estimates of litter weight

$\bar{s}$ = average litter weight at weaning of the sire progeny

$\bar{P}$ = average of litter weight at weaning excluding the sire's litter weight at weaning which is under evaluation.

The selection will be done using the above formula. The criteria would be litter weight at weaning. h^2 is estimated by intra sire regression of daughters on dam.

The step for intra sire regression are as follows:

(a) The dam litter weight at weaning will be the independent variable (X)

(b) The progeny litter weight will be the dependent variable (Y)

(c) Intersire covariance between X and Y will be calculated as under.

2. $$COV_{XY}\frac{k}{i=i}\left[(\Sigma Xi\,Yi)\_\frac{(\Sigma Xk)(\Sigma Yi)}{n_i}\right]$$

will give the intrasire covariance. The sire number (i) varies from 1 to K.

(c) The variance of X will be calculated by intrasire regression using the following formula.

$$\frac{k}{i=1}\left[(\Sigma X^2 i) - \frac{(\Sigma X_i)^2}{Ni}\right]*$$

*(Johanson, I. and Rondel J. (1968). Genetic and Animal Breeding O.W.H. Freeman and company, San Francisco).

(d) Therefore the regression will be equal to $\frac{b}{c}$, which will be half the additive genetic variance for the trait.

(e) Thus the h^2 by intrasire regression of daughter on dam will be 2 x regression value.

The index for each sire will be calculated using the formula same as above. They will be ranked for selection, whose male piglets only be selected for future breeding.

Dam's index

Selection index has to be developed using its litter weight at weaning and dam's body weight at 24 weeks of age. For the construction of selection index the following parameters have to be calculated.

(a) h^2A = Twice the intrasire regression of dam's body weight at 24 weeks. This will be done as per procedure suggested in sire selection programme.

(b) h^2B = Twice the intrasire regression of litter weight at weaning of progeny which has already been calculated in sire selection programme.

(c) r_{GAB} = Genetic correlation between traits B (litter weight at weaning) and trait A (dam's body weight at 24 weeks of age) is calculated by following formula using intrasire regression method.

$$rG_{AB} = \frac{\text{Covariance}\,G_{AB}}{\sqrt{\text{Variance}\,G_A} \times \sqrt{\text{Variance}\,G_B}}$$

The phenotypic correlation *i.e.* $r/_{AB}$ and SD of A and B are calculated by using standard statistical procedure.

(d) The genetic SD of A and B are also calculated and by using variance for A and variance for B by using sire component of variance.

(e) The values of h^2 estimates for dam's weight at 24 weeks (A) and litter weight at weaning (B) as reported in literature ($h^2_A = 0.2$ and $h^2B = 0.29$) were used. Similarly the genetic correlation between A and B was used ($rG_{AB} = 0.46$). The phenotypic correlation among AB and B will be calculated from the experimental data ($r_{AB} = -0.313$). Similarly the phenotypic standard deviation observed during experiments are $\sigma A = 4.41 \text{ and } \sigma B = 14.85$. Using these parameters a selection index for the dam's (I_D) were calculated for ranking the dams in each generation. The construction of selection index for dam's weight at 24 weeks and its litter weight at weaning a logical procedure is to first derive predication equation based on casual paths.

$$h^2_B = 0.29, r_{AB} = -0.313, \sigma_B = 14.85$$

Thus the equation for predicting the breeding value of A (dam weight at 24 weeks) and B (litter weight at weaning) will be:

$$b_{G_A} AB + r_{AB}\, b_{GA} BA = r_{G_{AA}}$$
$$rABB_{G_A} + b_{G_A} BA = rG_{AB}$$

In matrix notation it can be written as

$$\begin{bmatrix} 1 & -0.313 \\ -0.313 & 1 \end{bmatrix} \begin{bmatrix} \bar{b}_{G_A} A.B \\ \bar{b}_{G_A} B.A \end{bmatrix} = \begin{bmatrix} 0.447 \\ 0.46 \end{bmatrix}$$

$$b_{G_A} AB = \frac{\begin{bmatrix} 0.447 & -0.313 \\ 0.46 & 1.0 \end{bmatrix}}{\begin{bmatrix} 1.0 & -0.313 \\ -0.313 & 1.0 \end{bmatrix}} = \frac{0.447 - (-0.313 \times 0.46)}{1 - X(-0.313)(-0.313)}$$

$$B = \frac{\begin{bmatrix} 0.447 & -0.313 \\ 0.46 & 1.0 \end{bmatrix}}{\begin{bmatrix} 1.0 & -0.313 \\ -0.313 & 1.3 \end{bmatrix}} = \frac{0.477 - (-0.313 \times 0.46)}{1 \times X(-0.313)} = \frac{0.59095}{0.09203} = 0.6552$$

Next, the partial regressions are obtained from the standard partial regression coefficients:

$$b_{GA}\ AB = \bar{b}_{GA}\ AB \frac{\sigma_{GA}}{\sigma_{GA}} = 0.6552 (\frac{\sigma_{GA}}{\sigma_{A}})$$

$$b_{GA}\ B.A = \bar{b}_{GA}\ B.A. \frac{\sigma_{GA}}{\sigma_{GA}} = 0.6651 - (\frac{\sigma_{GA}}{\sigma_{B}})$$

The prediction equation for breeding value for dam's body weight at 24 weeks will be

$$G_A = \bar{G}_A + b_{GA} A.B (A - \bar{A}) + b_{GA} (B.\bar{B})$$
$$= -\bar{G}A + 0.6552 (\frac{\sigma_{GA}}{\sigma_A})(A - \bar{A}) + 0.665 \frac{\sigma_{GA}}{\sigma_B} (B - \bar{B})$$

To obtain the equation for prediction the breeding value of litter weight at weaning (B) from Dam's weight at 24 weeks (A), the same procedure is followed:

$$rGB\ A.B + r_{AB}\quad rGB\ B.A = r_{GBA}$$
$$rAB\ bGB\ A.B + rGB\ B.A = r_{GBB}$$

$$b'G_B\ A.B = \left[\frac{\begin{matrix} 0.46 & -0.313 \\ 0.539 & 1.0 \end{matrix}}{\begin{matrix} 1.0 & -0.313 \\ -0.313 & 1.0 \end{matrix}}\right] = \frac{0.6287}{0.90203} = 0.697$$

$$\bar{b}_{GB}\ B.A1 = \left[\frac{\begin{matrix} 1.0 & 0.46 \\ -0.313 & 0.539 \end{matrix}}{\begin{matrix} 1.0 & -0.313 \\ -0.313 & 1.0 \end{matrix}}\right] = 0.539 - \left(\frac{-0.313 \times 0.46 = 0.7572}{0.00203}\right)$$

The partial regression coefficients are obtained from the above standard partial regression values as follows

The equation for prediction of the breeding value for litter weight at weaning (B) is then

$$\bar{G}_B \; B\bar{G}_B \; \sigma_{GB} \; A.B(A-\bar{A}) \pm b_{GB} \; B.A(B-\bar{B})$$

$$= G_B + 0.696\frac{\sigma_{GB}}{\sigma_A}(A-\bar{A}) + 0.7572\frac{\sigma_{GB}}{\sigma_B}(B-\bar{B})$$

Thus the above equations can be written as

$$G_A = 0.6552\,\frac{\sigma_{GA}}{\sigma_A} \times A + 0.665\,\frac{\bar{\sigma}_{GA}}{\sigma_B} \times B$$

$$G_B = 0.697\,\frac{\sigma_{GB}}{\sigma_A} \times A + 0.7572\,\frac{\sigma_{GA}}{\sigma_B} \times B$$

The relative economic value of two traits A and B are to be formed to develop an index

$$I = \begin{bmatrix} E\ G \\ A\ A \end{bmatrix} + \begin{bmatrix} E\ G \\ B\ B \end{bmatrix}$$

If they are of equal economic importance, as standard deviation of dams weight at 24 weeks (A) is worth as much as a SD of litter weight (B). The standard deviation of dam's weight at 24 weeks (A) is 4.14 units while the litter weight at weaning (B) is 14.85. The standard deviation of litter weight at weaning (B) is approximately 3.37 times that of weight at 24 weeks. The weight B.W. at 24 weeks and litter weight at weaning equally, the prediction equation of dam weight at 24 weeks has to be multiplied by 3.37 thus

$$I = 3.37\left[0.6552\frac{\sigma_{GA}A}{\sigma_A}\right] + 3.37\left[0.665\frac{\sigma_{GA}B}{\sigma_B}\right] + 0.697\left[\frac{\sigma_{GA}A}{\sigma_A}\right] + 0.7572\left[\frac{\sigma_{GB}B}{\sigma_B}\right]$$

and

$\sigma_{GA}, \sigma_{GB}, \sigma_A$ and σ_B,

$$ID = \frac{\sigma_B}{\sigma_A}\left[0.6552\frac{\sigma_{GA}}{\sigma_A}\right]A + \frac{\sigma_B}{\sigma_A}\left[0.665\frac{\sigma_{GA}}{\sigma_B}\right]B + 0.697\left[\frac{\sigma_{GB}}{\sigma_A}\right]A + 0.7572\left[\frac{\sigma_{GB}}{\sigma_B}\right]B$$

Selection of gilts will be done using the IS and ID which is averaged and each gilt is ranked accordingly.

$$I\,gilt = \frac{IS + ID}{2}$$

Summary

To select the male piglet, sire index is calculated. The h^2 estimates of litter weight at weaning as per standard literature is 0.29, (Edwards and Omtvedi, 1971), will be used. The other parameters *i.e.*

n = which is weighted average size of full sibs family, will be calculated by each unit using the following formula. $n = \frac{\Sigma n}{N}$ for each sire

S = average litter weight at weaning of particular sire whose index is being calculated.

P = average litter weight at weaning of the population excluding, sire which is under evaluations.

The formula for selection of sire will be

$$I_S = \frac{0.5n \times 0.29}{1 + (n-1)\ 0.25 \times 0.29\ (S - P)}$$

For selections of the gilts, formula will be

$$I_G = \frac{I_S + I_D}{2}$$

The method of calculation of I_S has been given above while I_D will be calculated by the following procedure.

The standard parameters for the h^2 estimates of both the traits and the genetic correlation between two traits will be used.

h^2 of body weight of the dam at 24 weeks: $h^2A = 0.20$

h^2 of litter weight at weaning of the dam $h^2 = 0.29$ trait.

I_g between weight at 24 weeks and litter weight at weaning of the dam rg AB = 0.46

The other parameters which will be calculated from the data pertaining to the pig farm.

σ A = phenotypic standard deviation of the dam's body weight at 24 weeks.

σ B = phenotypic SD of litter weight at weaning of the dam

r_{AB} = phenotypic correlation of the above two traits.

σ_{GA} = genetic standard deviation of trait A (weight at 24 weeks) Calculated from the genetic variance (σ_G) using Sire component.

σ_{GB} = This will be for trait 'B' *i.e.* litter weight of the dam at weaning, calculated from sire component.

The formula (I_D) will be

$$I_D = \frac{\sigma_B}{\sigma_A}\left[0.6552\frac{\sigma_{GA}}{\sigma_A}\right]A + \frac{\sigma_B}{\sigma_A}\left[0.665\frac{\sigma_{GA}}{\sigma_B}\right]B + 0.697\left[\frac{\sigma_{GB}}{\sigma_A}\right]A + 0.7572\left[\frac{\sigma_{GB}}{\sigma_B}\right]B$$

Example for calculation of I_S

Sire No.	Dam no.	Progeny Litter wt. at weaning	No. of Progeny per Litter.	Av. litter wt at weaning for each sire
1.	1	23.2	6	
	2	19.0	4	
	3	19.0	7	
	4	24.5	4	
	Total	**85.7**	**n_1 = 5.5**	**15.38**
2.	5	22.8	8	
	6	28.0	2	
	7	20.5	8	
	Total	**71.3**	**n_2 = 7.33**	**9.73**
3.	8	15.0	8	
	9	23.4	6	
	10	16.75	8	
	11	22.0	6	
	Total	**77.15**	**n_3 = 7.14**	**10.8**

$$n_1 = \frac{6^2 + 4^2 + 7^2 + 4^2}{21} = 5.57$$

$$n_2 = \frac{8^2 + 2^2 + 8^2}{18} = 7.33$$

$$n_3 = \frac{8^2 + 6^2 + 8^2 + 6^2}{22} = 7.14$$

$$IS_1 = \frac{0.5(5.57)\times 0.29}{1=(5.57-1)\,0.25\times 0.29}\left[15.38-\frac{(9.73+10.8)}{2}\right]$$

$$= \frac{0.808}{1+0.33}(15.38-10.26) = 0.61(5-12) = 3.12$$

$$IS_2 = \frac{0.5(7.33)\times 0.29}{1+(7.33-1)\,0.25\times 0.29}\ (9.73) = (\frac{15.38+10.8}{2})$$

$$= \frac{1.063}{1+0.459}(9.73-12.09) = 0.720(-3.36) = -2.45$$

$$IS_3 = \frac{0.5(7.14)\times 0.29}{1+(7.14)\,0.25\times 0.29}\ (10.8-(\frac{15.35+9.73}{2})$$

$$= \frac{1.005}{1+0.445}\ (10.8-12.55)$$

$$= 0.716(-1.75) = -1.253$$

The ranking for above sire litter

Ranks	1st	2nd	3rd
Sire nos	S_1	S_3	S_2

The male piglets which are to be retained for breeding will be selected as per their sire ranking,The future mating should be done in such a way that the male piglets are of the same sire.

Selection of female piglets

Assume that the different parameter proposed to be calculated are such that,

$$\sigma_A = 4.41 \sigma_B = 14.85 r_{GB} = 0.313$$

$$\sigma_{GA} = 3.0, \sigma_{GB}\ 12.0,$$

Dam$_1$ litter weight at weaning (B) = 15.2

Dam$_1$ weight at 24 week (A) = 25.0

Then

$$ID_1 = \frac{14.85}{4.41}\left(0.6552\times\frac{3.0}{4.41}\right)25.0$$

$$+\frac{14.85}{4.41}\left(0.6652\times\frac{3.0}{14.85}\right)15.2$$

$$+ 0.697 \left(\frac{12.0}{4.41}\right) 25 + 0.7572 \left(\frac{12.0}{14.85}\right) 15.2$$

$$= 3.37 (0.45) 25.0 + 3.37 (0.13) 15.2$$

$$+ 0.657 (2.72) 25.0 + 0.7572 (0.81) 15.2 = 98.32$$

6.3 Methods of Selection

The amount of progress made, regardless of the method used, depends upon the size of the selection differential (selection intensity), the heritability of the trait, the length of the generation interval and some other factors. The net value of an animal is dependent upon several traits that may not be of equal economic value or that may be independent of each other. For this reason, it is usually necessary to select for more than one trait at a time. The desired traits will depend upon their economic value, but only those of real importance need to consider. When too many traits are selected for at one time, less improvement, in any particular one is expected. Assuming that the traits are independent and their economic value and heritability are almost the same, the progress in selection for any one trait is only about 1/n times as effective as it would be if selection were applied for that trait alone. When four traits are selected at one time in an index, the progress for one of these traits would be on the order of $^1/_2$ (not $^1/_4$) as effective as if it were selected for alone. For the selection of superior breeding stock, several methods can be used for determining which animal should be saved and which should be rejected from breeding purposes. Three of these methods which are generally used are given as below.

6.3.1 Tandem (individual) selection method

In this method, selection is practiced for only one trait at a time until satisfactory improvement has been made in this trait. Selection efforts for this trait are then relaxed and efforts are directed toward the improvement of a second, then a third traits and so on. This is the least efficient of the three methods practiced in respect of the amount of genetic progress made for the time and effort spent by the breeder.

The efficiency of this method depends a great deal on the genetic association between the traits selected for. When there is desirable genetic association between the traits, improvement in one by selection results in improvement in the other trait not selected for, the method could be quite efficient. If there is little or no genetic association between the traits, the efficiency would be less. Since a very long period of time would be involved in the selection practiced, the breeder might change his goals too often or become discouraged and not practise selection that was intensive and prolonged enough to improve any desirable trait effectively. A negative genetic association between two traits, in which selection for an increase in desirability in one trait results in a decrease in the desirability of another, would actually nullify or neutralize the progress made in selection for any one trait indicating a low efficiency of the method.

6.3.2 Independent culling method

In this method, selection may be practiced for two or more traits at a time, but for each trait a minimum standard is set that an animal must meet in order to be selected for breeding purposes. The failure to meet the minimum standard for any one trait causes that animal to be rejected for breeding purposes. Let us assume that pig A was from a litter of 9 pigs weaned, weighed 77 kg at 5 months, and had 1.3 inches of backfat. For pig B, let us assume that it was from a litter of 5 pigs weaned, weighed 94 kg at 5 months, and had 0.95 inches of backfat at 84 kg. If the independent culling method of selection were used, pig B would be rejected, because it was from a litter of only five pigs. However, it was much superior to pig A in its weight at five months and in backfat thickness, and much of this superiority could have been of a genetic nature. Thus in practice, there is likelihood to cull some genetically very superior individuals when this method is used.

The independent culling method of selection has been widely used in the past, especially in the selection of cattle and sheep for show purposes, where each animal must meet a standard of excellence for type and conformation regardless of its status for other economic traits. It is also used when a particular colour or colour pattern is required. It is still being used to a certain extent in the production of show buffalo/cattle and sheep. It does have an advantage over the tandem method, when selection is practiced for more than one trait at a time. Sometimes, it is also advantageous, because an animal may be culled at a young age for its failure to meet minimum standards for one particular trait, when sufficient time to complete the test might reveal superiority in other traits.

6.3.3 Selection index

This method is based on the separate determination of the value for each of the traits selected for and the addition of these values to give a total score for all the traits. The animals with the highest total scores are kept for breeding purposes. The influence of each trait on the final index is determined by how much weight that trait is given in relation to the other traits. The amount of weight given to each trait depends upon its relative economic value, since all traits are not equally important in this respect, and upon the heritability of each trait and the genetic associations among the traits.

The selection indices is more efficient than the independent culling method, as it allows the individuals which are superior in some traits to be saved for breeding purposes even though they may be slightly deficient in one or more of the other traits. If an index is properly constructed, taking all factors into consideration, it is a more efficient method of selection than either of the other two described earlier, because it should result in more genetic improvement for the time and effort made for its use.

Selection indices seem to be gaining in popularity in livestock breeding. The kind of index used and the weight given to each of the traits is determined to a certain extent by the circumstances under which the animals are produced. Some indices are used for selection between individuals, others for selection between the progeny of parents from different kinds of mating, such as line-crossing and crossbreeding, and still others for the selection between individuals based on the merit of their relatives, as in the case of dairy bulls, where the trait cannot be measured in that particular individual.

6.3.3.1 *Selection indexes*

Selection index is a number intended to be proportional to an individual's breeding value and therefore usable as a criterion for selecting or rejecting that individual. It is made by combining credits for the individual's merit and penalties for the defects.

Needs for a selection index

An individual's net merits depends upon many things. If selection for each of these traits is practiced separately, two things happen, which reduce considerably the effectiveness of the selection.

First one is some inadvertently emphasized some traits more and other less than intended or than he thinks he is doing.

Culling independently for different things gives no opportunity to let unusually high merit in one trait offsetting slightly low merit in other.

It is economically unwise or even impossible to select for one thing alone, since the usefulness and economic value of the individual plant or animal always depend on several things.

Construction of a selection index

If the observed values of the characters that are desired to be selected is denoted by X1, X2, X3....... etc, and the underlying genetic basis for each as G1, G2,

G3.... etc. respectively, then an additive function (the simplest possible) of the G's with the appropriate economic weights will give the "breeding worth" (denoted as H) of each animal. Thus H will equal a1G1 + a2G2 + a3G3 + ..., where a's are the relative economic weights.

Since G's cannot be observed directly, only X's are observed, the index I is constructed as a linear function of X's such that the correlation between I and H (*i.e.* RIH) is the maximum. Thus I = b1X1 + b2X2+b3X3________, where b's are so determined that RIH is maximum requires the use of multiple regression technique.

The relation among the X's, the G's and the H can be illustrated by the path coefficient diagram as given below.

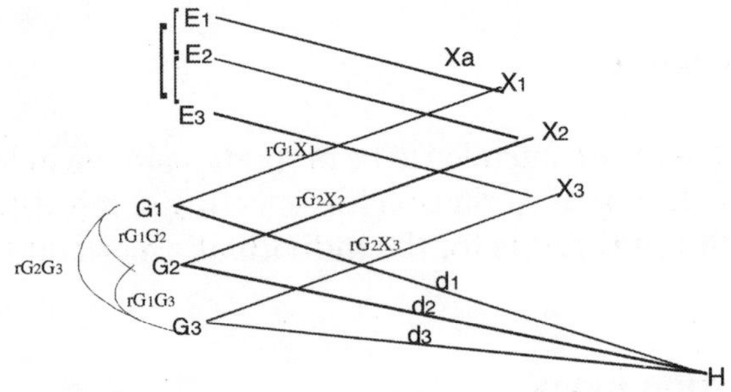

Here di = ai σ gi/ $_H$

ai = relative economic wt of the character

σ_{ai} = genetic standard deviation of the character

σ_H = standard deviation of H and

r Gi Xi = square root of the heritability of the character

Now the information that is needed for constituting a selection index can be summerized as follows:

1. Relative economic value of each trait choosen for improvement
2. Estimates of certain parameters
   - (i) Phenotypic
   - (a) Standard deviation for each trait
   - (b) Coefficient of linear correlation between each pair of traits
   - (c) Standard deviation for each trait
   - (d) Heritability of each trait

Selection index value for an animal is obtained by substituting the observed values of the animal in the formula for I.

Young (1961), has discussed in detail the relative response under these three methods.

Tandem method is by far the least efficient among the selection methods. If selection is made for an independent and equally important traits, with the same h^2 and variance, then the average response per generation for each traits, when the Tanden method is used, will diminish the response to a great extent when selection is for only one trait. On the above premise, the selection index method is $\sqrt{n}$ as efficient as tandem method.

When selection is based on independent culling levels, the selection intensity for a single traits is reduced as the number of traits to be considered increases. The selection intensity against the individual trait will thus be proportional to the function n $\sqrt{v}$ where the n is the number of traits and v is the fraction which must be saved for breeding.

Selection based on independent culling levels is more efficient than the tandem method, but less efficient than the selection index. In the latter case, usually high merit in one trait is allowed to compensate for slight inferiority in others.

Young (1961) extended the comparison of three selection methods to cover cases where the traits had unequal variances, h^2 and economic values. Factors such as selection intensity, the number of traits under selection and their relative importance (*i.e.* the product of economic weight, h^2 and phenotypic standard deviation) were found to influence the relative efficiency of the methods. Index selection is never less efficient than independent culling though in some cases it is not more efficient. Independent culling is never less but in some cases no more efficient than Tandem selection. With increasing the number of traits the superiority of the index method increases and its superiority is at a maximum when the traits are of equal importance. With increasingly intense selection, the superiority of index selection over independent culling decreases while its efficiency, as compared with that of Tandem method remain unchanged. The outcome of the three methods is strongly influenced by the phenotypic correlation between the traits under selection. The relative efficiency of index selection is higher when the phenotypic correlation is low or negative.

The greatest difficulty in selecting for two or more traits at the same time is that the possibility of strong negative genetic correlation may occur. It has been shown that h^2 for a combination of n negatively correlated traits with the same phenotypic and genetic variance approaches zero as the mean genetic correlation between all possible pairs of characters approaches –1/(n-1). Selection will then become ineffective.

6.4 Factors Affecting Selection Efficiency

Some factors which determine selection efficiency are: (1) Object in selection– definite goal, no change in objectives in a year or two. (2) Accuracy of the breeder in selecting superior stock will be increased if he compares all breeding animals under a standard environment. (3) Correction must be made for such factors such as age of dam and sex to increase accuracy. (4) In addition, he will be more accurate if he uses scales, rulers and other measuring devices whenever possible. (5) Accurate and detailed records are essential for increasing the accuracy.

6.4.1 Amount of selection pressure applied

The amount of selection pressure applied for a particular trait is known as the selection differential. In general, larger the selection differential, the more progress one can expect to make in selection.

6.4.2 Number of factors which affect the size of selection differential

(1) Number of animals that can be culled in the process of selecting breeding animals or the number of animals that needs to be kept in replacement purposes. Selection differential for males is almost always larger than that for females, since fewer males are needed for breeding purposes and they can be more extreme individuals. (2) Number of traits selected will have a tendency to reduce the size of selection differential for any one trait. Reason is that it is more difficult to find one individual who is outstanding for several traits than it is to find one that is outstanding for only one. (3) Level of performance: if the selection for a trait has been practiced for many years and the average of the herd for that particular trait is very high, it becomes more difficult to find individuals for breeding purposes that greatly exceed this average. On the other hand if there has been no selection for improvement in a particular herd and average in the trait is low, it becomes much easier to find individuals from a herd where the level of performances is very high.

6.4.3 Heritability of the traits

Selection for a trait that is lowly heritable will make little progress; selection for the trait that is highly heritable, should result in more progress in improving this trait. When the heritability of the trait is high, we expect a large portion of the selection differential to be due to heredity and less to environments. When the heritability of the trait is low, most of the selection differential may be due to environmental factors.

When heritability estimates together with the selection differential may be used to calculate the progress, we can expect to make in selection for certain traits.

Generation interval

It is the average age of the parents when their offspring are born. The generation internal in swine can be reduced to one year, if pig are selected from the first litters of gilts bred to boars of the same age. When this is practiced gilts can be bred when they are 7 to 8 months old and will produce litter by the time they are one year of age. If sows as well as boars are progeny tested before they are used to produce breeding or replacement offspring, the generation interval may be two years or even longer.

In four years time we should have the opportunity to produce from generations with first selection system, but with record only two generations would have been produced. It is obvious that the h^2 of the trait is the same, so we would expect to make more progress in selection in four than in two generations.

6.4.4 Genetic correlations among traits

Even if the heritability of the trait is as high as 70% no progress will be made in selection if the selection differential is zero. Furthermore, no progress will be made if the selection differential is large and the heritability of the trait is close to zero.

6.4.5 Heredity and environment interaction

The interaction of heredity and environment means that animals of certain genotype may perform more satisfactorily in one environment than they do in other. In other words, one environment permits the expression of genetic characters in a breed or strain, while another does not.

The Poland pigs were 10.5 kg heavier at 154 days of age than were the inbred Landrace pigs when both were fed the ration on pasture, but the difference was only 4.5 kg when they were fed in dry lot. Thus, the Landrace pigs grew faster in comparison to the Poland pigs in dry lot, than in pasture, which is another way of saying that the dry lot condition permitted the gene involved to achieve more complete expression. This seems reasonable since the Landrace breed was originally developed under dry lot conditions whereas Poland china were developed to a greater extent on pasture.

Breeders should be interested in knowing that genetic environmental interactions are important, and such knowledge should help answer the question of whether or not selection of animals for improvement in one set of conditions would also result in genetic improvement in another.

Importance of heredity and environment

It has been frequently discussed whether heredity or environment is the more important in the expression of economic traits. Such a discussion would be of little value, because it is now recognized that both are of very great importance. The best possible inheritance will not result in a superior herd or flock unless the proper environment is also supplied, so that the animals can attain the limit set for their inheritance. Half starved and neglected purebreds are truly a disappointment to livestock men in their appearance as well as their performance. Nevertheless, the best possible environment will not develop as superior herd or flock unless the proper inheritance is also present in the animals. To make the most possible use of good inheritance, we must select breeding animals which are superior because they possess more desirable genes or combination of genes. Superiority due to genes is the only thing that is transmitted from parent to their offspring. Superiority due to environment will not be transmitted by the parents. Thus superior environment must be provided for the offspring if they are to be the equal of their parents.

All of the phenotypic variations in a trait is due to hereditary (σ^2w) and to environment (σ^2e). The portion of the variation due to heredity would be equal to the hereditary variance divided by the total variance or percent hereditary variation

$$\left(\frac{\sigma^2_H}{\sigma^2_H + \sigma^2_e}\right)100$$

Let us assume that σ^2H is equal to 20 units and σ^2e is equal to 20 units. Thus % of the variance due to heredity would be

$$\left[\frac{20}{(20+20)}\right]\times 100 = 50\%$$

Suppose, however, that we are able to reduce the environment variation to an extent of only 10 units. In such a case, the portion of the variance due to heredity would be:

$$\left[\frac{20}{10+20}\right]\times 100 = 67\%$$

When we correct weaning weights for every piglet in a herd to the same age and same sex, as well as to the same age of dam, we are actually reducing the environmental variations between individuals in that herd and a larger proportion of the remaining variance should be due to heredity. Thus, the superior individuals after such connections are made would be more likely to be genetically superior, because we would increase the accuracy of picking those which possessed the more desirable genes or combination of genes.

For genetic reasons it is best to select and breed animals in the environments in which they have to perform.

In general, research results show up to the present that G × E is not very important in dairy, beef cattle and is more important in sheep, pig and poultry.

6.4.6 Complications of selection

(a) Genetic complication, (b) Operational complication

(a) Genetic complications

1. Heredity and environment

Most characteristics in animals are controlled by many genes, the same traits are also greatly influenced by environment. An animal with fast growth rate, raised in a deficient diet in an otherwise faulty environment, may end with same growth rate as an animal that has a poor genetic constitution for rate of growth, but was raised in good environment. Thus effect of environment can be responsible for mistake in selection. Both heredity and environment are responsible for the development of the character. The important thing for the breeder is to recognize the difference are heredity and thus increase accuracy of the selection.

2. Genotype and phenotype

The genotype of the animals is the animals' genetic constitution. It is more than the sum of all its genes, for it also includes the particular combination and arrangement of those genes. The particular gene will have different effects in different gene combinations. The genotype of an animal can therefore be referred to as its genetic environment. The genotype remains constant for an animal throughout its life.

The phenotype of an animal is the result of the interaction of the genotype and the environment in which the animal is developing. The phenotype, unlike genotype changes with time. This affects selection process.

Difficulties in selection arise because we can not identify the genotype of an animal accurately enough. If we know exactly the transmitting abilities of animals, progress from selection will surely follow.

3. Heritability

Most selection processes are based on phenotypic difference. Although we select on a phenotypic basis, our aim is to effect genotypic changes. The amount of change that selection is able to bring about is dependent on the relationship of phenotypic variation to genotypic variation. If the phenotype accurately reflects the genotype, selection will be quote accurate. If most of the phenotypic variation is environmental, progress form selection will be slow.

The larger the additively genetic portion of the phenotypic variance, the more accurately will a heritability estimation serve to identify the genotype. For this reason selection will be more effective in herds and for character were the h^2 is high. Heritability estimates are ratios expressed in percent and are usually designated by h^2. Like all ratios, the estimates will vary as their component vary.

The hereditary variation can be reduced through inbreeding and increased by an outcrossing or by a more complete control of environment. In a herd in which inbreeding of the animals is advancing the h^2 will decrease. After an outcross, the genetic variability, and therefore, the h^2 will be increased.

When the animals in herd are not raised under similar conditions, much of their phenotypic variation will be environmental. This will have the effect on reducing h^2. In our fraction E will be large and h^2 will be reduced. Where such a situation exists, many mistakes in selection will be made.

It can be seen that h^2 is based on the variation in a particular trait in a particular time and under particular environment.

A comparison of the variation between these parents-offspring-full sib relatives and variation between less closely related animals in the herd is the basis of all h^2 estimates.

Regression to the mean

Many breeders have been frustrated by the observation that the offspring of the animals that they selected had a tendency to regress to the average of the breed from which they were selected. This regression can now be explained fairly easily. When we get animals that are outstanding in characteristics, it is probably because these animals happened to get a favourable combination of genes and a satisfactory environment for these genes to express themselves. When these animals in turn reproduce, new combinations of genes are formed through segregation and independent assortment and these usually will be more like those of the average of the breed.

The genetic part of the regression can be at least partly avoided by increasing the homozygosity or genetic purity through inbreeding. The more nearly pure an animal is genetically, the less segregation there will be naturally. Where the heterozygote is superior to the homozygote, it will not be possible to fix this superiority. In most cases, the systematic crossing of line is the only way to restore superiority.

The environmental part of the regression can be lessened a great deal by keeping the same environment as far as possible from year to year.

Types of gene action

The fact that gene act differently in different combinations may make accurate selection more difficult. A simple case of dominance where A is dominant over to a, AA and Aa individuals will be of the same phenotype. They will be selected with equal preference, but AA will breed true where as Aa will segregate.

In case of over dominance, Aa will produce a larger effect than AA or aa. Here selection will favour Aa which can never be fixed.

Where there are many alleles in a series, combination of some of them will produce more favourable effects than others. For example in a series A1, A2, A3 and so on, A1 and A3 may produce a more favourable effect than any other combination. The job of the breeder is to increase the frequency of favourable alleles and to discard the less favorable ones. Selection with inbreeding should accomplish this.

In interactions of genes that are non alleles, a gene may complement or inhibit the action of another gene on group of genes. We do not know ways which gene interact to produce an effect, nor do we know the frequency of such interactions. We do know that the net effect of non-additive gene action is that the breeders cannot hope to continue to all the desirable effects in one super line of breed.

The breeder will do better to develop numerous lines that produce relatively well and then systematically cross those lines that produce the highest performing crossbred. Developing successfulness and finding suitable combinations for crossing can go on through the type and frequency of gene interaction. The methods are known and the results are gratifying.

Correlation of traits

Some characters are genetically correlated. For example it has been shown that a rapid rate of gain in swine positively correlated with efficiency of gain. Other characters are negatively correlated. In the case of positive correlations between desirable traits, selection is made somewhat easier, because selection for one is automatically works for the other.

Negative correlation between two desirable traits or positive correlation of desirable with undesirable traits have the same effect. They tend to lessen the effectiveness of selection.

Whenever possible, undesirable associations should be broken up by crossing, inbreeding and selection. A knowledge of the correlations between various characteristics should be a great help in avoiding mistake in selection.

Effects of inbreeding

It is generally known that a decline in all the attributes of vigour usually accompanies inbreeding. Hence many breeders hesitate to practice inbreeding. Inbreeding however, is necessary to introduce gene regeneration and to fix desirable gene contributions.

(b) Operational complication of selection

Object in selection

Many failure of selection in livestock can be attributed to lack of definite objectives. Selection will be more effective when the breeder has a definite objective for which to strive. The objective must be defined by measurements.

Number of traits

Selection becomes increasingly complex as the number of traits under selection increase. When single trait is subjected to selection it is simple matter to rank the population in order of their merit for that trait. This becomes more difficult as the number of traits is increased. The number of traits must be kept as small is practicably possible. The traits put under selection must be those with the greatest value from the stand point of utility.

Foundation stock

Selection may be ineffective because of an unfortunate or unwise choice of foundation stock. If the foundation animals are genetically poor, no one has yet demonstrated that selection pressure will be effective in bringing about improvement within a reasonable and workable period of time. Selection merely sorts genes and permits the better ones to be saved and the poor ones to be discarded. If the genes that we are looking for, are not in the foundation animals or are of very low frequency, they will have to be introduced by crossing or selection will be ineffective.

Selection can act only when there is variability. Genetic variability is caused by heterozygosity, and can be increased by out-breeding. Selection is ineffective for loci that are already homozygous.

Level of performance

Some time selection may be effective for a while and then it plateaus and no further progress taken place. For example in AI centres where proved dairy sires are used, it is easy to raise milk production in herds with low production. After several generation, as the level of performance of these herd is raised, further progress will be less andless. Selection here will loose effectiveness not because the quality of the sire is lower, but because the level of performance of herds has become higher.

When the level of performance of a line is already high further progress by selection will be slow, unless it is accompanied by a system of mating that will bring about new gene contribution.

Systems of selection

Too much rigidity in the system of selection may be a handicap to progress in an animal breeding programme. The system of selection should be flexible enough to allow the maximum selection pressure to be applied where the need is at any particular time. A fixed standard of selection, such as minimum record of performance, also has definite complication.

Length of time

In order to effect improvement in livestock through selection, a breeder must be prepared to continue his project for a relatively long period of time.

Number of animals

When the number of animals in a line or herd is small selection is severely restricted, because small herds or flocks offer very little opportunity for genetic segregation. There can be little selection in such cases. Even in less extreme cases, selection is likely to be handicapped through a lack of numbers.

6.4.7 Correlated characteristics

Correlated characters are of interest for 3 chief reasons.

(1) In connection with the genetic causes of correlation through the pleiotropic action of genes.

(2) In connection with the change brought about by selection it is important to know how the improvement in one character cause simultaneous changes in other character.

(3) In connection with natural selection the relationship between a matric character and fitness is important.

Genetic correlation

Genetic correlation is the correlation of breeding values. The genetic cause of correlation is pleiotropy through linkage is a cause of transient correlation. For example, genes that increase growth rate increase both stature and weight, so that they tend to cause correlation between these two characters. The degree of correlation arising from pleiotropy express the extent to which two characters are influenced by the same set of genes. But the correlation resulting from pleiotropy is the overall or net effect of all the segregating genes that effect both characters.

Environmental correlation

Environmental correlation is not strictly speaking the correlation of environmental deviations. Environmental correlation is so far two characters influenced by the same difference of environmental conditions. Again the correlation resulting from environmental causes is the overall effect of all the environmental factors that vary so may tend to cause a positive correlation or negative one. If both the characters have low heritability then phenotypic correlation is determined chiefly by the environmental correlation. If they have high heritability then genetic correlations is more important.

Phenotypic correlation

The association between two characters that can be directly observed is the correlation of phenotypic values or phenotypic correlation. This is determined by measurements of two characters in a number of individuals of the population in the same environment.

The genetic and environmental correlation are often very different in magnitude and sometimes different even in sign. A difference in sign between two correlations shows that genetic and environmental sources of variation affect the characters through different physiological mechanisms.

If highly inbred lines are available the environmental correlations can be determined directly from the phenotypic correlation with the lines or preferably within F_1's of crosses between the lines.

Estimate of genetic correlation are usually subject to rather large sampling errors and therefore seldom very precise. If it is low, then characters are to great extent different and high performance require a different set of genes.

If the genetic correlation is high then the two characters can be regarded as being substantially the same, if there are no special circumstances for offspring the h^2 or the intensity of selection will make little difference in which environment the selection is carried out.

If genetic correlation is low, then it will be advantageous to carry out the selection in the environment in which the population is determined. A character showed in two environment is to be regarded not as one character but as two. The physiological mechanisms are to some extent different and consequently the genes required for high performance are to some extent also different.

By regarding performance in different environments as different characters with the genetic correlation between them, we can in principle, solve the problem out lined above from a knowledge of heritabilities of the different characters and the genetic correlation between them.

If the genetic correlation is high, then performance in two different environment represents very nearly the same character, determined by very nearly the same set of genes.

6.4.8 Genotype environmental interaction

It means that the best genotype in one environment is not the best in another environment. Example: that breed of cattle, with the highest yield in temperate climate is unlikely to have the highest yield in tropical climate.

These matters have an important bearing on breeding policy. If selection is made under good conditions of feeding and management in the best farms at experimental stations, the improvement achieved be carried over when the later generation are transferred to poor conditions of management and feeding.

The idea of genetic correlation provide the basis for a solution of these problems in the following way:

Correlated response to selection

Response for a correlated character can be predicted if the genetic correlation and the h^2 of the two characters are known. With a low genetic correlation the expected response is small, and is liable to be occurred by random drift. Also if the genetic correlation is to any great extent caused by linkage, it is likely to diminish in magnitude through recombination, with a consequent dissemination of the correlated response.

Genetic correlation and selection limit

Just as the h^2 are expected to change after selection has been applied for some time, so also are the genetic correlations. If the selection has been applied to two characters simultaneously the genetic correlation between them is expected eventually to become negative for the following reasons.

Those pleiotropic genes that affect the two traits will be strongly acted on, by selection and brought rapidly towards fixation. They will then constitute little to the variance or the covariance of the two characters. The pleiotropic genes that affect one character favourably and other adversely will, however, be much less strongly influenced by selection and will remain longer at intermediate frequencies. Most of the remaining covariance of the two characters will, therefore, be due to these genes and the resulting genetic correlation will be negative.

The consequences of negative genetic correlation, whether produced by selection in this way or the two characters may each show a h^2 that is far from zero, and yet when selection is applied to them simultaneously, neither responds.

6.4.9 Response to selection

The response to selection is the change in the population mean after selection and is denoted by R. The difference in the mean of the selected parents from the mean of the population as a whole is known as selection differential (S). The intensity of selection (i) is calculated as selection differential divided by the standard phenotypic deviation of the trait. The response to selection is predicted from the heritability and the selection differential as:

$R = h^2 S$, which is popularly known as breeder's equation

Alternatively response due to selection, R can be calculated as $R = i\sigma_p h^2$

Considering average selection intensities for male and female, i_m and i_f, the response due to selection can be modified as

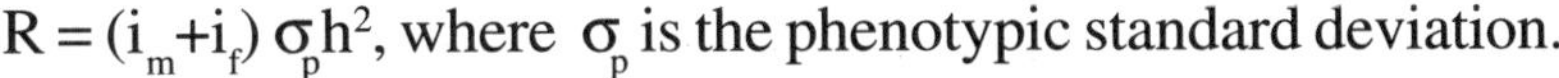

$R = (i_m + i_f)\, \sigma_p h^2$, where σ_p is the phenotypic standard deviation.

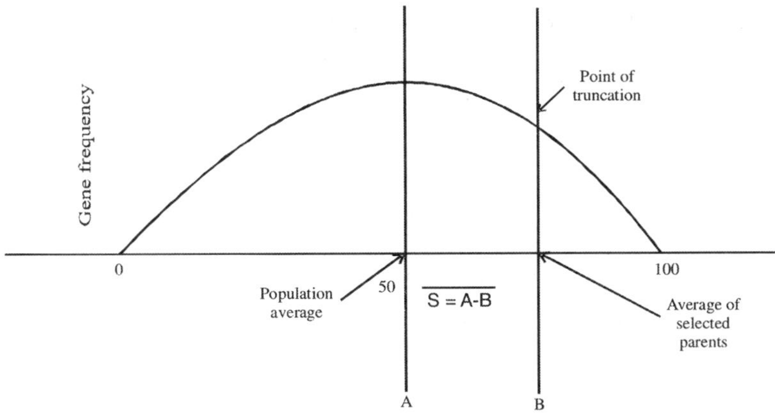

Fig 6.1. Standard deviation graph

Genetic gain (Δ G) per generation

Genetic gain (Δ G) is calculated from intensity of selection, accuracy of selection, genetic standard deviation (genetic variability) and generation interval. Thus, Δ G can be calculated as,

Genetic gain (Δ G) = intensity of selection × accuracy of selection × genetic standard deviation generation interval

p	i	p	i	p	i
1.00	0.0	–	–	–	–
0.90	0.20	0.09	1.80	0.008	2.74
0.80	0.35	0.08	1.85	0.006	2.83
0.70	0.50	0.07	1.91	0.004	2.96
0.60	0.64	0.06	1.98	0.002	3.17
0.50	0.80	0.05	2.06	0.001	3.38
0.40	0.97	0.04	2.15	0.0008	3.43
0.30	1.14	0.03	2.27	0.0006	3.51
0.20	1.40	0.02	2.42	0.0004	3.61
0.10	1.76	0.01	2.67	0.0002	3.79

Calculation of selection intensity, i based on 'i' proportion of animals (P) selected for breeding (Adapted from Population Genetics by Knud Christensen following the URL. http://www.husdyr.kvl.dk/htm/kc/popgen/genetics/8/1.htm)

Response under small amounts of inbreeding

When the amount of inbreeding is small enough that changes in the covariances between relatives are negligible, its main effect is inbreeding depression. Consider a population of modest size undergoing random mating, where the amount of inbreeding generated by genetic drift at generation t is $f_t H \Delta_t/2N_e$, provided t << N_e and f(0) = 0. If no epistasis is present, then inbreeding depression, "I is equal to bt/2N_e. The response due to inbreeding can be calculated for small amounts of inbreeding (f <0.05 – 0.1) as:

$$R(t)\square\ t.\bar{r}.h^2.\sigma_z - \frac{bt}{2N_e} = t.\bar{r}.(h^2\sigma_z - \frac{b}{2N_{e.}\bar{r}})$$

(After Walsh and Lynch, 2000. Selection under Inbreeding, University of Arizona) For f = t = (2Ne) > 0.1, the change in genetic variances from base population value must be taken into account (Walsh and Lynch, 2000).

Non-uniform response to selection

The response to selection is non uniform and can vary in selected line of animals since large number of factors determines the h^2 and S. Genetic variance is reduced by selection, in proportion to the reduction of phenotypic variance of the parents' relative to their entire generation, which is known as Blumer effect. Factors modulating response to selection can be summarized as:

Natural selection

Strong tendencies for and against natural selection for some of the traits favourably or adversely affect gene frequencies.

Linkages and correlated responses

The effect of one gene may be correlated with effect of other genes or linked, which can change predicted response due to selection.

Dominance and epistasis

Genetic interactions between and among loci such as dominance and epistasis can result in different gene effects at different frequencies, thus changing response to selection.

Genetic drift

Drift can cause the cumulative response in one direction to be greater than the other. Selection experiments are normally done with relatively small population sizes, therefore the chances of drift are high.

Change in environmental variance

Environmental conditions vary from season to season causing variation in responses.

Error in measurement of traits

Measurement error can result in substantially variable results. For example during selection on behavioral traits, or other traits with low repeatability, there can be error or variability in measurement of trait parameters.

Scale effects

The variance may change as a function of the mean, drifting towards one direction causing change in response to selection.

Inbreeding depression

Intense selection some times leads to development of inbred line, causing the gene frequencies to move in one direction. The directional dominance can result in biased selection.

Dam effects

The dam effects can act as an agent by which correlated effects are made, and can therefore cause asymmetry as well.

Genetical asymmetry and genes with large effects

When the average allelic frequencies at gene loci affecting a trait are different from $p = q = 0.5$, then the response to selection can tend to become asymmetric. Asymmetric responses can result if genes with large effects are present. For example, genes with large effect on fecundity such as X-linked Inverdale gene, $FecX^I$ in sheep.

Non-linear interactions between genes and environment

Selection is based on the principle that interactions between genes and environment are linear. Varying interaction of genes with environment can result in variation in response to selection.

6.4.10 Effectiveness of selection

When the actual performance of a whole herd or flock is improved over a period of years, it is taken to mean that selection has been effective in raising the level of performance. The actual production figures, however, give no indication of how much selection was practiced, or how much improvement was due to better nutrition and management.

If an increase in performance is due to management or some other environmental factors, the performance of the animal will slip back whenever the favourable environment does not prevail. If improvement in performance results from genetic improvement, the change is more of permanent nature.

Selection is generally considered to be effective when (i) it is successful in raising the level of production of herd, (ii) it is also effective when level of performance is maintained while the degree of purification (inbreeding) is increased. In other words, selection is effective, if it can offset the decline that usually accompanies inbreeding.

Technique for the appraisal of effectiveness of selection

Amount of selection pressure applied

The amount of selection pressure applied for a particular trait is known as the selection differential. The selection differential is usually defined as the average difference between the herd average of the individuals within the herd that are kept for breeding purpose or difference between the average of the selected animals and average of the groups from which they were selected. Example: Let us assume that three gilts A, B and C are selected from a group of gilts that average 80 kg at 154 days of age. A weighed 86 kg, B weighed 84 kg and C weighed 91 kg at 154 days. When these gilts farrowed, they raised 3, 2 and 1 gilts at 154 days of age respectively.

The weighted average of three gilts are

$3 \times 86 = 258$ kg
$2 \times 84 = 168$ kg
$1 \times 91 = 91$ kg
$6 \times 86 = 517$ kg
517 divided by 6 = 86 kg
Selection differential is 16
Average selected gilts = 86 kg
Average of all the gilts = 80 kg
thus, Selection differential = 6 kg

Some of the 6 kg superiority of the related gilts is due to environmental causes and some of the superiority will probably be due to favourable interactions of the particular genes of these gilts. Because of segregation and recombination, the offspring of these gilts will not have exactly the same genes as the gilts and consequently will not have the same gene interactions. Now how much superiority can be justly expected in the offspring of these gilts. If the h^2 of 154 day weight is 20%, 20% of the 6 kg or 1.2 kg superiority in the offspring of these gilts with respect to 70 kg weight is the maximum that can be expected.

If it is desired to know how much progress can be expected per year from the selection practiced, it is necessary to change the 1.4 kg per generation to per year basis. To do this we must know the generation interval of the herd.

The average generation interval of these pigs is 17.2 or 1.43 years.

The improvement of 0.95 kg at 154 days of age needs at least one further correction. It is possible, by statistical techniques to determine, how much decline results in 154 day weight with each percent increase in inbreeding. Let us assume that inbreeding of 0 to 1, 1 to 30% over 10 years period. This is equivalent to a 3% increase in inbreeding per year. It was found that 154 days weight decrease 300 g for each percent increase in inbreeding. The inbreeding in our example results in a decline of 3 times 300 g or 895 g at 154 days. The net effect of selection in the particular example amounts to 2.1 minus 895 g or a 59 g increase in 154 days weight per year. This is called annual expected genetic gain. It is the result of selecting pigs that weight 6.8 kg heavier than the average of the generation when due allowance is made for h^2, generation interval and effect of inbreeding.

The expected genetic gain per year can be expressed as the amount of selection (selection differential) times the accuracy of selection (h^2) divided by generation interval. From this amount is subtracted the annual decline from inbreeding.

Symbolically:

$m = (sh/g) - bI$
where, m = expected gain per year
s= the relative difference, selection differential
h = h^2 of the trait
g = generation interval
b = effect of inbreeding (on each percent)
I = the average increase in inbreeding per year
Accordingly genetic gain/yr can be calculated as:
$m = (15 \times 0.20/1.43) - (3 \times 0.658) = 2.1 - 1.97 = 0.13$

If now we obtain regression coefficient of the average annual 154 day weight on time, we can compare it with our expected gain. The regression coefficient is really a measure of the trend of the actual performance of our pigs with respect to 154 day weights. If 154 day weight have increased over the 10 years of our experiment, the regression coefficient will have a plus sign. If the 154 day weights have decreased the sign of the regression will be minus.

For convenience, let it be assumed that in the above example the agreement between the expected and the actual gain was very good. What does this mean? It means, first of all that the actual gain was as much as could reasonably be expected on the basis of the selection practised. It means further, that through selection it was possible to maintain the performance of pigs with respect to 154 day weights and sometime, increase the genetic purity of the pigs from no inbreeding to an average inbreeding coefficient to 30%. Selection would indeed be effective if this could be accomplished in short period of 10 years.

6.4.11 Effective breeding value (EBV)

EBV is the genetic merit of an individual, defined as twice the average deviation of its offspring from the population mean when mated randomly to an infinite population. This is known as true EBV. The estimated EBV is calculated from one or more measurements of performance, using phenotypic values of the individual animal or from relatives or contemporary animals. The net genetic worth of the animal, that is true overall breeding value computed by adding true breeding values for each selection objective (traits) with each true breeding value multiplied by the relevant net economic value. The actual genetic merit of an animal is its breeding value, which is the overall effect of all the genes. Selected animals transmit one-half of their genes; it is considered that the animal contributes one-half of its breeding value, to each progeny. Therefore, the expected difference between the progeny of a selected animal and the original population is half of the breeding value of that animal. EBV can be calculated using several methods based on the information available on animal or its relatives, contemporary group animals. EBV can be calculated when only one record on the individual pig is available as

$$EBV = a\,(X_1 - X_2)$$

where a is the weighting factor such as heritability or percent of the variation in the trait under genetic control, economic merit etc, X_1 is the record of individual animal for the trait and X_2 average of contemporary group of animals for the trait.

From EBV expected progeny differences (EPD) may be calculated since EPD is half of the animal's EBV (EPD = 1/2 EBV). The EPD predicts progeny performance relative to the group or population average. The EPD for progeny is the sum of the EPDs of the sire and dam. The EBV or index can be calculated for large number of measurements incorporating a number of animals, traits, heritability and relationship as shown by Christensen (2009) as under

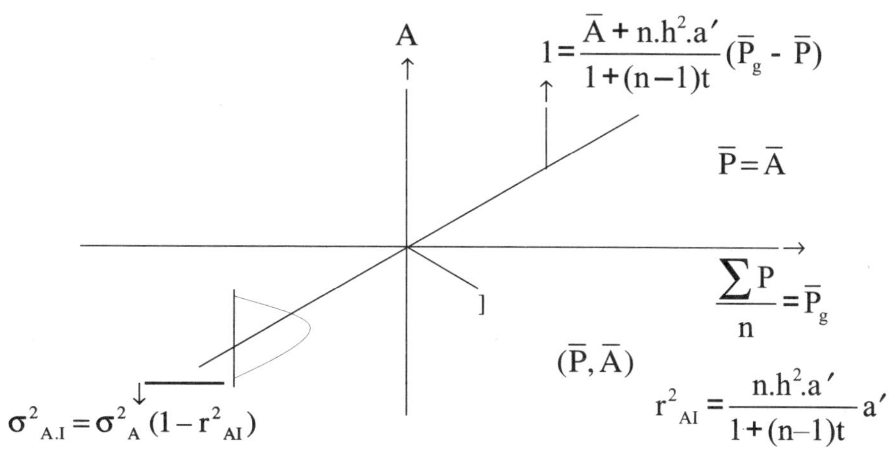

P's = phenotype values of the trait
n = number of measurements
Pg bar = average of a uniformly related group of P's
a' = degree of relationship between the P's and the animal being estimated for the index
a = degree of relationship between P's
P bar = average of the population
A bar = average breeding value of the population = P bar
h^2 = heritability
c2 = common environmental factor for the P's
$t = ah^2 + c2$

(Adapted from Population Genetics by Knud Christensen following the URL. http://www.husdyr.kvl.dk/htm/kc/popgen/genetics/7/2.htm)

All the available information on EPD of parents, ancestors, progenies etc may be analyzed through a linear mixed models statistical technique Best Linear Unbiased Prediction (BLUP).

Best linear unbiased prediction (BLUP)

The best linear unbiased prediction (BLUP) is used in linear mixed models for the prediction of random effects. BLUP was derived by H.C Henderson in 1975. Best linear unbiased predictions (BLUPs) of random effects are equivalent to best linear unbiased estimates (BLUEs) of fixed effects (Henderson, 1975). The effectiveness of genetic prediction using the sire-maternal grandsire mixed model analysed using BLUP is found to give an unbiased prediction of one-half of the true breeding value of a sire and one-fourth of the true breeding value of the maternal grandsire even when the population size is less.

Advantages and uses of BLUP

- BLUP provides true BV of the animal and improvement probable in progenies through breeding programmes.
- It is possible to apply BLUP for wide range of applications.
- Path analysis of inheritance from ancestors and relatives
- BLUP can be used to supplement to other evaluations and selection methods

The BLUP is normally distributed with an average at 100, *e.g.* the average animal being evaluated has a BLUP value of 100. The BLUP values above 100 means the probability that the progeny will be above average is more and vice versa. The BLUP values are calculated for all the characters separately under evaluation.

CHAPTER 7

BREEDING

7.1 Systems of Breeding

The art of breeding lies in the proper application of principles of heredity to animal improvement. The problem of animal improvement may be approached in two ways:

(i) Modification of environment as better feeding, management and disease control.

(ii) Genetic improvement, which is permanent, *e.g.* selection and mating systems.

Systems of breeding do not create any new gene. They sort out old genes into new patterns. Success, therefore, depends upon the proportion of favourable genes present in the foundation stock. Genes that are not present in the foundation animals can sometimes be found in other strains or populations and can be introduced through crosses.

Breeding

Inbreeding	Outbreeding
i. Close breeding	i. Crossbreeding
ii. Line breeding	ii. Outcrossing
	iii. Top crossing
	iv. Back crossing
	v. Grading
	vi. Species hybridization

7.2 Inbreeding

Inbreeding is the mating of males and females that are related. We consider animals to be related only when they have one or more ancestors in common in the first 4 to 6 generations of their pedigree. The intensity of inbreeding depends on the degree of relationship, *e.g.* mating of son to dam or brother sister mating are called close breeding in contrast to cousin mating or those which are not closely related.

Measurement of relationship between individuals helps us to understand the intensity of inbreeding. Two animals no nearer related than the average of their breed have a relationship of zero. Two animals with exactly similar genotype have a relationship of 100. They are alike in 100% of the genes, *e.g.* identical twins. The degree of relationship therefore, ranges from 0 to 100.

Relationship may be of two kinds, direct and collateral. You are directly related to your father. That is you and your father has more genes in common (50%) than do unrelated members of the human population. Similarly, one half of your genes are identical with those of your mother. You and your cousins are collateral relatives, because you both have some ancestors in common. Your cousin probably has some identical genes that came to each of you from your common grand parents.

7.2.1 Coefficient of inbreeding

When animals which are related, in other words those having genes in common, are mated, more homozygosity results in the offspring in relation to average animals of the same breed in the foundation stock. Inbreeding, therefore, increases homozygosity or decreases the heterozygosity in individuals.

The average percentage increases in homozygosity or decreases in heterozygosity in an inbreed animal in relation to an average animal of the same breed of the foundation stock is known as the coefficient of inbreeding. It is obtained by multiplying the relationship among parents by ½; since the new generation produced is once further removed from the common ancestors and a further halving of the genetic material occurs. The formula for the coefficient of inbreeding of individuals (Fx) is

$$Fx = \sum 0.5^{n1+n2+1}(1+FA)$$

n_1: The number of generations or halving from the common ancestor to the father.

n_2: The corresponding number for the mother, $\sum$: summation

The coefficients are not absolute but relative measures. It measures the probable similarity of germ cells. It is useful for study of breeds and lines within a breed and analysis and comparison of individuals, groups and breeds for the part inbreeding plays in their respective developments.

Effect of inbreeding

Inbreeding is the mating of animals, which are related or have more number of similar genes. Hence, it increases the likelihood of similar genes becoming paired. In other words it increases the percentage of homozygotes and reduces the proportion of heterozygotes.Inbreeding makes the genes, favourable or unfavourable, homozygous. When the animals are homozygous for a number of genes, the regularity of inheritance is assured, *i.e.* it fixes the characteristics. A high degree of homozygosity increases the prepotency of the inbred individuals *i.e.* the ability of a parent to impress its characteristics uniformly on its offspring.

Reasons for inbreeding

1. To promote genetic purity and thereby increase prepotency.
2. To bring undesirable recessives to light and give the breeder an opportunity of culling them from the stock. When a sire is mated to 20 of its daughters, if it does not throw out any recessive character, it may be reasonably stated that the sire is not heterozygous for character under question.
3. To develop inbred lines for nicking ability.
4. For regrouping the genetic material.

It is generally believed that inbreeding reduces vigour. The reasons for this are:

(i) The recessive genes become homozygous during inbreeding.

(ii) If over dominance exists, where a2 a2 is superior, inbreeding diminishes the quality of trait and a2 a2 becomes a1 a1 and a2 a2 during inbreeding. To offset the bad affects of inbreeding, it is desirable that it is practiced only in herds that are better than average *i.e.* where frequency of desirable genes is more. It may be practical in herds where an outstanding sire has been used. It is also necessary that the breeder should know the merits and demerits of this system before he practices it. Inbreeding should not be practiced in grades or in commercial herds below average for the sake of economy in a single sire herd.

7.2.2 Line breeding

Line breeding is a form of inbreeding, but so directed as to keep the relationship of the individuals very close to as in admired ancestor. The admired ancestor is usually a male since it can give more offspring during its life time than a female. When we say that an animal is a line breed, the question immediately arises, line bred to what?

When and why line breeding? When a sire, used on good dams produce offspring better than their dams, the breeder should line breed at once strongly to this sire, while the animal is yet alive. It can be used on its daughters and grand daughters generation after generation. Often an animal is old or dead before its superiority is recognized. If its sons and daughters are mated to unrelated individuals, within three to four generations, the influence of the outstanding animals is so scattered that no one descendent is like the original individual. So, line breeding holds the expected amount of inheritance from the admired ancestor as a constant level instead of letting it to be halved every generation. If at the time of death, there are no relatives more closely related than 50%, we cannot produce animals more close to it than that, but it may be possible by inbreeding to keep and maintain that level.

Line breeding builds up homozygosity and prepotency. It tends to hold the gain made by selection while attempt is made to make further gain.

Line breeding is specially useful where there is much epistasis, where a desired characteristic depends on a combination of genes and where the combination tends to get scattered at each generation. These genes can be made homozygous in different lines and lines can be crossed for their combining or nicking ability.

When progress by inbreeding comes to a standstill, line breeding makes additional progress possible.

Dangers of line breeding

Line breeding tends to make frequency of desirable or undesirable genes homozygous rapidly. Hence choosing of the ancestor (sire) to line breed is very important. Those that are definitely superior should alone be selected. Besides rigid selection, culling of the undesirable recessives is highly essential. Line breeding should be practiced only in herds distinctly superior to the general average of the breed.

7.2.3 Prepotency

Prepotency is the ability of an individual to stamp its characteristics on its offsprings to such an extent that they resemble their parents more closely than is usual. It is the property of the characteristic and not the individual breed or sex. When two individuals are mated, one may have more influence than the other on the offspring. Similarly, some lines and breeds are more prepotent than others. However, prepotency cannot be passed on from one generation to another unless it is possessed by both sires and dams.

A high degree of homozygosity and the possession of a high percentage of dominant genes are the inherent qualities that will enable an animal to stamp its own characteristics on majority of its offspring. A perfectly homozygous animal produces only one kind of gametes and all its offspring will receive exactly the same gene from it. Any genetic difference between the offspring would depend entirely on their having received different genes from the other parents. If the parent is homozygous for several dominant genes, all the offspring will resemble it, irrespective of what they received from the other parent. Here, prepotency is the maximum.

Measure of prepotency

Inbreeding and increase of homozygosity is the only means of making animals prepotent for characteristic. The more the animals are inbred the more they become homozygous for a number of genes. The inbreeding coefficient then is the best estimate of an animal's prepotency. Prepotency, however is not transmissible from parent to offspring.

7.2.4 Physiological basis of inbreeding effect

Many adverse effects of inbreeding are due to several pairs of recessive genes each of which have only a slightly detrimental effect on the same trait.

(a) Probably the action of most, if not all such genes is through the failure to produce required enzymes or through production of abnormal proteins or other compounds.

(b) The adverse effects of inbreeding may be due to some physiological inefficiency and perhaps to a deficiency or lack of balance of hormones of endocrine system.

(c) Reduction in the reproductive efficiency.

7.2.5 Additive gene action

In additive gene action, there are no dominant or recessive genes, nor are there interactions between the various alleles or pairs of genes. Inbreeding would cause both plus and neutral genes to become more homozygous, but if selection were not practised, there would be no decline in the trait as inbreeding increased.

Effect of Inbreeding on different kinds of gene action dominance and recessiveness

Decline in vigor which accompanies inbreeding is due to the uncovering of detrimental recessive genes through increased homozygosity. These recessive genes are hidden by dominant genes in the non-inbred population. If complete homozygocity were attained and this is not very likely, there would be further decrease in the values because there would no further uncovering of recessive genes.

This type of gene action can also be over dominance, responsible for adverse effects of inbreeding in farm animals as homozygsity increases. In this case if we select superior animals for breeding purposes, however, there would be tendency to favour those that are more homozygous and cull those that are more heterozygous. This would result in a slower increase in the degree of heterozygous in the population, than one might expect.

7.2.6 Inbreeding experiment done in pigs

Inbreeding of the pigs affects their performance directly because of their genetic constitution, whereas the inbreeding of the dams affects the pigs through the maternal environment provided to them from conception to weaning. It will be noted that the genetic constitution of the pigs or their own inbreeding caused a decrease in litter size at birth 21, 56 and 154 days, with the effects becoming progressively less as the pigs grew older. This indicates that the vigour of the pigs was adversely affected by inbreeding. The inbreeding of pigs had little or no effect on their growth rate up to 56 days of age, but at 154 days there was 1.56 kg less weight per pig for each 10% increase in inbreeding. Inbreeding seemed to affect rate of gain less than it affected survival rate.

Inbreeding also affected the performance of the sow. Increased inbreeding of the sow resulted in a reduction in litter size and to a lesser extent, the weights of the pigs. Since litter size up to the time of birth is determined by ovulation rate and embryonic death losses, the results show that these factors were adversely affected by inbreeding. Maternal influences on pig weight after birth and up to 154 days of age are a reflection of milking and mothering ability of the sows. Inbred sows were inferior to non-inbred sows in this respect.

Inbreeding also delays the onset of sexual maturity in gilts and in boars. The inbred boars do not perform as satisfactorily as non-inbred boars because of a lack of mating desire or libido. Inbred gilts generally produce fewer eggs during oestrus and farrow smaller litters than those which are not inbred. The influence of inbreeding on carcass quality seems to be very small or nonexistent.

Line breeding should be used only in a purebred population of a high degree of excellence. Line breeding does not seem to have the traditional fears that are associated with inbreeding. Line breeding is often used when there is a high likelyhood of reducing the merit of the herd when outside sires are introduced. Thus only breeders with superior herds can really justify line breeding.

7.3 Outbreeding

Outbreeding is the mating of animals distinctly less closely related to each other than the average of the population, *i.e.* those that have no common ancestors in the preceding 4 to 6 generations of their pedigree. It is just the opposite of inbreeding. It promotes the pairing of unlike genes by mating animals that belong to different families, breeds or species. Thus it increases heterozygosity and variability. The chief reasons for outbreeding are:

(i) To bring about an increase in vigour. Vigour includes almost anything that pertains to desirability, *e.g.* rate of gain, efficiency of gain, fertility, general strength etc.
(ii) To make full use of dominance of characteristics.
(iii) To introduce new genes in a closed population. If a certain breed or family is deficient in a certain trait, the quickest and most certain method of improving that trait is to introduce genes through crossbreeding to some stock known to be superior in that trait.
(iv) To start new breeds with a broad genetic background.
(v) To produce market animals making use of heterosis.

Outbreeding includes:

(a) crossbreeding,
(b) out crossing,
(c) back crossing,
(d) top crossing,
(e) grading and
(f) species hybridization.

7.3.1 Crossbreeding

Crossbreeding is the mating of two animals, which are pure bred but belong to different breeds. It is widely practiced in swine, sheep, and poultry and less so in cattle and horses. The main purpose in this is to produce commercial stock where the individual merit for economic traits is promoted. The breeding value of the individual, however, is lowered.

When the crosses are used for breeding purposes, their offspring are more variable than the crossbreds and generally average, somewhat lower in individual merit, below their purebred grand parents.

Judicious crossing of breeds that complement each other might result in increased vigour. The economy of crossbreeding, therefore, depends upon whether the increase in production is more than enough to balance the possible confusion regarding the breeding value of the crossbred individuals and also increase in cost of replacement of pure bred stock under a cross breeding system. It is more profitable where fertility is highest and females can be kept for long and the cost of their replacement is lowest. Mainly for these reasons, it is mostly practiced in swine, poultry and sheep.

In a crossbreeding system, the males are to be discarded because of their lowered breeding value. The heterosis in females can, however, be utilized by crossing it with a third different breed and rotating the same in a systematic manner. This is known as triple crossing or rotational crossing.

Three breeds are used in this system. The females or crosses are used on sire of pure breeds in rotation. The crossbreds will soon come to have $^{4}/_{7}$ of inheritance of the breed of immediate sire, $^{2}/_{7}$ from the breed of maternal grand sire and $^{1}/_{7}$ of the hereditary material of the other pure breed. Heterosis is thus continuously maintained.

Crisscrossing is another method proposed for utilizing heterosis in dams, without incurring the full decline in average individual merit which usually occurred when crossbreds are mated.

7.3.1.1 *New breeds from crossbreeds*

Crossbreeding has been utilized for developing several new breeds of livestock. It offers a broad genetic basis from which, by a process of selection and inbreeding, new gene combinations can be made for specific purposes.

The question of development of new breeds arises only when we are not satisfied with the existing breeds as regards their utility value and when we feel that their value can be enhanced by making new gene combinations from two breeds which is likely to complement each other in different traits. Many new breeds have been developed by crossing different exotic breeds. Some of the new breeds are Beltsville, Lacombe, Marryland etc. The detail characteristics have been discussed in Chapter 4.

7.3.2 Outcrossing

Outcrossing is the mating of the animals that are members of the same breed, but show difference in the herd. Intense inbreeding makes the genes homozygous and at the same time closes the door for further improvement. It fixes the deleterious genes also. An outcrossing brings into the herd new genes and gives an opportunity for selection and further improvement of the herd.

Outcrossing is a useful procedure where it is desired to change the type of the herd rather drastically, when necessitated by market demands.

7.3.3 Top crossing

Top crossing is the mating of a male of a certain family to females of another family of the same breed. It is the same in principle as grading up, except that top crossing is usually applied to different families within a pure breed, whereas grading up is applied to continued use of sires of one pure breed starting with foundation females which are of another breed or mongrel stock.

7.3.4 Back crossing

Back crossing is the mating of a crossbred animal back to one of the pure parent races which were used to produce it. It is commonly used in genetic studies, but not widely used by breeders. When one of the parents possesses all or most of the recessive traits, the back cross permits surer analysis of the genetic situation than an F_2 does.

A heterozygous individual of the F_1, when crossed with a member of the homozygous recessive parent race, the offspring group themselves into a phenotypic ratio of 1:1; on the other hand if the individual of the parent race were to be homozygous dominant, all the offspring will be phenotypically alike.

7.3.5 Grading up

Grading up is the continuous use of purebred sires on females of another breed or mongrel stock, to raise them quickly to the level of the purebred sires. When the purebreds are relatively scarce, this is the only quickest way available for improving the mongrel stock. In grading up, generally, the first cross shows a marked improvement over the original stock. Further improvement by each successive cross is progressively less. In a purebred, which is stationary in level, the mongrel stock by the seventh generation almost reaches that of the purebred.

7.3.6 Species hybridization

By crossing two different species sometimes we get good fertile individuals. The mule is a good example of a commercially important species hybrid. *e.g.,* Mare × Jack = Mule; She Ass × Stallion = Hinny. Male mules are always sterile as far as is yet known. A few cases of fertile mare mules have, however been reported. But these are very rare. Hinny is generally inferior to mule as work animal. It is also sterile.

Horse having 32 pairs of chromosomes and ass 31 pairs, the mule comes to possess 63 single chromosomes in all. The mare mules have given birth to mule foals and horse foals when bred to jack and stallion respectively. The inference is that the mare rules essential function as mares as far as the genetics of their eggs is concerned.

If all the horse chromosomes were extruded in the polar body, these mules will function genetically as asses. But no case of this sort has been reported. True breeding of mules as such seems also theoretically impossible.

European cattle and American bison when crossed, produce sterile males and fertile females. By back crossing the females to bison and cattle, attempts are being made to form a new breed of cattle, the cattalo.

7.4 Heterosis or Hybrid Vigour

Heterosis or Hybrid vigour is a phenomenon in which the crosses of unrelated individuals often result in progeny with increased vigour much above their parents. The progeny may be from the crossing of strains, varieties, or species. One of the explanations for this increased vigour is that genes favourable to production are usually dominant over their opposites. As a species or breed develops, it becomes homozygous for some dominant genes. They also have few unfavourable recessive genes. When one breed is crossed with the other one parent supplies a favourable dominant gene to offset the recessive one supplied by the other and vice versa.

The offspring, therefore, has a larger number of dominant genes than does either parent and is likely to be more vigorous.

Another explanation for hybrid vigour is overdominance, where a heterozygous condition is much more superior to any of the homozygous conditions.

Heterosis is much employed to produce commercial stock where the individual merit is promoted, but the breeding value is lowered. The successful exploitation of heterosis depends upon how superior the crosses are over the purebreds and whether it is worth the confusion caused in lowering the breeding value of the individual and the cost of replacement of purebred stock. For these reasons it is more commonly practiced in poultry, swine and sheep where the fertility is high and the cost of replacement of purebred stock is likely to be low.

How to estimate heterosis

Heterosis is expressed by some traits but some feels that the best measure is the amount that the F_1 exceeds the other higher parents. Others feel that hererosis is best measured by comparing the near of offspring with that of the purebred parents by the following formula.

$$\text{Percent heterosis} = \frac{\text{Mean of the } F_1 \text{ offspring} - \text{Mean of parent breed} \times 100}{\text{Mean of parent breed}}$$

Genetic explanation of heterosis

Heterosis is caused by heterozygosity involving genes with non-additive effects. Non additive gene action includes dominance, over dominance and epistasis.

Dominance

(a) Gene frequency of one allele may be much higher than that of the other.

(b) One breed could be homozygous dominant for several pairs and homozygous recessive for another (AABBccdd) while the second could be respectively homozygous recessive and homozygous dominance to those pairs (aaBBccDD when individuals of the two breeds are crossed the F_1 AaBBCCDd) would be superior to both parents in that particular trait having at least one dominant gene each pair.

Heterosis should be theoretically possible to capture the superiority in the single line by making individual homozygous dominant for all pairs of genes. For instance, few individual in F_2 were AABB. If animals of this genotype were mated inter-se, there offspring would all have the same genotype. However, these homozygous dominant would be difficult to distinguish, for they would rescissible the heterozytes in phenotype.

Even though it is theoretically possible to get a strain that is homozygous dominant for several genes it is not practically possible over dominance.

With this kind of gene action, it would never be possible to fix heterosis in a single pure strain because the gene action is entirely dependent up on heterozygosity.

7.5 Fundamental Rules of Breeding

Certain fundamental principles have to be kept in mind no matter whatever breeding system is adopted:

(i) Breeding stock should not be over fatty although it should be well fed so as to capable of resisting diseases.
(ii) Pig feed should contain sufficient green feed of good quality and about 50% grain and grain byproducts and vitamin E, essential for fertility, should be present in feed in adequate quantity.
(iii) Boars should not be raised along with sow in the same paddock.
(iv) Mating of boars and sows should only be carried out when they attain maturity, which is normally achieved at 7–8 months of age.
(v) Sows should be usually mated on second day of heat period which normally lasts for 2–3 days till pregnancy is achieved.
(vi) Young boars in the first year should only be used once in a fortnight to avoid over work. After 4–5 years of service, the boars should be considered for replacement.
(vii) Sows in the herd be considered for replacement by young gilts after 5 years of age.
(viii) Selected boars should be 6–7 months of age with strong constitution, meaty and well filled hams with well attached testicles and with at least 12 rudimentary teats. It should have good evidence of production record of sire and dam.
(ix) The sows should look feminine and possess neat head, long middle, straight back and belly line, meaty and well filled hams having 12 evenly spaced teats starting well forward. They should come from a dam which was proved a prolific mother with good milking capacity so that it can wean good sized litters.

Table 7.1 Expected Advantages of Crossbred over Purebred Pigs

	Expected advantage of crossbreds as % of purebreds	
	First cross Boar purebred Sow crossbred	**Multiple cross Boar purebred Sow crossbred**
Litter size at farrowing	0	5
Survival	7	12
Litter size at weaning	10	20
Live weight at 154 days	11	14
Total litter weight at 154 days	22	30

Source: Louca and Robinson (1967).

Table 7.2 The Relationship between Heritability and the Expression of Hybrid Vigour in some Production Traits of Pigs

Production trait	Heritability	Hybrid vigour
Litter size at weaning	*	***
Litter weight	*	***
Survival ability	*	***
Rate of gain	**	**
Efficiency of food conversion	**	*
Percentage lean in carcass	***	
Back fat thickness	***	
Body length	****	

Note: The relative degree of heritability or hybrid vigour is expressed by the number of asterisks.

7.6 All India Coordinated Project on Pigs

Introduction

Government of India developed several programmes in later part of 1960's to improve the livestock sector. One of the important programmes, using bilateral assistance with several European countries, launched development of bacon factories along with a large pig development farm based on exotic pigs, which would provide pure bred stocks for entrepreneurs who would like to take up pig farming with exotic stock as a vocation and entrepreneurship. It was believed that with incomes generated by these farmers by raising crossbreds using modern methods of extension with govt. supports, could utilize the market provided by the bacon factory and this would act as a catalyst to upgrade their skills and work towards total replacement of indigenous breeds by improved pigs from developed countries. Several project complexes were created by the animal Husbandry Department of Government of India in collaboration with State Government *viz*. Aligarh in Uttar Pradesh; Alwar in Rajasthan; Kalyani in West Bengal; Gannavaram in Andhra Pradesh etc. This was consistent with the general policy framework that poultry and pigs being fast growers could replace local populations much faster than other livestock and at a much lower cost, to improve the livestock sector in general and livelihood of small and marginal farmers in particular.

Indian Council of Agricultural Research (ICAR) in 1970 revamped its research programme on pigs production based on review of the bacon factory development programme of the Animal Husbandry Department, which would provide improved breeding material of developed breeds through genetic improvement and adaptability under India's eco-climatic conditions. It would also focus on studies on nutrition of these breeds and develop economically sustainable low cost rations using conventional and non-conventional feed ingredients. The third focus in its objective was to study the disease portfolio and develop a system of disease control so that the small and marginal farmers would benefit from the technology. Four research stations were initially identified to undertake this research namely, IVRI, Izatnagar; JNKVV, Jabalpur; APAU, Gannavaram; and AAU, Khanapara, Assam and three breeds, Large White Yorkshire, Middle White Yorkshire and Landrace were taken up for experimentation. After the initial problems of acclimatization etc., by 1975 it was realized that unlike poultry, where investments from private sector in integration and organization of large poultry units became the order of development, no such development was forthcoming in the pig sector. And therefore, there was a strategic shift in the process of development of piggery in the country. It was realized that the small and marginal farmers did not incorporate better health care and nutrition in their entrepreneurship and continued to use scavenging model as the main system of raising pigs. However, there was significant improvement raising crossbred pigs. In view of this, it was decided that the project is to be remodeled and two major innovations need to be considered (i) the genetic merit, buffering capacity and adaptability of the indigenous breeds under the improved nutritional and health care management, and (ii) the possibility of developing new synthetic breeds from crossbred segregating populations. The project design was recast to include (i) study of indigenous breed populations, (ii) their crossbreds at 50 and 75% genetic levels in a forward cross and also to interbreed at these two levels. The programme on nutrition and health care management was broadened to these genotypes. The project progress was reviewed and as per the recommendations of the Midterm Review Committee ICAR established a National Research Centre on Pig at Rani, Guwahati in 2002. During the XI th five year plan period the total number of AICRP centres were increased to ten. The current AICRP on Pig centres are in position at Assam Agricultural University, Khanapara, Guwahati, Jawaharlal Nehru Krishi Viswa Vidyalaya, Jabalpur, Birsa Agricultural University, Kanke, Ranchi, College of Veterinary and Animal Science, Kerala Agricultural University, Mannuthy, Tamilnadu Veterinary Animal Science University, Kattupakkam, Sri Venketeswara Veterinary University, Tirupati, Indian Veterinary Research Institute, Izatnagar, ICAR Research Complex for Goa, Goa, College of Veterinary Sciences and Animal Husbandry, CAU, Aizawl, Mizoram, SASARD, Nagaland University, Medziphema, Nagaland.Since inception the AICRP has produced thousands of piglets, identified several non-conventional feed materials and generated location specific technologies related to swine husbandry.

Table 7.3 Average performance and Carcass Characteristics of Local Pigs (foundation stock) at Four Centres of AICRP on Pigs

Izatnagar	
No. of service/conception	1.25
Farrowing %	100.00
Average litter size	7.80
Average birth weight (kg)	0.79
Average litter size at weaning	6.25
Average litter weight at weaning	38.62

Table 7.3 (*Contd...*)

Slaughter weight (kg)	47.89 ± 1.13	(on barrows)
Hot carcass weight (kg)	35.54 ± 0.96	(on barrows)
Dressing %	72.02 ± 0.32	(on barrows)
Carcass length (cm)	59.25 ± 0.51	(on barrows)
Backfat thickness (cm)	2.25 ± 0.20	(on barrows)
Lion eye area (sq. cm)	14.60 ± 0.45	(on barrows)
Chilled carcass weight (kg)	34.36 ± 0.93	(on barrows)
Chilling loss (kg)	1.16 ± 0.0	(on barrows)
Jabalpur		
No. of service/conception	2.10	
Farrowing %	100.00	
Avg. litter size at birth	6.74	
Avg. birth weight (kg)	0.70	
Avg. litter size at weaning	5.52	
Avg. litter wt. at weaning	38.62	
Slaughter weight (kg)	44.75 ± 1.379	(at 40 weeks age)
Hot carcass weight (kg)	30.841± 1.58	(at 40 weeks age)
Dressing %	68.69 ± 0.83	(at 40 weeks age)
Carcass length (cm)	53.85 ± 0.89	(at 40 weeks age)
Backfat thickness (cm)	2.06 ± 0.11	(at 40 weeks age)
Loin eye area (sq. cm)	19.27 ± 2.02	(at 40 weeks age)
Chilled carcass weight (kg)	No data	
Chilling loss (kg)	No data	
Khanapara, Guwahati		
No. of service/conception	2.65	
Farrowing %	100.00	
Avg. litter size at birth	4.84	
Avg. birth weight (kg)	0.90	
Avg. litter size at weaning	3.10	
Avg. litter wt. at weaning	21.32	
Slaughter weight (kg)	22.91 ± 1.05	(at 34 weeks age)
Hot carcass weight (kg)	17.77 ± 1.17	(at 34 weeks age)
Dressing %	71.25 ± 0.65	(at 34 weeks age)
Carcass length (cm)	44.11 ± 1.76	(at 34 weeks age)
Backfat thickness (cm)	2.25 ± 0.20	(at 34 weeks age)
Loin eye area (sq. cm)	9.34 ± 0.35	(at 34 weeks age)
Chilled carcass weight (kg)	16.74 ± 0.54	(at 34 weeks age)
Chilling loss (kg)	0.95 ± 0.11	(at 34 weeks age)
Gannavaram, Tirupati		
No. of service/conception	1.04	
Farrowing %	100.00	
Avg. litter size at birth	6.67 ± 0.13	
Avg. birth weight (kg)	0.78 ± 0.04	
Avg. litter size at weaning	5.31 ± 0.12	
Avg. litter wt. at weaning	43.92 ± 1.05	
Slaughter weight (kg)	48.98 ± 1.24	(on barrows)
Hot carcass weight (kg)	37.35 ± 1.76	(on barrows)
Dressing %	77.50 ± 2.24	(on barrows)

Table 7.3 *(Contd...)*

Carcass length (cm)	54.68 ± 1.40	(on barrows)
Backfat thickness (cm)	2.35 ± 0.08	(on barrows)
Loin eye area (cm^2)	19.16 ± 1.01	(on barrows)
Chilled carcass weight (kg)	No data	
Chilling loss (kg)	No data	

Table 7.4 Performance of Local Breeds, 50% Crossbred and 75% Crossbred at AICRP on Pigs during 1988–89

Traits	Local	50% crossbred	75% crossbred
Avg. litter size at birth	04.77 ± 0.41	05.20 ± 0.86	07.00 ± 0.30
Avg. wt at birth	00.89 ± 0.017	00.94 ± 0.01	01.27 ± 0.04
Avg. litter size at weaning	04.15 ± 0.52	04.80 ± 1.02	05.57 ± 0.72
Avg. litter wt at weaning	34.31 ± 3.98	51.12 ± 9.35	66.34 ± 7.75
Farrowing %	81.25	50.00	63.64
Weaning weight	08.26 ± 0.16	10.65 ± 0.20	11.91 ± 0.41
No. of service per conception	02.60	02.62	02.42
Mortality %			
(a) Preweaning	02.95	00.74	
(b) Post weaning			
(c) Adult			
Slaughter weight (kg)	16.91 ± 1.75	45.47 ± 1.34	
Hot carcass weight (kg)	10.12 ± 1.06	28.18 ± 0.81	
Chilled carcass weight	09.61 ± 1.01	26.62 ± 0.75	
Dressing %	59.74 ± 0.83	62.07 ± 0.85	
Carcass length (cm)	41.73 ± 1.55	60.89 ± 0.72	
Ham weight (kg	02.44 ± 0.28	06.34 ± 0.20	
Shoulder weight (kg)	03.33 ± 0.27	07.61 ± 0.17	
Belly weight (kg)	01.55 ± 0.21	05.39 ± 0.24	
Loin weight (kg)	02.29 ± 0.29		
Backfat thickness (cm)	00.88 ± 0.07	01.78 ± 7.66	
Loin eye area (sq. cm)	00.50 ± 0.03		
Chilling loss (kg)	00.50 ± 0.03	01.56 ± 7.66	

As a result of systematic implementation of crossbreeding programme in the AICRP, there was a definite increase in growth rate of the crossbred pigs and the average slaughter weight also increased by more than 2.5 times. There was an increase in average litter size from 4.77 to 5.20 and 7.00 in 50% and 75% crossbred respectively along with increase in birth weight (from 0.89 kg to 0.94 kg in 50% and 1.27 kg in 75%). Average litter size in weaning also increased from 4.15 to 4.80 in 50% and 5.57 in 75% crossbreds. Average litter wt at weaning also increased from 34.3 to 51.12 kg in 50% and 66.34 in 75%. Although, there was no increase in conception rate in 50% but there was significant improvement in 75% crossbred pigs. There was decrease in farrowing percentage from 81.25 to 50% and 63.64 in 75% crosses. There was definite improvement in all the carcass characteristics of 50% crossbred pigs including slaughter weight due to increase in body weight in crossbred population.

Table 7.5 Carcass Characteristics of Indigenous Breeds

Sl.No.	Traits	Izatnagar	Khanapara	Jabalpur	Tirupati
1.	Hot carcass weight (kg)	35.54	19.56	22.60	37.64
2.	Chilled carcass weight	34.36	18.73	–	–
3.	Dressing %	72.02	71.25	66.09	78.39
4.	Carcass length (cm)	59.25	40.95	49.21	54.84
5.	Loin eye area (sq. cm)	14.60	10.34	13.65	18.95
6.	Backfat thickness (cm)	2.57	4.98	1.59	2.77
7.	Shoulder weight (kg)	–	4.78	7.20	5.12
8.	Loin weight (kg)	–	4.01	–	10.28
9.	Ham weight (kg)	7.11	–	5.56	7.72
10.	Shoulder %	–	23.67	32.96	13.32
11.	Ham %	20.80	–	24.46	19.51
12.	Loin %	–	20.47	–	25.74

Table 7.6 Carcass Characteristics of Exotic Breeds

Sl.No.	Traits	Landrace		Large White Yorkshire	
		Izatnagar	Khanapara	Jabalpur	Tirupati
1.	Hot carcass weight (kg)	55.90	52.84	56.84	59.00
2.	Chilled carcass weight	54.20	–	57.52	56.96
3.	Dressing %	75.43	61.49	71.44	79.03
4.	Carcass length (cm)	74.59	84.76	70.94	74.98
5.	Loin eye area (sq. cm)	31.89	18.21	25.36	24.42
6.	Backfat thickness (cm)	2.24	2.26	2.60	2.70
7.	Shoulder weight (kg)	10.04	14.06	11.08	12.36
8.	Ham weight (kg)	64.0	13.18	10.45	13.04
9.	Loin weight (kg)	–	14.57	8.97	14.46
10.	Shoulder %	18.38	21.53	20.23	21.31
11.	Ham %	23.50	25.35	19.29	22.39
12.	Loin %	–	27.47	16.44	24.38

7.7 National Research Centre (NRC) on Pig

ICAR established a NRC at Guwahati, Assam to undertake basic, applied and strategic research on pigs and to act as information repository on swine husbandry for policy planning.The NRC is coordinating AICRP on pigs as well as a new programme, Megaseed project on pig, initiated in XI plan. The NRC maintains local strain (Meghalaya local, Nagaland local and Ghungroo) and exotic breeds of pigs (Hampshire and Duroc). The research centre implements its research projects and currently emphasize is being laid upon to characterise indigenous germplasm of pigs, development of feeding standards, development of reproductive technologies that are applicable under field conditions, popularisation of artificial insemination, candidate marker approach for economically important traits, important diseases of swine and development of pig health calender.

CHAPTER 8

HERITABILITY AND REPEATABILITY ESTIMATES

8.1 Heritability Estimate

The most important function of the heritability in genetic study of metric characters has not yet been mentioned namely its predictive role, expressing the reliability of the phenotypic value as a guide to the breeding value. Only the phenotypic values of individuals can be directly measured, but it is the breeding value that determines their influence in the next generation. Therefore, if the breeder or experimenter chooses individuals to the parents according to their phenotypic values, his success in changing the characteristics of the population can be predicted only form the knowledge of the degree of correspondence between phenotypic value and breeding values. This degree of correspondence is measured by the heritability.

The heritability is defined as the ratio of additive genetic variance to phenotypic variance.

$$h^2 = VA/VP$$

h^2 stands for the heritability itself and not for its square, VA stands for additive variance and VP for phenotypic variance

An equivalent meaning of the heritability is the regression of breeding value on phenotypic values. $h^2 = bAP$.

It is important to realize that the heritability is a property not only of a character but also of the population and the environmental circumstances to which the individuals are subjected to. Since the value of the heritability depends on the magnitude of all components of variance, a change in any one of these will affect it. All the genetic components are influenced by the gene frequencies and may, therefore, differ from one population to other. In particular, a small population maintained long enough for an appreciable amount of fixation to have taken place are expected to show lower heritability than large populations. The environmental variance is dependent on the conditions of culture or management. More variable conditions reduce the heritability, where as more uniform conditions increase it. So, whenever value is stated for heritability of a given character, it must be understood to refer to a particular population and particular conditions.

On the whole, the characters with lowest heritability are those that are most closely related with reproductive fitness, while the characters with highest heritability are those that might be judged on the biological ground to be the least important as determinant of natural fitness.

8.1.1 Methods of estimating heritability

8.1.1.1 *Identical twin method*

One egg twins are derived from the same egg and thus have the same genetic make up. Any difference in such twins should be of an environmental nature. Fraternal twins or two egg twins develop form two different eggs and should not be more alike genetically than full brothers–full sisters which are not twins. Variations in the two egg twins would be due to both heredity and environment. Therefore, comparison of two egg and one egg twins would give an estimate of the relative influence of the heredity and environment on a particular trait. It appears that heritability from identical twin data are too high and are not a true indication of the progress and so, one would expect to make in selection for a particular trait.

8.1.1.2 *Isogenic method*

In the case of isogonic lines method the estimate of a heritability obtained is of the broad sense. However, this method is seldom used in livestock since twins are the only isogonics lines possible in animals and they are rare.

Usually heritability is approximated in the broad sense since all the dominance and epistatic effect contribute a similarity among gene monozygotic twins. Positively maternal and contemporary environmental factors also could initiate heritability value.

8.1.1.3 *Intra sire regression of offspring on the dam*

A heritability estimate determined by this method is largely heritability in the narrow sense, which is mostly additive genetic action. Since each male is mated to several females, the regression of offspring on mid parent is inappropriate and since there are usually few male parents, the simple regression on one or other parents are both unsuitable. The heritability can however, be satisfactorily estimated from average regression of offspring on dams and calculated within his groups. That is to say regression of offspring on dam is calculated separately for each set of dams mated to one sire, and the regression of each set pooled in a weighted average. This method is commonly used for the estimation of heritability in farm animals. The intra-sire regression of offspring on dam estimates half the heritability and the offspring receives a sample half of the genes of the parent and covariance between parent and offspring is expected to include one half of the additively genetic variance for the trait. The variance of the parents' measurements is obtained and the regression of offspring on parents represents one half of heritability. Since relationship is only one of the two parents involved, the regression must be multiplied by two to compute heritability.

8.1.1.4 *Regression of offspring on mid parent*

A complication in the use of regression of offspring on mid parent arises if the variance is not equal in the two sexes. The genetic covariance of offspring on mid parent is equal to half of the additive variance on condition that the sexes are equal in variance. If this is not so, the regression on mid parent cannot, strictly speaking, be used and heritability must be estimated separately for each sex from the regression of daughters or mothers and of sons or fathers. If the heritability is found to be equal in two sexes, then joint estimate can be made from the regressive or mid parent, by taking the near value of the offspring as the weighted mean of the males and females.

8.1.1.5 *Half sib analysis*

It is a common form in which following data are obtained with animals:

(i) The number of males (sires) each mated to several females (dams).
(ii) The males and females are randomly chosen and randomly mated. A number of offspring from each female are measured to provide data. The individual measured thus form a population of half sibs and full sib families.

An analysis of variance is then made by which phenotypic variances is divided into observational components attributed to differences between progeny of different males, between sire component ($\sigma 2S$) to difference between the progeny of females mated to the same male between dams (within sire component ($\sigma 2D$) and differences between individual offspring of the same female (within progenies component ($\sigma 2w$). The form of analysis is given in the table 8.1.

Table 8.1 Form of Analysis of Half Sib and Full Sib Families

Source	d.f	Mean square	Component of mean square
Between sires	S-1	MSs	$\sigma^2w + K\sigma^2d + dk\sigma^2s$
Between dams (within sire)	S (d-1)	MSD	$\sigma^2w + K\sigma^2d$
Within progeny	Sd (K-1)	MSw	σ^2w

S No. of sire; D No. of dams per sire; K No. of offspring per dam.

There are supposed to be S sires, each mated to D dams, which produce K offspring each. The value of mean square are denoted by MSs, MSD, MSw. The mean square within progeny is itself the estimate of the within progeny variance component $\sigma 2w$, but the other means squares are not variance components. The composition of the mean squares in terms of observational components of variance are shown in the right hand column of the table, consideration of which will show how the variance component are to be estimated. The dam mean square, for example is made up of the within progeny component together with k times between dam component, so the between dam component is estimated as $\sigma^2d = (1/k)$ (MSD–MSw). Similarly, the between sire component is estimated as $\sigma^2s = (1/dk)$ MSs–MSD) where dk is the number of offspring per sire. If there are unequal numbers of offspring from the dams in the sire groups, the mean value of k and d can be used with little error provided the inequality of numbers is not very great. The next step is to deduce the connection between the observational components that have been estimated from the data and casual components, particularly the additive genetic variance, the estimation of which is the main purpose of analysis.

Table 8.2 Observational Components of Variance

Observational components	Covariance	Casual components estimated
Sires σ^2s	Cov (hs)	¼ VA
Dams σ^2d	Cov (ps)-Cov (HS)	¼ VA+ ¼ VD + Vec
Progenies σ^2w	VP – cov (FS)	½ VA+ ¾ VD + Vew
Total $\sigma^2T+\sigma^2s+\sigma^2d+\sigma^2w$	VP	VA + VD+Vec + Vew
Sires + dam $\sigma^2s+\sigma^2d$	Cov (FS)	½ VA + ¼ VD + Vec

Interpretation of between one component (σ^2s) is the variance between the means of half sib families and it, therefore, estimates the phenotype covariance of half sibs cov (HS), which is ¼VA. Thus σ^2s = ¼VA. Next consider the within progeny component, σ^2w. Since any between group variance component is equal to the covariance of the members of the groups, it follows that within group component is equal to total variance minus the covariance of members of the groups. The progenies of the dam are full sib families and the within progeny variance is estimated as VP = cov (Es). This leads to the interpretation of σ^2w = ½ VA+ ¾ VD + VEW. Finally, there remains the between dam components and what it estimates can be found by subtraction as follows:

$$\sigma^2_D = \sigma^2_T - \sigma^2 s + \sigma^2 w = cov\ (FS) - cov\ (HS) = ¼\ VA + ¼\ VDT + VEc.$$

Consideration of the between sire and between dam components will show that their sum gives an estimate of full sib covariance cov (FS) but this provide no new information for estimating the casual components.

The problems of experimental design are first the choice of method and second, the decision of how many individuals in each family are to be measured. We shall consider only a limitation of total number of individuals measured that is to say, we shall assume the total number of individual measured to be the same for all methods and all experimental designs. What we have to do then, is to consider each method on this basis and see what design and which method will give an estimate of heritability with lowest sampling variance.

8.1.1.6 *Offspring parent regression*

The regression on one parent must be doubled to give the estimate of heritability, but regression an mid parent is itself the estimate. The estimate based on mid parent values has a considerably less sampling variance. A regression on mid parent values, in general, yield more precise estimate of heritability for a given total number of individuals measured.

8.1.1.7 *Sib analysis*

Sib analysis are considered estimates obtained from the inter-class correlation of full sib or half sib families. Half sib analysis should generally be designed with families between 20 and 30. Thus other things being equal, an estimate from full sib families is twice as precise as from half sib families.

Regression method is preferable for estimating moderately high heritability and sib correlation method is preferably on low heritability. In the absence of prior knowledge of the heritability the analysis should be planned with 3–4 dams per sire and 10 offspring per dam.

8.1.1.8 *The precision of estimates of heritability*

It is of the greatest importance to know the precision of any estimate of heritability. When one estimate has been obtained, one wants to be able to indicate its precision by the standard error. And when an experiment aimed at estimating heritability of a trait is being planned, one wants to choose the method and possible precision within the limitations imposed by the scale of the experiment. The precision of an estimate depends on its sampling variance, the lower the sampling variance, the greater the precision, and the standard error is the square root of the sampling variance. Estimates of heritability are derived from estimates of either a regression coefficient or an interclass correlation coefficient.

8.2 Repeatability Estimates

Repeatability is the measure of the expression of the same character at different times in the life of the same individual. Thus, there is no chance for segregation or independent assortment of genes. Milk production is repeated in 3 or 4 years. Wool is repeated twice a year for 2–3 years.

If the repeatability is very high, the animal can be kept or culled based on the first record of observation. On the other hand, if the repeatability is very low, more than one observation on the same character is necessary, before a decision on the life time production of that animal can be made.

Repeatability is essentially an upper limit for heritability. Heritability cannot be more than repeatability. Repeatability is due to the genetic constitution of the individual plus environmental factors which have a similar effect in the different periods in which a record is made. For instance a good cow does better than the rest of the herd in the second lactation due to better genetic composition it has for milk production and also due to better feeding and management condition she has in the first lactation and in the second lactation.

The repeatability is $\sigma^2g + \sigma^2_{Pe}$ σ^2g is a component due to the genetic constitution and σ^2_{pe} component is due to permanent environment. Thus one can see that repeatability will always be more than heritability which is only σ^2g component.

8.2.1 Use of repeatability

Repeatability estimates also give an idea how many records should be obtained on an individual before it may be culled from the herd or flock.

Repeatability estimates tells us something about how to allot animals on a feeding trial. If the repeatability of the trait is high, it becomes increasingly important to divide the offspring of each sire or each dam evenly among the different lots. Otherwise, if the offspring from each one parent were in one lot and those of another parent in the second lot, lot difference ascribed to treatment might actually be largely due to hereditary differences. The procedure is of less importance if the repeatability of the trait is low.

Since neither the genes nor gene combinations influencing the successive expression of a trait change, repeatability should be at least as large as heritability in a broad sense. It may be larger since certain permanent environmental influences may be included in the numerator of the repeatability fraction but they would, of course, be non genetic.

8.2.2 Method of calculating repeatability

Repeatability can be computed as a regression of future performance on past performance. It may be derived from analysis of variance as an interclass correlation among records or observations of traits on same individual

Repeatability can be expressed as:

$$\sigma^2 g + \sigma^2 d + \sigma^2 i + \sigma^2 pe \,/\, \sigma^2 g + \sigma^2 d + \sigma^2 i + \sigma^2 pe + \sigma^2$$

Where $\sigma^2 pe$ represents variance associated with permanent environmental influences which make for differences in the particular expression of a particular trait for the several individual in the population. For example a cow may accidentally permanently loose an quarter and this would influence milk yield in all future lactations.

Good animals ought to be able to perform well each time and the extent to which a record can be repeated is termed the repeatability.

R = Genetic variation + general environmental variation/total phenotypic variance

$$R = \sigma^2 A + \sigma^2 D + \sigma^2 I + \sigma^2 Eg / \sigma^2 A + \sigma^2 D + \sigma^2 I + \sigma^2 Eg + \sigma^2 Es - \text{specific effect}$$

The character which have a high specific effect will have low repeatability and those with low specific effect will be highly repeatable. Repeatability indicates the correlation between records and it also an upper limit in the heritability.

Approximate values of the repeatability of various characteristics in pigs:

Litter size at birth 10–20
Litter size at weaning 10
Litter weight at birth 25–40
Weaning weight 10–15
Adult weight 35
Litter weight at 8th week 5–15.

CHAPTER 9

SELECTION OF HERD

9.1 Factors to be Considered in Selecting the Herd

At the outset, it should be recognized that the vast majority of the swine producers keep pigs one for profit and more importantly as a livelihood option. Large population of producers from weaker sections of society keeps them for the reasons of livelihood, to supplement the family income and for a source of nutrition. The new generations of entrepreneurs are now raising pigs to generate maximum profit from their herds for both the groups. The factors which need to be considered are same, though some of the parameters may differ. In chapter 20, the economics and profitability has been described in detail. For maximum profit and satisfaction in establishing the herd, the individual pig producer must give consideration to the type, breed, size of herd, uniformity, health, age, price, and suitability of the farm. Gilts from improved breeds can be selected from the breeding herds at 4–5 months of age, when they should weigh 68–90 kg whenever it is possible. They should be selected on the basis of records to ensure that they do not possess any inherited defects that they come from families having large litter size and early sexual maturity. The problems of genetic disease in pigs are discussed by Done and Wijeratne (1972). Pigs should be healthy, possess sound feet, be well grown, have at least 14 prominent teats, a good carcass conformation, rapid weight gain and good feed conversion efficiency at the time of selection.

(a) Type: Meat vs Bacon; and Large, Medium, or Small

With reference to pigs, the word type is used in a dual capacity: (i) to denote the final product for which breed is selected, such as meat, bacon, or lard; and (ii) to denote the difference in form and general conformation within a breed on the basis of rangy, medium or chuffy type.

Historically, there were two types of pigs; the lard type and the bacon type. The lard type was a thick-bodied hog carrying a large amount of fat, while the bacon type was developed to meet the demand for high quality lean bacon.

(b) Selection of breed

No one breed of pigs can be said to excel others in all points of pigs production and for all conditions. It is true, however, that particular breed characteristics may result in a certain breed bring better adapted to given conditions; for example, pigs of light colour are subject to sunburn. In the end, therefore, the selection of a particular breed is most often a matter of personal preference, and usually the breed that the individual producer likes is the one with which he will have the greatest degree of success, where no definite preference exists, however, it is well to choose the breed that is most popular in the community. The producer should also give some thought to the local market demands and initial costs.

(c) Size of herd

Pigs multiply more rapidly than any other class of farm animals. They also breed at an early age, produce twice each year, and bear litters. It does not take long, therefore, to get into the pig business. The eventual size of the herd is best determined by the following factors: (1) size of farm, (2) available grains and pastures, (3) confinement facilities, (4) kind and amount of labor, (5) disease and parasite situation, (6) probable market, and (7) comparative profits from pigs and other types of enterprises.

(d) Uniformity

Uniformity of type and ancestry gives assurance of the production of high quality pigs that are alike and true to type. This applies both to the purebred and the grade herd. Uniform offspring sell at a premium at any age, whether they are sold as purebreds for foundation stock, as feeder pigs or as slaughter pigs. With a uniform group of sows, it is also possible to make a more intelligent selection of the herd boar.

(e) Health

Breeding animals that are in a thrifty, vigorous condition and that have been raised under a system of swine sanitation should have a decided preference. Tests should be made to make certain of freedom from swine brucellosis or contagious abortion. In fact, all purchases should be made subject to the animals being free from contagious diseases.

(f) Age

In establishing the herd, the beginner may well purchase a few gilts that are well grown, uniform in type and of good ancestry and that have been mated to a proven sire. Although less risk is involved in the purchase of tried sows, the cost is likely to be greater in relation to the ultimate value of the sows on the market.

With limited capital, it may be necessary to consider the purchase of a younger boar. Usually a wider selection is afforded with this procedure, and in addition, the younger animal has longer life of usefulness ahead.

(g) Price

The beginner should always start in conservative way. However, this does not mean purchase of poor animals because they are cheap.

(h) Suitability of the farm

Some of the things that characterize successful major enterprises are:

(1) Swine knowledge, interest, and skill of the operator
(2) A plentiful supply of grains or other high energy feeds in the immediate area.
(3) Available labour, skilled in caring of swine, especially at farrowing time.

(i) Selection based on type or individuality

Selection based on type or individuality implies the selection of those animals that approach the ideal or standard of perfection most closely and the culling out of those that fall short of these standards.

(j) Meatiness–measuring backfat

Thickness of back fat, which has a heritability of 50%, has long been recognized as an important measure of meatiness in pigs.

Today, three mechanical methods are available and may be used by producers in determining back fat on live hogs; namely, the probes, the lean meter and ultra sonic. Each of these methods requires hog restraint.

The probe and the lean meter were developed for the purpose of obtaining objective measures of back fat. Ultra sonic is used to determine loin eye areas as well as back fat.

(k) Selection based on pedigree

In the selection of breeding animals, the pedigree is a record of the individual's heredity or inheritance. If the ancestry is good, it lends confidence in projecting how well young animals may breed. If pedigree selection is to be of any help, one must be familiar with the individual animals listed therein.

The boar should always be purebred, which means that he is of known ancestry. This alone is not enough, for he should also be a good representative of the breed selected; and his pedigree should contain an impressive list of noted animals. Likewise, it is important that the sows be of good ancestry, regardless of whether they are purebreds, grades, or crossbreds. Such ancestry and breeding give more assurance of the production of high quality pigs those are uniform and true to type.

(l) Selection based on show ring winnings (relevant to developed economies)

Swine producers have long looked favorably upon using show ring winnings as a basis of selection. Perhaps the principal value of selections based on show ring winnings lies in the fact that shows direct the attention of the amateur to those types and strains of pigs that at the moment are meeting with the approval of the better breeders and judges.

(m) Selection based on production testing

No criterion that can be used in selecting an animal is as accurate or important as past performance. It is recommended, therefore, that one should purchase tried sows and a proven boar when such animals of the right kind can be secured at reasonable prices.

(i) Performance

In general, the fastest growing gilts which are from large litters should be saved for replacement gilts.

(ii) Backfat

Replacement gilts should be lean, having 3 cm or less of back fat.

(iii) Feed efficiency

Feed efficiency is favored by selecting fast growing, low backfat gilt.

(iv) Well developed underline

Replacement gilts should possess a sufficient number of functional teats to nurse a large litter. Six functional and uniformly spaced teats on each side is a good standard. Gilts with inverted or scarred nipples should not be saved.

(v) Reproductive soundness

Most anatomical defects of the reproductive system are internal and hence not visible. However, gilts with small vulva are likely to possess infantile reproductive tracts and should not be kept.

(vi) Feet and legs

Gilt should have legs that are set wide out on the corners of the body and the legs should be heavy boned with a slight angle to the pasterns.

9.2 Selecting Boars

When selecting boars, some traits are apparent from the records of relatives, while other traits are apparent, and may be selected for, from the boar's own record. In selecting boars, the following traits or standards should be examined:

(a) Behaviour

Behavioral traits are those characteristics that express themselves as docileness, temperament, sex characteristics, maturity, and aggressiveness. These are associated with reproductive potential.

(b) Sow productivity

Sow productivity traits include such things as reproductive ability, litter size, milking ability, and mothering ability. The number of pigs farrowed and weaned and the average pig birth weight in litter are the most common measures. Litter weight at 21 days is probably the best single measure of sow productivity. Boars should be selected only from those litters of 10 or more pigs farrowed and 8 or more pigs weaned.

Behavioral and sow productivity traits are very important in the financial returns of the swine enterprise. Therefore, when selecting a boar or a gilt for these traits, use records of the sire and dam, litter records, records of other relatives and any records available on the animal being selected. A crossbreeding programme will maximize improvement of these two traits.

(c) Performance

Performance traits include (a) growth rate measured as gain per day from weaning to market, or age at 230 lb (104 kg), and (b) feed conversion. These traits are above average in economic value. When selecting for these items, one should place more emphasis on the boar's own record and less emphasis on records of relatives.

(d) Backfat

Carcass merit is probably best evaluated by taking measurements such as backfat thickness, loin eye area, or estimating the percent muscle in the animal. Of these measurements, backfat is the single most important and best measure of leanness. These traits have very high heritability values. When selecting for carcass merit (backfat), place most emphasis on the boar's own record.

(e) Reproductive soundness

Characteristics associated with soundness include: the spacing, number and presentation of the teats; genetic abnormalities such as hernia and cryptorchidism and mating ability. Boars should possess 12 or more well spaced teats. Genetic abnormalities and mating ability traits have a very high economic importance. For these traits, insist that relatives of these selected boars be free of these defects and rely on the breeder's integrity. Physical soundness of the feet and legs, and bone size and strength, are also important. Feet and legs should demonstrate medium to large bone; wide stance both front and rear; free in movement; good cushion to both front and rear feet; and equal size toes.

(f) Conformation

This includes body length, depth, height, and skeletal size; muscle size and shape; masculinity characteristics and testicular development. Conformation traits such as length and height have high heritability values. It is important to select boars on the basis of their own records for these characteristics.

Boars should be selected and purchased at 6 to 7 months of age and should be used at a minimum of 8 months of age. It is recommended that all replacement boars be purchased at least 60 days before the breeding season. This allows them to be isolated and checked for health, conditioned, and test mated or evaluated for reproductive performance.

The primary consideration of producers is to select those boars only that will maintain the present production level of the herd and at the same time lessen weaknesses in the herd.

9.3 Judging Swine

The parameters of judging swine depend upon the purpose for which they are raised *viz.* market hogs, breeding or as stud boar.

For proper judging it is essential to have thorough knowledge of ideal physical characteristics of swine of different breeds, age group, reproductive characteristics etc.

For selection of market hogs emphasis should be given on: (a) Muscle–expression, shape and dimension of forearm, blade, loin and ham. (b) Lean growth considering fast growing and efficient growing, leanness–absence of fat deposition (c) Structure and movement- whether stands squarely and correctly on its feet and legs and is at proper angles to shoulder, hock and pastern. (d) Skeletal width and dimension–internal volume, depth of rib, outward shape to rib etc. From the side it should be looked for type, balance, length, depth, bulge and firmness of ham. From rear view it should be looked for correct turn; thickness, meatiness and muscling in the ham, loin and hump. From the front view it should be observed for cleanness over shoulder.

For selection of breeding swine, emphasis should be given on: (a) functionality –structure and movement similar to market hogs, internal dimension and condition like depth of rib, shape to rib cage, ability to maintain condition; (b) growth and performance; (c) balance and eye appeal, muscle, feminity/masculinity.

From carcass point of view the swine should have higher percent lean, more kilos of lean product, rank higher on fat free lean index, higher cutability, greater lean value, more shapely carcass and less fat opposite the tenth rib.

CHAPTER 10

REPRODUCTION IN PIG

10.1 Female Reproductive System

Anatomy

The female reproductive system is composed of ovaries, oviducts, uterus, cervix, vagina and vulva. The ovary may be referred to as the female gonad; the vulva and the clitoris as the external genitalia and the other organs as the internal genitalia.

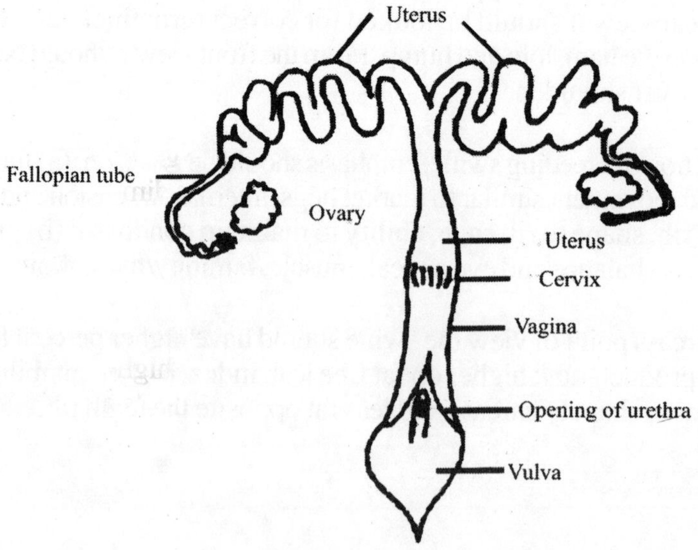

Fig. 10.1. Female reproductive system

10.1.1 The ovary

The ovaries are paired organs and serve a double role similar to that of the testes, that is gametogenic and endocrine function. This dual function is complementary, since gametogenesis requires certain changes in the reproductive tract to complete reproduction. The response of the reproductive tract is caused by the gonadal steroid hormones.

The ovaries remain in the abdominal cavity and there are some differences in functional capacity of the two ovaries in various species. A functional ovary in sow may have several follicles or corpora lutea giving an appearance of a cluster of grapes.

The ovary is suspended with the help of mesovarium and is composed of cortex and medulla surrounded by germinal epithelium. The ovarian medulla consists of irregularly arranged fibroblastic connective tissue and extensive nervous and vascular system that reach the ovary through the hilus. Beneath the germinal epithelium is the tunica albuginea and then large mass of follicles.

a. Primary follicles

Originate from the germinal epithelium and consist of single layer of follicular cells surrounding the oocyte.

b. Growing follicles

These are the follicles which have left the resting stage of primordial follicles and began to grow, but do not develop the thecal layer or antrum and can have two or more layers of follicular cells.

c. Graafian follicles (vesicular follicles)

The mature tertiary follicle, which appears as a fluid-filled blister on the surface of the ovary contains *liquor folliculi. Liquor folliculi* is rich in estrogens and inhibin (Fig. 10.2).

With rupture of the follicle, bleeding occurs and a blood clot forms at the ovulation site. The ruptured follicle with its blood-filled cavity is called a *corpus hemorrhagicum*. The corpus hemorrhagicum is replaced by the *corpus luteum,* which forms rapidly by proliferation of a mixture of theca external, theca internal, and granulosa cells. Granulosa cells, which form the main component of the corpus luteum, enlarge and acquire a large amount of mitochondria and other intracellular structure involved in synthesis and secretion of progesterone. The corpus luteum is a solid, yellowish body which is only ovarian source of the progestins. When the corpus luteum regresses, the yellowish color is lost and eventually appears as a small white scar on the surface of the ovary (*corpus albicans*). If the animal is pregnant, the corpus luteum will not regress until late pregnancy for most species, including pigs.

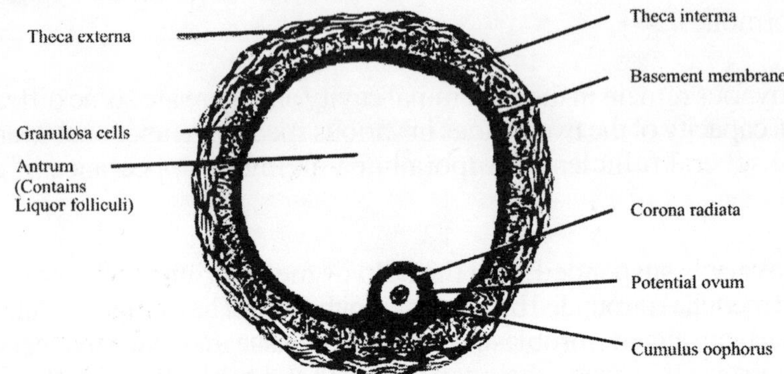

Fig. 10.2. Functionally important features of a Graafian follicle, (Redrawn from Hafez. 1974. *Reproduction in Farm Animals.* (3rd edn.) Lea and Febiger.)

d. Atretic follicles

These follicles result from Graafian follicles that do not ovulate.

Tubular genitalia

It consists of fallopian tubes or oviducts, uterus and vagina.

10.1.2 Oviduct

The oviducts (also called fallopian tubes) are a pair of convoluted tubes extending from near the ovaries to and becoming continuous with the tips of the uterine horns. Their functions include transportation of ova and spermatozoa, which must be conveyed in opposite directions. In addition they are the site of fertilization and the early cell divisions of the embryo. An oviduct, which is from 20 to 30 cm long for most farm species, is divided into three segments. The funnel-shaped infundibulum which is separate from the ovary in the pigs. The ampulla, the middle segment, is from 3 to 5 mm in diameter and accounts for about half of the total length of the oviduct. The mucosal lining of the ampulla has longitudinal folds, which increase the surface area of the lumen. The ampullary-isthmic junction is difficult to locate anatomically and is distinguished by having a thicker smooth muscle layer than the ampulla and less number of mucosal folds. A higher ratio of secretary to ciliated cells is characteristic of the isthmus.

Functions of oviduct

(1) The oviduct has the unique function of conveying the eggs and spermatozoa in opposite directions almost simultaneously.
(2) The fimbriae transport ovulated eggs from the ovarian surface to the infundibulum.
(3) The eggs are transported through the mucosal folds to the ampulla where fertilization and early cleavage of fertilized eggs take place.
(4) The embryos remain in the oviduct for 4 days before they are transported to the uterus.
(5) The utero-tubal junction controls, in part, the transport of sperms from the uterus to the oviduct.
(6) The oviductal fluid provides a suitable environment for fertilization and cleavage of fertilized ova.

10.1.3 Uterus

The uterus extends from the uterotubal junctions to the cervix. For the cow, sow and mare the overall length may range from 35 to 60 cm. In the sow, doe, ewe and cow the uterine horns account for 80 to 90% of the total length. The major function of the uterus is to retain and nourish the embryo or fetus. The *bicornuate uterus* is found in the sow and is characterized by a small uterine body just anterior to the cervical canal and two long uterine horns. The sow has longer uterine horns than the cow.

The endometrium provides a mechanism for attachment of the extraembryonic membranes, the placenta. The placenta form the conduit for transport of nutrients from maternal blood to embryonic or fetal blood and waste products back for elimination through the maternal systems. The nature of the placental attachment differs among species. The sow has a diffuse (surface) placental attachment which is histologically epitheliochorial.

Functions of uterus

Uterus serves a number of functions. The endometrium and its fluid play a major role in the reproductive process.

(1) It facilitates sperm transport from the site of ejaculation to the site of fertilization in the oviduct. As spermatozoa are transported through the uterine lumen to the oviduct, they undergo "capacitation" in the endometrial secretion.

(2) It regulates the function of the corpus luteum through prostaglandin secretion.

(3) It secretes "uterine milk" for nourishing the zygote.

(4) It initiates implantation of the embryo.

(5) During pregnancy, uterus nourishes the foetus for its development through adequate blood supply within the endometrium.

(6) Uterus initiates parturition and facilitates expulsion of foetus through its contractile activity at the time of parturition.

10.1.4 Cervix

Cervix is thick-walled and inelastic, the anterior end being continuous with the body of the uterus while the posterior end protrudes into the vagina. For most farm species the length will range from 5 to 10 cm with an outside diameter of 2 to 5 cm. The primary function of the cervix is to prevent microbial contamination of the uterus; however, it also may serve as a sperm reservoir after mating. Semen is deposited into the cervix during natural mating in sows.

The cervical canal in the sow is funnel-shaped, with ridges in the canal having a corkscrew configuration which conforms to that of the glans penis in the boar. The internal longitudinal folds present in the cervix of pig are known as sulcus pelvinae. Cervical mucus seals the female genital tract during the pregnancy.

Function of cervix

(1) Facilitation of sperm transport through the cervical mucus to the uterine lumen.

(2) It acts as sperm reservoir.

(3) It plays a role in the selection of viable sperms, thus preventing the transport of non-viable and defective sperms.

(4) Cervix acts as "physiological barrier" against entry of micro-organisms into the uterus.

(5) Cervix secretes thin mucus during oestrus and thick mucus during pregnancy.

(6) At the time of parturition a relaxed cervix permits expulsion of foetus.

10.1.5 Vagina

The *vagina* is tubular in shape, thin-walled and quite elastic and about 10 to 15 cm in length in the sow.

Functions of vagina

(1) It serves as copulatory organ in female.
(2) It acts as conduit for foetus at parturition.
(3) It protects the reproductive tract from invading micro-organisms.
(4) It is also a site for sperm antigen-antibody reaction.

10.1.6 Vulva

The *vulva,* or external genitalia, consists of the vestibule with related parts and the labia. The *vestibule* is that portion of the female duct system that is common to both the reproductive and urinary systems.

Even though the female reproductive tract may be partially resting on the floor of the pelvis, the *broad ligament* is considered the principal supporting structure. The *ovarian arteries*, also called utero-ovarian arteries, branch and supply blood to the ovaries, oviducts and a portion of the uterine horns.

10.2 Puberty

Puberty is defined as the time when the male and female gonads become capable of producing and releasing gametes. In the female, this would be associated with oestrus and ovulation. In the male, release of the first spermatozoa from the seminiferous tubule is the indication of puberty.

Boars frequently reach puberty before they are 7 months of age. The production of spermatozoa is usually not evident until they reach 10 to 12 months of age, while the gilts initially come in to estrus at 6 to 8 months of age.

Sexual maturity

An animal reaches maturity when its gametes become capable of fertilization. In domestic animals, several additional weeks must pass after puberty to reach sexual maturity. The first released gametes are incapable of fertilization, where as several weeks later, when sexual maturity is reached, the gametes become capable of fertilization.

Physiological events leading up to puberty

1. Decreased negative feedback of oestradiol
2. Maturation of the hypothalamic area of brain

3. Increased frequency of release of luteinizing hormone (LH) pulses
4. Enhanced development of ovarian follicles
5. Enough oestradiol produced to induce behavioural oestrus and a preovulatory surge of gonadotrophins

10.2.1 Factors affecting the age at puberty

Many factors such as breed, body weight, social and climate, season, nutrition, sex etc. affect the onset of puberty.

(a) Breed and body weight

In general, the smaller the breed of a particular species, experience puberty at an earlier age. The smaller breeds frequently reach puberty several months earlier than the larger breeds.

(b) Social and climate

The presence of a boar hastens puberty in pig. Climate includes an interaction of temperature, humidity, diurnal variation and daylight. The consensus is that climatic conditions in tropic favours early.

(c) Plane of nutrition

In non-seasonal breeding animals, a high plane of nutrition favors an early puberty, and a low plane of nutrition delays puberty. Growth is accelerated by over feeding, but retarded by under feeding, hence delays attainment of puberty.

(d) Sex

Females of all species reach puberty at an earlier age than male.

10.2.2 Oestrous cycle

Unlike the other large farm animal species such as the cow or the mare, that have only one or occasionally two offspring, the pig is the only farm animal species with multiple offsprings. During each oestrous cycle more than 15–20 ova are usually ovulated.

The basic mechanism of the oestrous cycle in pigs is very similar to other farm species. However, there are some differences that are important to know for good management of reproduction in a pig farm.

It is the rhythmic sexual behavioural pattern developed in female animals. This behavioural change, the oestrus (from Greek "Oestrus meaning mad desire", and referred to by a lay term "heat") characterized by sexual receptivity occurs at a regular interval, if the animal does not conceive. The combination of physiological events which begins at one oestrus and end at the next, is called "oestrous cycle". Although each species has its own peculiarities in respect to the pattern of oestrous cycle, basically all are similar.

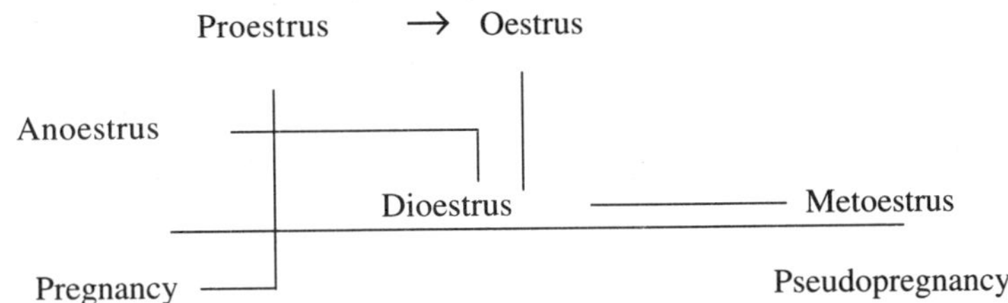

Fig. 10.3. Oestrous cycle of domestic animals

Oestrus often does not accompany the first follicular development and ovulation. On the other hand, "silent heat" prevails, which means there are no external manifestation of oestrus on the part of the female even though the ovarian follicle has matured and ovulated.

10.2.2.1 *Phases of oestrous cycle*

The oestrous cycle is commonly divided into four phases: Proestrus, oestrus, metoestrus and dioestrus. However, based on the hormonal status of the animals this cycle/period can be divided conveniently into two, *viz*, (a) oestrogenic or follicular phase comprising oestrus and proestrus, and (b) progestetional or luteal phase, comprising metoestrus and dioestrus.

Proestrus

It is an ill defined period which in different species stays for different length of time. It is accompanied with the following:

1. Growth of Graafian follicles under the influence of FSH,
2. Theca interna secretes increasing amount of oestradiol,
3. Increased serum oestradiol,
4. Increased excretion of oestrogen in urine,
5. The corpus luteum undergoes rapid degeneration both in morphology and in function, and

6. Decreasing serum progesterone.

The increasing amount of oestrogen in circulation causes the following

1. Oedema of vulva,
2. Increased vascularity of secondary organs (uterus, vagina etc.),
3. Gradual relaxation of cervix,
4. Increased secretion of mucus from cervix and vagina,
5. Increased vascularity of the endometrium
6. Growth of cells and cilia lining the oviduct, and
7. The animal tends to show interest in male at the late stage.

Oestrus

It is a well defined period characterized by the following.

1. The desire of the female to accept boar
2. The changes of the proestrus are intensified, *i.e.*

   a. the Graafian follicle becomes large and mature,
   b. the ovum undergoes maturation changes,
   c. the oviduct becomes tonic and exhibits the spontaneous movements,
   d. epithelium of the oviduct matures and the cilia become active,
   e. fimbriae of the oviduct arrange themselves close to the Graafian follicle,
   f. blood supply to the entire reproductive tract is increased – mucous membrane of the vagina exhibits congestion,
   g. mucous membrane of the oviduct grows rapidly and secretes increasing amount of fluids,
   h. increased migration of leucocytes to the uterine lumen,
   i. secretion of mucus is increased,
   j. vulva becomes relaxed and oedematous and stringe of mucus may hang out,
   k. relaxation of cervix,
   l. in most species, ovulation occurs towards the end of oestrus,

Metoestrus

It is an ill defined period under the influence of progesterone and exhibits the following changes in the reproductive organs.

1. Presence of corpus haemorrhagicum on the ovarian surface.
2. Transformation of corpus haemorrhagicum to corpus luteum under the influence of LH/LTH from the anterior pituitary.
3. Corpus luteum (CL) thus formed synthesises increasing amount of progesterone and resulting in higher serum progesterone.
4. Vaginal epithelium undergoes desquamation, a process assisted by heavy invasion of leucocytes so that the epithelium returns to its normal size.
5. Mucus secretion is decreased.
6. Uterine glands undergo further growth.
7. Uterine and oviductal tone gradually ceases and they become soft and pliable.
8. This period is followed by dioestrus, and in seasonally monoestrous animals, it is followed by either a period of anoestrus or pseudopregnancy if pregnancy does not occur.

Dioestrus

It is the longest period of oestrous cycle, and is characterized by presence of mature corpus luteum and high level of circulating progesterone. The reproductive tract exhibits the following.

1. Differentiation of endometrium and hypertrophy of uterine glands.
2. Cervix becomes constricted and cervical and vaginal mucus scant.
3. Mucous membrane of vagina becomes pale.
4. At the later stage of dioestrus, CL may show regression accompanied by regressive changes in the endometrium and uterine glands.
5. If pregnancy occurs, CL continues to function throughout pregnancy period.
6. If pregnancy does not occur, this period is followed by appearance of proestrus in polyoestrous animals.

Table 10.1 Reproductive Cycle in Pig

Age at puberty (months)	Average length of		Time of ovulation in relation to oestrus	Length of gestation (days)	Post-arturient occurence of oestrus (days)
	oestrous cycle (days)	oestrus (in hr)			
6–7	20–22	72	Before end of oestrus	114	25

10.2.3 Detection of oestrus

Observation of the sow for failure to return to estrus after mating is the most common pregnancy detection method. This technique is based on the premise that pregnant sows rarely exhibit estrus during gestation, and that non pregnant sows will return to oestrus within 17 to 24 days after breeding. The pig rearer's ability to detect signs of oestrus is improved if the sow's behavior is observed in the presence of a boar. Oestrus detection can be used as a means of pregnancy evaluation if gestation facilities are designed to allow daily fenceline contact between boars and sows, or if the boar and sow can be placed in the same pen each day for boar parade so that sows in heat are detected.

Few studies have investigated the applicability of oestrus detection for diagnosis of pregnancy. Accuracies of 39% to 98% were reported for the detection of return to estrus as a pregnancy diagnostic technique. It was concluded that daily estrus detection throughout gestation provided the best indicator of farrowing rate. False positive tests are obtained when sows become persistently anestrus due to cystic ovarian degeneration or to inactive, acyclic ovaries or become pseudopregnant. When the design of gestation facilities does not allow daily boar exposure to the bred sow, the likelihood of detecting returns to estrus is dramatically reduced. Because of the requirement for special facilities and increased labor, oestrus detection often is not a favored method of pregnancy diagnosis.

10.2.4 Formation of corpus luteum

Following ovulation, the cells that developed within the follicle undergo a differentiation process by action of pituitary hormones. This process is called luteinization and gives rise to the second ovarian structure, the corpora lutea (pl. corpus luteum). This structure is often referred to simply as the CL and has the important function of secreting the hormone progesterone.

The CL goes through a maturation and regression cycle much the same as the follicle. A blood clot-type structure known as a corpora hemorrhagicum forms in the cavity left by the ruptured follicle and is transformed into a CL by day 5 of the cycle (day 0 = oestrus). The CL is fully functional from day 5 to day 15 of the cycle and then begins to regress if the female does not become pregnant. The CL regresses and no longer secretes progesterone as the follicle of the next oestrous cycle begins to develop. As the CL regresses further, it becomes known as the corpus albicans and remains visible on the ovary for several subsequent cycles.

1. Reduction of the blood flow in the corpus luteum (CL)

A rapid decrease in luteal blood flow has been recently proposed as one of the main luteolytic actions of PGF_2 alpha. It was demonstrated that the reduction in luteal blood supply 8 hr after prostaglandin injection was coincident with the onset of structural luteolysis, the first significant decrease in CL volume was seen.

2. Direct action on luteal cells

A direct action of prostaglandin on the luteal cells, resulting from both the decrease in cyclic adenosine monophosphate (cAMP) synthesis normally produced in response to LH and the inhibition of the steroidogenic action of cAMP. These effects would be further amplified by a reduction in the number of receptors for LH.

This theory is further supported by the result of a study that demonstrated that a prostaglandin-induced decrease in plasma progesterone concentrations occurs before a detectable decrease in both the volume of the CL and the luteal blood flow.

10.2.5 Fertilization

It is the union of a male and female gametes to form the genome of a new diploid organism. Fertilization consists of series of steps that begins when egg and sperm first come into contact and end with the fusion of haploid genomes. Prior to fertilization, the two gametes must become fully mature and be transported to the oviduct where fertilization occurs.

The events occur prior to fertilization are:

(a) Sperm transport

Semen is ejaculated and deposited initially into the uterus in pigs and it has been observed that despite these differences in deposition site and significant differences in the number of sperm ejacuated, there is little variation among species in the total number of sperm that reach the oviducts. Typically, a few hundred to a few thousand sperms reach the oviducts following a single mating, which usually represent far less than one percent of the sperm in the ejaculate.

(b) Egg transport

Mammalian eggs consist of the oocyte embedded in a cluster of follicle cells. In order to reach the site of fertilization, the ovulated egg must be picked up and transported into oviduct through an opening called the ostium. In pigs the ovarian end of the oviduct forms into a funnel-shaped structure called the fimbria, which is positioned to partially cover the ovary. Once an oocyte enters the oviduct, it is propelled by ciliary motion down into the ampulla, where fertilization takes place. The oviduct provides the appropriate environment not only for fertilization, but for early embryonic development and it is important that the embryo remain there for a period of about three days before uterus is ready for hosting the embryo.

(c) Sperm capacitation

Freshly ejaculated sperms undergo a series of changes known collectively as capacitation before it can fertilize ova. Capacitation is associated with removal of adherent seminal plasma proteins, reorganization of plasma membrane lipids and proteins. It also seems to involve an influx of extracellular calcium, increase in cyclic AMP and decrease in intracellular pH. The molecular details of capacitation appear to vary somewhat among species.

(d) Sperm-zona pellucida binding

Binding of sperm to the zona pellucida of ova is a receptor-ligand interaction with a high degree of species specificity. The carbohydrate groups on the zona pellucida glycoproteins function as sperm receptors. The molecules that binds this receptor has been only partially characterized.

(e) The acrosome reaction

The acrosome reaction is an enzyme mediated process leading to fusion between the plasma membrane and outer acrosomal membrane. Membrane fusion (actually an exocytosis) and vesiculation expose the acrosomal contents, leading to release of acrosomal enzymes from the sperm head. As the acrosome reaction progresses, the sperm passes through the zona pellucida. Sperm that lose their acrosomes before encountering the oocyte are unable to bind to the zona pellucida further and cannot fertilize another ova. Therefore, the acrosomal integrity of ejaculated sperm is commonly used for testing functional capacity of sperm during semen analysis.

(f) Penetration of the zona pellucida

The constant force from the forward movement of sperm flagellating tail, in combination with acrosomal enzymes, allows the sperm move through the zona pellucida.

(g) Sperm-oocyte binding

Once a sperm penetrates the zona pellucida, it binds to and fuses with the plasma membrane of the oocyte at the posterior (post-acrosomal) region of the sperm head.

(h) Egg activation and the cortical reaction

Prior to fertilization, the egg is in a quiescent state, arrested in metaphase of the second meiotic division. Upon binding of a sperm, the egg rapidly undergoes a number of metabolic and physical changes that collectively are called egg activation. Prominent effects include a rise in the intracellular concentration of calcium, completion of the second meiotic division and the so-called cortical reaction.

The cortical reaction refers to a massive exocytosis of cortical granules seen shortly after sperm-oocyte fusion. Cortical granules contain a mixture of enzymes, including several proteases, which diffuse into the zona pellucida following exocytosis from the egg. The entire enzymatic reaction involving proteases which alter the structure of the zona pellucida is known as the zona reaction.

(i) The zona reaction

The zona reaction refers to an alteration in the structure of the zona pellucida catalyzed by proteases from cortical granules. The critical importance of the zona reaction is that it represents the major block to polyspermy in most mammals. This effect is the result of two measurable changes induced in the zona pellucida namely, hardening of *zona pellucida* and destruction of *sperm receptors*. Therefore, any sperm that have not yet bound to the zona pellucida will no longer be able to bind.

(j) Post-fertilization events

Following fusion of the fertilizing sperm with the oocyte the chromatin material undergoes decondensation and form pronuclei.

The product of fertilization is a one cell embryo, zygote with a diploid complement of chromosomes.

10.2.6 Pregnancy

In mammals, it is the interval between fertilization and birth. It covers the total period of development of the offspring, which consists of a preimplantation phase (from fertilization to implantation in the mother's womb), an embryonic phase (from implantation to the formation of recognizable organs), and a fetal phase (from organ formation to birth).

Duration of pregnancy

For most species, the amount a foetus grows before birth determines the length of the gestation period. Smaller species normally have a shorter gestation period than larger animals. The length of gestation in pigs is about 114 days. However, growth does not necessarily determine the length of gestation for all species, especially for those with a breeding season. Species that use a breeding season usually give birth during a specific time of year when food is available.

Various other factors can come into play in determining the duration of gestation. For humans, males normally gestate several days longer than females and multiple pregnancies gestate for a shorter period. Ethnicity may also lengthen or shorten gestation. Some events, such as preterm birth, can greatly shorten the length of gestation.

10.2.6.1 *Pregnancy diagnosis*

Early and accurate identification of pregnant and non-pregnant sows and gilts improves reproductive efficiency in commercial swine farms. Detection of returns to estrus after mating, ultrasound devices and other methods has been used for pregnancy diagnosis. Presently, there is not an ideal pregnancy detection technique that is commercially available. The sensitivity (ability to detect pregnant animals and represents the proportion of pregnant animals that test positively), specificity (ability to detect non-pregnant animals and represents the proportion of non-pregnant animals that test negatively) and positive predictive value (proportion of pregnant animals among those that test positively) are used to assess accuracy.

Hormone concentrations

Serum concentrations of prostaglandin-F_{2} (PGF), progesterone and estrone sulphate have been used as indicators of pregnancy. These hormone concentrations are dynamic and considerable knowledge regarding endocrine changes in pregnant and non-pregnant sows is required prior to using these techniques for pregnancy diagnosis. Presently, determination of serum progesterone concentrations is the only test with any commercial application.

The progesterone pregnancy test has an overall accuracy of >88%. It has >97% sensitivity, but has a specificity of 60 to 90%. False positive results are common when non-conceiving sows and gilts have delayed or irregular returns to oestrus, and when non-pregnant sows and gilts are anoestrus due to cystic ovarian disease. In contrast, false negative results are rare with this technique which is consistent with the postulate that progesterone is required for pregnancy maintenance in swine. The method's limitations include the necessity of collecting blood and, until recently, the need of a laboratory for analysis. The advent of commercially available enzyme linked immunosorbent assays to measure blood concentrations of progesterone in swine makes the test more practical. These kits have not been well evaluated for use in swine.

The overall accuracy of the estrone sulphate test in the evaluation of pregnancy status has been found to range from 82 to 100%. As with other early tests of pregnancy, animals may be correctly diagnosed as pregnant, but fail to farrow if the fetuses die after the test has been conducted. Quantitative commercial assay kits for the determination of estrone sulphate concentrations in serum from swine are not available. The need to collect blood (or urine) samples limits the practical application of this technique for pregnancy diagnosis in swine.

Physical methods

(i) Rectal palpation

It has been demonstrated that pregnancy diagnosis by rectal palpation of the sow was practical and reasonably accurate. The disadvantages of the technique were that the pelvic canal and rectum were often too small for the procedure to be used on low parity sows. False negative results, presumably due to errors in palpation technique or palpation too early, were more common than false positive diagnoses. Despite the potential application of this technique, it has not gained popularity in North America.

Other physical methods of pregnancy diagnosis include radiography, laparoscopy, and vaginal biopsy. These methods are not practically feasible under commercial swine production system.

(ii) Ultrasound techniques

Mechanical ultrasound devices are commonly utilized because they are easy to use, are commercially available, and perceived as being accurate. Three types of ultrasound equipment are available for pregnancy diagnosis in swine.

(iii) Doppler ultrasound

The Doppler ultrasound instruments utilize the transmission to and reflection of ultrasound beams from moving objects such as the fetal heart and pulsating umbilical vessels or uterine arteries. Blood flow to the uterine artery in the pregnant sow and gilt is detected as a regular 50 to 100 beats/minute while blood flow in the umbilical arteries is detected at 150 to 250 beats/minute.

Two types of transducer probes currently are available for use with the Doppler instruments: an abdominal probe and a rectal probe. The abdominal probe is positioned on the flank of the animal, lateral to the nipples, and aimed at the sow's pelvis area. The ultrasound waves are emitted and received by transducers and are converted to an audible signal. The rectal probe functions similarly, with the obvious exception of the positioning of the transducer. There were no differences between the accuracies of the rectal and abdominal probes.

Both false positive and false negative diagnoses can be obtained when using either the abdominal or rectal probes. There was increased likelihood of false positive diagnoses if examinations were done when sows and gilts were in proestrus or estrus, or if sows and gilts had active endometritis. False negative diagnoses were obtained if examinations were conducted prior to approximately 30 days, if examinations were conducted in a noisy environment, or if faeces became packed around the rectal probe.

Amplitude depth (a mode pulse echo) ultrasound

Amplitude depth machines utilize ultrasonic waves to detect the fluid filled uterus. A transducer is placed against the flank and oriented toward the uterus. Since the contents of the gravid uterus differ in acoustic impedance from that of adjacent tissues, some of the emitted ultrasonic energy is reflected to the transducer and is converted to an audible signal, a deflection on an oscilloscope screen, or illumination of a light (diode) or series of lights.

Pregnancies were not confirmed prior to 20 days, but from approximately 30 days until 75 days after breeding, the overall accuracy was commonly >90%. The percentage of false negative and uncertain determinations increased from 75 days until farrowing. These changes in accuracy parallel alterations in volume of allantoic fluids and fetal growth. Some models of amplitude depth instruments were more severely affected by inaccuracies. Errors in the placement of the transducer resulted in the detection of a fluidfilled urinary bladder, which yielded a false positive diagnosis. False positive results were obtained when sows were affected with endometrial oedema from zearalenone toxicosis, pyometra, or when the litter died and was neither aborted nor resorbed. False negative results also were obtained when the test was made before 28 days of gestation or after day 80.

Real-time ultrasound scanning

Portable real-time ultrasound scanners were used to evaluate the reproductive tracts of mares, heifers, bitches and for pregnancy diagnosis in sows and gilts.

Besides pregnancy detection, there are other potential applications of real time ultrasound. Pseudopregnant sows and gilts with uteri containing mummified foetuses were differentiated from pregnant sows. Sows experiencing difficult and/ or prolonged farrowing can be examined for piglets retained in the uterus. In addition, sows and gilts with endometritis often were identified and distinguished from females in later stages of pregnancy; however, they are difficult to distinguish from sows that were at 18 to 21 days of gestation.

10.2.7 Parturition (Farrowing)

To appreciate the intricacies of the farrowing process it is necessary to understand the anatomy of the pelvis and the reproductive tract at farrowing. As farrowing approaches the vulva becomes enlarged, together with the vagina that leads to the cervix or opening into the womb. The neck of the cervix opens into the two long horns of the womb that contain the piglets. The umbilical cord of the piglet terminates at the placenta which is attached to the surface of the womb. Nutrients pass from the blood of the sow across the placenta and into the developing piglet. The placenta also extends around the piglet as a sac which contains fluids and waste materials, produced by the piglet during its growth. The placenta and the sac are referred to as the afterbirth.

This is an intriguing mechanism activated by the piglet once it reaches its final stage of maturity, at approximately 115 days after mating. The sequence of events is depicted in Fig. 10.4. The piglet activates its pituitary and adrenal glands to produce corticosteroids. These hormones are then carried via its blood stream to the placenta. The placenta then produces prostaglandins, which are circulated to the sow's ovary. As have been seen earlier, the corpora lutea in the ovaries are responsible for the maintenance of pregnancy. Prostaglandins cause them to regress, thus terminating the pregnancy and allowing the hormones that initiate farrowing to commence.

10.2.7.1 *Length of pregnancy*

The mean length in the sow is between 114–115 days with a range from 111–120. Gilts tend to have a shorter pregnancy. The variation within the range is influenced by the herd, environment, breed, litter size (it tends to be shorter in larger litters and longer in smaller litters) and the time of year.

10.2.7.2 *The farrowing process*

This can be considered in three stages, the pre-farrowing period, the farrowing process and the immediate post-farrowing period when the afterbirth is expelled.

Stage 1. The pre-farrowing period

The preparation for farrowing starts some 10 to 14 days prior to the actual date, with the development of the mammary glands and the swelling of the vulva. At the same time teat enlargement occurs and the veins supplying the udder stand out prominently. The impending signs of farrowing include a reduced appetite and restlessness, the sow standing up and lying down and if bedding is available chewing and moving this around in her mouth. If she is loose-housed on straw she will make a bed. Within 12 hr of actual delivery of piglets, milk is secreted into the mammary glands and with a gentle hand and finger massage it can be expressed from the teats. This is one of the most reliable signs of impending parturition. A slight mucous discharge may be seen on the lips of the vulva. If a small round pellet of faeces is seen in the mucous and the sow is distressed, farrowing has started and it is highly likely the first piglet is presented backwards. This small pellet is the meconium or first faeces coming from the rectum of the piglet inside. An internal examination is immediately required. The final part of stage 1 is the opening of the cervix to allow the pigs to be pushed out of the uterus, through the vagina and into the world.

The initiation of farrowing

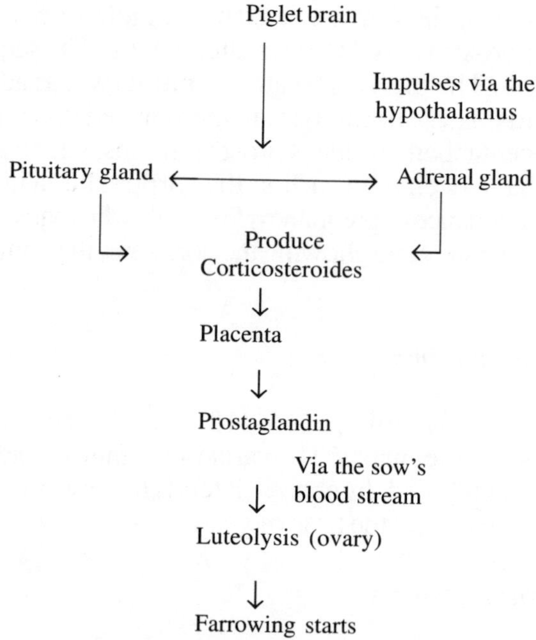

Fig. 10.4. Sequence of events at farrowing

Stage 2. The farrowing process

This can range from 3 to 8 hr and piglets are usually delivered every 10 to 20 minutes but there is a wide variation. Consult the sow and litter card to see if there have been any previous problems at farrowing. For example if a sow has had high stillbirth rates, monitor her more closely and take any necessary actions. There is often a gap between the first and second piglet of up to three quarters of an hour. The majority of pigs are born head first but there are more pigs presented backwards towards the end of the farrowing period. Immediately prior to the presentation of a pig the sow lays on her side, often shivering and lifting the upper back leg. This is an important point to take note of because it may indicate the presence of a stillborn pig. Twitching of the tail is seen just as a pig is about to be born.

Stage 3. Delivery of the placenta

This usually takes place over a period of 1 to 4 hr and is an indication that the sow has finished farrowing although some afterbirth will sometimes be passed during the process of farrowing. Once the sow has completed the farrowing process there are certain signs that should be observed.

- She appears at peace, grunts and calls to the piglets.
- The shivering and movement of the top hind leg ceases. If this is still occurring it is likely that a piglet is still present.

After the placenta has been delivered there will be a slight but sometimes heavy discharge for the next 3 to 5 days. Provided the udder is normal, the sow is normal and eating well ignore it, it is a natural post-farrowing process. Occasionally a pathogenic organism enters the uterus causing inflammation (endometritis). This may cause illness, requiring treatment.

10.2.8 Reproductive efficiency in pig

Reproductive efficiency of a species can be defined as the relative capacity to reproduce itself under optimal conditions. Sound livestock management should always include some method to monitor continuously the reproductive performance of any herd or flock. Comprehensive procedures for assessing efficiency should measure the total number of viable offspring produced by all mature females in the breeding unit over a suitable time period. Since most domesticated species have inter parturition intervals of less than twelve months, the calculations usually made to cover on a yearly basis. A number of procedures exist to tabulate and summarize breeding data with microcomputers which is now the preferred tool for updating, analyzing and storing results. Regardless of the method used, it will only provide useful information if accurate records are maintained, the figures are calculated properly, the results are reviewed regularly to determine whether goals are being met and the outcome of these assessments used as a basis for management decisions.

Producers occasionally omit primiparous females or animals that are kept for long periods but fail to get pregnant. Such summaries, based only on those animals that reproduce successfully, produce inflated results. Thus, performance figures calculated with such omissions might look impressive but do not illustrate the true reproductive performance.

For sows, true productivity must combine litter size, piglet survival and farrowing interval, so a useful measure is *piglets weaned per sow per year*. Many producers delude themselves, by omitting culled females and gilts from the calculation. However, all members of the breeding herd are part of the inventory and overhead, so each should be included in any efficiency calculation. Gilts may be added to the monthly inventory as soon as they pass market weight and are retained as potential breeding stock or when they are first mated. When the performance is calculated on an annual basis, proportional figures can be used for females kept for periods less than a full year. Litter sizes are sometimes calculated and even published in official production statistics with gilt litters excluded. Since gilts produce 30 to 50% of total litters in commercial piggeries, such figures may look impressive but are quite misleading. Realistic goals for intensive producers who include all mature female members of the breeding herd in their calculations are 18 to 22 piglets weaned per sow per year.

Another simple way to monitor productivity of a large continuously farrowing sow unit is by calculation of the *piglets weaned or finished animals marketed per sow per month*. This is simply the actual number weaned or marketed divided by the total number of mature females on the herd inventory for every month of the year. A monthly average of 1.5 corresponds to 18 piglets per sow per year and this can be maintained or exceeded in most problem free operations.

The piglets weaned per sow per year are a reliable measure of performance over the long term, but even if the calculation is revised monthly, it could take some time before a problem becomes obvious from this summary. The calculation of *piglets weaned per sow per month* should indicate any reduction in reproductive efficiency much sooner.

10.2.8.1 *Factors affecting reproductive efficiency*

The factors which influence the breeding efficiency of sow are as follows:

1. Number of ova
2. Percentage of fertilization
3. Embryonic death
4. Age at first pregnancy
5. Frequency of pregnancy
6. Longevity
7. Optimum nutrition

10.2.8.2 *Management practises to improve breeding efficiency*

1. To keep accurate breeding records of dates of heat, service and parturition. Use records in predicting the dates of heat and observe the females carefully for heat.
2. Breed sows near the end of heat period.
3. Have females with abnormal discharges examined and treated by veterinarian.
4. Get the females checked for pregnancy at the proper time after breeding.
5. Buy replacements only from healthy herds and test them before putting them in herd.
6. Have the females give birth in isolation, preferably in a farrowing room and clean up and sterilize the area once parturition is over.
7. Follow a programme of disease prevention, test and vaccination for diseases affecting reproduction and vaccinate the animals against such diseases.
8. Practice a general sanitation programme.
9. Supply adequate nutrition.
10. Detect silent or mild heat, by using a vasectomised boar.
11. Provide suitable shelter management.

10.2.9 Sexual behaviour of sow

Courtship behaviour lasts only a short time when a boar is placed in a small pen with an oestrus female. The sow plays the critical role of meeting sexual partners as boars show equal choice between an oestrus and an anoestrus sow. The male sniffs the female, noses sides, flanks and vulva, and emits a 'mating song' of soft guttural grunts (6–8 seconds). He foams at the mouth and moves his jaw from side to side as the female poses and bites the male's ears gently. When the sow becomes stationary the boar mounts. Androsterone within boar saliva aids in eliciting the standing response in the sow. Some sows are more attractive to boars than others and occasionally a sow may avoid and refuse to stand for a specific boar. Rearing females in isolation from males delays the standing response of the females once they are introduced to boars.

Pheromones in boar saliva and preputial secretions induce oestrus in gilts and sows (this is known as the boar effect). The presence of stimuli from boars (namely odour) will induce earlier puberty in gilts than if no other stimuli were present. Because an oestrus sow will stand near the boar, penning breeding females adjacent to a boar makes identification of oestrus sows easy. The social environment that boars have been raised in, influences their levels of sexual activity. Boars that are raised individually with no visual contact with immature females, but who can hear and smell the females, have reduced copulation frequency and shorter average duration of ejaculation compared to those raised in all-male or male-female groups. Boars that engage in more courting activity, especially nosing of the sow's flanks before mating, have higher conception rates. This study suggested that extra flank-nosing might stimulate oxytocin release from the sow's pituitary gland and this could increase sperm transport and the number of sperm in the oviduct and so increase the chances of fertilization. Dominant boars cover the markings of subordinate animals with urine that is often contaminated with preputial secretions.

10.3 Male Reproductive System

The organs of the male reproductive system are specialized for the following functions:

1. To produce, maintain and transport sperm (the male reproductive cells) and protective fluid (semen)
2. To discharge sperm within the female reproductive tract
3. To produce and secrete male sex hormones

10.3.1 Testes

Testes are considered primary because they produce male gametes (spermatozoa) and male sex hormones (androgens). Germ cells, located in the seminiferous tubules, undergo continual cell divisions, forming new spermatozoa throughout the normal reproductive life of the male.

Testes also differ from ovaries in that they do not remain in the body cavity. They descend from their site of origin, near the kidneys, down through the inguinal canal into the scrotum. Descent of the testes occurs because of an apparent shortening of the *gubernaculum,* a ligament extending from the inguinal region and attaching to the tail the epididymis. This apparent shortening occurs because the gubernaculum does not grow as rapidly as the body wall. The testes are drawn closer to the inguinal canal and intra-abdominal pressure aids passage of the testes through the inguinal canals into the scrotum. Both gonadotropic hormones and androgens regulate descent of the testes. In some cases one or both testes fail to descend due to a defect in development .If neither descends, the animal is termed a *bilateral cryptorchid*. Bilateral *cryptorchid* are sterile. If only one testis descends he is a *unilateral cryptorchid.* The unilateral *cryptorchid* is usually fertile due to the descended testis. The *cryptorchid* condition can be corrected by surgery, but this is not recommended for farm animals. The condition can be inherited; therefore, surgical correction would result in the perpetuation of an undesirable trait.

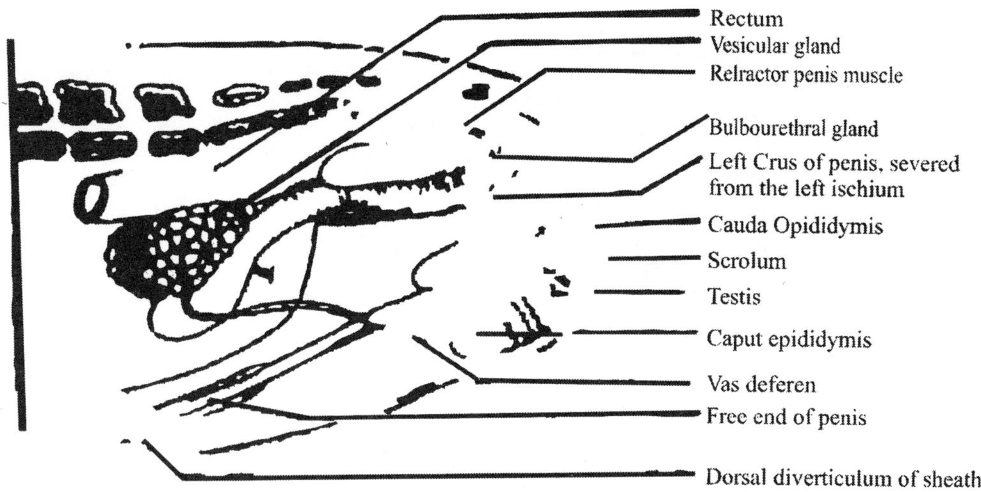

Fig. 10.5. Diagram of the reproductive system of the boar (Redrawn from Sorenson. 1979. Animal Reproduction: Principles and Practises McGraw-Hill.)

Functional morphology of testis

In all species testes are covered with the *tunica vaginalis,* a serous tissue, which is an extension of the peritoneum. This serous coat is obtained as the testes descend into the scrotum and is attached along the line of the epididymis. The outer layer of the testes, the *tunica albuginea testis,* is a thin white membrane of elastic connective tissue. Numerous blood vessels are visible just under its surface. Beneath the tunica albuginea testis is the *parenchyma*, the functional layer of the testes. The parenchyma has a yellowish color and is divided into segments by incomplete septa of connective tissue. Located within these segments of parenchyma tissue are the *seminiferous tubules*. Seminiferous tubules are formed from primary sex cords. They contain germ cells (spermatogonia) and Sertoli cells. Sertoli cells are larger and less numerous than spermatogonia. With stimulation by FSH, Sertoli cells produce both androgen binding protein and inhibin. Seminiferous tubules are the site of spermatozoa production. They are small, convoluted tubules approximately 200 nm in diameter. It has been estimated that the seminiferous tubules from a pair of bull testes, stretched out and laid, approach 5 km in length. They make up 80% of the weight of the testes. Seminiferous tubules join a network of tubules, the *rete testis,* which connects to 12 to 15 small ducts, the *vasa efferential,* which converge into the head of the epididymis (Fig. 10.6).

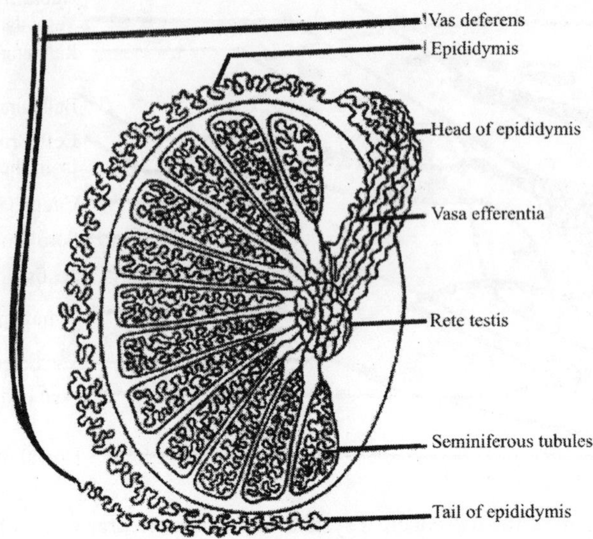

Fig. 10.6. Sagittal section of testis illustrating segments of parenchymal tissue which contain the seminiferous tubules, rete testes, vasa efferentia is, epididymis, and scrotal portion of the vas deferens.

Leydig (interstitial) cells are found in the parenchyma of the testes between the seminiferous tubules (Fig.10.7). LH stimulates Leydig cells to produce testosterone and small quantities of the androgens.

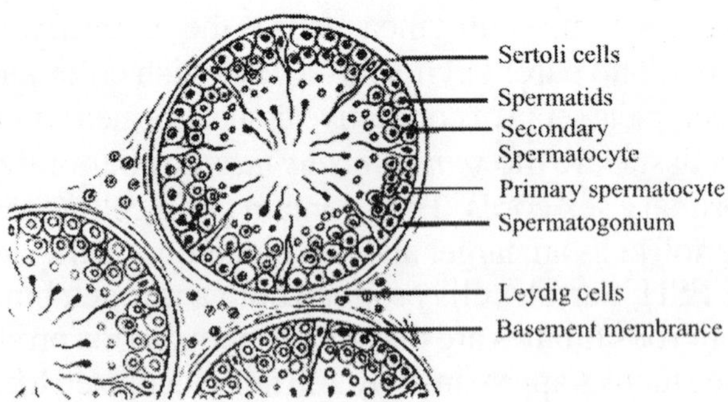

Fig. 10.7. Cross section of parenchymal tissue showing relationship between the seminiferous tubules and interstitial tissue containing Leydig cells.

Testosterone is needed for development of secondary sex characteristics and for normal mating behavior. In addition, it is necessary for the function of the accessory glands, production of spermatozoa and maintenance of the male duct system. Through its effects on the male, testosterone aids in maintenance of optimum conditions for spermatogenesis, transport of spermatozoa and deposition of spermatozoa into the female tract. Normal body temperature will not affect the function of the Leydig cells. *e.g*, bilateral cryptorchids develop secondary sex characteristics, have normal sexual vigor, and can do all things associated with reproduction except production of spermatozoa.

10.3.2 Scrotum and spermatic cord

The *scrotum* is two-lobed sac which encloses the testes. It is located in the inguinal region between the rear legs of most species. The scrotum has the same embryonic origin as the labia majora in the female. It is composed of an outer layer of thin skin with numerous large sweat and sebaceous glands. This outer layer lined with a layer of smooth muscle fibers, the *tunica dartos,* which is interspersed with connective tissue. The tunica dartos divides the scrotum into two pouches, and is attached to the tunica vaginalis at the bottom of each pouch.

The *spermatic cord* connects the testis to its life support mechanisms, the convoluted testicular arteries and surrounding venus plexus, and nerve trunks. In addition, the spermatic cord is composed of smooth muscle fibers, connective tissue, and a portion of the vas deferens. Both the spermatic cords and scrotum contribute to the support of the testes. Also, they have a joint function in regulating the temperature of the testes.

Temperature control of scrotum

Several examples can be given to illustrate the importance of temperature control of the testes. The higher temperature causes degeneration of the cells lining the wall of the seminiferous tubules. The bilateral cryptorchid is sterile, as the production of spermatozoa stops when the temperature inside the testes is as high as normal body temperature.

During cold weather, contraction of cremaster muscles causes the scrotum to contract and the spermatic cords to shorten, drawing the testes closer to the body. During hot weather, these muscles relax, permitting the scrotum to stretch and the spermatic cord to lengthen. Thus, the testes swing down away from the body. These muscles do not respond to changes in temperature until near the age of puberty. They must be sensitized by testosterone to respond to changing ambient temperature.

Actual cooling of testes occurs by two mechanisms. The skin of the scrotum has both sweat and sebaceous glands which are more active during hot weather.

Evaporation of the secretions of these glands cools the scrotum and thus the testes. The external scrotum has been observed to be 2° to 5 °C cooler than the temperature inside the testes. As the scrotum stretches during hot weather, more surface area is provided by relaxation of dartos for evaporative cooling. In addition to cooling occurs through heat exchange in the circulatory system, as arteries transporting blood at internal body temperature transcend the spermatic cord, their convoluted folds pass through a network of veins, the *pampiniform venous plexus,* transporting cooler blood back towards the heart. Some cooling of arterial blood occurs before it reaches the testes. The lengthening of the cord during hot weather provides more surface area for this heat exchange.

10.3.3 Epididymis

Epididymis, the first external duct leading from the testis, is fused longitudinally to the surface of the testis and is encased in the tunica vaginalis with the testis. The single convoluted duct is covered with an extension of the tunica albuginea testis. The *caput* (head) of the epididymis is a flattened area at the apex of the testis, where 12 to 15 small ducts, the vasa efferentia, merge into a single duct. The *corpus* (body) extending along the longitudinal axis of the testis is a single duct which becomes continuous with the *cauda* (tail). The total length of this convoluted duct is about 34 meters in the bull and longer in the ram, boar, and stallion. The lumen of the cauda is wider than the lumen of the corpus.

Functions of epididymis

a. Sperm transport

As a duct leading from the testes, the epididymis serves to transport spermatozoa.

b. Sperm concentration

A second function of the epididymis is concentration of spermatozoa. Spermatozoa entering the epididymis from the testis, they concentrate to about 4×10^9 (4 billion) spermatozoa per ml. Concentration occurs as the fluids, which suspend spermatozoa in the testes, are absorbed by the epithelial cells of the epididymis. Absorption of these fluids occur principally in the caput and proximal end of the corpus.

c. Storage of spermatozoa

Most are stored in the cauda of the epididymis where concentrated spermatozoa are packed into the wide lumen. The low pH, high viscosity, high carbon dioxide concentration, high potassium-to-sodium ratio, the influence of testosterone and probably other factors combine to contribute to a lower metabolic rate and extended life.

d. Maturation of spermatozoa

When recently formed spermatozoa enters the caput from the vasa efferentia they have ability for neither motility nor fertility. As they pass through the epididymis they gain the ability to be both motile and fertile. If the cauda is ligated at each end, those spermatozoa closer to the corpus increased in fertility for up to 25 days. During the same period, those closer to the vasa deferens exhibited reduced fertilizing ability. Therefore, it appears that spermatozoa gain ability to be fertile in the cauda and then start to age and deteriorate if they are not removed.

10.3.4 Vas deferens and urethra

The vas deferens is a pair of ducts with one leading from the distal end of the cauda of each epididymis.

The urethra is a single duct which extends from the junction of the ampullae to the end of the penis. It serves as an excretory duct for both urine and semen.

10.3.5 Accessory sex glands

The *accessory glands* (Figure 10.8) are located along the pelvic portion of the urethra, with ducts which empty their secretions into the urethra. They include the *vesicular glands,* the *prostate gland* and the *bulbourethral glands.* They contribute greatly to the fluid volume of semen. In addition, their secretions are solution of buffers, nutrients, and other substances needed to ensure optimum motility and fertility of semen.

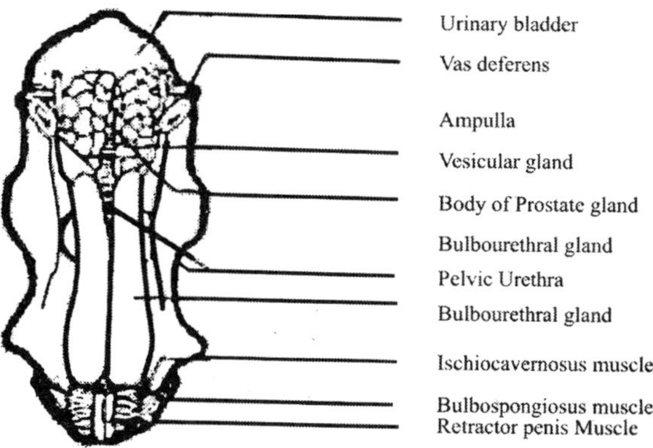

Fig. 10.8. Accessory glands of boar (Redrawn from Ashdown and Hancock. 1974. Reproduction in Farm Animals 3rd ed. Hafez. Lea and Febiger).

a. Vesicular glands

The vesicular glands (sometimes called seminal vesicle) are a pair of lobular glands that are easily identified because of their knobby appearance. They have been described as having the appearance of a "cluster of grapes." They are of similar length in the bull, boar and stallion (15 to 35 cm), but the width and thickness of the vesicular glands of the bull is approximately half than that of the boar and stallion.

b. Prostate gland

The prostate is a single gland located around and along the urethra just posterior to the excretory ducts of the vesicular glands. A prostate body is visible in excised tracts and can be palpated in bulls and stallions. However, some report that the contribution of the prostate gland is at least as substantial as that of the vesicular glands in boars. The prostate of the boar is larger than that of the bull.

c. Bulbourethral glands

The bulbourethral (Cowper's) glands are well developed pair of glands located along the urethra near the point where it exits from the pelvis. They are about the size and shape of walnuts in bulls, but are much larger in boars. They contribute very little to the fluid volume of semen. In boars, their secretions account for gel portion of semen which coagulates. This is strained from boar semen before it is used for artificial insemination. During natural service, the white lumps formed by coagulation may prevent semen from flowing back through the cervix into the vagina of sows.

10.3.6 Penis

It is the organ of copulation in males (Figure 10.9). It forms dorsally around the urethra from the point where the urethra leaves the pelvis, with the external urethral orifice at the free end of the penis. Bulls, boars and rams have a *sigmoid flexure,* an S-shaped bend in the penis which permits it to be retracted completely into the body. These three species and the stallion have *retractor penis muscles,* a pair of smooth muscles which will relax to permit extension of the penis and contract to draw the penis back into the body. These retractor penis muscles arise from the vertebrae in the coccygeal region and are fused to the ventral penis just anterior to the sigmoid flexure. The *glans penis* which is the free end of the penis, is well supplied with sensory nerves and is homologous to the clitoris of the female. In most species the penis is fibroelastic, containing small amounts of erectile tissue. The penis of stallions contains more erectile tissue than is found in bulls, boars, bucks and rams.

Erectile tissue is cavernous (spongy) tissue located in two regions of the penis. The *corpus spongiosum penis* is the cavernous tissue around the urethra. It enlarges into the penile bulb, which is covered with *bulbospongiosum* muscle at the base of the penis. The *corpus cavernosum penis* is a larger area of cavernous rods from the *ischiocavernosus muscle*, eventually fusing to form one cavernous area. As it proceeds toward excitement, cause extension of the penis (erection) and facilitating the final ejection of semen during ejaculation. Both the bulbospongiosum muscle and ischiocavernosus muscle are striated skeletal muscles, rather than the smooth muscle associated with most of the male and female tracts.

Fig. 10.9. Shape of the glans penis of boar (Redrawn from Ashdown and Hancock. 1974. Reproduction in Farm Animals. (3rd ed.). ed. Hafez. Lea and Febiger.)

10.3.7 Prepuce

The prepuce (sheath) is an invagination of skin which completely encloses the free end of the penis. It has the same embryonic origin as the labia minora in the female. It can be divided into a prepenile portion, which is the outer fold, and the penile portion, or inner folds. The orifice of the prepuce is surrounded by long and tough preputial hairs.

Endocrine control

The entire male reproductive system is dependent on hormones, which are chemicals that stimulate or regulate the activity of cells or organs. The primary hormones involved in the functioning of the male reproductive system are follicle-stimulating hormone (FSH), luteinizing hormone (LH) and testosterone.

FSH and LH are produced by the pituitary gland located at the base of the brain. FSH is necessary for sperm production (spermatogenesis), and LH stimulates the production of testosterone, which is necessary to continue the process of spermatogenesis. Testosterone also is important in the development of male characteristics, including muscle mass and strength, fat distribution, bone mass and sex drive.

10.4 Puberty

In the male, puberty can be defined less succinctly than in the female. Generally, it is considered the time when spermatozoa are in the ejaculate, the age will be 4 to

8 months for boars. However, spermatozoa are formed in the seminiferous tubules several weeks before they are seen in the ejaculate.

A number of other changes can be seen in males, starting several weeks before fertile spermatozoa are in ejaculate. These include changes in body conformation, increased aggressiveness and sexual desire, rapid growth of the penis and testes, and separation of the penis from the prepuce so that extension of the penis is possible. Timing of these events varies with species.

Development of testicular function is primary to the changes observed as puberty approaches. This development is regulated by the endocrine system. LH is necessary for the development of the Leydig cells and for their function. However, during the period around puberty synergistic effects from FSH and prolactin have been reported. FSH and prolactin appear to make the Leydig cells more responsive to LH in young males by increasing and maintaining receptor sites for LH. As the Leydig cells develop and become functional, increasing concentrations of testosterone will stimulate most other changes associated with approaching puberty. Synergistic effects from testosterone and FSH stimulate development of Sertoli cells, production of androgen binding protein and preparation of the seminiferous tubules for production of spermatozoa.

As with the female, puberty is not sexual maturity in the male. Some rams and boars are used for breeding and are highly fertile after about 6 months of age. However, the testes size and total production of spermatozoa increases until about 18 months of age. There is a high correlation between the size of the testes and total spermatozoa production.

All factors which affect age at puberty in females will affect age at puberty in males. Genetic effects on puberty are seen by comparing species or breeds within a species. Any adverse environmental factor which slows growth rate will delay puberty.

10.4.1 Spermatogenesis

Spermatogenesis is the process by which spermatozoa are formed. This process occurs in the seminiferous tubules. Output of spermatozoa per day has been reported to be 4 billion for beef bulls, 7 billion for dairy bulls, 8 billion for rams, 10 billion for stallions, and 15 to 20 billion for boars. Actual production of spermatozoa may be 50 to 100 times higher, because all that are produced cannot be collected. After formation in the seminiferous tubules, spermatozoa will be forced through the rete testis and vasa efferentia into the epididymis, where they are stored while undergoing maturation changes that make them capable of fertilization. After puberty, spermatogenesis will proceed as a continuous process and may occur due to ambient temperature in all species, and due to light in rams and bucks. Reciprocal action of FSH, LH and testosterone is necessary for the maintenance of spermatogenesis.

The process of spermatogenesis

Spermatogenesis can be divided into two distinct phases (Figure 10.10). The first is *Spermatocytogenesis,* a series of divisions during which spermatogonia form spermatids. The second is *spermiogenesis,* a phase where spermatids undergo a metamophosis, forming spermatozoa. As spermatogenesis proceeds, the developing gametes migrate from the basement membrane of the seminiferous tubules toward the lumen.

1. Spermatocytogenesis

Two types of cells are located along the basement membrane of the seminiferous tubules (Figure 10.10). Sertoli cells, which are larger and less numerous, are somatic cells which play a supporting role during both spermatocytogenesis and spermiogenesis. Spermatogonia, the small, rounded, more numerous cells, are the potential gametes.

After migrating to the embryonic testes, primordial germ cells will undergo a number of mitotic divisions before forming gonocytes. Before puberty gonocytes will differentiate into A_0, A_1 and A_2 spermatogonia and are located along the basement membrane of the seminiferous tubules. The A_2 spermatogonium will divide, forming a dormant (A_1) spermatogonium and an active (A_3) spermatogonium and starting a new generation of developing germ cells. The active spermatogonium will undergo four mitotic divisions in bulls and rams, eventually forming 16 primary spermatocytes. In rams, these mitotic divisions are completed in 15 to 17 days. During the next step, each primary spermatocyte will undergo a meiotic division forming two *secondary spermatocytes.* With this division, the chromosome complement in the nucleus is reduced by half so that nuclei in secondary spermatocytes contain unpaired (n) chromosomes. Within a few hr after their formation each secondary spermatocyte will again divide, forming two *spermatids.* Thus, four spermatids form from each primary spermatocyte, or 64 from each active (A_3) spermatogonium, in bulls and rams. Since A_1 spermatogonia divide by mitosis to from A_2 spermatogonia, the potential yield of spermatids is higher than is actually realized. Degeneration of spermatogonia during mitotic divisions accounts for this loss in efficiency.

Even though A_0 spermatogonia (reserve stem cells) will occasionally divide, forming new A_0 and A_1 spermatogonia, formation of dormant spermatogonia from spermatogonia is the key to maintaining the continuity of spermatogenesis and there by not diminishing the supply of potential gametes within the testes.

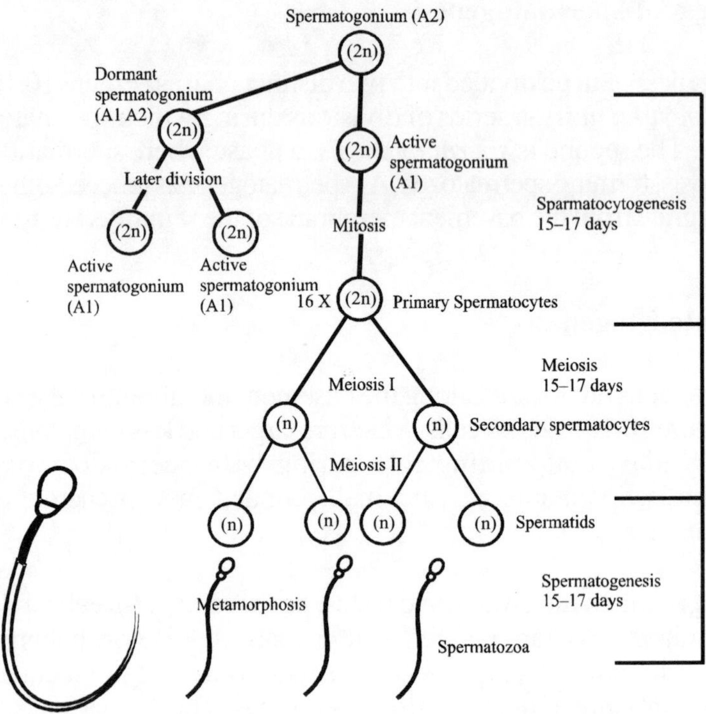

Fig. 10.10. Spermatogenesis indicating the sequence of events and time involved in spermatogenesis.

2. Spermiogenesis

During spermatogenesis spermatids are attached to Sertoli cells. Each spermatids undergoes a metamorphosis (change in morphology), forming a spermatozoon. During this metamorphosis the nuclear material will compact in one part of the cell, forming the head of the spermatozoon, while the rest of the cell elongates, forming the tail. The *acrosome*, a cap around the head of the spermatozoon, will form from the Golgi apparatus of the spermatids. As the cytoplasm from the spermatids is cast off during formation of the tail, a *cytoplasmic droplet* will form on the neck of the spermatozoon. The mitochondria from the spermatid will form in a spiral around the upper one sixth of the tail, forming the *mitochondrial sheath*. Newly formed spermatozoa will then be released from the Sertoli cell and forced out through the lumen of the seminiferous tubules into rete testis. Spermatozoa are unique cells in that have no cytoplasm, and after maturation possess the ability to be progressively motile.

Hormonal control of spermatogenesis

The endocrinology of reproduction has not been studied in males as extensively as in females. In bulls and rams there are 3 to 7 surges in LH per day followed by similar surges in testosterone. The principal role of LH in regulation of spermatogenesis appears to be indirect in that it stimulates the release of testosterone from the cells of Leydig. Testosterone and FSH then act on the seminiferous tubules to stimulate spermatogenesis. Testosterone is necessary for the regulation of this process. On the other hand FSH appears more dominant in regulating spermiogenesis. Both testosterone and FSH may exert their influence directly through germ cells and/or indirectly through Sertoli cells. FSH stimulates the Sertoli cells to secrete both androgen binding protein (ABP) and inhibin. ABP may simply be a carrier for testosterone, making it more readily available during spermatogenesis in the seminiferous tubules and transporting it through the rete testis, vasa efferentia to the epididymis. ABP is absorbed in the epididymis. Feedback control operating between the testes, hypothalamus and anteriorpituitary in regulating the release of gonadotropins (FSH and LH) and gonadal steroids (testosterone) are similar to those described for the female. Testosterone has a negative feedback effect on the hypothalamus and anterior pituitary. High concentrations of testosterone will inhibit the release of GnRH, FSH and LH, whereas low concentrations permit their release. It has been demonstrated that $PGF_{2}\alpha$ will stimulate the release of LH and testosterone. Therefore, $PGF_{2}\alpha$ may be involved in the feedback regulation between the hypothalamus, anterior pituitary and testes.

10.4.2 Semen characteristics

a. Volume

Semen volume varies between 100 and 500 ml; a large volume does not mean the total spermatozoa content is greater than that from smaller ejaculates. If the semen is to be used undiluted, its volume determines the number of inseminations possible from each collection. The dose per insemination is between 50 and 100 ml.

b. Motility

Subjective assessment of sperm cell motility is the best single way of estimating semen quality. Semen from obviously fertile boars need not be routinely evaluated unless dilution is required for inseminating a large number of sows.

Using the low power lens (×50) of a microscope, examine semen on a warm slide (30° to 35 °C) immediately after collection. Good quality semen shows a typical 'wave' motion and individual spermatozoa movement. A poor sample shows weak motility and 'clumping' of spermatozoa.

Motility depends on how much of each fraction is collected. If the collection is mainly sperm-rich fraction, there is higher motility and wave motion than when accessory fluids dilute the semen.

c. Density

Accurate estimates of semen density are needed only with high dilution rates done at animal breeding centres. Density varies from 100 to 500 million sperm cells per ml with 60 to 80 billion in a normal ejaculation. Density is also influenced by how much post-sperm fraction is collected.

d. Abnormalities

Abnormalities include detached heads, bent and coiled tails, broken tails and twin heads or tails. There are more immature sperm cells evident by protoplasmic droplets on their tails when the boar has been overworked. If total abnormalities are less than 25%, semen quality is satisfactory.

10.4.3 Semen evaluation

Determining the initial quality of a boar ejaculate is the first step in semen processing and should ensure that prior to further processing, a high quality artificial insemination dose of semen will be produced. Effective screening methods for ejaculates prior to processing are necessary for improving on farm reproductive performances. Ideally, ejaculates that are thoroughly evaluated prior to processing help identify poor quality semen. Daily evaluations of gross motility and morphology of stored semen sample will help ensure that due to some unforeseen reason, deteriorated semen doses are not used at the farm level.

With this in mind, the objectives of Table 10.2 is to describe boar semen evaluation methods and outline specific guidelines for acceptance and rejection of boar ejaculates upon laboratory entry.

a. Concentration

Generally, there are four basic parameters that are measured to evaluate boar semen quality: Concentration, motility, morphology and acrosome integrity. Of these, concentration and motility are perhaps most routinely used for sorting ejaculates prior to processing since they require the least amount of time and are required to calculate semen doses/ejaculate. Measuring semen concentration or total numbers of spermatozoa is not a component of semen quality evaluation, but more so, as a tool to monitor the health and productive output of the boar and as the primary feature in processing boar ejaculates for optimizing the genetic potential of a single individual. Accurate assessment of sperm numbers is not the only factor for increasing semen doses per ejaculate and boar stud efficiency in terms of semen output.

b. Gross motility

Gross ejaculate motility appears to an important aspect of semen evaluation. A recent study that evaluated an inseminated split ejaculates shortly (<24h) after collection suggests that farrowing rates and litter sizes will decrease when initial semen motility is recorded and used at levels below 62.5%. However, it is important to considering that semen from commercial studs, unless hand delivered after processing, is seldom used within this time period. Because semen motility decreases during storage, the minimum motility rates during initial evaluation of semen at the boar stud should be higher than 60%, and many stud farm have established a motility cutoff level between 70–80%. The minimum motility rate for processing a semen collection at each facility centre needs to be based on the projected storage length before use and expected motility rate decline over this period of storage time. Boar studs must also recognize that semen storage conditions and handling are perceived to be poorer on the farm than in the stud. Therefore, retained semen samples in the stud farm for daily quality monitoring will most likely have better motility rates than the homogeneous samples that were sent to the farm. Communication between the sow farm and boar stud farm in regard to this discrepancy, will enable the stud to select an initial motility rate acceptance level that helps ensure that when shipped semen is actually used, motility rates are above 60 %.

Visual estimates of the percentage of motile spermatozoa by light microscopy are the most widely used and acceptable method. Technician's skill and experience greatly influence the relative accuracy of this procedure. Briefly, a very small drop of diluted spermatozoa (dilution rate must be standard for all evaluations) is placed on a warmed microscopic slide and overlaid with a cover slip. The sample should be dilute enough to view individual sperm cells at 400 × power. Although gross estimates can be derived from viewing groups of sperms, technicians should be trained by first giving a gross estimate and then counting 10 cells in 5 different fields and averaging the % of motile cells (only those with forward motility) from the count for determining overall gross motility.

Table 10.2 Minimum Procedures and Equipment for Semen Quality Evaluation of Boar Ejaculates Following Collection and Prior to Processing

Evaluation procedures		Equipment needed[a]
1a.	Visual and olfactory assessment of ejaculate	None
1b.	Determine semen volume and sperm concentration	Balance and a hemacytometer or photospectometer
2 Motility		
a.	Prepare a 1:10 dilution of semen with semen extender	
b.	Gently rotate the semen	
c.	Remove a small sample (5 to 10 ml) and place in a clean glass test tube.	
d.	If, necessary, warm it to 36 to 37 degrees centigrade (body temperature)	Small water bath
e.	Place a small drop on a pre-warmed slide and gently place a cover slip over the drop.	Slide warmer
f.	Immediately examine the sample at 100 × and then at 400 ×	Self illuminating microscope capable of 100 ×, 400 ×, magnification and glass slides with coverslip
g.	Estimate the percentage of sperm in field that are progressively	
h.	Examine several fields and establish an average.	
i.	Record your estimate to the nearest 5 or 10% units.	Small, disposable plastic pipette
3. Morphology		
a.	After the motility estimate is complete, allow the slide to cool. Motility will slow or stop and individual sperm cells can be observedorPrepare a stained semen sample using step 4a, with a mixture (1:1) of morphology stain and formal saline.	Self-illuminating microscope capable of 100 × and 400 × and 1000 × (oil) magnification; glass slides and immersion oil.
b.	Switch to the 400 × objective and observe individual cells in several fields.	Eosin-nigrosin stain
c.	Estimate, in several fields, the percentage of cells that are "normal". (see example pictures)	
4. Acrosome integrity		Self illuminating phase contrast microscope capable of 100 ×, 400 ×, 1000 × (oil) magnification;
a.	From the same semen sample in step 1a, prepare a 1:1 dilution of semen and a mixture (1:1) of formal saline and Acrosome stain on a glass slide.	Formal saline: 6.19 g $Na_2HPO_3 2H_2O$: 2.54 g KH_2PO_4: 4.41g NaCL: 125 ml 38% formaldehyde: 1000 ml distilled water. naphthol yellow or erythrocin stain
b.	Place one or two drops of semen and 1–2 drops of the stain mixture on a glass slide and mix gently with the tip of the pipette. Use the edge of a second slide to draw the mixture across the flat slide to produce a thin layer. Allow the slide to air dry. Place a drop of microscope immersion oil under the slide and view first at 10 × to focus, and then switch to either 40 × or 100 × and view individual cells. (Be sure that you donot get oil on non-oil lens)	
d.	Estimate, in several fields, the percentage of cells that are "normal".	

d. Morphology

Sperm morphology and acrosome integrity are also effective tools to estimate semen viability and can also provide more information about the ejaculate in terms of its quality than is possible with just a motility evaluation. Both of these criteria are important to use, along with motility, as a determinant for keeping or discarding ejaculates.

Because motile sperm may be morphologically abnormal, poorly motile sperm may can fertilize eggs, and sperm without intact acrosomes cannot fertilize eggs, boar stud farm which do not evaluate all three of these semen quality components likely to underestimate the true fertility potential and quality of an ejaculate.

Like motility, it appears that a certain percentage normal morphological spermatozoa are needed in an AI dose to optimize fertility rates. Research data supports routine evaluation of spermatozoa for morphological normality. Semen collections with less than 70% normal morphological sperm can be identified as inferior collections if the semen is used at or below this level. Since the rate of morphological deterioration during storage is probably highly variable between boars, the initial processing level for normal spermatozoa is probably higher than 70% normal morphology when semen is used after extended storage lengths (>24 hr).

A rough morphological examination can be easily performed at the same time as semen motility, however, ideal morphological examinations are conducted with phase contrast microscopy that allows for a greater distinction of sperm membranes and parts. A precise evaluation will be obtained by performing separate counts for sperm head morphology, droplets and tail morphology. Morphology counts should be immediately conducted under phase contrast microscopy (400 × or 1000 ×) using 1–2 drops of semen diluted 1:10 with semen extender. If semen cannot be immediately analyzed (<30 min), fix or preserve the semen drops on the slide with 0.5–1 ml of normal saline. In addition to preserving the sample, sperm will be immobilized, and thus, much easier to view. Samples can be viewed wet or dry mounted and viewed under oil immersion after fixing. Deformities in head shape, tail formation and cytoplasmic droplets (proximal-near the head; distal-middle of tail) should be counted as abnormal spermatozoa.

10.4.4 Semen processing and extension

The semen collected on-farm can be used: undiluted fresh or diluted chilled.

a. Undiluted fresh semen

Procedures for microscopic evaluation, extension and storage of semen can be complex and are rarely needed in small within-herd AI programs. Keep the semen at the temperature collected (34 °C) and use within a few hours. Best results are obtained when 100 ml of gel-free fraction is used. Before dividing the semen into doses, gently rock the container to disperse sperm cells settled on the bottom. Results with fresh semen compare with those from natural mating. Undiluted semen from young boars can be used to inseminate two or three sows while from older boars, four to six inseminations are possible from one collection.

b. Diluted semen

A simple skim milk diluent can be used but it protects semen for only a few hr. More complex diluents allow chilling of semen and extend 'shelf-life' to several days. Diluents are available from animal breeding centres and can be freeze stored until required. They are preferable to the milk diluent. For on-farm AI programs, one part gel free semen is added to four or five parts diluent; *e.g.* 20 ml of good quality semen is made up to 80 or 100 ml with diluent. Higher dilution is possible when accurate estimates of total live and normal spermatozoa are made.

10.5 Semen preservation

For efficient use of semen in artificial insemination programme, it is require preserving the semen in a manner that the semen maintains the normal integrity. There are different methods of semen preservation:

A. Preservation of semen at ambient temperature (room temperature)

The semen at room temperature should be stored in a vial so that the vial should contain only one dose of semen. The vials should be wrapped with cotton so that it cannot be exposed to sunlight.

B. Preservation of semen at refrigerated temperature (5 °C)

For preservation of semen at refrigerated temperature Egg Yolk Citrate (EYC) extender is used. After dilution with extender, the semen is transferred to a vials and this sealed with metallic wax. The vials are placed in a tray or petridish and kept at refrigerator temperature at 5 °C. The semen preserved at 5 °C remains fit for insemination for 3–4 days.

C. Deep freezing of boar semen

Although cryopreserved boar semen has been available since 1975, a major breakthrough in commercial application has not yet occurred. There is ongoing research to improve sperm survival after thawing, to limit the damage occurring to spermatozoa during freezing and to further minimize the number of spermatozoa needed to establish a pregnancy. Boar spermatozoa are exposed to lipid peroxidation during freezing and thawing, which causes damage to the sperm membranes and impairs energy metabolism

The addition of antioxidants or chelating agents, *e.g.* catalase, vitamin E, glutathione, butylated hydroxytoluene or superoxide dismutase, to the still standard egg-yolk based cooling and freezing media for boar semen effectively prevented this damage. In general, final glycerol concentrations of 2–3% in the freezing media, cooling rates of -30 to -50 °C/minute, and thawing rates of 1200 to 1800 °C/ minute resulted in the best sperm survival.

However, cooling and thawing rates individually optimized for sub-standard freezing boars have substantially improved their sperm quality after cryopreservation. With deep intrauterine insemination, the sperm dose has been decreased from 6 to 1×10^9 spermatozoa without compromising farrowing rate or litter size.

Minimizing insemination-to-ovulation intervals, based either on estimated or determined ovulation, have also improved the fertility after AI with cryopreserved boar semen.

With this combination of different approaches, acceptable fertility with cryopreserved boar semen can be achieved, facilitating the use of cryopreserved boar semen in routine AI programs.

Commercial frozen semen is available either in pellet form or in straws. Thawing of frozen semen in pellets or straws is done by eventually adding the semen to extenders. This is done to ensure a large enough volume is being inseminated with the sperm cells.

Boar semen that has been subjected to cryopreservation, or even cooling below 15 °C show many signs of being capacitated. Consequently, they show signs of reduced longevity, tend to undergo spontaneous acrosome reaction soon after rewarming, as if they have become capacitated.

The cooling process appears to accelerate sperm development more than fresh semen, therefore longevity in the female reproductive tract is reduced.

The researcher's challenge is to determine at what degree freezing/cooling mimics capacitation, and whether it can be prevented or reversed. They tackle this challenge by comparing sperm that have been incubated in a capacitating medium with those that have been cooled, then rewarmed. The capacitation similarities recorded in the comparison are temperature dependent. The proportion of live spermatozoa showing a capacitation like change was related inversely to the final temperature reached in the range from 0 ° to 24 °C. The capacitation effect was seen when rewarmed to 39 °C, but could be prevented by rewarming to only 24 °C. Not only does this indicate an enzyme-related response, but it offers some clues that it is possible to extend the viability of frozen semen.

The preliminary conclusions are that the cooling response indeed resembles capacitation and it does not represent bypassing the need for capacitation.

10.5.1 Semen extenders

Properties of good semen extender

a) To be isotonic with semen
b) To have buffering capacity
c) To protect sperm from cold shock injury during cooling from body temperature to 5°C: lecithin and lipoproteins from egg yolk or milk
d) To provide nutrients for sperm metabolism: egg yolk, milk and some simple sugars
e) To control microbial contaminants : antibiotics
f) To protect sperm from injury during freezing and thawing: glycerol
g) To preserve sperm life with a minimum drop in fertility

Most porcine semen extenders come packaged in a powdered form. When buying powdered extenders in bulk, they should be broken down and re-packaged in tightly sealed containers that will make the desired volume of liquid extender. If not mixed in the powdered extender, preservative antibiotics should be added the day the powdered extender is reconstituted with water. Purchased extenders should have production dates, be kept in a frost-free refrigerator, and be used within six months of purchase.

10.5.1.1 *Function of extender*

1. To allow multiple inseminations from a single ejaculate.
2. The extender must provide temperature protection for sperm while reducing the metabolic rate of sperm cells in cool storage.

3. The extender functions to provide membrane stabilization in cool temperatures, energy sources for sperm metabolism, pH buffering from sperm cell waste, ions for membrane and cell balance, and antibiotics to prevent growth of microbes that can cause disease and compete for nutrients.

Handling of semen should be done carefully to avoid cold shock, overheating and contamination with urine, dust and water, exposure to direct sunlight should be avoided.

10.5.1.2 *Extender preparation*

To prepare the extender, weigh out the specified amount to make 1 liter. Using a graduated cylinder or a scale, pour one liter of distilled water into a mixing container. Add the extender powder to the water and mix well until powdered or particulate material is dissolved. The extender should be made up 1–2 hr in advance in order to let the pH and ions achieve equilibrium. The extender can be made up in advance and stored frozen. However, if it will not be frozen immediately, it should be utilized within 24 h of mixing. This is to prevent the antibiotics, which are effective for a limited amount of time, from losing their potency. Once the semen sample has been collected and evaluated, and the extender has equilibrated at room temperature, the semen temperature should be measured and the extender temperature adjusted to within 1 °C of the semen sample.

10.5.1.3 Determining extension rate

1. Sperm cells/cc × % motile × % normal × ejaculate volume = total sperm.
2. *e.g.* 60 billion sperm × 90% motile × 95% normal × 150 cc = 51 billion sperm.
3. To get a desired 3 billion sperm/80cc insemination dose:

   340×10^6 sperm/cc (determined sperm concentration) × 150 cc (ejaculate volume) = 51×10^9 total sperm.
4. 51×10^9 sperm ÷ 3.0×10^9 sperm/dose = 17 doses
5. Multiply doses (17) × volume of semen dose (80 cc) = 1360 cc final volume.
6. 1360 cc – 150 cc = 1210 cc
7. Add 1210 cc of extender to 150 cc semen sample.
8. Gently put 80 cc of extended semen into 17 bottles, tubes, or bags for 3.0×10^9 sperm/dose.

10.5.1.4 *Extending semen*

Total numbers of sperm per dose of semen tend to range from 2-6 billion (sperm concentration of 25 to 80 × 10^6 cells/ml). A dose of semen should contain at least 60 ml and no more than 120 ml total volume; 65–85ml being the most common volumes for a dose of extended porcine semen. The final dilution rate of semen into extender should be dependent upon initial ejaculate quality, extender type, and anticipated duration of storage time. Some facilities employ an arbitrary extension ratio of 1 part semen (sperm-rich fraction) to 7–11 parts extender when storing and using semen within 24–72 hr. Optimum extension ratios for each type of extender have yet to be established by the industry; therefore, this current practice remains questionable. If boar semen is to be extended by the volume ratio method, a conservative dilution of 1 part semen (whole ejaculate) to 4 parts extender should be followed, with the extended product used within 24 hr of extension. Problems that can occur when using the volume ratio method are: (i) semen is under diluted, allowing for exhaustion of available energy substrates and buffers over a shorter period of time, and (ii) semen is over diluted, potentially causing reduced sperm viability and fertility. In addition, the optimum number of doses of semen is not obtained; therefore, an economic and genetic loss occurs because the use of sperm cells is not maximized. The freshly collected semen and extender should be at same temperatures for mixing. The mixing of semen and extender can be accomplished by adding either semen into the extender or vise versa. Semen is diluted with extender using either a 1 (*i.e*., add all of the calculated volume of extender at one time) or 2 (*i.e*., adding one-half the calculated volume of extender to semen, allowing it to equilibrate for 5 to 10 minutes, then adding the remaining extender to achieve final volume) step technique. Since the one step process is easier and less time consuming, it is the method preferred by many laboratories.

10.5.1.5 *Precautions during semen extention*

For extending semen, the following precautions should be adopted

1. Dilutors should be prepared aseptically with analytical grade of chemicals.
2. Cleaned and sterilized equipments should be used.
3. Handling of semen should be done carefully to avoid cold shock, overheating and contamination with urine, dust and water, exposure to direct sunlight should be avoided.

10.5.1.6 Semen transportation

Care to be taken for transportation of chilled semen

a. Moisture should not enter the tube.

b. Semen tube/vials should be prevented from breaking.

c. Jerk should be minimum.

d. The semen vials should not come into close contact with ice.

The transportation of the chilled liquid semen may be done either in thermos flask or insulated carton. The thermos flask should have a layer of crushed ice at the bottom and a good layer of cotton/wool. Then the bottle containing chilled semen may be put.

Frozen semen is transported dipped in liquid nitrogen. The following precaution should be taken during transportion of frozen semen.

1. The level of liquid nitrogen should not go down and the semen straws should remain dipped in liquid nitrogen.
2. Jerk should be minimum.

10.6 Sexual behaviour in boars

10.6.1 *Confinement sexual behaviour*

When boars and gilts in oestrus placed together in a small pen, courtship rituals and sexual behaviors are abbreviated. Sexual behaviour under confinement rearing condition usually involves a short time period. The male mount and ejaculate quickly without an elaborate demonstration of courtship behaviour.

10.6.2 *Free-range sexual behaviour*

In free living swine, elaborate courtship and sexual behaviour patterns are demonstrated.

Boars frequently reach puberty before they are 7 months of age. The production of spermatozoa is usually not evident until they reach 10 to 12 months of age. Behavioral puberty or sexual activity generally coincides with the age of spermatozoa production.

The estrus female goes to the boar and sniffs his anal and preputial areas. If the estrus female runs from the boar, he pursues her, attempting to her to a stand still position. During pursuit, he noses her sides, flank and vulva. Additionally, he emits a series of soft guttural grunts. This vocalization has been entitled as "matting song". When the estrus female remains stationary, the boar presses his nose lightly against her head, shoulder or flank. Next, he proceeds towards her genital and anal region with increasingly more vigorous muzzling. Often he places his head between her rear legs and with a quick upward jerk raises her hindquarter. At this point the boar grinds his teeth, moves his jaws from side to side, foams at the mouth and grunt continually. The boar emits urine in rhythmic manner during this stage of sexual arousal. Once the copulation begins, it continues for a period of 4 to 5 minute. The female generally stands immobile during this time. A boar may repeat mating with an oestrus sow 4–8 times over an interval ranging from 12 minute to 15 hr.

10.6.3 *Climatic effect on sexual behaviour on pigs*

Number of observations have been made to know the effect of climate and temperature on the sexual behaviour of swine. When white pigs are exposed to tropical sunlight even for a short period, their ejaculation time has increased and sperm concentration and motility has decreased without affecting the libido (Egbunike and Dede (1980). Steinbach (1972a) observed that sexual libido appears to be affected by the climate. He stated that refusal to mount and ejaculate was positively related to the effective mean monthly temperature. In Nigeria, it has been observed that boars need more time to ejaculate during hottest months of the year. It appeared that climate does not affect the ovulation rate of female pigs. However, it affects the oestrous cycle. According to Steinbach (1972b), oestrus in pigs lasts longer during cooler months and the incidence of missed heat increases when the ambient temperature rises above 23° C. This situation was confirmed by Serres (1992) (Table10.2). It shows that an increase in the ambient temperature from 27° to 33° C, increases the number of sows failing to show heat after mating. Edwards *et al.* (1968) has shown that extreme heat may increase embryonic death. This observation has been confirmed by an experiment conducted by Omtvedt *et al.* (1971). They investigated the effect of heat stress at 37.8° C for 17 hr and 32.2 °C for the remaining 7 hr of the day, as against a controlled temperature of 23.3 °C. They found that heat stress on first litter gilts during the first 8 days after oestrus reduced the pregnancy rate at 30 days after oestrus by 43%, whereas heat stress from the 8th to 16th day reduced the pregnancy rate by 21%. It was also observed that the number of viable embryos was significantly lower in the stressed group in both periods. Although heat stress did not appear to affect the gilts during mid-pregnancy significantly, it had a very significant effect towards the end of the pregnancy (102–110 days). Two gilts died in the stressed group. The number of piglets born alive and stillborn were 6.0 ± 0.76 and 5.2 ± 0.62 in the stressed group and 10.4 ± 0.74 and 0.4 ± 0.62 in the control group, respectively. Of the piglets born alive 71.7% survived to 21 days in the stressed group and 88.5% in the control group. Therefore, it has been suggested (Tomkins *et al.,* 1967) that in the practical management of breeding pigs, at the time of service and for some time afterwards, and again towards the end of the gestation period, it is particularly important to protect sows and gilts from extreme heat stress. There is some experimental evidence that the gestation period may be slightly shortened in hotter months.

Fecundity is about the same as it would be in the temperate zone, but weaning weights are low and stillborn and piglet mortality rates are high.

Table 10.3 The Effect of Ambient Temperature on Reprodutive Performance of Pigs

	Ambient temperature (°C)		
	26–27°	30°	33°
No. sows	74	80	80
No. sows on heat	74	78	73
No. anoestrus	0	2	7
No. returning	2	8	8
% pregnant sows	90.5	84.8	77.5

Source: Serres, (1992).

Some reproductive data for a sow herd in Ibadan in Nigeria are shown in Table 10.4.

Table 10.4 Data from the Sow Herd at Ibadan in Nigeria for the Years 1967–69

Trait	Breed	
	Large white[a]	Landrace[a]
Conception rate (%)	67	61
Farrowing interval (days)	176	177
No. of litters per sow per year	2.1	2.1
Litter size		
No. at birth	8.9	9.2
No. at weaning	7.0	7.2
Stillborn (%)	4.5	7.5
Piglet mortality (%) birth to weaning	21.4	22.2
No. of pigs sreared per sow per year	14	15
Litter weight (kg)		
At birth	11	13
At weaning[b]	42	44
Total weaning[b] weight per sow per year (kg)	86	91

[a] Descendants of Large White and landrace foundation stock imported from the United Kingdom and Sweden, respectively

[b] Age at weaning, 35 days

10.7 Artificial insemination (AI) in pig

Artificial insemination (AI) of pigs has been used in Australia since the early 1970s but only became popular after 1981 when boars and frozen semen were imported. The genetic influence of imported boars has been more widely spread by the use of AI.

Advantages of AI

1. Semen from a range of top-performing tested boars of several breeds is available from AI centres.
2. The genetic influence of good boars can be spread more widely.
3. AI is a safe, cheap method of introducing new genes into pig herds, especially those classified as specific patbogen-free, minimal disease or high health status.
4. There is less risk introducing exotic diseases with AI than in the importation of live pigs.
5. AI overcomes size differences between boars and sows.
6. It may be used during temporary shortages of boars from death, lameness or failure to work.

Disadvantages

1. Reduced farrowing rate (50%) with frozen semen.
2. Lower than average results with chilled semen stored longer than 72 hr.
3. Disappointing results where AI is poorly timed or done incorrectly.

A. Collecting semen

Use a shady, draught-free area when collecting semen since exposure to ultra violet light, sudden temperature changes and water contamination, lower the spermatozoa's viability.

When the boar is mounted on the dummy, grasp the spiral end of his penis with the hand (gloved or bare, it must be clean, dry and warm). Allow the boar to thrust through the clenched hand a few times before applying pressure. Hand pressure on the spiral part of the penis, imitates that of the oestrus sow's cervix, stimulating ejaculation. With experience, it becomes obvious that some boars prefer more pressure than others. The long hairs around the boar's prepuce should be clipped to minimize injury.

When the penis is 'locked' in the hand and the boar relaxes, a four-phase ejaculation follows in a few seconds, taking 5 to 10 minutes to complete. The first phase, called the pre-sperm fraction, has clear seminal fluid, some gel, dead sperm cells and is heavily contaminated with bacteria. It should be discarded.

The next phase is the sperm-rich fraction, easily recognized by its creamy-white colour. Although only 50 ml in volume, it contains the greatest density of spermatozoa. Because spermatozoa are very sensitive to rapid temperature change, a warm, dry, insulated collecting flask is required to safeguard semen fertility. The third fraction, greyish because of lower density of spermatozoa, accounts for about 80 ml of the collection. Fractions two and three only are collected when semen is to be diluted for storage over a few days.

The fourth phase or post-sperm fraction provides the large semen volume peculiar to pigs. Up to 250 ml of clear seminal plasma free of spermatozoa, plus gel is secreted from the accessory glands. The gel portion apparently has no physiological significance. It is separated from the collection by several layers of gauze or a similar filter fastened over the collection flask. Filtering particles of gel from the semen prevents catheter blockage during insemination. Hair, skin or dust particles from the boar or dummy must also be excluded from the flask during collection.

Large amounts of gel signal the end of ejaculation. When it is clear that the boar's erection is fading, a second ejaculation can be stimulated with brief, firm, pulsating hand pressure applied to his penis.

B. Collection frequency

To ensure high sperm cell concentration and semen volume, do not collect boars more than three times a week. Collect twice a day or on two successive days to re-inseminate females bred 8 to 12 hr before. Excessive collection in a short time reduces the quality and quantity of semen and the boar's sex drive.

C. Succeeding with AI

Whether using semen collected on-farm or buying it from an AI centre, successful insemination hinges on:

1. Detecting oestrus in the sow
2. Timing of insemination
3. Using the right technique
4. Correct storage and handling of semen

1. Detecting oestrus

During her 50 to 60 hr oestrus or 'heat' period the sow will mate but she is only highly fertile for 24 to 32 hr (Fig. 10.11).

To experienced stock persons, the signs of impending oestrus are obvious. In a group, the sow mounts others while swelling and reddening of the vulva (in gilts especially) gives early warning. There is little mucus secreted from the vagina at this stage, afterwards it is plentiful and has better lubricating qualities than that seen at the beginning of oestrus. The colour of the mucous changes from clear to greyish-looking at this stage.

In the last 12 to 24 hr, the sow 'stands' for the boar but increasingly less for the stock person. Insemination at this time also gives poor results.

2. Timing of insemination

During ovulation, the ovaries shed eggs for 40 hr (range 36 to 50 hr) after the onset of oestrus. Spermatozoa in freshly collected or chilled semen have to mature or capacitate for 2 to 3 hr in the female's oviduct before being capable of fertilization, but frozen semen requires no time for capacitation.

Spermatozoa can be found in the oviducts of naturally mated sows for 24 hr or more after copulation while the viable life of frozen and then thawed spermatozoa is only about 8 to 10 hr in the oviducts. The eggs or ova have a much shorter life, being viable for less than 6 hr in the oviduct. Ideally, the eggs should be fertilized within hr of being shed; embryos from aged eggs tend to die more readily.

Inseminations with fresh or chilled semen achieve optimum conception about 12 hr before ovulation. Since ovulation follows onset of heat after about 40 hr the best insemination time is 28 hr (about a day) after onset. As the exact time of oestrus onset is difficult to pinpoint on farm, inexperienced persons particularly, should perform two inseminations 12 to 16 hr apart to achieve best results.

Some sows may 'honk', lose their appetite and appear nervous, thus signalling approaching oestrus. This period before proper oestrus lasts 2 to 4 days.

The sow shows onset of oestrus by accepting the boar's sexual advances. The vulva is still red and swollen and watery mucus is often seen after sexual stimulation. At this time only the boar gets the sow to 'stand' and insemination results in poor fertility.

Peak fertility (for AI and natural mating) is in the middle 24 hr when both boar and inseminator can get a strong 'standing' reaction (starts about 12 hr after onset of oestrus). The swollen red vulva has noticeably subsided by this time.

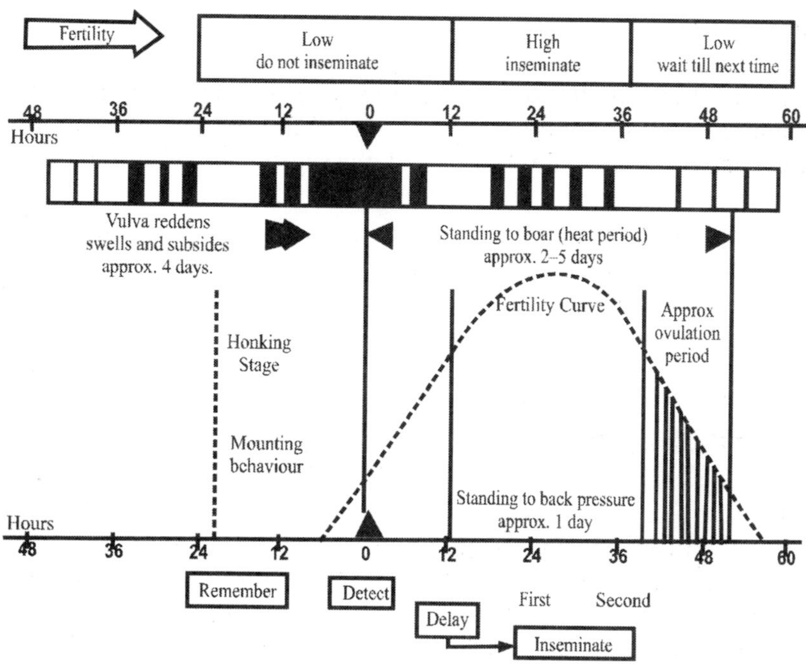

Fig. 10.11. Sow oestrus and timing of insemination

Fig 10.12. Testing the 'standing' reaction with a boar nearby

3. Insemination technique

The third important element with AI is using the right technique. The technique should imitate natural mating as closely as possible.

Before inserting the catheter, clean the sow's vulva and the area around it with a fresh, damp cloth or paper towel. This reduces the chances of introducing infectious material into the uterus and also provides sexual stimulation.

Manually stimulate the sow in the presence of a mature boar to induce the mating stance. This is done by applying pressure between her shoulder and mudpack, knee pressure to her flanks and massaging her vulva and udder.

The uterine horns of a mature sow are up to 1.5 meters in length and sperm cells cannot travel this distance on their own. Oxytocin secreted from the sow's brain in response to stimulation cause smooth muscles in the tract to contract and push the spermatozoa towards the oviduct. This process is critical for a successful insemination.

a. Inserting the catheter

Lubricate the tip of the catheter (Melrose type only) with a small amount of semen. Part the lips of the vulva and gently insert the catheter upwards into the vagina ensuring that the tip does not penetrate the urethra (the bladder opening on the vaginal floor). Accidental penetration allows urine to flow from a distended bladder through the catheter. As urine harms sperm cells, a fresh catheter should then be used.

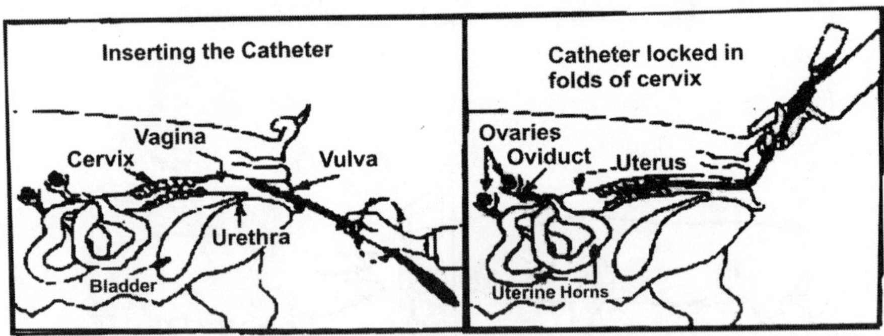

Fig 10.13. Diagram of the sow's reproductive organs with catheter positioned for insemination.

The catheter is gently pushed through the vagina until resistance is felt at the opening of the cervix. Spiral tipped catheters are rotated anti-clockwise and steadily pushed to lock into the cervix. As the funnel shaped cervix of a sow in oestrus is firm and well lubricated the inseminator can apply positive pressure while 'locking in' the catheter. Its shape helps to direct the catheter so that insertion is seldom difficult. In some gilts the hymen membrane may cause resistance in the first 10 cm.

The catheter is inserted as far as it will go. The flange on the non-spiral type can be felt passing over the cervical folds while the spiral type is 'locked-in' when a gentle pull fails to free it.

(b) Attaching the semen bottle

With semen kept within the desired temperature (15° to 20 °C), spermatozoa lose their motility and settle to the bottom of the bottle. Twice per day and before use, the bottles or tubes must be gently agitated to re-suspend the cells. Thirty minutes before use, remove the chilled semen bottle from its pack and allow the semen to warm up. This can be done by placing the bottle in a warm place, for example, a shirt pocket. Immediately before use, gently rock the insemination bottle to redistribute the spermatozoa through the diluent. Cut the tip of the inseminating bottle with a clean, sharp knife. Grip the bottle by its cap and firmly fit the tip into toe inserted catheter. The bottle is raised over the sow's back and squeezed a little to remove the airlock in the catheter. With the aid of gravity, semen is drawn into the sow by wave-like muscular contractions in her uterine horns. Sexual stimulation should be continued to promote the contractions.

If semen does not flow freely from a soft-walled inseminating bottle or when gentle pressure is applied to a less pliable one:

1. The opening at the catheter's tip may be blocked by a fold in the cervix; withdraw the catheter a little way or slightly rotate the spiral tipped catheter.
2. The sow may not be contented or sufficiently stimulated. The hormone oxytocin is responsible for uterine contractions during insemination. Aggressive handling, especially of nervous sows, triggers the release of adrenaline, a hormone that inhibits oxytocin. Boar presence, firm hand rubbing between the sow's shoulder and midback area, knee pressure in the flank and massaging the vulva and udder stimulate the sow. Periodic movement of the catheter also helps.
3. There is a blockage evident by resistance when the bottle is firmly squeezed. Remove the bottle and if necessary, the catheter to check for blockage (*e.g.* from gel particles or faulty equipment or too small an opening in the semen bottle's spout).

(c) Insemination

When the sow is well stimulated, insemination should only take 5 to 10 minutes. It is possible to let the semen gravity-feed into the catheter when using soft walled bottles; gentle pressure may be used with any type of bottle, particularly the firmer types.

If the catheter is not far enough in or the inseminator is too hasty, back-flow may occur. When this happens, disconnect the bottle, reposition the catheter and delay insemination for a minute or two. Semen is often squeezed out by abdominal pressure when the sow moves suddenly, for example, when coughing or struggling. A small back-flow is not unusual and not detrimental provided enough semen with an adequate sperm count is used.

After semen enters the sow's reproductive tract, sexual stimulation should continue for several minutes. This ensures that the uterus actively forces the spermatozoa towards the oviducts, where fertilization of the eggs takes place. The catheter is left in place to assist with stimulation.

Stimulation is important in the sow because spermatozoa have to travel the 1.0 to 1.5 m length of the uterine horns. When removing the catheter, ensure that the free end is higher than the vulva or semen may siphon out. Allow the sow to remain undisturbed following insemination.

Equipments for AI

a. Catheter

The catheters commonly used for AI are the reusable rubber 'Melrose' with spiral tip or disposable plastic catheters having several types of tip. All are easy to use and achieve good results. Many inseminators prefer catheters with spiral tips that 'lock in' like the boar's penis, reducing back-flow during insemination. A drawback with reusable catheters is the high standard of cleaning and hygienic storage needed between uses.

b. Semen bottles or tubes

Chilled semen bought from AI centres comes in ready-to-use inseminating bottles or tubes. But if semen is collected and used fresh or diluted on-farm, a supply of clean plastic inseminating bottles will be needed.

Care of equipments

Since boar semen is an excellent medium for growing bacteria, all AI equipment must be kept clean. Immediately after use, soak reusable equipment in cold water so that semen or other material is easily removed later.

Do not use soaps or detergents because they affect sperm viability. Particles of gel can be removed with a brush.

Rinse, then boil rubber 'Melrose' catheters in distilled water for 10 to 20 minutes before reuse. Tap water must not be used for rinsing because it leaves mineral deposits on the equipment.

Store equipment in a dust-free cabinet or when completely dry, in a sealed plastic bag.

Hints for successful AI

1. When handling the sow, be firm but not aggressive.
2. If a sow or gilt first 'stands' to back pressure in the morning, when near a boar, inseminate late in the afternoon of the same day and again the following morning.
3. If she stands in the afternoon, inseminate her the following morning and again late that afternoon.
4. If only doing a single insemination, time it to coincide with the second period as above.
5. If mucous is present, it should be creamy in colour rather than clear. If clear, then insemination is too early.
6. Use clean equipment for every insemination. Ensure the sow is adequately stimulated before insemination.
7. Carefully lock the catheter into the cervix.
8. Be patient, allow the semen to flow slowly.
9. Continue stimulating the sow during insemination. Handle the sow firmly but not aggressively.
10. Watch for catheter blockages and semen back-flow.

CHAPTER 11

GROWTH

11.1 Introduction

Growth simply means increase in weight or dimension of the body. Growth is anabolism reduced from catabolism. Growth has been defined as irreversible time change in measured dimension and function. The growth may be accomplished by increase in size of cells (hypertrophy), number of cells (hyperplasia) and incorporation of materials inside the cell. It can be defined as correlated increase in the mass of the body at definite intervals of time in way characteristics of species. Several major changes occur as an animal passes from the zygote to its mature form and size. Perhaps the most obvious change in size and mass has been termed growth. In addition to these, there are fundamental changes in shape and body composition which have been termed differentiation. Because growth and differentiation are inseparable their combination is called development.

Von Bertalanffy's equation: $dW/dt = a.Wb - c.Wd$

Body weight increment = anabolic increment – catabolic loss, w being body weight, a and c are constants.

The body cells can be classifies into 3 types based on growth.

Renewing cell population

The cells which are constantly destroyed and replenished by proliferation of stem cells to a generative zone, *e.g.* epidermis and derivatives, endometrium.

Expanding cell population

Cell division continues until the adult size of an organ is reached, *e.g.* Liver, kidney and glandular cells.

Static cell population

Static tissues are expanding in which cell division is restricted to the early stages of development, although cellular hypertrophy may occur later. *e.g.* neurons, muscle fibres

11.2 Types of growth

1. Compensatory growth

When a part of organ stops growing or whole part of its experimentally removed, compensatory growth takes place which increases functional capacity of the tissues. The tissues capable of compensatory growth are liver, testis, adrenal and ovary. The tissues such as nervous, muscular, limbs, teeth, sense organs etc cannot undergo compensatory growth in mammals.

2. Differential growth

Differential growth centers become active at different times and exhibit different rates of activity. These growth centers are coordinated to produce predetermined form and characteristic of the species.

Growth can be described in animals into two phases namely, prenatal growth and postnatal growth.

11.2.1 Prenatal growth

Prenatal growth occurs between formation of zygote and birth of the animal. During this intrauterine life, the continuous growth rate can be arbitrarily divided as stage of embryo and stage of fetus. In the first phase the zygote undergoes cleavage to form cluster of cells called as blastomeres. The multicelled structure, called morula undergoes a process called as gastrulation to form gastrula. The morula forms itself into a layer of cells called the trophoblast surrounding a fluid filled space, the blastocoele. The trophoblast contributes to the placenta and is lost at birth. The inner cell mass together with the trophoblast is called as blasotocyst. The embryo developing from the inner cell mass becomes roofed over by amniotic folds that later fuse to form the amnion. While the newly formed animal is developing its various types of tissues (ectodermal, endodermal and mesodermal tissues) it is called an embryo but, after these tissues are acquired and until birth or hatching, it is called a foetus.

11.2.2 Postnatal growth

The post natal growth can be divided into prepubertal, pubertal, reproductive and senescence phases.

11.2.3 Growth curve

If dimensions or weight of the animal from conception to senescence is plotted against time, the curve follows an S structure, the curve is sigmoid. The growth curves of meat animals raised under commercial conditions may appear as relatively flat slopes (the middle segment of the flat "S"), and the sigmoid shape may only become apparent if the data include very young animals or animal beyond a typical market weight due to various factors affecting growth. In other words, growth velocity is approximately constant during the commercial growing period. In adult animals the sigmoid curve tends to become j shaped.

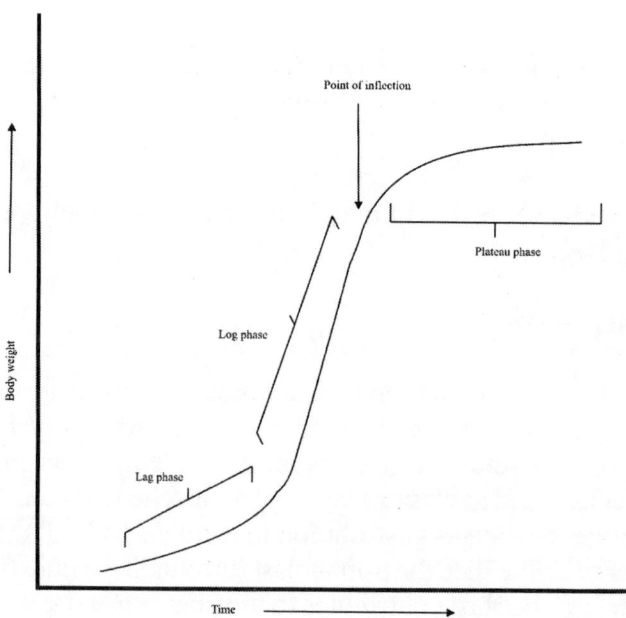

Fig 11.1. Sigmoid growth curve

The point of inflection is the point of maximum growth with respect to age. It is the point at which the animal comes to puberty and lowest mortality. The point of inflection is also known as point of physiological equivalence.

11.3 Factors affecting Growth Rate in Pigs

A plethora of factors affect growth rate in pigs. The following is the list of factors which affect growth in pigs.

Genotype

One of important factors affecting growth rate in pigs is its innate genetic capacity to grow at a given rate.

Sex and age of the animal

The growth pattern is determined by age as well as sex of the animal. In male animals, the growth is sluggish in comparison to females in the initial phase, however, in later stages, a higher growth is achieved in males.

Size and age of dam

Pigs farrowed from sows were about four pounds heavier at weaning than pigs from gilts (Nordskog *et al,* 1944).

Plane of nutrition

The plane of nutrition profoundly influences growth at all stages. In the initial phases, nutrient deficiencies may cause irreversible changes in the growth of animal. In other wards, the effect of nutrition depends upon the age of the animal and the extent of deficiency or excess to which it is subjected to.

Environment

A major component of the physical environment is the climate. Its effect on growth is brought about by a complex of interacting components resulting in both direct and indirect effects upon the growing animal. The indirect effects are usually the more important for the ruminant they include ambient temperature and factors controlling the level of soil moisture available for plant growth, thus affecting the quantity and quality of available nutrients. Much the same complex of factors determines the microclimate of many micro-organisms and their vectors, these playing a major role in the dynamics of parasitic organisms and hence the levels of disease (Charles, 1985).

Environmental variance common to each litter was greatest at 21 days and accounted for 37% of the total variance. On gains after weaning it decreased from 24%, 28 days post-weaning, to 7%, 112 days post-weaning. Environmental effects peculiar to individual pigs accounted for approximately one-half of the total variance in the periods of gain and weights studied (Nordskog *et al,* 1944).

Hormones

The entire growth is precisely regulated by various hormones affecting cell growth, nutrient uptake and utilization. Growth hormone, insulin, corticosteroids and thyroid hormones in general play an important role in determining animal growth.

11.4 Growth Factors

Several growth factors besides hormone have been identified to influence animal growth. The following is a non exhaustive list of growth factors.

1. Bone morphogenetic proteins (BMPs)
2. Epidermal growth factor (EGF)
3. Erythropoietin (EPO)
4. Fibroblast growth factor (FGF)
5. Granulocyte-colony stimulating factor (G-CSF)
6. Granulocyte-macrophage colony stimulating factor (GM-CSF)
7. Growth differentiation factor-9 (GDF9)
8. Hepatocyte growth factor (HGF)
9. Hepatoma derived growth factor (HDGF)
10. Insulin-like growth factor (IGF)
11. Myostatin (GDF-8)
12. Nerve growth factor (NGF) and other neurotrophins
13. Platelet-derived growth factor (PDGF)
14. Thrombopoietin (TPO)
15. Transforming growth factor alpha(TGF-á)
16. Transforming growth factor beta (TGF-â)
17. Vascular endothelial growth factor (VEGF)

11.5 Allometric Growth in Pigs

Sir Julian Huxey defined the size relationship between the whole body and its organ mathematically as

$$Y = bX^k$$

Taking log on both sides,

$$\log y = \log b + k \log x$$

The slope of the resulting regression is called the allometric growth ratio, often designated as k. The k represent specific growth rate coefficient for a particular tissue. When the growth rate of the organ and body are same, it is termed as isometric. The isometric growth may be positive or negative. The equation can be used for growth of different organs of the body. When $k = 1$, both components are growing at the same rate and When $k < 1$, the organ/ structure represented on the Y is growing more slowly than the on the X axis and vice versa is true when $k > 1$.

Allometric growth of carcass muscles in pig

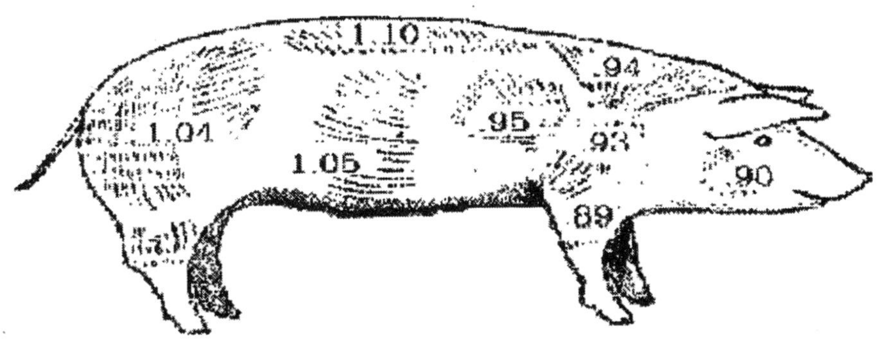

Fig. 11.2. Allometric growth ratios for muscle groups of the pig.

(Adapted from Swatland, H.J. http://www.aps.uoguelph.ca/~swatland/ gasman.html)

Allometric growth ratio may be used to categorize muscles into one of three monophasic categories high, average and low impetus depending on whether their allometric growth ratio is greater than, equal to, or less than a value of 1 respectively.

The readers may refer to original texts on growth, its regulation and manipulation as detailed description will be beyond the scope of the book.

CHAPTER 12

PHYSIOLOGY OF DIGESTION

12.0 Physiology of Digestion

A brief note to physiology of digestion is provided in the following sections as an introductory to subsequent chapters on nutrition of swine.

12.1 The Digestive Tract of the Pig

The pig has a digestive system which is classified as monogastric or nonruminant. The monogastrics differs from that of a polygastric or ruminant digestive system found in cattle and sheep. Due to the differences in the digestive systems, cattle can utilize different types of feeds than pigs. Cattle and sheep can suevive on coarse grains, while pigs must eat feed that can be digested more easily.

Digestion is the breakdown of food occurring along the digestive tract. The digestive tract may be thought of as a long tube through which food passes. As food passes through the digestive tract, it is broken down into smaller and smaller units. These small units of food are absorbed as nutrients or pass out of the body as urine and faeces.

The digestive tract of the pig has five main parts: the mouth, oesophagus, stomach and small and large intestines.

12.1.1 Mouth

The mouth is where food enters the digestive tract and where mechanical breakdown of food begins. The teeth chew and grind food into smaller pieces. Saliva, produced in the mouth, acts to soften and moisten the small food particles. Saliva also contains an enzyme which starts the digestion of starch. The tongue helps by pushing the food toward the esophagus.

12.1.2 Oesophagus

The oesophagus is a tube which carries the food from the mouth to the stomach. A series of muscle contractions push the food toward the stomach. Swallowing is the first of these contractions. At the end of the esophagus is the cardiac valve, which prevents food from passing from the stomach back into the oesophagus.

12.1.3 Stomach

The stomach is the next part of the digestive tract. It is a reaction chamber where chemicals are added to the food. Certain cells along the stomach wall secrete hydrochloric acid and enzymes. These chemicals help break down food into small particles of carbohydrates, protein and fats. Some particles are absorbed from the stomach into the bloodstream. Other particles which the stomach cannot absorb pass on to the small intestine through the pyloric valve.

12.1.4 Small intestine

The small intestine is a complex tube which lies in a spiral, allowing it to fit in a small space. Its wall has many tiny finger-like projections known as villi, which increase the absorptive area of the intestine. The cells along the small intestine's wall produce enzymes that aid digestion and absorb digested foods.

At the first section of the small intestine called the duodenum, secretions from the liver and pancreas are added. Secretions from the liver are stored in the gall bladder and pass into the intestine through the bile duct. These bile secretions aid in the digestion of fats.

Digestive juices from the pancreas pass through the pancreatic duct into the small intestine. These secretions contain enzymes that are vital to the digestion of fats, carbohydrates, and proteins.

Most food nutrients are absorbed in the second and third parts of the small intestine, called the jejunum and the ileum. Undigested nutrients and secretions pass on to the large intestine through the ileocecal valve.

A "blind gut" or cecum is located at the beginning of the large intestine. In most animals, the cecum has little function. However, in animals such as the horse and rabbit, the cecum is very important in the digestion of fibrous feeds. The last major part of the digestive tract, the large intestine, is shorter, but larger in diameter than the small intestine. Its main function is the absorption of water.

12.1.5 Large intestine

The large intestine is a reservoir for waste materials that make up the feces. Some digestion takes place in the large intestine. Mucous is added to the remaining food in the large intestine, which acts as a lubricant to make passage easier. Muscle contractions push food through the intestines. The terminal portion of the large intestine is called the rectum.

The anus is an opening through which undigested food passes out of the body. Food that enters the mouth and is not digested or absorbed as it passes down the digestive tract is excreted through the anus as faeces.

12.2 Uptake and Mastication of Feed

Whenever ground concentrates mixed up with any liquid like water or skim milk is offered to the pig, it dips its snout to the bottom of the trough and sucks the mixed feed with the help of the tongue. Longer particles like grass or beets are chewed by the molar teeth. Since the angles of the mouth are situated far back, part of the mouth will always remain above the surface of the trough. The portion which remains above the surface sucks air together with the mixed feed. The suction of air into the mouth and rapid chewing movements with open lips cause the characteristic smacking and slurping sounds. After the uptake of feed in the mouth, it is thoroughly mixed up with saliva. The amount of saliva depends upon the type of feed. Less the moisture content of the feed, more will be the secretion of saliva.

There are three salivary glands (a) parotid glands, lying in the space below the ear and behind the border of the lower jaw; (b) the sub maxillary glands, lying just within the angle of the lower jaw and (c) the sublingual glands, which lies at the side of the root of the tongue. Each of these gland are paired, so that actually there are six glands. The parotid glands produce serous, an alkaline fluid, which normally contains small amounts of the enzyme amylase, aids in hydrolyzing the carbohydrates. This activity is many times less than that found in human saliva. The secretion from the other salivary glands contains no enzyme. The carbohydrate portion of the diet is thus subjected to amylase action in the mouth which is continued for some time in the stomach until the acid in gastric juice has inactivated the enzyme.

12.3 Digestion in Stomach

The wall of the stomach is full of glands. The ducts of these gastric glands open into the stomach cavity, so that as the gastric juice is produced it pours directly into the stomach cavity. Three types of cells have been described in the gastric glands. They are (1) parietal cells which secrete HCl, (2) neck cells which secrete mucin, a mucus substance; (3) chief cells which produce enzyme pepsinogen which later on changes into pepsin. Thus the gastric juice is composed of water, mucin, pepsin and HCl.

Even though food does not remain long enough in the mouth for amylase to complete the carbohydrate digestion, the action of the enzyme continues long after the food has entered the stomach. The pH, optimum for amylase, is about 7; however, it will continue to act until the pH has fallen to 4.5. The first phase of gastric secretion is called the amylolytic phase. Bacterial carbohydrases and carbohydrases of the feed will also show activity during the amylolytic phase of gastric digestion. Through the action of these enzymes starch and other polysaccharides are broken down to soluble carbohydrates such as, erythrodextrin and to some extent to maltose.

Once the stomach juice penetrates the swallowed feed, the pepsin starts to break down the feed proteins. There will then be the phase with both amylolytic and proteolytic activity, the amyloproteolytic phase of gastric digestion. Through the action of pepsin and HCl the feed protein are broken down to peptides. The optimum pH for pepsin is about 2. When the pH falls below 4, the amylase activity is completely inhibited and the only enzyme acting is pepsin. It is to be noted that the various phase of digestion in the stomach go on concomitantly.

No lipases are secreted from the stomach glands. Gastric juice may show weak lypolytic activity due to lipases reflexed from the small intestine into the stomach. These lipases may liberate some free fatty acids from ingested lipids. However, the main change of feed lipids in the stomach is a result of gastric motility which by churning and kneading turn the lipids into a coarse emulsion.

12.4 Digestion in Small Intestine

The liver and pancreas are the two large glands connected to the first part of the small intestine. By peristaltic movement of the small intestine the content is mixed and transferred through three parts, *viz.*, duodenum, the jejunum and the ilium. In the small intestine the digestion is continued through the action of the enzymes from the intestinal glands and from pancreatic secretion.

There is a continuous formation and secretion of bile in the liver. A portion of synthesized bile is stored in the gall bladder and rest flows directly through the duct to the small intestine. In liver bile the percentage of dry matter is about 3 whereas bladder bile has about 16% dry matter. The higher dry matter content in the bladder is due to absorption of water from the gall bladder. The major organic components of bile are the bile pigments and the bile salts with significant amounts of lipids as phospholipids (mainly lacithin and lyso-lecithin) and some cholesterol are present. Bile salts are sodium and potassium salts of glycocholic or taurocholic acid. The reaction is usually weak alkaline with a pH of about 7 to 8.

12.4.1 Pancreas

Pancreas is both an endocrine and exocrine gland. The endocrine secretions, insulin and glucagon are not considered here. There are three proteolytic zymogens in pancreatic juice namely, trypsinogen, chymotrypsinogen and a procarboxypeptidase. All three zymogens are then rapidly converted into active enzymes such as trypsin, chymotrypsin and carboxypeptidase respectively. Apart from proteolytic enzymes the pancreatic juice, which is clear and distinctly alkaline, contains several other lypolytic and amylolytic enzymes. The pig secretes about 7 to 15 liters pancreatic juice per day.

The duodenal mucosa contains branched coiled tubular glands called *Brunner glands,* that give an abundant secretion. It is viscous, sticky fluid with an alkaline reaction. The pH of duodenal juice from pigs ranges from 8.4 to 8.9. The alkali and mucin or mucin like substances in the secretion protect the intestinal mucosa from injury by the acid chyme coming from the stomach. There might be some digestive enzymes in the duodenal secretion.

When the acid chyme from the stomach is mixed with the alkaline secretion from the liver, the pancreas and the intestinal gland, the acid is partly neutralized and the pH of the intestinal content rises slowly as the content is passed down through the tract.

The lipolytic activity of pancreatic juice is due to a specific enzyme, pancreatic lipase which removes only fatty acid residues linked to primary hydroxyl groups (a groups) of triglycerides. The rate of lipase action is increased by emulsifying agents such as bile salts, lecithin and lysolecithin.

The digestion of starch initiated by salivary amylase is continued by the action of pancreatic amylase, resulting starch into maltose and isomaltose. Ingested carbohydrates with different glucosidic linkages are attacked by other carbohydrases present in the intestinal secretions. The products of the action of these carbohydrates are different disaccharides. The epithelial cells of the small intestine contain four enzymes, *viz.* lactase, sucrase, maltase and isomaltase, which are capable of splitting the disaccharides lactose, sucrose, maltose and isomaltose respectively, into their constituent monosaccharide. There is much reason to believe that these enzymes are located in the brush border of the cell lining the lumen of the intestine and that the disaccharides are digested as they come in contact with this border. The digested products are then immediately absorbed into the portal blood.

However, it is important to note that raw potato starch and cellulose are not digested by the intestinal enzymes in the pig. Part of these substances is digested by bacterial fermentation in the large intestine including caecum.

12.5 Digestion in Caecum and Colon

It has been known for a long time that the pig can digest crude fibre to some extent and that this digestion is entirely dependent upon bacterial fermentation in the caecum and colon. The digestible coefficient of crude fibre of normal swine varies from 10 to 90%. The variations are probably due to (1) changes in the intestinal flora which again varies with the type of diet. In normal pigs there are 10^8 to 10^9 micro-organisms per gram of caecal contents. The predominating species are lactobacilli and streptococci, and (2) to the amount of cellulose in the ration. It has been estimated that the optimum level of crude fibre in swine ration should be about 6 to 7%.

The products of bacterial fermentation of cellulose are volatile fatty acids (VFA) with acetic acid as the predominant acid. The average composition of mixed VFA from the caecum of pigs is as follows: 62% acetic acid, 28% propionic acid and 10% butyric acid.

VFA produced in the caecum and colon of pigs are rapidly absorbed in the blood system and are readily utilized by the animal.

Substantial amount of digestible proteins are absorbed after hydrolysis in the small intestine. Some amount of amides and non-essential amino acids are catabolised and due to deamination results in the formation of ammonia which is readily absorbed through caecum and colon. Since ammonia is a toxic substance, it is converted into urea and excreted again.

Microorganisms are capable of synthesizing several water soluble vitamins and thus play a role in the normal supply of these elements.

The digestive tract acts to digest and absorb nutrient necessary for maintenance of cells and growth. Efficient absorption of nutrients depends on each segment of the digestive system functioning to its maximum capacity. The genetic differences between exotic breeds and indigenous breed in respect to efficient utilization of different feed stuffs are not known and need to be investigated.

CHAPTER 13

NUTRITION AND FEEDS RESOURCES

13.0 Pig Nutrition

13.1 Principles of Pig Nutrition

Efficient and profitable pig production depends upon an understanding of the concepts of genetics, environment, herd health, management and nutrition. These factors interact with each other and their net output determines the level of production and profitability. Feed represents 60–75% of the total cost of pork production. Therefore, amino acids, carbohydrates, vitamins, minerals and water must be provided and balanced to meet pig's requirements. Thus, a thorough knowledge of the principles of pig nutrition is essential in order to maintain a profitable pig enterprise.

In organized pig production unit, the main objective of feeding strategy and diet formulation is to maximize profits. Therefore, the aim of swine feeding is to maximize the economic efficiency; the indispensable nutrients are fed as close as possible to requirements. The formulation of economic, nutritionally adequate pig ration requires the knowledge of: (i) the nutrient requirements of different categories of pigs; (ii) the nutrient contents and presence of toxic principles of feed ingredients; and (iii) the availability of the nutrients in feed ingredients.

For development of such optimum feeding strategies, consideration of factors such as genetics, environment, availability and variability of feed ingredients, non conventional feed ingredients and stability of nutrients in feed ingredients, interactions among the nutrients and non-nutritive factors, etc. is necessary. In addition, there must be an effective means to incorporate all the necessary information to formulate efficient diets in a convenient and economical manner.

Several factors affect a pig's requirement of specific nutrient. These factors influence feed intake, which will require changing the concentration of the nutrient in the diet to meet the pig's requirement on an amount-per-day basis. Some of the factors are:

a) Environmental temperature or weather
b) Breed, sex and genetic background of pigs
c) Health status of the herd
d) Presence of moulds, toxins or inhibitors in the diet
e) Availability and absorption of dietary nutrients
f) Variability of nutrient content in the feed
g) Level of feed additives or growth promoters
h) Energy concentration of the diet
i) Level of feeding, such as restricted feeding vs. *ad libitum*

Environmental temperatures and housing conditions play an important role in determining the pig's nutrient needs for maintenance. Pigs housed in outside dirt lots are exposed to greater temperature changes than those housed in confinement facilities and may have greater maintenance needs. In addition, research has indicated that pigs of different sex, breeds or genetic background may have different capacities for production, thus different nutrient requirements. It is reasonable to expect that a sow weaning 27 pigs per year would have higher requirements than one weaning 15 pigs per year.

Pig is a single stomach animal and cannot utilize coarse fodder. A pig must therefore be fed maximum of concentrate and minimum of roughages. The major goal of a swine feeding program is to provide them the proper amount of necessary nutrients in a palatable form at the lowest possible cost. The response pigs make to the feed depends on (a) their genetic ability to convert feed efficiently, (b) their freedom from disease and parasites, and (c) the proper combination of essential nutrients in the ration. Feed represents about 65–80% of the cost of producing hogs. Thus, successful pig production requires a carefully planned and efficient feeding program.

In the temperate zone practical rations for pigs are usually based on a daily feed allowance, as pig farmers aim to achieve the fastest possible growth without excessive deposition of fat. The protein in the feed must be adequate, not only in total amount, but also in the amount of individual essential amino acids, as pigs cannot synthesise a number of these amino acids, the most important being lysine and tryptophan. As the majority of pigs are pen fed, special attention must also be paid to the mineral and vitamin content of the daily feed.

The effect of high environmental temperatures on the nutrient requirements of the pig has not been fully explored. At ambient temperatures above the thermal neutral zone for medium sized pigs (20–25 °C) there is a progressive decrease in total feed intake. This decreases total essential amino acid, mineral and vitamin intake and it has been suggested that the crude protein, mineral and vitamin contents per unit of feed should be increased to compensate. Additional fat in the diet may be desirable as may be the amino acid supplementation of low quality protein diets (Serres, 1992). The evidence with regard to requirements for some vitamins is controversial. *e.g.* it has been suggested by some researchers and refuted by others that riboflavin requirement decreases as ambient temperature increases above the thermal neutral zone. Additional water will certainly be required by pigs as ambient temperatures rise.

13.2 Characteristics of Good Ration

In order to plan a feeding program for pig, it is important to become familiar with the characteristics of a good ration for this kind of livestock. This makes it possible to plan rations of suitable feeds which provide the essential nutrients in proper amounts at the lowest possible cost.

It should be balanced

By definition, a balanced ration is one which furnishes all the required nutrients in such proportion, amount and form as will, without waste, properly nourish a given animal or group of animals to which it is fed. In truly balanced rations, both the amount and the proportion of nutrients vary with each age group of animals and with the purpose for which they are fed.

The ration which most nearly meets the requirements of the animals will be the most productive to the extent that it utilizes readily available feeds, also it is likely to be the cheapest when measured by results.

It should be well adapted to the age and individuality of the animal

The recommended nutrient allowances for swine provide a higher level of protein for young pigs making rapid growth than for older animals.

Suitable for the category of pigs for which it is being fed

Growing pigs and fattening pigs require nutrients in quite different proportions, while the requirements of brood sows may differ considerably from both of these. This is due to the varying uses the different classes of pigs make of the nutrients they eat. Pigs that are growing rapidly and sows at the height of lactation have great need for protein, which makes up a large share of the produce they yield. On the other hand, mature pigs and those in the later stages of fattening, require relatively little protein and a great deal of carbohydrates. The greatest need for calcium and phosphorus is also during early growth and lactations. A large amount of fiber is a detriment in rations being fed for rapid production but plays a useful role in restricted feeding.

It should be palatable

Feeds which pigs eat rapidly are said to be palatable to them. A balanced ration is usually more palatable and therefore is consumed in greater amounts than an unbalanced one. The fineness of grind will affect palatability; the amount of mineral and high fiber feeds in the ration will also affect palatability. Certain feeds and nutrients added in the ration will increase palatability and other will decrease it.

It should not contain anti-nutritional factors

Do not use those feeds containing antinutritional factors because they may result in health losses. When such feeds are included in the ration, swine frequently eat less of these feeds and therefore gain much more slowly. This means that one needs a knowledge of injurious substances within the feed ingredients which may containing fluorine or raw rock phosphate, selenium in grains from selenium contain areas, gossypol in cotton seed, ergot in seeds such as in barley when infected and lanolin in citrus seed meal are examples of some these substances.

It should be adapted to the system of farming

In the corn surplus areas corn is included in swine rations because of its availability and cheapness. In regions better adapted to small grains than to corn, swine rations include more barley, oats, wheat, or sorghum grain (according to the region) and less corn. Mill by-products make up a considerably larger proportion of the total feed of swine.

It should be economical

Balanced rations are more likely to be profitable than unbalanced ones. The relative economy of rations will vary with almost any marked change in price ratios of feeds. Therefore, it is important to know the relative nutritive values of feeds similar in nature in order to take advantage of price changes. Price variations among grains and protein feeds afford the alert producer an opportunity for increased profits if he knows the relative values of feeds.

A good ration does not produce an inferior product. Ground soybeans, distillery slop, rice bran, rice polish, sesame seed, flaxseed, sunflower seed and corn wheat germ meal produce soft pork when fed in liberal amounts to fattening pigs. These feeds should be avoided or fed in small quantities and pig should be fed hardening feeds after they weigh 50 to 60 kg.

13.3 Nutrient Requirement of Pigs and Utilization

Pigs are simple stomached animals and they can not be expected to consume coarse feeds like straws, stovers and inferior quality forages. Therefore, the most important considerations in pig's ration is the percentage of crude fibre. In growing-fattening pigs 6–8% fibre may be used while the sow ration may contain 10–12% fibre. It has been reported that an increase in dietary fibre by 1% depressed the digestibility of gross energy by about 3.5%.

Pigs require all the 10 essential amino acids in addition to the 16 vitamins. The amino acid requirement of growing pigs increases as the levels of the dietary energy and protein increase. The vitamin C and K are not essential since they are synthesized. Essential fatty acids like linolic and arachidonic are considered necessary including about 13 or so mineral elements. Pigs are the most rapidly growing livestock and suffer from nutritional deficiencies than the other livestock. This means that nutritional needs of the pig for carbohydrates, fats, proteins, minerals and vitamins must be met fully for profitable and efficient production.

The practical rations for pigs are usually based on a daily feed allowance, as pig farmers aim to achieve the fastest possible growth without excessive deposition of fat. Milk of sow is deficient in iron and copper hence care is to be taken during early growth to supply these nutrients. Pigs, probably more than any meat animal, is influenced by the kind of fat fed to them. Unsaturated fats produce oily carcass and thus its excess feeding should be avoided. The NRC, ARC as well as ICAR has formulated feeding standards for the pigs. The nutrient requirements for each function are computed separately to arrive at final requirement of the nutrient.

13.3.1 Energy

Energy is produced when organic molecules undergo oxidation. Energy is either released as heat or is trapped in high-energy bonds for subsequent use for the metabolic processes in animals.

The energy content of the feed is the most important concern and the ration should provide adequate energy to the pig. In pigs the energy requirement and energy content of feedstuffs is generally defined in terms of digestible energy (DE) or metabolizable energy (ME), rather than gross energy (GE). The use of DE accounts for the feed energy losses due to variation in digestibility of feeds and enhances the accuracy of diet formulation. ME takes into account the energy lost in the urine and combustible gases produced in the digestive tract. Net energy (NE) is the difference between ME and heat increment. Because it is the energy ultimately required by the animal for maintenance or production, the NE would be the best measure of the energy that is available to the animal. The estimation of the NE is, however, difficult and imprecise and affected by many factors (NRC,1988). Therefore, the use of NE might be too sensitive to be of practical use and unlikely to provide any greater precision in formulating diets or predicting responses compared with the ME or DE system.

The amount of energy lost in the urine is not a constant, and the ME content of the diet decreases with poor quality and excess protein relative to the pig's needs because of the increased excretion of nitrogen (N) as urea. The ME can be calculated as follows:

ME=DE [96–(0.202% CP)]/100, wherein the CP is the crude protein content of the feed.

The published ME to DE ratios range from 0.92 to 0.98 (ARC, 1981; NRC, 1988). For practical purposes, a commonly accepted value of 0.96 is regarded as an appropriate factor for a wide range of ingredients and diets. The loss of energy as combustible gases in pigs is generally ignored because the losses are negligible and difficult to measure (NRC, 1988). Furthermore, the variation in the relationship between DE and ME is more of a function of the animal than of the feed ingredient itself. In addition, DE values for the pig are available for most of the commonly used feed ingredients. For these reasons, it is preferable to use DE values, which can be determined much more easily and precisely, rather than the ME, to express the requirements of pigs and describe the energy value of feed ingredients and diets.

Energy requirements

Maintenance

The ME requirement for maintenance (ME_m) includes the need of all body functions and moderate activity. These requirements are usually expressed on a metabolic body weight basis, which is defined as body weight raised to the 0.75 power ($BW^{0.75}$). Other exponents have been suggested as more appropriate: 0.67 (Heusner, 1982); 0.60 (Noblet *et al.,* 1989b); 0.42 (Noblet *et al.,* 1994).

During gestation, 60 and 80% of the total energy requirement is used for maintenance. The National Research Council (1988) concluded from the available literature that the daily requirement for maintenance of pregnant sows was 106 kcal of ME or 110 kcal of DE/kg of BW $^{0.75}$/day. Noblet *et al.* (1990), on the basis of recent estimates, concluded that the daily requirement was 105 kcal of ME/kg of BW $^{0.75}$ for primiparous and multiparious sows. Beyer *et al.* (1994) reached a similar conclusion from the literature (103 kcal of ME/kg of BW $^{0.75}$/ day) for primiparous sows but reported data to indicate an increase from 93 kcal of ME/kg of BW $^{0.75}$ in the fourth parity.

Growth

Estimates for the energy costs of protein retention (ME_{Dr}) range from 6.8 to 14.0 Mcal of ME/kg, with a mean of 10.6 Mcal of ME/kg (Tess *et al.*, 1984). Literature estimates of the energy costs of fat deposition (ME_f) range from 9.5 to 16.3 Mcal of ME/kg, with a mean of 12.5 Mcal of ME/kg (Tess *et al.*, 1984). Although the mean energy costs/kg of protein or fat deposited are approximately equal (Wenk *et al.*, 1980), 1 kg of lean muscle tissue is only 20 to 23% proteins, whereas 1 kg of adipose tissue is 80 to 95% fats. Therefore, the energy cost for muscle tissue production is considerably less than that for fat tissue deposition.

Pregnancy

The feed and energy requirements of the pregnant sow will vary with body weight, target body weight gain during pregnancy, and other management and environmental parameters. The Agricultural Research Council (1981), Cole (1982), Seerley and Ewan (1983), and Aherne and Kirkwood (1985) reviewed the effects of energy intake during gestation on sow weight gain and reproductive performance. Aherne and Kirkwood (1985) and Willams *et al.* (1985) suggested that sows should be fed and managed so that they gain 25 kg of maternal tissues throughout pregnancy for at least the first three or four parities. The weight of the placenta and other products of conception should be approximately 20 kg for a total of 45 kg of gestational weight gain of the sow (Verstegen *et al.*, 1987; Noblet *et al.*, 1990).

Lactation

The long term reproductive efficiency of the sow is best served by minimizing weight loss during lactation (Dourmand *et al.*, 1994). Such a strategy requires only minimal restoration of weight in the next pregnancy. The daily energy requirements during lactation include a requirement for maintenance 7(ME_m) and a requirement for milk production. The energy requirement for milk production can be estimated from the growth rate of the suckling pig and the number of pigs in the litter (Noblet and Etience, 1989).

Determination of the energy value

The energy of the TDN component of a ration should be the largest fraction since the pig is not equipped to digest fibrous feeds. Starch and fat in the feed form the main source of energy. Rapid gains cannot be obtained if pigs are fed grain rations containing less than 70% of TDN. Thus, pigs less than 40 kg live weight, when fed heavy rations containing mainly maize, sorghum, oat, barley, wheat or rice give best performance. During the finishing period the use of lighter rations having 65 to 70% total digestible nutrients will often lead to marked improvement in grades.

Energy and total feed requirements are closely related. Under good conditions, pigs in the period following weaning require about 1.2 kg feed per kg of live weight gain. This high efficiency naturally declines until 1.8 to 2.8 kg feed is consumed for every 1 kg of gain towards the latter part of the finishing period. This progressive change in feed efficiency underlines the importance of using good rations in the growing period. At this stage, feed efficiencies are high because the development of muscles has priority over fat in the growth of the pig.

Feed requirements increase from 0.9 to 1.2 kg per day for an 8 week old piglet to about 3.6 kg per day for a pig weighing 90 kg. Mature, pregnant sows require about 3.1 or 3.6 kg per day, and more in cold weather. Lactating sows may exceed 5.5 kg of feed daily depending on the size of the litter.

Cereals and millets are the main ingredients of swine rations. The byproducts of the cereal grains may also be fed. They have higher fibre, protein and phosphorus content, but are deficient in calcium, and poor in the quality of protein and specific vitamins. Generally, the fibre content in swine rations should be around 5 or 6%. If this is exceeded, there is a reduction in feed utilization efficiency.

13.3.2 Proteins and amino acids

Protein generally refers to crude protein, which is defined for mixed feedstuffs as the nitrogen content × 6.25. This definition is based on the assumption that, on average, the nitrogen content is 16 g of nitrogen/100 g of protein. Proteins are composed of amino acids, and it is actually the amino acids that are the essential nutrients. Therefore, the dietary provision of amino acids in correct amounts and proportions determinestheadequacyofadietary protein concentrate. Supplemental non-protein nitrogen, such as urea, has not produced beneficial responses in swine that were fed practical diets (Hays *et al.,* 1957; Kornegay *et al.,* 1965; Wehrbein *et al.,* 1970).

Animals continually use proteins, either to build new tissues, as in growth and reproduction, or to repair worn-out tissues. Thus swine require a regular intake of protein. If adequate proteins are lacking in the diet, the swine suffer a reduction in growth or loss of weight. Ultimately, protein will be withdrawn from certain tissues to maintain the functions of the more vital tissues of the body as long as possible. Protein is needed for milk, meat, hide, hoof, hair, hormones, enzymes, blood cells etc. Thus, protein affects every body function. It has been shown also that animals are more resistant to infections if they are fed an adequate protein ration. The elements in the blood stream which resists diseases are proteins. So, adequate protein in the diet is one way of keeping animals in disease fighting trim.

Proteins are made of many amino acids combined with each other. These amino acids are put together in various combinations to form proteins; they are many times referred to as the building blocks of proteins. Every protein has a definite amino acid composition and no two are alike.

Amino acids contain nitrogen combined with carbon, hydrogen, oxygen and sometimes sulphur and phosphorus. The nitrogen is in the form of an amino group (NH_2); it is from this that the name of the amino acid is derived.

Essential amino acids

Animals can make certain of the amino acids or other nutrients in the ration. These are called non-essential amino acids. Other amino acids, however, can not be made in the body from other substances or cannot be made fast.

Amino acid classification for the pig:

Essential amino acids	Non-essential amino acids
Lysine	Glycine
Tryptophan	Serine
Methionine	Alanine
Valine	Norleucine
Histidine	Aspartic acid
Phynylealanine	Glutamic acid
Leucine	Hydroxyglutamic acid
Isoleusine	Cystines
Threonine	Citrulline
Arginine	Proline
	Hydroxyproline
	Tyrosine

The pig requires all the 10 essential amino acids for maximum growth.

Quality of protein

Feeds which supply proper proportions and amounts of the various essential amino acids supply are called good quality protein. Those feeds which furnish an inadequate amount of any essential amino acids have poor quality protein.

If any one essential amino acid is lacking in proper amount, it will limit the utilization of other amino acids in the ration. This means that one serious amino acid deficiency will cause the entire ration to be inadequate. For this reason, it is very necessary that feeds low in one or more essential amino acids are not fed alone; otherwise, swine will make poor use of the protein supplied by the feed in performing the body function which requires protein.

Lysine, tryptophan, and methionine are three of the essential amino acids which are apt to be in borderline in certain rations.

Time factor in protein feeding

For efficient use, protein must be present at the right time, in the right amount and balance, as well as in the correct form. A pig cannot consume an excess of amino acids today to take care of tomorrow's needs as amino acids are not stored or carried over. Actually, animals do better if they eat all the required essential amino acids at the same time. The lack of any one essential amino acid will cause the waste of all the others. Proteins are not utilized in the body as such. First they are broken down into amino acids, which are then recombined to form the body's own protein. For protein synthesis, therefore, it is necessary that all the required amino acids be present simultaneously.

Excess protein

Excess protein is deaminized as the nitrogen is removed as ammonia and urea. The remainder of the protein molecule serves as a source of energy or is stored as fat through complex mechanisms in the body.

Amino acids

Lysine

Lysine is apt to be borderline or deficient in swine ration, since corn and other cereal grains are generally low in lysine. A lysine deficiency results in reduced appetite, loss of weight, poor feed efficiency, rough, dry hair coat, and a general emaciated condition. The addition of 2.0% DL-lysine to the ration of deficient pigs causes an immediate improvement in growth, appetite, and thriftiness. It has been recommended minimum of lysine level of 0.9% associated with a crude protein level of 18.5% in dry matter in pigs up to 50 kg and 0.7% lysine associated with crude protein level of 15% in live weight ranging 50–90 kg. These are equivalent to 4.9% and 4.7% lysine in crude protein respectively.

Tryptophan

Tryptophan is another essential amino acid which is apt to be low or borderline in certain swine rations, since corn is low in tryptophan. A lack of tryptophan causes a loss in weight, poor feed consumption, depraved appetite, rough hair coat, and symptoms of inanition in the pig. Adding tryptophan to the ration of deficient pigs caused an immediate response and recovery.

An interrelationship exists between niacin and tryptophan. Pigs can use tryptophan to synthesize niacin. Niacin, however, cannot be converted back to tryptophan. Consequently, an adequate level of niacin in the ration spares tryptophan, since tryptophan will not be used to make niacin. If the tryptophan is at a high level in the diet, animals will probably not develop a nicotinic acid deficiency. It was found that 0.17 to 0.19% tryptophan was needed in 20% protein diets and 0.16 in 16% protein diets.

Methionine

Pigs do not require cystine in ration if they have sufficient methionine to meet their requirements for both methionine and the synthesis of cystine. This is important because, when an adequate amount of cystine is included in the ration, methionine is no longer converted to cystine. Instead, it is used as methionine to form new tissue and carry out the other functions for which methionine is essential.

An interrelationship exists between methionine and choline. Methionine can furnish methyl groups for choline synthesis. In a diet which is mildly deficient in both, adding either one will improve growth. Methionine is effective both in correcting a methionine deficiency and in promoting the synthesis of choline. The baby pig requires choline at a level of approximately 0.1% of the dry matter in the diet.

Histidine

Histidine is usually supplied in adequate amounts in good swine rations. A deficiency of histidine in the ration resulted in decreased growth and efficiency of feed utilization. When histidine was added to the diet of deficient pigs, they showed greatly increased appetites and resumed growth almost immediately. The histidine requirement of weaning pigs should be 0.2% with a 13% protein ration.

Phenylalanine

Phenylalanine is usually adequately supplied in good swine rations. A deficiency of phenylalanine in the pig resulted in decreased growth and efficiency of feed utilization. Tyrosine is synthesized from phenylalanine. This means that if the ration contains sufficient tyrosine, phenylalanine will not be used up for the synthesis of tyrosine. If the ration is short of tyrosine, then phenylalanine will be used for tyrosine synthesis.

Arginine

Arginine is not apt to be deficient in practical swine ration. A deficiency of arginine causes a lower growth and lowered efficiency of feed utilization with the pig. Arginine can be synthesized at a rate sufficient to permit about 60% of normal growth. Pigs need a dietary source of arginine and a level of 0.20% L-argninine in the ration or 1.77% of the protein, is adequate to meet the pig's need along with the amount synthesized by the pig.

Leucine

Leucine is present in sufficient quantity in feeds to meet the requirements of the pig. A deficiency of leucine in the pig causes a decrease of appetite, feed efficiency and rate of gain.

Isoleucine

Isoleucine is not apt to be deficient in good swine rations. Lack of isoleucine in the ration, decreases growth rate, efficiency of feed utilization and nitrogen retention. They also found that weaning pigs fed as a 22% protein ration required 0.7% L-isoleucine.

Threonine

Threonine is present in sufficient quantity in feeds to meet the requirements of the pig. A deficiency of threonine decreases feed consumption, rate of gain and efficiency of feed utilization. Suckling pigs fed a simulated milk diet containing approximately 25% protein, requires approximately 0.9% L-threonine in the ration.

Valine

Valine is not apt to be deficient in practical swine rations. A deficiency of valine in the ration decreased daily feed consumption, rate of growth and feed efficiency. The weaning pigs fed a 12.8% protein ration require 0.4% L-valine in the diet.

Bioavailability of aminoacids

All the amoinoacids present in the diet is not available to the animals since most proteins are not fully digested, the amino acids are not fully absorbed, and not all absorbed amino acids are metabolically available. Diets vary considerably in the proportions of their amino acids that are biologically available. The amino acids in some proteins such as milk products, fish meal and egg proteins are almost fully bioavailable, whereas those in other proteins such as certain plant seeds are available in much less quantities at tissue level. Expressing amino acids requirements in terms of bioavailable requirements is, therefore, desirable. However, it means that to formulate swine diets, the bioavailable amino acid content of the ingredients being considered must be known.

The bioavailability of amino acids in the protein of dietary ingredients can be determined by various ways. The primary method to determine bioavailability has been to measure the proportion of a dietary amino acid that has been absorbed from the gastrointestinal tract when digesta reach the terminal ileum region of small intestine (illeal digestibilities). Furthermore, unless a correction is made for endogenous amino acid losses, the complete terminology is 'apparent illeal digestibility'.When apparent digestibilities are determined, feedstuffs with low protein content are undervalued relative to feedstuffs with high protein content because of the relatively greatercontribution of endogenous amino acids. In addition, because of the way in which ideal protein patterns were determined, these patterns reflect true illeal digestibility rather than apparent illeal digestibility.

Protein requirement of pigs

Young pigs require more protein in the ration than older animals, which are sharing less protein and more fat in the bodies. The protein levels recommended by National Research Council are:

Table 13.1 Requirements of Protein

Weight of pig	% crude protein
Birth to 34 kg	14
Weight 34 to 56 kg	12
Weight 56 to 90 kg	10

This quantity of protein is needed in the ration properly fortified with antibiotics, vitamins and minerals.

Table 13.2 Ideal Ratios of Amino Acids to Lysine for Maintenance, Protein Accretion, Milk Synthesis, and Body Tissue

Amino acid	Maintenance[a]	Protein accretion[b]	Milk synthesis[c]	Body tissue[d]
Lysine	100	100	100	100
Arginine	-200	48	66	105
Histidine	32	32	40	45
Isoleucine	75	54	55	50
Leucine	70	102	115	109
Methionine	28	27	26	27
Methionine + cysteine	123	55	45	45
Phenylalanine	50	60	55	60
Phenylalanine + tyrosine	121	93	112	103
Threonine	151	60	58	58
Tryptophan	26	18	18	10
Valine	67	68	85	69

[a] Maintenance ratios were calculated based on the data of Baker *et al.* (1996 a,b), Baker and Allee (1970), and Fuller *et al.* (1989). The negative value for arginine reflects arginine synthesis in excess of the needs for maintenance.

[b] Accretion ratios were derived by starting with ratios from Fuller *et al.*, (1989) and then adjusting to values that produced blends for maintenance + accretion that were more consistent with recent empirically determined values (Baker and Chung, 1992; Baker *et al.*, 1993); Hahn and Baker, 1995; Baker, 1997).

[c] Milk protein synthesis ratios were those proposed by Pettigrew (1993) based on a survey of the literature; the value of 73 for valine proposed by Pettigrew was modified to 85.

[d] Body tissue protein rati.os were from a survey of the literature (Pettigrew, 1993).

The ratios for protein accretion were derived by starting with the ratios proposed by Fuller *et al.* (1989). However, these ratios were adjusted to values that produced blends for maintenance and accretion which were more consistent with recent empirically determined values (for a discussion, see Baker and Chung, 1992; Baker *et al.*, 1993; Hahn and Baker, 1995; Baker, 1997). The ratios for milk production were from the review of Pettigrew (1993) except that the value of 73 for valine was modified to 85. The ratios for body tissue protein were also from the review of Pettigrew (1993). Although it is recognized that the amino acid composition of body protein changes as a pig matures (Kyriazakis *et al.*, 1993), a fixed pattern was used.

13.3.3 Lipids

The term "lipid" includes both fats and oils. Originally, linoleic acid and arachidonic acids were both identified as essential fatty acids (EFA) that must be supplied in the diet (Cunnane, 1984). Now it is recognized that these fatty acids are members of N-6 series of EFA and that arachidonic acid can be derived in vivo from lilnoleic acid. It is difficult to produce overt signs of an EFA deficiency in pigs. Enser (1984) has reported normal growth in pigs from weaning to slaughter weight when they are fed diets containing only 0.1% linoleic acid. The Agricultural Research Council (1981) suggested the EFA requirements are 3.0% of dietary DE for pigs up to 30 kg and 1.5% of dietary DE from 30 to 90 kg. These are equivalent to about 1.2 and 0.6% of the diet

Fat has a bearing on quality of carcass, soft fats renders carcass unpalatable and give unpleasant appearance and are difficult to be cured, they loose weight and show discolouration. Best quality fat is needed for quick growth.

The value of adding fat to the diets of weaning pigs is uncertain. Pettigrew and Moser (1991) summarized data involving 92 comparisons of fat additions for pigs from 5 to 20 kg. In this weight range, addition of fat reduced growth rate and feed intake while it improved gain-to-feed ratio. The response of growth rate was small (0.01 kg) and variable, with similar numbers of positive (37) and negative (38) responses. Inconsistent responses to added fat may be a result of a number of factors, including the age of the pig at the start of the experiment, the amount of fat added, the type of fat, and the method by which the fat was added. Pettigrew and Moser (1991) reported responses for studies in which a constant protein-to-energy ratio was maintained and found no response in growth rate, a reduction in feed intake, and an improvement in gain-to-feed ratio when fat was added.

For growing-finishing swine (20 to 100 kg), the summary by Pettigrew and Moser (1991) indicated consistent improvement in growth rate, reduction in feed intake, improvement in gain-to-feed ratio, but an increase in back-fat thickness in response to addition of fat to swine diets.

The age of the pig, chain length of the fatty acids in the fat, free fatty acid concentration, and unsaturated-to-saturated (U:S) fatty acid ratio influence the apparent digestibility of fat (Stahly, 1984).

Evidence suggests that the addition of fat to the diets of sows during late gestation or lactation increases the milk yield, fat content of colostrums and milk, and survival of pigs from birth to weaning, especially for lightweight pigs (Moser and Lewis, 1980; Coffey *et al.,* 1982; Seerley, 1984; Pettigrew and Moser, 1991).

13.3.4 Fibre

The fibre content in pig feed should be low, for obtaining efficient conversion of feed and maximum growth chaffed hay, silage, oat and brans which are high in fibre content should not be high in rations of growing pigs.

13.3.5 Minerals

Pigs have a dietary requirement for certain inorganic elements. These include calcium, chlorine, copper, iodine, iron, magnesium, manganese, phosphorus, potassium, selenium, sodium, sulfur, and zinc. Chromium is now recognized as an essential mineral (National Research Council, 1997), but a quantitative requirement has not been established. Cobalt also is required in the synthesis of vitamin B_{12}. Pigs may also require other trace elements (*i.e.,* arsenic, boron, bromine, fluorine, molybdenum, nickel, silicon, tin, and vanadium) which have been shown to have a physiological role in one or more species (Underwood, 1977; Nielsen, 1984). These elements are required at such low levels, however, that their dietary essentiality has not been proven.

The functions of these inorganic elements are extremely diverse. They range from structural functions in some tissues to a wide variety of regulatory functions in other tissues. Most pigs are now raised in confinement, without access to soil or forage; this rearing environment may increase the need for mineral supplementation. Meeting the mineral requirements will be influenced by the bioavailability of minerals in feed ingredients. Several minerals, including antimony, arsenic, cadmium, fluorine, lead, and mercury, can be toxic to swine (Carson, 1986).

The vital functions of minerals

Minerals perform important functions in the animal body. Besides being constituents of bone and teeth, mineral element serve the body in many other ways. Nearly every process of the animal body depends on one or more of the mineral elements for proper functioning. Minerals are just as essential for growth, reproduction and lactation as are proteins, fats, carbohydrates and vitamins.

A lack of minerals in the ration may cause any of the following deficiency symptoms reduced or poor appetite, expensive, poor gains, rickets, soft or brittle bones, bending of the ribs, stiffness or malformed joints, posterior paralysis 'going down in the back', goiter, unthrifty looking, born hairless, failure to come in heat regularly, poor milk production, weak or dead young and many other ailments.

Supplying mineral needs

Supply those elements that are most apt to be lacking in pig rations. Give particular attention to calcium, phosphorus, salt and iodine. To prevent nutritional anemia in suckling pigs kept on concrete or board floors, supply iron and traces of copper and cobalt. All of the other minerals listed above will usually be supplied in rations that are balanced in all other respects.

Not providing iron and copper to prevent nutritional anemia is responsible for the death of many suckling pigs that do not have contact with soil. Pigs are born with small stores of iron and copper in their bodies and sow's milk is very deficient in these elements. Consequently, if new born pigs are not given outside sources of these elements during the 3 or 4 weeks and they subsist entirely on sow's milk, the hemoglobin level of the blood may fall so low that they die. The fastest growing pigs are the first to suffer because rapid gains require the greatest supply of these elements. Pigs that do not become too anemic by the time they start eating solid feed usually recover, because most solid feeds contain enough iron and copper to permit recovery.

Anemic pigs show a paleness of the unpigmented skin and mucous membranes. In severe cases they develop a thickened, wrinkled condition of the skin of the neck and shoulders. They become listless and inactive, which increases they danger of their being crushed by the sow. They frequently develop thumps (laboured breathing) which may lead to their death due to suffocation (lack of oxygen carrying capacity of the blood).

The most direct way to prevent anemia is to paint the udder of the sow daily with a solution of 0.5 kg of ferrous sulphate in 1.7 litres of water. This compound contains the trace of copper required. As the pigs nurse they get these elements. Continue the treatment for 3 or 4 weeks or until the pigs are eating generously of solid feeds.

A pile or low box of clean soil in the corner of the pen, in which the pigs will root, also affords protection from anemia. The solution of ferrous sulphate may be sprinkled on the clean soil as an added precaution if so desired. Get clean soil from the garden or roadside where no pigs have been there.

Feeding iron and copper salts to sows nursing pigs is not effective in preventing anemia in pigs, as the udder of the sow does not allow these mineral elements (iron and copper) to pass into the milk. Pigs having access to soil will not need a supplemental source of iron and copper.

13.3.5.1 *Major or macro minerals*

Calcium and phosphorus

Calcium and phosphorus play a major role in the development and maintenance of the skeletal system and perform many other physiologic functions (Hays, 1976;

Peo, 1976, 1991; Kornegay, 1985). Peo (1991) indicated that adequate calcium and phosphorus nutrition for all classes of swine is dependent upon: (1) an adequate supply of each element in anavailable form in the diet, (2) a suitable ratio of available calcium and phosphorus in the diet, and (3) the presence of adequate vitamin D. A wide calcium-to-phosphorus ratio lowers phosphorus absorption, resulting in reduced growth and bone calcification, especially if the diet is marginal in phosphorus (Peo *et al.,* 1969; Vipperman *et al.,* 1974; Doige *et al.,* 1975; van Kempen *et al.,* 1976; Reinhart and Mahan, 1986; Hall *et al.,* 1991; Wilde and Jourquin, 1992; Eeckhout *et al.,* 1995; Qian *et al.,* 1996). An adequate amount of vitamin D is also necessary for proper metabolism of calcium and phosphorus, but a very high level of vitamin D can mobilize excessive amounts of calcium and phosphorus from bones (Hancock *et al.,* 1986; Jongbloed, 1987).

The levels of calcium and phosphorus that result in maximum growth rate are not necessarily adequate for maximum bone mineralization. The requirements for maximizing bone strength and bone-ash content are at least 0.1% unit higher than the requirements for maximum rate and efficiency of gain (Cromwell *et al.,* 1970; Mahan *et al.,* 1980; Crenshaw *et al.,* 1981; Kornegay and Thomas, 1981; Mahan, 1982; Maxson and Mahan, 1983; Koch *et al.,* 1984; Combs *et al.,* 1991 a, b).

The dietary calcium and phosphorus requirements, expressed as a percentage of the diet, may be slightly higher for gilts than for barrows (Thomas and Kornegay, 1981; Calabotta *et al.,* 1982). Feeding of dietary levels of calcium and phosphorus, sufficient to maximize bone mineralization in gilts during early growth and development improved reproductive longevity in one study (Nimmo *et al.,* 1981a, b) but not in other studies (Arthur *et al.,*1983a, b; Kornegay *et al.,* 1984). The calcium and phosphorus requirements of the developing boar are greater than those of the barrow and gilt (Cromwell *et al.,* 1979; Hickman *et al.,* 1983; Kesel *et al.,* 1983; Hansen *et al.,* 1987).

Pigs possessing a high lean growth rate do not seem to have a higher dietary requirement for calcium and phosphorus as compared with pigs having a moderate lean growth rate, according to a study by Bertram *et al.* (1994). However, when the lean growth rate is increased by treating pigs with porcine somatotropin, the dietary requirement, expressed as percentage of the diet, increases due to the reduced daily feed intake resulting from porcine somatotropin treatment (Weeden *et al.,* 1993a,b; Carter and Cromwell, 1998a,b). There is also strong evidence that porcine somatotropin treated pigs require greater daily amounts of calcium and phosphorus to maximize growth performance, bone mineralization, and carcass leanness than untreated pigs (Carter and Cromwell, 1998a, b).

During pregnancy, the physiological requirements for calcium and phosphorus increase in proportion to the need for fetal growth and reach a maximum in late gestation. During lactation, the requirements are affected by the level of milk production by the sow. Generally, the requirements for calcium and phosphorus are based on a feeding level of 1.8 to 2.0 kg of feed/day during gestation and 5 to 6 kg of feed/day during lactation. If sows are fed less than 1.8 kg of feed during gestation, the diet should be formulated to contain sufficient concentrations of calcium and phosphorus to meet the daily requirements.

Calcium and phosphorus are required for bone and teeth formation, nerve function, muscle contraction, blood circulation, cell permeability, as well as essential for milk production. Phosphorus, a component of phospholipids is important in lipid transport and metabolism and cell membrane structure. In energy metabolism a component of RNA and DNA, the vital cellular constituents, are required for protein synthesis. Phosphorus is a constituent of several enzyme systems.

Sources of Ca and P are ground lime stone or oyster shell flour. Where both Ca and P are needed, use monocalcium phosphate, dicalcium phosphate, tricalcium phosphate, deflourinated phosphate or bone meal monosodium phosphate, di-sodium phosphate, sodium tripolyphosphate, ammonium phosphate solution, or feed grade phosphoric acid.

1.5 to $^{2}/_{3}$ in phytate form is not available to swine; although fairly good utilizaiotn of phytate P is achieved through action of enzyme phytate (S) in the intestine. Most favorable Ca:P ratio is between 1:1 and 1.4:1. Sow's milk contains a Ca:P ratio of 1.3 :1.

Magnesium (Mg)

Magnesium is a cofactor in many enzyme systems and is a constituent of bone. The magnesium requirement of artificially reared pigs fed milk based semi purified diets is between 300 and 500 mg/kg of diet (Mayo *et al.,* 1959; Bartley *et al.,* 1961; Miller *et al.,* 1965 c,d). Milk contains adequate magnesium to meet the requirement of suckling pigs (Miller *et al.,* 1965 c, d). The magnesium requirement of weanling-growing-finishing swine is probably not higher than that of the young pig. The magnesium in a corn-soybean meal diet (0.14 to 0.18%) is apparently adequate (Svajgr *et al.,* 1969; Krider *et al.,* 1975), although some research suggests that the magnesium in natural ingredients is only 50 to 60% available to the pig (Miller, 1980; Nuoranne *et al.,* 1980).

In order of appearance, signs of magnesium deficiency include hyperirritability, muscular twitching, reluctance to stand, weak pasterns, loss of equilibrium, and tetany followed by death (Mayo *et al.,* 1959; Miller *et al.,* 1965 c). The toxic level of magnesium is not known. The maximum tolerable level for swine is approximately 0.3% (National Research Council, 1980).

Magnesium is essential for normal skeletal development as constituent of bone and enzyme activation in many enzyme systems.

Potassium (K)

Potassium is the third most abundant mineral in the body of the pig, surpassed only by calcium and phosphorus (Manners and McCrea, 1964), and is the most abundant mineral in muscle tissue (Stant *et al.,* 1969). Potassium is involved in electrolyte balance and neuromuscular function. It also serves as the monovalent cation to balance anions intracellularly, as part of the sodium potassium pump physiological mechanism.

The dietary potassium requirement of pigs from 1 to 4 kg body weight is estimated to be between 0.27 and 0.39% (Manners and McCrea, 1964); from 5 to 10 kg, 0.26 to 0.33% (Jensen *et al.,* 1961; Combs *et al.,* 1985); at 16 kg, 0.23 to 0.28 % (Meyer *et al.,* 1950); and from 20 to 35 kg, less than 0.15% (Hughes and Ittner, 1942; Mraz *et al.,* 1958). No estimates are available for finishing or breeding pigs. The content of potassium in most practical diets is normally adequate to meet these requirements for all classes of swine. The potassium in corn and soybean meal is 90 to 97% available (Combs and Miller, 1985).

Signs of potassium deficiency include anorexia, rough hair coat, emaciation, inactivity, and ataxia (Jensen *et al.,* 1961). Electrocardiograms of potassium-deficient pigs showed reduced heart rate and increased electrocardial intervals (Cox *et al.,* 1966). Necropsy of affected pigs revealed no unique gross pathology.

Corn contain 0.27% potassium and other cereals contains 0.42–0.49% potassium.

Sodium and Chlorine

Sodium and chlorine (chloride) are the principal extra cellular cation and anion, respectively, in the body. Chloride is the chief anion in gastric juice.

The dietary sodium requirement of growing-finishing pigs is no greater than 0.08 to 0.10% of the diet (Meyer *et al.,* 1950; Alcantara *et al.,* 1980; Cromwell *et al.,* 1981a; Froseth *et al.,* 1982a; Honeyfield and Froseth, 1985; Honeyfield *et al.,* 1985; Kornegay *et al.,* 1991). The dietary chlorine (chloride) requirement is less well defined but is probably no higher than 0.08% for the growing pig (Froseth *et al.,* 1982a; Honeyfield and Froseth, 1985; Honeyfield *et al.,* 1985).

Mahan *et al.* (1996a, b) reported that weaning pigs fed diets containing dried whey or dried plasma (both are relatively high in sodium) responded to added sodium as sodium chloride or sodium phosphate and to added chloride as hydrochloric acid. Their results indicate that early-weaned pigs require more sodium and chlorine than previously thought. Thus, the estimated dietary sodium and chloride requirements have been increased to 0.25% of each from 3 to 5 kg, to 0.20% of each from 5 to 10 kg, and to 0.15% of each from 10 to 20 kg body weight.

The sodium and chlorine requirements of breeding animals are not well established. The results of one study suggested that 0.3% dietary sodium chloride (0.12 % sodium) was not sufficient for pregnant sows (Friend and Wolynetz, 1981). In a regional study, pig birth weights and weaning weights were reduced when sodium chloride was reduced from 0.50 to 0.25% during gestation and lactation for two or more parities.

Sodium and chlorine are available in salt in loose form. A good deal of salt is provided by tankage and fish meal. In iodine deficient areas, stabilized iodized salt should be used. When pigs are salt starved, precaution should be taken to prevent overeating of it.

13.3.5.2 *Trace or micro minerals*

Cobalt

Cobalt is a component of vitamin B_{12} (Rickes *et al.,* 1948). There is no evidence that pigs have an absolute requirement for cobalt, other than for its role in vitamin B_{12}. Cobalt can substitute for zinc in the enzyme carboxypeptidase and for part of the zinc in the enzyme alkaline phosphatase. Hoekstra (1970) and Chung *et al.* (1976) have shown that supplemental cobalt prevents lesions associated with a zinc deficiency.

Copper (Cu)

The pig requires copper for the synthesis of hemoglobin and for the synthesis and activation of several oxidative enzymes necessary for normal metabolism (Miller *et al.,* 1979). A level of 5 to 6 ppm in the diet is adequate for the neonatal pig (Okonkwo *et al.,* 1979; Hill *et al.,* 1983a). The requirement for later stages of growth is probably no greater than 5 to 6 ppm.

Copper salts with high biological availabilities include the sulphate, carbonate and chloride salts (Miller, 1980; Cromwell *et al.,* 1998).

Copper is an essential element in a number of enzyme systems and necessary for synthesizing hemoglobin and preventing nutritional anemia. Hemoglobin serves as a carrier of oxygen throughout the body.

Sources of copper are copper sulphate, copper carbonate and copper oxide and are equally effective.

Beyond the sucking period, natural feedstuffs usually contain enough copper. Apart from the role of copper as an essential trace element, much higher levels (65.8 to 113.6 mg/lb, or 125–250 mg/kg) in the diet have been shown to support increased rate and efficiency of gain of pigs to breeding age.

Iodine (I)

The majority of the iodine in swine is present in the thyroid gland, where it exists as a component of mono-, di-, tri-, and tetraiodothyronine (thyroxine). These hormones are important in the regulation of metabolic rate. Hart and Steenbock (1918) and Kalkus (1920) demonstrated that hypothyroidism existed in swine raised in the northwestern United States and the Great Lakes region because of iodine-deficient feedstuffs produced on low-iodine soil.

Calcium iodate, potassium iodate, and pentacalcium orthoperiodate are nutritionally available forms of iodine and are more stable in salt mixtures than are sodium iodide or potassium iodide (Kuhajek and Andelfinger, 1970). The incorporation of iodized salt (0.007% iodine), at a level of 0.2% of the diet, provides sufficient iodine (0.14 ppm) to meet the needs of growing pigs fed grain soybean meal diets.

A severe iodine deficiency causes pigs to be stunted and lethargic and to have an enlarged thyroid (Beeson *et al.,* 1947; Braude and Cotchin, 1949; Sihombing *et al.,* 1974). Sows fed iodine-deficient, goitrogenic diets farrow weak or dead pigs that are hairless, show symptoms of myxedema, and have an enlarged, hemorrhagic thyroid (Hart and Steenbock, 1918; Slatter, 1955; Devilat and Skoknic, 1971).

Iodine is needed by the thyroid gland for making thyroxin and triodothyonine, the iodine containing hormones, which control the rate of body metabolism or heat production.

Iron (Fe)

Iron is required as a component of hemoglobin in red blood cells. Iron also is found in muscle as myoglobin, in serum as transferrin, in the placenta as uteroferrin, in milk as lactoferrin, and in the liver as ferritin and hemosiderin (Zimmerman,

1980; Ducsay *et al.,* 1984). It also plays an important role in the body as a constituent of several metabolic enzymes. Pigs are born with about 50 mg of iron, most of which is present as hemoglobin (Venn *et al.,* 1947). A high level of iron fed to sows during late gestation (Brady *et al.,* 1978) or parenteral administration of iron dextran to sows in gestation (Rydberg *et al.,* 1959; Pond *et al.,* 1961; Ducsay *et al.,* 1984) does not substantially increase placental transfer of iron to fetuses.

The suckling pig must retain 7 to 16 mg of iron daily, or 21 mg of iron/kg of body weight gain to maintain adequate levels of hemoglobin and storage iron (Venn *et al.,* 1947; Braude *et al.,* 1962). Sow's milk contains an average of only 1 mg of iron per liter (Brady *et al.,* 1978). Thus, pigs receiving only milk rapidly develop anemia (Hart *et al.,* 1930; Venn *et al.,* 1947).

Miller *et al.* (1982) suggested a requirement of 100 mg of iron/kg of milk solids for pigs raised in a conventional or germ-free environment. The iron requirement of pigs fed a dry, casein-based diet is about 50% higher per unit of dry matter than for those fed a similar diet in liquid form (Hitchcock *et al.,* 1974).

The post weaning dietary iron requirement is about 80 ppm (Pickett *et al.,* 1960). In later growth and maturity, this requirement diminishes as the rate of increase in blood volume slows. Natural feed ingredients usually supply enough iron to meet post weaning requirements.

The hemoglobin concentration of blood is a reliable indicator of the pig's iron status and it is easy to determine. Hemoglobin levels of 10 g/dL of whole blood are considered adequate. A hemoglobin level of 8 g/dL suggests borderline anemia, and a level of 7 g/dL or less represents anemia (Zimmerman, 1980). The type of anemia resulting from iron deficiency is hypochromic-microcytic anemia. Anemic pigs show evidence of poor growth, listlessness, rough hair coats, wrinkled skin, and paleness of mucous membranes. Fast-growing anemic pigs may die suddenly of anoxia. A characteristic sign is labored breathing after minimal activity or a spasmodic jerking of the diaphragm muscles, from which the term "thumps" arises. Necropsy findings include an enlarged and fatty liver; thin, watery blood; marked dilatation of the heart; and an enlarged firm spleen. Anemic pigs are more susceptible to infectious diseases (Osborne and Davis, 1968).

Iron is necessary for formation of hemoglobin, an iron containing compound which enables the blood to carry oxygen. Iron is also important to certain enzyme systems.

Practical sources are ferrous sulfate or ferric ammonium citrate.For the prevention or treatment of anemia in young pigs (i) place a little uncontaminated sod (topsoil from an area where pigs have not run for years) in corner of the pen daily, (ii) inject a suitable iron preparation at a level of 150–200 mg into baby pigs at 1–3 days of age, (iii) swab the sow's udder with iron solution, (iv) give iron copper pill or (v) swallow access to oral iron preparations. In addition, the pig should be encourage to eat a grain ration as soon as they are old enough.

New born pigs require an average 47 mg of iron. Iron has a detoxifying effect when added to gossypol containing diet. Add iron from soluble source to free gossypol at a weight ratio of 1:1. Milk is deficient in iron (sow's milk contains an average of 1 mg of iron/litter), but feeding high level of iron to sows does not seem to increase the iron level in their milk.

Manganese

Manganese functions as a component of several enzymes involved in carbohydrate, lipid, and protein metabolism. Manganese is essential for the synthesis of chondroitin sulfate, a component of mucopolysaccharides in the organic matrix of bone (Leach and Muenster, 1962).

Long-term feeding of a diet containing only 0.5 ppm of manganese results in abnormal skeletal growth, increased fat deposition, irregular or absence of estrous cycles, resorbed fetuses, small, weak pigs at birth, and reduced milk production (Plumlee *et al.,* 1956).

Seleninum (Se)

Selenium is a component of the enzyme glutathione peroxidase (Rotruck *et al.,* 1973), which detoxifies lipid peroxides and provides protection of cellular and subcellular membranes against peroxide damage. Thus, the mutual sparing effect of selenium and vitamin E stems from their shared antiperoxidant roles. High levels of vitamin E, however, do not completely eliminate the need for selenium (Ewan *et al.,* 1969; Bengtsson *et al.,* 1978a, b; Hakkarainen *et al.,* 1978). Selenium has been shown to have a function in thyroid metabolism, because iodothyronine 5'-deiodinase has been identified as a selenoprotein (Arthur, 1994).

The dietary requirement for selenium ranges from 0.3 ppm for weaning pigs to 0.15 ppm for finishing pigs and sows (Groce *et al.,* 1971, 1973a, b; Ku *et al.,* 1973; Mahan *et al.,* 1973; Ullrey, 1974; Young *et al.,* 1976; Glienke and Ewan, 1977; Wilkinson *et al.,* 1977a,b; Mahan and Moxon, 1978a, b, 1984; Piatkowski *et al.,* 1979; Meyer *et al.,* 1981). The requirement for selenium is influenced by dietary phosphorus level (Lowry *et al.,* 1985b) but not dietary calcium level (Lowry *et al.,* 1985a). Several forms of selenium, including selenium-enriched yeast, sodium selenite, and sodium selenate, are effective in meeting the dietary requirement (Mahan and Magee, 1991; Suomi and Alaviuhkola, 1992; Mahan and Parrett, 1996; Mahan and Kim, 1996). The selenium status of the dam influences reproductive performance and the selenium requirement of suckling and weanling pigs (Van Vleet *et al.,* 1973; Mahan *et al.,* 1977; Piatkowski *et al.,* 1979; Chavez, 1985; Ramisz *et al.,* 1993).

Zinc (Zn)

Zinc is a component of many metalloenzymes, including DNA and RNA synthetases and transferases, many digestive enzymes, and is associated with the hormone, insulin. Hence, this element plays an important role in protein, carbohydrate, and lipid metabolism.

The classic sign of zinc deficiency in growing pigs is hyperkeratinization of the skin, a condition called parakeratosis (Kernkamp and Ferrin, 1953; Tucker and Salmon, 1955). Zinc deficiency reduces the rate and efficiency of growth and levels of serum zinc, alkaline phosphatase, and albumin (Hoekstra *et al.,* 1956, 1967; Luecke *et al.,* 1957; Theuer and Hoekstra, 1966; Miller *et al.,* 1968, 1970; Prasad *et al.,* 1969, 1971; Ku *et al.,* 1970). Gilts fed zinc-deficient diets during gestation and lactation produce fewer and smaller pigs, which have reduced serum and tissue zinc levels (Pond and Jones, 1964; Hoekstra *et al.,* 1967; Hill *et al.,* 1983a,b,c). The zinc concentration in the milk from these dams is also reduced (Pond and Jones, 1964). Zinc deficiency retards testicular development of boars and thymic development of young pigs (Miller *et al.,* 1968; Liptrap *et al.,* 1970).

Source of zinc are zinc carbonate, zinc sulfate, zinc chloride or zinc oxide.

13.3.5.3 *Vitamins*

1. Fat soluable vitamins

Vitamin–A

Vitamin A is essential for vision, reproduction, growth and maintenance of differentiated epithelia, and mucus secretions. Except for its role in vision (Wald, 1968), the exact role of vitamin A in these functions is undefined (Goodman, 1979, 1980). Recent evidence, however, suggests that vitamin A may be involved in gene expression.

Pigs are less efficient in converting carotenoid precursors to vitamin A. This conversion occurs primarily in intestinal mucosa (Fidge *et al.,* 1969).

Chew *et al.* (1982) and Brief and Chew (1985) have suggested that β carotene plays a role in reproduction that is independent of vitamin A. Swine are able to store vitamin A in the liver, which makes the vitamin available during periods of low intake. Vitamin A deficiency in swine results in reduced weight gain, incoordination, posterior paralysis, blindness, increased cerebrospinal fluid pressure, decreased plasma levels, and reduced liver storage (Guilbert *et al.,* 1937; Braude *et al.,* 1941; Hentges *et al.,* 1952a; Frape *et al.,* 1959; Hjarde *et al.,* 1961; Nelson *et al.,* 1962, 1964). Practical sources of vitamin A are forages. One mg of beta carotene from natural feedstuffs is equal to 200–500 IU of vitamin A available for swine.

Vitamin–D

Vitamin D and its hormonal metabolites act on the mucosal cells of the small intestine, causing the formation of calcium-binding proteins. These proteins facilitate calcium and magnesium absorption and influence phosphorus absorption. The actions of vitamin D metabolites, together with parathyroid hormone and calcitonin, maintain calcium and phosphorus homeostasis. Braidman and Anderson (1985) have reviewed the endocrine functions of vitamin D.

Bethke *et al.* (1946) indicated that vitamins D_2 and D_3 were equally effective in meeting the vitamin D needs of swine. Horst *et al.* (1982), however, demonstrated that pigs discriminate in their metabolism of the two forms of vitamin D. Additional research is needed in swine to quantify the differences in absorption and utilization of these forms.

The vitamin D_2 requirement of the baby pig fed a casein glucose diet is 100 IU/kg of diet (Miller *et al.,* 1964, 1965). The requirement is higher if isolated soy protein is fed (Miller *et al.,* 1965; Hendricks *et al.,* 1967). Vitamin D deficiency reduces retention of calcium, phosphorus, and magnesium (Miller *et al.,* 1965). Bethke *et al.* (1946) suggested a minimum requirement of 200 IU/kg of diet for growing pigs. In other studies, however, vitamin D supplementation did not improve weight gain (Wahlstrom and Stolte, 1958; Combs *et al.,* 1966).

Vitamin D aids in assimilation and utilization of calcium and phosphorous and necessary in the normal bone development of animals including the bones of the fetus.

Vitamin D_2 (ergocalciferol) and vitamin D_3 (cholecalciferol) are similar for biological activity in swine. Irradiated yeast, exposure to sunlight, sun cured hay, 10% alfalfa in the total ration will normally supply sufficient Vitamin D.

Grains, grain by-products and high protein feed stuffs are practically devoid of vitamin D. Therefore, unless swine are exposed daily to the ultraviolet rays of the sun, the diet should be fortified with vitamin-D. When animals are exposed to direct sunlight, the ultraviolet light produces vitamin D from traces of cholesterol in the skin. The vitamin D requirement is less when a proper balance of calcium and phosphorus exists in the ration. One IU vitamin D is defined as the biological activity of 0.25mg of Crystalline vitamin D.

Vitamin–E (tocopherol)

For many years the primary source of vitamin E in feed was the tocopherols found in green plants and seeds. Oxidation, which is accelerated by heat, moisture, rancid fat, and trace minerals, rapidly destroys natural vitamin E. Therefore, predicting the amount of vitamin E activity in feed ingredients is difficult. Vitamin E losses of 50 to 70% can occur in alfalfa stored at 32 °C for 12 weeks; losses of 5 to 30 % can occur during dehydration of alfa-alfa (Livingstone *et al.,* 1968). Storage of high-moisture grain or its treatment with organic acids greatly reduces its vitamin E content (Madsen *et al.,* 1973; Lynch *et al.,* 1975; Young *et al.,* 1975, 1978).

Inclusion of high levels of vitamin E in the diet may increase the immune response (Ellis and Vorhies, 1976; Tiege, 1977; Nockels, 1979; Peplowski *et al.,* 1980; Wuryastuti *et al.,* 1993), although Bonnette *et al.* (1990) found no evidence of an increased humoral or cell-mediated immune response in young pigs fed high levels of vitamin E.

Vitamin E functions as an antioxidant at the cell membrane level, and it has a structural role in cell membranes. There are vitamin E deficiency diseases that respond to vitamin E, selenium, or antioxidants.

Information is available on the vitamin E requirements for reproduction (Hanson and Hathaway, 1948; Adamstone *et al.,* 1949; Cline *et al.,* 1974; Malm *et al.,* 1976; Young *et al.,* 1977, 1978; Wilkinson *et al.,* 1977a; Nielsen *et al.,* 1979; Piatkowski *et al.,* 1979; Mahan, 1991, 1994). Placental transfer of tocopherol from dam to fetus is minimal, so the offspring must rely on colostrum and milk to meet their daily needs. The content of vitamin E in sow colostrum and milk is dependent on the vitamin E content of the sow's diet (Mahan, 1991).

Many dietary factors affect the vitamin E requirement, including levels of selenium, unsaturated fatty acids, sulfur amino acids, retinol, copper, iron, and synthetic antioxidants. Michel *et al.* (1969) prevented deaths in pigs fed a corn soybean diet containing 5 to 8 mg of vitamin E/kg and 0.04 to 0.06 mg of selenium/kg by supplementing the diet with 22 mg of vitamin E/kg. Studies of corn soybean meal diets fed to growing-finishing pigs suggest that 5 mg of vitamin E/kg and 0.04

mg of selenium/kg are inadequate for growing-finishing pigs and may result in deficiency lesions and mortality.

This vitamin acts primarily as a lipid soluble antioxidant. It is required for growth and normal reproduction.

Practical sources are high quality green feeds, whole cereal grains (corn, oats, wheat, rye etc.) and germ of cereal grains, alpha tocopherol. Tocopherols differ in their biological activity, with d alpha-tocopherol being the most active. One IU of vitamin E is equivalent in biopotency of 1 mg d-alpha-tocopherol acetate.

Vitamin–K

Although it was the last of the four fat-soluble vitamins to be discovered, the metabolic role of vitamin K has been more clearly defined than that of vitamins A, D, and E (Suttie, 1980; Kormann and Weiser, 1984). Vitamin K is essential for the synthesis of prothrombin, factor VII, factor IX, and factor X, which are necessary for the normal clotting of blood. These proteins are synthesized in the liver as inactive precursors. The action of vitamin K converts them to biologically active compounds (Suttie and Jackson, 1977; Suttie, 1980). This activation occurs by enzymatic -carboxylation of specific glutamate residues. The resulting carboxyglutamate residues are strong chelator of calcium ions, which are essential for blood coagulation. A deficiency of vitamin K or the presence of anticoagulation compounds reduces the number of carboxyglutamate residues, resulting in a loss of activity and prolonged bleeding times. In addition to its role in blood clotting, there is evidence that vitamin K dependent protein and peptides may be involved in calcium metabolism (Suttie, 1980; Kormann and Weiser, 1984).

Vitamin K deficiency increases prothrombin and clotting times and may result in internal hemorrhages and death (Schendel and Johnson, 1962; Brooks *et al.*, 1973; Seerley *et al.,* 1976; Hall *et al.,* 1986, 1991). Schendel and Johnson (1962) reported a requirement of 5 µg of menadione sodium phosphate/kg of body weight for 1- and 2-day-old pigs fed a purified liquid diet.

Muhrer *et al.* (1970), Osweiler (1970), and Fritschen *et al.* (1971) reported an occurrence of hemorrhagic conditions in pigs under field conditions. Mycotoxin contaminated ingredients were suspected in these incidents, and vitamin K supplementation (2.0 mg of menadione/kg of diet) prevented the hemorrhagic syndrome.

Under practical conditions, the vitamin K requirement is met by vitamin K in feed stuffs and by intestinal synthesis. Best natural sources include legumes and other green forages. Most common synthetic sources are menadione sodium bisulfite and menadione dimethyl primidinol bisulfite. Vitamin K is commonly added to rations.

2. Water soluble vitamins

Folacin (folic acid)

Folacin includes a group of compounds with folic acid activity. Chemically, folacin consists of a pteridine ring, paraaminobenzoic acid (PABA), and glutamic acid. Animal cells neither synthesize PABA, nor can they attach glutamic acid to pteroic acid. A deficiency of folacin causes a disturbance in the metabolism of single carbon compounds, including the synthesis of methyl groups, serine, purines, and thymine. Folacin is involved in the conversion of serine to glycine and homocysteine to methionine.

Except for the studies of Matte *et al.* (1984a,b; 1992) and Lindemann and Kornegay (1986; 1989), results have indicated that the folacin contribution of ingredients commonly fed to swine when combined with bacterial synthesis within the intestinal tract, adequately meets the requirement for all classes of swine.

Practical swine rations and intestinal synthesis are believed to contain adequate amounts of folacin.

Niacin (Nicotinic acid)

Niacin or nicotinic acid is a component of the coenzymes nicotinamide-adenine dinucleotide (NAD) and nicotin-amide-adenine dinucleotide phosphate (NADP). These coenzymes are essential for the metabolism of carbohydrates, proteins, and lipids.

Niacin is required by all living cells and an essential component of important metabolic enzyme systems involved in lipid carbohydrate protein metabolism.

The richest source of Niacin are fish meals, condensed fish soluble, dried corn from distilleries, soluble corn gluten feed, yeast, liver meal, peanut oil meal, wheat bran, wheat shorts, wheat standard middling, rice bran and rice polishing. Tankage, meat scraps, meat and bone scraps, alfalfa hay and other legume hays or meals, soybean oil meal, barely, or other cereal grains (except corn) are only fair sources of niacin. Corn contains less niacin than the other cereal grains. The niacin requirement can be met easily by a wise choice of commonly used feeding stuffs.

Niacin occurs in corn wheat and milo in bound form, hence, it may be unavailable to the pig. The tryptophan level affects the niacin requirement because of the conversion of tryptophan to niacin.

Pantothenic acid

The pantothenic acid requirement of 2 to 10 kg pigs fed synthetic diets was 15.0 mg/kg (Stothers *et al.,* 1955); and for 10 to 50 kg pigs, estimates range from about 4.0 to 9.0 mg/kg of diet (Luecke *et al.,* 1953; Barnhart *et al.,* 1957; Sewell *et al.,* 1962; Palm *et al.,* 1968). Requirement estimates for pigs weighing between 20 and 90 kg have varied from 6.0 to 10.5 mg of pantothenic acid/kg of diet (Pond *et al.,* 1960; Davey and Stevenson, 1963; Palm *et al.,* 1968; Meade *et al.,* 1969; Roth-Maier and Kirchgessner, 1977).

Pantothenic acid deficiency signs include slow growth, anorexia, diarrhoea, dry skin, rough hair coat, alopecia, reduced immune response, and an abnormal movement of the hind legs called goose stepping (Hughes and Ittner, 1942; Wintrobe *et al.,* 1943b; Luecke *et al.,* 1948, 1950, 1952; Wiese *et al.,* 1951; Stothers *et al.,* 1955; Harmon *et al.,* 1963).

Pantothenic acid plays an important role in oxidative, carbohydrate and fat metabolism.

Dried milk products, condensed fish solubles and alfalfa meal, green pastures, cane molasses, yeasts, rice bran, wheat bran, wheat grey shorts, peanut and oil meal are sources of pantothenic acid. It is widely distributed and occurs in practically all feed stuffs. However, the quantity present may not always be sufficient to meet the needs of the pig.

Thiamin (B-1)

Thiamin is essential for carbohydrate and protein metabolism. The coenzyme, thiamin pyrophosphate, is essential for the oxidative decarboxylation of á–keto acids. Thiamin is very heat-labile. Therefore, excess heat or autoclaving can reduce the thiamin content of dietary components, particularly when reducing sugars are present.

Miller *et al.* (1955) estimated a thiamin requirement of 1.5 mg/kg for pigs weighing about 2 kg initially and fed to approximately 10 kg of body weight. Pigs weaned at 3 weeks and fed to about 40 kg of body weight required about 1.0 mg of thiamin/kg of diet (Van Etten *et al.,* 1940; Ellis and Madsen, 1944).

Thiamin-deficient pigs exhibit loss of appetite; a reduction in weight gain, body temperature, and heart rate and occasionally, vomiting. Other effects observed in thiamin deficiency are heart hypertrophy, flabby heart, myocardial degeneration, and sudden death because of heart failure. Animals deficient in thiamin also have elevated plasma pyruvate concentrations (Hughes, 1940b; Van Etten *et al.,* 1940; Follis *et al.,* 1943; Wintrobe *et al.,* 1943a; Ellis and Madsen, 1944; Heinemann *et al.,* 1946; Miller *et al.,* 1955). Most of the cereal grains used in swine diets are rich in thiamin. Hence, grain oilseed meal diets fed to all classes of swine are considered adequate in this B-vitamin, and it is not generally included as a supplement for swine diets.

It acts as a coenzyme in energy metabolism. It promote appetite and growth, required for normal carbohydrate metabolism and reproduction.

Practical sources of vitamin are thiamin hydrochloride, green pastures, well cured green leafy hays, cereal grains, peas, brewers yeast. Thiamin content of normal feeds is usually sufficient

Riboflavin (B-2)

A component of two coenzymes, flavin mononucleotide (FMN) and flavin adenine dinucleotide (FAD), riboflavin is important in the metabolism of proteins, fats, and carbohydrates. In feedstuffs, most of the riboflavin activity exists as FAD.

Estimates of the riboflavin requirement for pigs weighing 2 to 20 kg range from 2.0 to 3.0 mg/kg of synthetic diet (Forbes and Haines, 1952; Miller *et al.,* 1953). Riboflavin requirement estimates range from 1.1 to 2.9 mg/kg for growing pigs fed synthetic diets (Hughes, 1940a; Krider *et al.,* 1949; Mitchell *et al.,* 1950; Terrill *et al.,* 1955), whereas the estimates vary from 1.8 to 3.1 mg/kg of diet when practical diets are fed (Krider *et al.,* 1949; Miller and Ellis, 1951). Riboflavin deficiency has led to anestrus (Esch *et al.,* 1981) and reproductive failure in gilts (Miller *et al.,* 1953; Frank *et al.,* 1984).

Signs of riboflavin deficiency in young growing pigs include slow growth, cataracts, stiffness of gait, seborrhea, vomiting, and alopecia (Wintrobe *et al.,* 1944; Miller and Ellis, 1951; Lehrer and Wiese, 1952; Miller *et al.,* 1953). In severe riboflavin deficiency, researchers have observed increased blood neutrophil granulocytes, decreased immune response, discolored liver and kidney tissue, fatty liver, collapsed follicles, degenerating ova, and degenerating myelin of the sciatic and brachial nerves (Wintrobe *et al.,* 1944; Krider *et al.,* 1949; Mitchell *et al.,* 1950; Forbes and Haines, 1952; Lehrer and Wiese, 1952; Miller *et al.,* 1953; Terrill *et al.,* 1955; Harmon *et al.,* 1963).

Practical sources are synthetic riboflavin, green pastures, milk and milk products, meat and fish meal. High quality hay or meat, yeast, dried corn, distillery solubles and dried linseed meal, cotton seed meal and pea nut oil meal.

Vitamin–B-6 (Pyridoxine, Pyridoxal, Pyridoxamin)

Vitamin B_6 plays a crucial role in central nervous system function. It is involved in the decarboxylation of amino acid derivatives for the synthesis of neurotransmitters and neuroinhibitors.

Miller *et al.* (1957) and Kösters and Kirchgessner (1976a,b) suggested a dietary requirement of 1.0 to 2.0 mg/kg of diet for the pig weighing initially about 2 kg and fed to 10 kg of body weight. Requirement estimates for the 10 to 20 kg pig range from 1.2 to 1.8 mg of vitamin B_6/kg of diet (Sewell *et al.,* 1964; Kösters and Kirchgessner, 1976 a,b).

A deficiency of vitamin B_6 reduces appetite and growth rate. Advanced deficiency will result in an exudate development around the eyes, convulsions, ataxia, coma, and death.

Practical sources of vitamin B-6 are cereal grains and their by-products, rice bran, green pastures, well cured alfalfa hay, yeast, milk products, meat and fish products. Vitamin B-6 content of normal feeds is usually sufficient.

Biotin

Biotin is important metabolically as a cofactor for several enzymes' function.

With sows, biotin supplementation has been reported to improve hoof hardness and compression, compressive strength, and the condition of skin and hair coat, as well as to reduce hoof cracks and footpad lesions (Grandhi and Strain, 1980; Webb *et al.,* 1984. Lewis *et al.* (1991) reported that adding 0.33 mg/kg of biotin to a corn-soybean meal diet for sows during both gestation and lactation increased the number of pigs weaned but did not improve foot health.

Biotin deficiency signs include excessive hair loss, skin ulcerations and dermatitis, exudate around the eyes, inflammation of the mucous membranes of the mouth, transverse cracking of the hooves, and the cracking or bleeding of the footpads (Cunha *et al.,* 1946 and 1948; Lindley and Cunha, 1946; Lehrer *et al.,* 1952).

Practical sources are feed stuffs and intestinal synthesis. Alfalfa meal, green forages, milk products and cereal grains and their by-products are good sources.

Very young pigs (under about 3 weeks of age) do not produce enough biotin in their intestine until they develop microflora capable of synthesizing it.

Choline

Choline remains in the B-vitamin category even though the quantity required far exceeds the "trace organic nutrient" definition of a vitamin. It is generally added to swine diets as choline chloride, which contains 74.6% choline activity (Emmert *et al.,* 1996).

Choline-deficient pigs have reduced weight gain, rough hair coats, decreased red blood cell counts and hematocrit and hemoglobin concentrations, increased plasma alkaline phosphatase, and unbalanced and staggering gaits. Livers and kidneys exhibit fat infiltration. In a severe choline deficiency, kidney glomeruli can become occluded from massive fat infiltration (Wintrobe *et al.,* 1942; Johnson and James, 1948; Neumann *et al.,* 1949; Russett *et al.,* 1979b).

Choline is involved in nerve impulses. It is a component of phospholipids. It helps in lipid transport. It is essential for growth and proper function of liver and kidneys.

Practical sources are meat meal, soya bean oil meal, fish meal condensed fish solubles, liver meal and dried distilleries solubles.

Choline content of normal feeds is usually sufficient. But studies have shown that more live pigs are born and weaned when sows received supplemental choline throughout gestation. Methionine can replace the choline needed for donation of methyl groups.

Vitamin–B-12

Vitamin B_{12}, or cyanocobalamin, contains the trace element cobalt in its molecule, which is a unique feature among vitamins. Vitamin B_{12} as a coenzyme is involved in the de novo synthesis of labile methyl groups derived from formate, glycine, or serine, and their transfer to homocysteine to form methionine. It is also important in the methylation of uracil to form thymine, which is converted to thymidine and used for the synthesis of DNA.

Vitamin B_{12} supplements are produced commercially by microbial fermentation and are usually added to grain soybean meal diets.

Pigs that are deficient in vitamin B_{12} display reduced weight gain, loss of appetite, rough skin and hair coat, irritability, hypersensitivity, and hind leg incoordination. Blood samples from deficient pigs indicate normocytic anemia and high neutrophil and low lymphocyte counts (Anderson and Hogan, 1950b; Neumann and Johnson, 1950; Neumann *et al.,* 1950; Cartwright *et al.,* 1951; Richardson *et al.,* 1951; Catron *et al.,* 1952).

It serves numerous metabolic functions and essential for normal growth and reproduction in swine. It is required for the maturation of red blood cells.

The richest source of Vitamin B_{12} are condensed fish solubles, liver and glandular meals, fish meal. Meat and bone scraps, tankage and milk products are good sources.

Vitamin B_{12} is liable to be lacking in swine ration. Synthesis of vitamin B_{12} by intestinal flora may supplement dietary sources. B_{12} contains the trace element cobalt, hence synthesis of B_{12} in the intestines is dependent on the presence of cobalt in the feed. This may be the major, if not the only, function of cobalt as an essential nutrient.

Vitamin–C (ascorbic acid)

Vitamin C is a water-soluble antioxidant that is involved in the oxidation of aromatic amino acids, synthesis of norepinephrine and carnitine, and in the reduction of cellular ferritin iron for transport to the body fluids. Ascorbic acid is also essential for hydroxylation of proline and lysine, which are integral constituents of collagen. Collagen is essential for growth of cartilage and bone. Vitamin C enhances the formation of both bone matrix and tooth dentin.

Using pigs weaned at 3 to 4 weeks of age, Brown *et al.* (1975), Yen and Pond (1981), and Mahan *et al.* (1994) reported that weight gains were improved by supplementing the diet with vitamin C. In pigs weighing 24 kg initially, Mahan *et al.* (1966) observed an improvement in weight gain from parenteral dosing and feed supplementation with vitamin C. In two of three trials, growing pigs (15 to 27 kg) fed to about 90 kg of body weight responded to vitamin C supplementation (Cromwell *et al.,* 1970).

Table 13.3 Recommended Nutrient Allowances for Pigs[a]

Component		Breeding sows			Growing pigs live weight (kg)[b]		
Component	Unit	Pregnancy	Lactation		Porkers (20–65 kg)	Bacones (50–90kg)	Heavy pig (65–125 kg)
			3 weeks weaning	5–8 weeks weaning			
	Feed (kg/day)	2.0	5.2	5.2	1.2	2.2	2.4
Digestible energy	(MJ/kg)	13.0	13.0	13.0	14.0	13.5	13.0
Crude protein	(g/kg)	130	170	160	220	180	140
Lysine	(g/kg)	4.5	8.0	7.0	13.6	10.4	6.6
Methionine/cystine	(g/kg)	3.1	4.4	3.8	6.8	5.2	5.3
Threonine	(g/kg)	0.7	1.5	1.3	1.9	1.5	1.0
Calcium	(g/kg)	8.5	8.5	8.5	9.8	8.1	7.8
Phosphorus	(g/kg)	6.5	6.5	6.5	7.0	6.1	5.9
Salt (NaCl)	(g/kg)	3.0	3.0	3.0	3.2	3.1	3.0
Iron	(mg/kg)	60.0	60.0	60.0	62.0	59.0	57.0
Zinc	(mg/kg)	50.0	50.0	50.0	56.0	49.0	47.0
Manganese	(mg/kg)	16.0	16.0	16.0	11.0	11.0	11.0
Iodine	(mg/kg)	0.5	0.5	0.5	0.15	0.15	0.15
Selenium	(mg/kg)	0.15	0.15	0.15	0.15	0.15	0.15
Vitamin A	(IU/kg)	8000	8000	8000	8000	6000	6000
Vitamin D	(IU/kg)	1000	1000	1000	1000	750	750
Vitamin E	(IU/kg)	15	15	15	15	15	15
Riboflavin	(mg/kg)	3.0	3.0	3.0	3.0	3.0	3.0
Pantothenic acid	(mg/kg)	10.0	10.0	10.0	10.0	10.0	10.0
Vitamin B_{12}	(mg/kg)	0.015	0.015	0.015	0.01	0.01	0.01
Biotin	(mg/kg)	0.30	0.30	0.30	0.2	0.2	0.2

[a] Assuming an average litter size of nine piglets. Only feed quantity varies with the number of piglets in the litter

[b] Assuming a growth rate of 0.7 kg/day

Source: McDonald *et al.*, 1988

13.3.5.4 *Water*

Water is the main constituent of the animal's body, constituting 50 to 80% of the live weight, depending on age and degree of fatness plays several physiological functions. In neonatal pig the water content is about 82% of the empty body weight whereas in adult pig it is about 53%. An animal can lose almost all of its fat and about 50% of its body protein and survive. However, the loss of 10% of its body water can be fatal. A good water supply is defined both in terms of quantity and quality of the water. Total water intake is directly related to feed dry matter (DM) intake.

The main functions of water in the body are:

- Provides turgidity to the cell and plays important role as structural element
- Movement of nutrients from cells of tissues and removal of waste from cells.
- As solvent for variety of substances and hence facilitates chemical reactions, transport of materials within the body
- Major component of body fluids (blood, CSF, lymph etc), interstitial fluid and secretions (milk, sweat, semen etc)
- to help eliminate waste products of digestion and metabolism
- to regulate blood osmotic pressure
- in the body's thermoregulation as affected by evaporation of water from the respiratory tract and from the skin's surface

Pigs fulfill their water needs from three major sources:

- drinking water
- water contained in feed (about 10–12%
- metabolic water produced as a result of oxidation of carbohydrates (0.56l/kg), protein (0.45l/kg) and fat (1.19l/kg).

The major sources of water loss is through: (a) respiration (b) sweat (c) urine and (d) faeces.

Water consumption requirements depend on factors such as:

- age of pig
- rate and composition of body weight gain per day
- pregnancy
- lactation
- type of diet
- level of dry matter intake
- level of activity
- quality of water

- temperature of the water offered
- surrounding air temperature

Table 13.4 Water Requirement of Various Categories of Pig

Swine type	Weight range (kg)	Water requirement range (L/day)
Weaner	7–22	01.0–03.2
Feeder pig	23–36	03.2–04.5
	36–70	04.5–07.3
	70–110	07.3–10.0
Gestating sow/boar	–	13.6–17.2
Lactating sow[c]	–	18.1–22.7

Pigs consume more water in hot weather to regulate the body temperature. If they are denied that extra water, the feed intake declines and the growth decreases. The animals should receive all they want to drink, at least two to thee times a day.

The tolerance to total salts dissolved in drinking water depends upon animal species, with poultry being most sensitive, hogs moderately sensitive and ruminant animals least sensitive. In general, a total soluble salt content of less than 1000 mg/litre is considered a low level of salinity suitable for all types of livestock. Salt contents between 1000 mg/litre and 3000 mg/litre are satisfactory for all types of livestock but may cause watery droppings in poultry or diarrhea in livestock not accustomed to this salt level. Salt levels above 3000 mg/litre are not recommended for poultry and are more likely to result in cases of livestock refusal. Salt levels above 5000 mg/litre are not recommended for lactating animals. Avoid levels above 7000 mg/litre for all livestock.

13.4 Computation of Different Types of Rations

13.4.1 Computation of ration

Exact feeding value of different feeds available and mixed in the ration are taken into consideration for working out Total Digestible Nutrients of starch equivalent and digestible protein. Quality of feed is best expressed in terms of Nutritive ration which is the ratio of protein to carbohydrate food constituents. The nutritive ratio and feed intake and anticipated gain in weight are given below for assessment of the performance of pigs.

The protein content of ration must contain essential amino acids at one or other stage of pig's life for good performance and optimum growth.

Farmer has to ensure the nutritive ratio and nutritive content of food in case feeds being computed by him and ready made balanced pig feed is not available in feeding trough. Rapid growth in early stages till they attain 50 kg body weight leads to formation of muscular tissues and beyond this leads to deposition of fat. To avoid obesity restricted feeding at later stages is necessary. Pig should be fed regularly and changes from one ration to the other should always be resorted gradually.

Table 13.5 Assessment of Performance of Different Ration

Sl. No.		Type of pig and approximate body weight	Nutritive ratio of ration	Approx daily feed intake (kg)	Anticipated average gain in wt. (kg)
1		Piglets (birth to 15 kg) Creep feed for weaners	1.0 to 4.0	0.14 to 0.7	0.32
2		Weaners (15 to 25 kg)	1.0 to 4.5	0.7 to 1.4	0.29
3		Growers (25 to 50 kg)	1.0 to 5.5	1.4 to 2.0	0.64
4		Fatteners (50 to 90 kg)	1.0 to 6.0	2.3 to 2.7	0.84
5		Adults			
	(a)	Pregnant sows	1.0 to 5.0	1.0 to 5.0	–
	(b)	Suckling sows	1.0 to 5.0	5.4	–
6		Boars			
	(a)	Below 15 months	1.0 to 5.0	2.7	–
	(b)	Above 15 months	1.0 to 5.0	2.3	–

13.5 Replacement of Some Feed Ingredients with Locally Available Cheap Feeds in Computation of Rations

As we already know, cost of feed is the largest component comprising about 80% of the total cost of production and for profitable pig enterprise efforts should be made by pig keepers to reduce the cost of feed by making more efficient utilisation of locally available cheap feed ingredients such as agricultural and industrial by-products and waste products, which are not used by human being, may be used for replacement. Nutritive values of these ingredients have to be kept in view. For feeding of newly born piglets for first three days after farrowing colostrum be given, which is the first milk of the sows which has just farrowed and this colostrum protests young piglets from diseases later. For feeding young piglet skimmed milk, butter milk, dried whey, cheese rind, and burnt baby feed etc., can partly replace some of the feed ingredients in creep ration.

Similarly slaughterhouse and meat packing plants by-products can easily be converted and processed to form cheap and highly nutritive diet for pigs. The by-products like blood in the form of blood meal, bone meal, meat meal, clean entrails residues, can be easily converted into suitable products, and mixed in pig feed replacing expensive ingredients, where they are locally available.

In places where poultry dressing plants or hatcheries are available, the waste of these units can be used for replacing some feed ingredients.

Fish meal and fish residue meal, provide clean highly nutritive feed supplement which can be partly used for replacement in feed, wherever they are locally available. They are good source for providing high quality protein, calcium and phosphorus. It can be used up to 5% of the ration.

Hotel and community kitchen waste, with little processing, cooking and boiling, can be used for feeding pigs by replacement of some ingredients. Guar meal which is the residue left after removal of germ from seeds can be used for replacement in feed rations for growing and finishing pigs. It is rich in protein and contains 40 to 45% protein with better amino-acids; guar meal can also be utilized to a limit of 12% to 15% for replacing 65% of ground-nut cake in finishing ration of pigs.

Brewery waste products can be included in pig ration up to 15% and as a replacement of rice bran to the extent of 75%.

13.6 Feed Resources and their Nutritive Value-Cereals

Maize

It is rich in carbohydrate and carotene or vitamin-A. It can be used up to 85% of the ration for growing pigs and at little lower rate in pregnant sows. It should be coarsely ground or crushed after which it should be mixed with other feed ingredients. Maize is an excellent source of linoleic acid but deficient in lysine and tryptophan.

Barley

It has approximately 90% of the feeding value and should be ground for mixing in ration and used in high levels in the rations. In feeding trials in India, it has proved very efficient in enhancing growth rate. The protein of barley is deficient in lysine and methionine and it contains b–D-glucose

Oats

Its not commonly used in India as a feed for pigs due to its low availability, it can only be used up to 30% in crushed or ground form in the ration due to its relatively high fibre content.

Sorghum and millet

It has 95% of feeding value of maize. In ration for growers and finishers this can be included to the extent of 50% and millet to the extent of 35%. They should be crushed are ground before mixing. It is deficient in lysine, methionine and arginine.

Damaged wheat

In India this is available from big wheat godowns and warehouses of FCI. The feeding value is equal or slightly inferior to maize, it should not be ground too fine.

Wheat bran

This is widely available in India for feeding of pigs and other livestock. Its feeding value is equivalent to 85% of maize, and has high fibre content, it is always desirable to use it in fatteners and breeding sows. It can be used upto 40% of the ration.

Root crops

Potato

This is available in surplus in some areas and can be mixed with grain in feed to the extent of 75%. It can also be used in form of flour up to 30% of the ration. Feeding value is 30% of the maize.

Sweet potato

It is a good source of energy (3500 kcal ME/kg) and protein. It contains antitrypsin. On drying, the antitrypsins are denatured. It can be used for older pigs. Sweet potato flour can replace grain in pig rations up to35 to 40%.

Sugarcain molasses

It is the cheapest carbohydrate feed available. It is mixed with meal and improved palatability of the ration, increased use causes scour. In young piglets not more than 5% be used.

Tapioca

This root is widely used for pig feeding and pigs fattened on this develop firm fat. Four parts of this root can replace one part of maize. Tapioca root meal is rich in energy. It contains cyanogenic glucosides (HCN 15-400 mg/kg). Drying of roots eliminates 8.5% HCN.

13.7 Other Sources of Protein

Soyabean meal

Soyabean meal contains 38–40% protein and 18–20% fat. It contains protease inhibitors which bind and render unavailable the enzyme trypsin and chymotrypsin. It is the source of allergenic proteins such as conglycinin and b–conglycinin that reduces the efficiency and caused scouring in young piglets. It is an excellent source of lysine, typtophan and threonine but is deficient in methionine.

Mustard or rapeseed meal

It has lower protein and energy than soyabean meal. The conditioning process destroys the enzyme myrosinase, which converts glucosinolates to goitrogenic compounds: oxalolidone-2-thione and isothiocynate.

Groundnut meal

G.N. meal contain about 35–40% protein and has a poor amino acid profile and is deficient in methionine, lysine and tryptophan. G.N. meal contains trypsin inhibitors and other protease inhibitors. The undesirable constituent often associated with G.N. meal is aflatoxin-produced by the fungus *Aspargillus flavus* that infest ground nuts.

Sunflower meal

Sunflower meal has about 25% protein but high levels of chlorogenic acid, a tannin like compound, inhibits activity of digestive enzymes (trypsin, chymotrypsin, amylase and lipase). Addition of methionine and choline are required to counteract the effect of chlorogenic acid.

Cotton seed meal

Cotton seed meal contains protein but may contain free gossypol (0.02 to 0.05%). The protein is deficient in lysine, methionine, thereonine and tryptophan. It also contains the cyclopropenoid fatty acids, malvalic and steraulic acid. Gossypol can bind with iron 1:4 and hence for the purpose of detoxification with ferrous sulphate 1.2% is optimum level.

Fish meal and fish residue meal

Fish meal and fish residue meal, provide clean highly nutritive feed supplement which can be partly used for replacement in feed, wherever they are locally available. They are good source for providing high quality protein (CP about 40–60%), calcium and phosphorus. It can be used up to 5% of the ration.

Blood meal

It can be used up to 3 to 5% in pigs ration.

Coconut meal

After expression of oil from coconut the available residue can be used and mixed up to 20 to 25% to the ration. It stimulates milk secretion and is good for lactating sows.

Linseed meal

Should not be used more than 5% in ration.

Meat and bone meal

The quality considerably varies depending on proportion of meat and bone meal or pure meat used. Depending on the economic price at which they are available its choice be made and mixed up to 5 to 10% of the ration.

13.8 Non Conventional Feed Ingredients

Some of the possible non conventional feed resources for pig are as under.

Tea waste

Tea waste, a tea industry by-product, is available to the tune of 906876 tonns annually. The factory tea waste contains CP of 19.5%. It contains about 4.9% tannic acid. It also contains higher level of all other essential amino acids more than cotton seed cake, methionine content exceeds that of whole egg protein. The material can be used up to the 15% level in the concentrate mixture of pigs.

Nahar seed meal

The seeds contain about 13% crude protein. The DCP and TDN values were found to be 12.84 and 78.06% respectively on DM basis. The expeller pressed Nahar seed meal can be used up to 15% level in growing pigs.

Ajar seed

The ajar seed has a CP of 10.03% and 3.06% tannic acid. Ajar seed kernal can be incorporated up to 15% level replacing maize grain (w/w) in growing pigs.

Wild colocasia (*Colocasia esculenta*)

The boiled colocasia contains 9.86% CP and 4150 Kcal GE/kg. Boiling reduces the oxalate content from 3.48 to 0.79 and tannic acid from 1.92 to 0.77%. Colocasia whole plant can replace 20% concentrate in pig ration (Baruah, personal communication).

Niger cake

Niger cake contains about 35% protein but is deficient in lysine, methionine and tryptophan.

Keranja meal

It contains 30% protein and 28% oil. The oil contains karanjine (1.47 mg /ml oil).

Rubber seed meal

It contains 26% protein and 20–40 mg/kg HCN which needs to the neutrilized before use.

Ambadi meal

Ambadi seed meal contains 28% protein and 0.35% tannins. The protein is low in lysine and methionine.

Mahua meal

Mahua seed meal contains about 20% protein. Mowrein, a saponins (19%) and tannins (1.5%) are present in mahua seed meal. It needs to be detoxified before use.

Guar meal

The guar meal contains about 40% protein and is deficient in methionine and lysine. The antinutrients present are trypsin inhibitors, HCN, haemagglutins and residual gums. These can be detoxified using current technologies before use.

Deoiled silkworm pupae meal

Deoiled silkworm pupae meal is good source of protein (about 65%) and phosphorus. The protein is rich in lysine, methionine, arginine, tryptophan and isoleocine, but low in threonine.

Leucaena

(*Leucaena leucocephala*) leaf meal contains 1500 Kcal ME /kg. It is a good source of protein (19%) and contains mimosine (3–5%), and tannins (0.95%). It can be used for short durations.

Coffee grounds

Spent coffee grounds which are a waste product of coffee industry can be incorporated in place of rice bran to the extent of 50% in grower and finishing ration of pigs.

Banana

Over ripe banana or dried bananas where available cheap can be used as a source of energy in pig feed and can be used for replacement to the extent of 20–30% of maize in ration of growing and finishing pigs.

Brewery and distillery grains

Both these items can be used for fattening pigs. Distillery grains have better feeding value.

Forages

Adult pigs can be given greed forage up to 4.5 kg per day. They can also be fed rations containing grass or legume meal, guinea, para and elephant grass, sweet potato tops or artificially dried leaves of legumes *viz* berseem and lucern. The feeing rate should not exceed 5% of total ration.

Table 13.6 Nutritive Value of different feeds

Name of feed ingredient	CP %	% of digestible crude protein	% of total digestible nutrients	DE (kcal/kg)	ME (kcal/kg)	Calcium %	Phosphorus %
Alfa-alfa meal	17.00	–	–	1830	1650	1.53	0.26
Bakery waste	10.80	–	–	3940	3700	0.13	0.25
Barley	11.30	10.00	77.70	3050	2910	0.06	0.35
Blood meal	77.71	–	–	2850	2350	0.37	0.27
Bone meal	50.00	45.00	79.20	–	–	29.00	12.60
Brewers grain	26.50	–	–	2100	1960	0.32	0.56
Buck wheat	11.10	–	–	2825	2640	0.09	0.31
Coconut cake	21.90	18.00	77.00	3010	2565	0.16	0.58
Cotton seed meal	39.00	32.80	72.50	2575	2315	0.19	1.06
Fish meal	50.0	45.00	65.00	2900	2695	5.36	3.42
Gingley oil cake	42.60	38.00	78.00	3350	3035	1.90	1.22
Groundnut cake	49.10	41.90	68.50	3415	3245	0.22	0.65
Hotel waste	–	2.20	23.50	–	–	–	–
Linseed cake	31.00	30.60	75.50	3060	2710	0.39	0.83
Maize	24.80	8.90	85.30	3100	2715	0.10	0.40
Meat–cum-bone meal	51.50	39.90	64.30	2440	2225	9.99	4.98
Meat meal	54.00	45.00	66.70	2695	2595	7.69	3.88
Molasses	4.80	1.00	53.70	–	–	0.66	0.08
Mustard oil cake	36.00	27.00	74.00	–	–	–	–
Oats	9.00	7.10	–	2770	2710	0.07	0.31
Poultry bye products	64.10	–	–	3090	2860	4.46	2.41
Rape seed	35.60	27.0	74.00	2885	2640	0.63	1.01
Rice bran	13.30	8.4	67.40	3100	2850	0.07	1.61
Rice polish	13.00	5.7	80.70	3770	3350	0.09	1.18
Skimmed milk	34.60	28.90	9.10	3980	3715	1.31	1.00
Sorghum	9.20	–	–	3380	3340	0.03	0.29
Soybean meal	43.80	42.0	78.10	3490	3180	0.32	0.65
Sunflower meal	33.00	23.00	71.00	2010	1830	0.36	0.86
Tapioca meal	3.30	2.00	64.00	3385	3330	0.22	0.13
Wheat bran	15.70	13.30	66.90	2420	2275	0.16	1.20

13.9 Feed Processing

Feed ingredients can be given in different forms, and mixed to form suitable feed for feeding of pigs. At present, considering improvement in feed efficiency, processing costs, incidence of gastric ulcers, an average particle size of 700 to 800 microns is recommended.

a. Grinding

Grinding is the most common method of feed processing for the swine producer and nearly all feed ingredients will be subjected to some type of particle size reduction. Particle size reduction increases the surface area of the grain, allowing for greater interaction with digestive enzymes, improving feed efficiency. It also improves the ease of handling and mixing characteristics.

Feed ingredient specially grain and cakes need to be grinded. Grinding increases the food value of the grains by about 20%, but feeds should not be grinded finely, as it involves higher cost and may adversely affect the digestive system.

b. Dry or wet feeding

Such feeding is carried out by mixing dry food with sufficient water in trough not to make it liquid like. This provides better food conversion and higher growth rate than feeding dry as digestibility and palatability of feed is increased.

c. Pellet form

Feed in form of pellets which are generally manufactured by organized feed mixing plants is more economical as wastage of feeds is reduced and digestibility is better.

d. Cooking

It is essential to adapt for feeding such ingredients like skim milk, butter milk, meat, blood meal and kitchen wastes etc. This is adapted to kill any pathogenic organisms present in these ingredients.

Common processing methods for pig feeds

There are many methods for processing pig feeds. In addition to grinding, the most common forms of feed processing are pelleting, extruding, and roasting.

Pelleting

Pellets can be made of different lengths, diameter, and degree of hardness. The ingredients of the diet will influence the hardness of the pellet and pellet quality. Various studies suggest a 3 to 10% improvement in growth rate and feed efficiency when pigs are fed pelleted diets rather than a meal. This appears to result from less feed waste with pelleted feeds. Pelleting appears to improve the nutritional valueof high fiber feed ingrediens to a greater extent than that of low fiber ingredients.This may be a result of increasing the bulk density of the feed. However, as energy costs increase, the economics of pelleting swine feeds may change. The increased diet cost must be offset by the improved feed efficiency or other productive measure of pigs feed in the pelleted form. Of future importance is the potential benefits that pelleting produces by sanitizing the feed. This aspect has yet to be examined in swine production and may play an integral part in future production systems.

Extrusion and roasting

Extrusion processing involves the application of heat, pressure, and or steam to an ingredient or diet. Extruders are sometimes used for on farm processing of soybeans. If properly heated, this is an easy way to add fat to swine diets and utilize home grown soybeans. Recent research shows that moist, extruded, soy protein concentrate is an excellent protein sources for baby pigs.

Because of volume and tonnage, extrusion of complete feeds is usually not economically justified based on performance of pigs fed extruded complete feeds. Furthermore, extrusion increases the bulkiness of the diet, making it more difficult for the pig to consume enough feed to meet its nutrient requirement.

Roasting can also be used to process home grown soybeans. This can also be an alternative method for adding fat to swine diets. However, roasting temperature and times must be checked to ensure adequate processing. The added cost of the extruded or roasted products must be the ultimate consideration in determining the feasibility of their use in swine diets.

Other processing methods

Several alternative processing methods are available to swine producers, such as steam flaking, micronizing. However, these processing methods often do not improve pig performance enough to justify the added expense of processing.

13.10 Feed Additives

13.10.1 Availability of feed additives

Feed additives are used by most swine producers because of their demonstrated ability to increase growth rate, improve feed utilization, and reduce mortality and morbidity from clinical and sub-clinical infection.

In general, additives available for swine producers fall into five classifications (i) antibiotics, (ii) chemobiotics or chemotherapeutics, (iii) anthelmintics or deformers, (iv) copper compounds, and (v) probiotics.

13.10.2 Selecting feed additive

There are many feed additives on the market, and they differ widely in chemical composition and mode of action. Selection of a specific feed additive and the level needed for optimal response will vary with the existing farm environment, management conditions, and the stage of the production cycle.

It is highly recommended that an accurate diagnosis and an antibiotic sensitivity test be performed to determine what compounds would be effective. In the long run, the initial expense of a sensitivity test will be great value because unnecessary drugs and inadequate levels will be avoided.

Producers who are planning to use a feed additive for treatment or prevention of a disease should consult their veterinarian or other professional who has training in the pharmacodynamics and efficiency of drugs. Some drugs (*e.g.* nitrofurazone is still approved but seldom of value in outbreaks of swine dysentery) are not as effective as others. Certain drugs will appear to be a good treatment based upon a sensitivity test, but will be unsatisfactory because the drug has limited absorption from the intestine. Furazolidone and neomycin often appear to be good drugs against organisms causing pneumonia, but neither are absorbed to any degree from the intestine. Rotation of antibiotics, evaluation of different antibiotics, or use with approved mixtures may be advisable, if the response to a feed additive appears to be diminishing.

Because antibiotics are expensive and their use is coming under grater scrutiny by health authorities and the public, their indiscriminate use should be avoided. Antibiotics should not be used to replace good management.

13.10.3 Recommended levels of feed additives

Level of usage depends upon the type of additive and the purpose of the compounds. Many additives have two levels, one for prevention and other for treatment. Always consult manufacturers direction before mixing. In addition, the food and drug administration has proposed that the sub-therapeutic use of certain antibacterial compounds in feed be restricted and withdrawal periods observed. Thus, it is important to recognize that approved usage of any feed additive and withdrawal periods are subject to change, and it is imperative to keep updated on any changes.

Certain feed additives must be withdrawn from the feed prior to slaughter at varying intervals to ensure residue-free carcasses.

Effectiveness of additives in stages of the production cycle

The response to feed additives is greatest in starter (4.5 to 22.6 kg) diets. The response to feed additives is less during the finishing period (55 kg to market weight) than it is at younger age. Their use in finishing diets is questionable.

Herds that have experienced problems with conception rates and litter size have often been helped by the addition of antibiotics to brood sow diets. However, the routine feeding of antibiotics to the breeding herd is discouraged, unless there is a history of reproductive problems.

13.10.4 Non nutritive feed additives

Non nutrient feed additives are commonly included in swine diets. Of these, the antimicrobial agents are the additives most commonly used. Antimicrobial agents, along with anthelmintics, are defined as "drugs" by the Food and Drug Administration (FDA). Thus, their usage levels, allowable combinations, and periods of withdrawal prior to slaughter are regulated by the FDA and are published annually in the Feed Additive Compendium (1998). In addition, certain other additives are sometimes included in swine diets. The association of American Feed Control Officials (1998) has established guidelines for the use of many of these products in animal feeds.

13.10.5 Antimicrobial agents

These are compounds that suppress or inhibit the growth of microorganisms. This class of compounds includes the antibiotics (naturally occurring substances produced by yeasts, molds, and other microorganisms) and the chemotherapeutics (chemically synthesized substances). They are added to feed at low (subtherapeutic) levels for growth promotion, improvement of feed utilization, reduction of mortality and morbidity, and improvement of reproductive performance. Antimicrobial agents also are used at moderate-to-high (prophylaxis) levels for the prevention of disease in exposed animals, and at high (therapeutic) levels for the treatment of certain swine diseases. Currently, 17 antimicrobial agents are approved for use in swine feed (Feed Additive Compendium, 1998). Of these, eight require withdrawal from the feed (on schedules ranging from 5 to 70 days) before animals are slaughtered, and nine do not require a withdrawal period.

Antibacterial agents also are effective in improving reproductive performance (Cromwell, 1991). A summary of nine experiments (1931 sows) indicated that farrowing rate was improved from 75.4% in controls to 82.1% in treated sows, and the number of live pigs born was increased from 10.0 to 10.4, respectively, when antimicrobials were included in the diet at the time of breeding. In 11 experiments (2105 sows), inclusion of antimicrobials in the lactation diet increased survival of pigs to weaning (84.9 versus 87.1% of pigs born alive) and pig weaning weights (4.65 versus 4.70 kg).

Although the mechanism of action of antimicrobials is not well understood, their effects are generally grouped into three categories: a metabolic effect, a nutritional effect, and a disease control effect. The first effect implies that these compounds directly influence certain metabolic processes in the animal (*e.g.*, increased rate of protein synthesis). The second effect implies that antimicrobials cause changes in the microbial population that result in increased utilization of nutrients by the host animal. This effect is supported by evidence that antimicrobials reduce intestinal wall thickness (thus improving absorption of nutrients), and that they reduce total gut mass (thus reducing heat loss from tissues with high metabolic activity). Most of the data support the disease control effect as the primary mode of action. This effect implies that antimicrobials suppress microorganisms that cause nonspecific, subclinical disease, thereby allowing the host animal to achieve a growth rate closer to its maximum potential. This suggested mechanism of action is supported by the greater response to antimicrobials that occurs in young versus older pigs, in a "dirty" versus "clean" environment, and in low-health versus high-health animals.

An antibiotic is a compound synthesized by living organisms, such as bacteria or molds, which inhibits the growth of another. Chemibiotics are compounds similar to antibiotics but they are produced chemically rather than microbiologically. Anthelmintics or dewormers are compounds added to swine diets, generally for short intervals, to help control of worm accumulation.

Antibiotics improve animal performances by:

- Ø Reduction or elimination of activity of pathogenic bacteria
- Ø Elimination of toxin producing microbes

- Ø Stimulation of growth of microorganisms that synthesize unidentified nutrients
- Ø Reduction of growth of microorganism that compete with host animal
- Ø Increased absorptive capacity of the intestine by reducing the thickness of the intestinal wall

Typically 20–25mg/kg of antibiotics are used, which has been demonstrated to increase growth rate and feed efficiency between 4–15% and 2–8% respectively.

13.10.6 Copper compounds (copper sulphate) have growth stimulating value similar to antibiotics. They are also effective as a therapeutic treatment for intestinal disorders that do not respond satisfactorily to antibiotics or chembiotics. Probiotics, which means 'in favour of life', have an opposite effect to antibiotics on the microorganisms of the digestive tract. It has been theorized that probiotics increase the population of desirable microorganisms instead of directly killing or inhibiting undesirable organisms. The young pigs up to 16 weeks may be fed with 175mg Cu/kg , which must be reduced to 100 mg/kg thereafter and to 35 mg/kg from 6 month of age onwards at the maximum.

Use of copper sulphate as a growth promoter

The use of copper sulphate as a growth promoting in swine has become widespread in the United States and Europe. Research has shown that when 125 to 250 ppm of actual copper (450 to 900 g of copper sulphate per tonne of feed) is added to starter pig diets, an improvement in growth and feed efficiency and a reduction in mortality is observed. When a combination of supplemental copper and antibiotics is fed in starter diets, improved performance is observed compared with the addition of antimicrobial agents alone. Copper sulphate can be added to starter pig diet but is not recommended for use in the finishing phase.

Cooper sulphate increase corrosion, thereby reducing the longevity of galvanized woven wire floors and feeders. It also has been shown to decrease the bacterial degradation of manure in lagoons. Copper, when fed in excess of 300–500 ppm (1 to 2.5 kg of copper sulphate per tonne), may be toxic, particularly if the diets are low in zinc and iron.

13.10.7 Probiotics

Probiotics are live microbial organism when fed to the animals produce a favorable microbial environment in the gut. It may be a directly fed microorganism or products of microbes. Probiotics are organisms promote proliferation of desirable organisms in the gut, thereby eliminating pathogenic microbes. Probiotics can be classified into two types, namely, live microbial cultures and nonviable fermentation products of microbes. The probiotics has been found to act in many ways in the animal body, mainly by:

- Preventing colonization of GI tract by pathogenic organisms
- Neutralization of endotoxins
- Exerting bactericidal activity by production of lactic acid and reduction of gut content pH.
- Prevention of amines which irritate the GI tract by coliforms
- Increased immunity of the host animal.

In pigs strains of Lactobacilli, Bacillus and Streptococci are used as probiotics.

CHAPTER 14

FEEDING OF VARIOUS CATEGORIES OF PIGS

The various categories of pigs are fed with rations depending upon age and functions such as gestation, lactation, semen production etc. The piglets are fed with creep ration, pre-starter ration for 1–3 weeks of age followed by starter ration up to about 8 weeks of age. The pigs after weaning are fed with grower ration up to a body weight of about 60 kg. Later, these pigs (finishers) are fed with a finisher ration till marketed. The lactation animals and breeding boars are fed with rations specifically designed suiting these physiological functions.

14.1 Computation of Ration

Exact feeding value of different feeds available and mixed in the ration are taken in consideration for working out total digestible nutrients of starch equivalent and digestible protein. Quality of feed is best expressed in terms of nutritive ratio which is the ratio of protein to carbohydrate food constituents. The nutritive ratio and feed intake and anticipated gain in weight are given below for assessment of the performance of pigs.

The protein content of ration must contain essential amino acids at one or other stage of pig's life for good performance and optimum growth. Farmer has to ensure the nutritive ratio and nutritive content of food in case feeds being computed by him and ready made balanced pig feed is not available in watering trough. Rapid growth in early stages till they attain 50 kg body weight leads to formation of muscular tissues and beyond his leads to deposition of fat. To avoid obesity restricted feeding at later stages is necessary. Pig should be fed regularly and changes from one ration to the other should always be resorted gradually.

Table 14.1 Assessment of Performance of Different Ration

Sl. No.	Type of pig and approximate body wt	Nutritive ratio of ration	Approx daily feed intake in kg	Anticipated average gain in wt in kg
1	Piglets (birth to 15 kg) Creep feed for weaners	1.0 to 4.0	0.14 to 0.7	0.32
2	Weaners (15 to 25 kg)	1.0 to 4.5	0.7 to 1.4	0.29
3	Growers (25 to 50 kg)	1.0 to 5.5	1.4 to 2.0	0.64
4	Fatteners (50 to 90 kg)	1.0 to 6.0	2.3 to 2.7	0.84
5	Adults			
	Pregnant sows	1.0 to 5.0	1.0 to 5.0	–
	Suckling sows	1.0 to 5.0	5.4	–
6	Boars			
	Below 15 months	1.0 to 5.0	2.7	–
	Above 15 months	1.0 to 5.0	2.3	–

14.2 Method of Feeding

In modern intensive pig production the method of feeding affects productivity. Feeding methods should be chosen according to the aim of the production of type of meat, rationing system, form of food and type of food available.

14.2.1 Complete diets

Complete diets are those in which all the ingredients are mixed together before feeding, usually in dry form. These diets can be fed dry, as pellets, or wet, either by the addition of water to the feed in the trough at feeding time or preparing in liquid form.

14.2.2 *Ad libitum* feeding

Diets which are based largely on bulky feeds can often be fed *ad libitum* because the pigs capacity will limit their energy intake. It is a labour saving operation and allows the pigs to fulfill their full growth potential. The *ad libitum* feeding may cause a lot of food wastage, and the efficiency of food conversion is somewhat low. The finished carcass tend to be rather fat. The young animals are fed *ad libitum* up to 45 kg body weight.

14.2.3 Restricted feeding

The animals are given a measured amount of feed once, twice of three times a day. In this feeding the growth is restricted, but the feed conversion efficiency is improved. The quality of animals becomes better by lean meat deposition. Ration feeding has a higher demand than *ad libitum* but there is a saving on wastage.

14.3 Feeding of Pigs

The feeding of pigs can be divided into different categories

1. Piglet ration (a) Pre starter, (b) Creep
2. Growers ration
3. Finisher ration
4. Gestation ration
5. Lactation ration
6. Feeding replacement stock
6. Boars

Requirement of essential nutrients varies in different type of pigs, varying with weight and breeding stage. Broadly following requirements should be kept in view.

14.3.1 Piglet ration

From birth to 3–4 weeks, pigs are maintained only on dams milk and all requirements are met through the milk. For first three days feeding of colostrum is essential, because colostrum contains the antibodies necessary for building up the baby pig's disease resistance. Problem only arises when the piglet becomes orphan due to loss of dam in post-parturient period. Various milk substitutes can be used and commonly milk mixed with glucose is artificially fed to pigs through special efforts.

Equalizing litters within 24–48 hr and transferring pigs so that litters contain pigs of similar weight can improve pig survival. Commercial milk replacers can also be used to provide supplemental milk during lactation or the first few days post-weaning. A good replacer should contain at least 24–28% protein and 8–10% fat.

14.3.1.1 *Pre starter ration*

Piglets are generally kept with dam till 8 weeks of age when they are weaned, from 4 to 8 weeks of age. Both dams milk and special starter ration are used (pre starter ration) , as piglets at this age have the maximum rate of growth as they are capable of utilizing maximum part of their feed nutrients for growth. In view of this, ration at this stage should be high in protein content and palatable.

14.3.1.2 *Creep ration*

It is a weaning diet, suitable for weaning the young, which are the only animals able to penetrate the creep. In addition to sows' milk, pigs need a creep feed to make maximum gain through weaning. A fresh creep feed should be provided at one week of age in a place where pigs can get away from the sow. Research shows that very little creep feed will be consumed before 3 weeks of age.

Creep feed would contain 20% high quality protein most of which should be of animal origin and 80% total digestible nutrients. Use of milk powder, gur, sugar makes feed palatable. The creep food should have low fibre content and not more than 1.2%. Use of good quality creep food is with 22% of protein, results in piglets gaining good body weight up to 12 to 20 kg at the time of weaning i.e. between 7 and 8 weeks of age.

A creep ration should be of high-quality, complete mixed feed that is eaten readily. Good creep rations can be purchased or mixed on the farm. When creep rations are formulated and mixed on the farm, particular care to be taken using a high energy palatable mixture that meets the pig's nutrient needs.

Getting pigs to eat adequate amounts of a creep ration is often a problem. Creep feeder should be placed in a warm, dry, well-lighted area. Feeding should be done in small amounts, and need to feed frequently to keep the ration fresh. Sprinkling feed on the floor or placing it in a shallow pan may help pigs start to eat. Pelleted feeds are usually eaten more readily than meal.

Table 14.2 Creep and Starter Rations (14 to 56 day after farrowing)

Ingredient	%	Ingredient	%
Maize	65	Wheat bran	10
Groundnut cake	14	Ground yellow maize	40
Molasses	5	Groundnut cake	10
Wheat bran	10	Til cake	10
Fish meal	5	Skimmed milk	10
(Antibiotics can be added)		Fish meal	6
		Molasses	10
		Brewers yeast	2
		Mineral mixture	2
		Vitamin A+B_2+D_3	10 g

Source: Dr. S.K. Ranjhan and co-workers

14.3.2 Growers ration

For 12 to 25 kg body weight

Usually piglets are weaned when they are of 12–50 kg body weight and they still have good capacity for weight gain at this stage. In a farrow-to-finish operation grower diets represents approximately 30–35% of the feed usage. The growing pig (22 kg) is still in the growth phase in which it is depositing lean tissue at a first rate. Therefore, high levels of lysine and other amino acids are necessary to promote maximum lean growth.

Starter rations having low fibre contents with high quality protein should be used for feeding these weaned piglets and should contain 18% protein and 80% of TDN and calcium should be 0.65% and phosphorus 0.50%.

For 25 to 40 kg body weight

At this stage should contain 16% protein, 75% TDN, with calcium and phosphorus @ 0.50%, crude fibre content should not exceed 3 to 4%. Animal protein should be at least 5%.

For 40 to 60 kg body weight

Ration should have 17% protein and 70% TDN and crude fibre not exceeding 5 to 6%.

For 60 to 80 kg body weight

Ration should have 14% protein and 70% TDN and about 3 kg of this is given in a day.

For 80 to 100 kg body weight

Ration should have 14% protein and 70% TDN and about 3.5 kg of this is given in a day.

Grower's rations are grouped on the basis of inclusion of cereals and non-cereal grains.

Table 14.3 Grower's Ration*

Ingredient		Ingredient	
Maize	30.0 kg	Maize	20.0 kg
Ground nut cake	20.0 kg	Groundnut cake	10.0 kg
Wheat bran	40.0 kg	Wheat bran 40%	40.0 kg
		Milo –	10.0 kg
Fish meal	7.5 kg	Fish meal	7.5 kg
Mineral mixture	2.5 kg	Mineral mixture and	2.5 kg
Vitamins supplement	10 g	Vitamins supplement	10 g

Source: Dr. S.K. Ranjhan and co-workers

*Note: Expected growth rate with the above rations (about 0.5 kg) per head per day with a feed efficiency of 4 kg of meal to 1 kg wt gain.

Table 14.4 Non-cereal Ration*

Ingredient		Ingredient	
Wheat bran	70.0 kg	Yellow maize	10.0 kg
Ground nut cake	20.0 kg	Groundnut cake	20.0 kg
Fish meal	6.5 kg	Wheat bran	60.0 kg
Mineral mixture	1.5 kg	Fish meal	6.5 kg
Common salt	0.5 kg	Common salt	0.5 kg
Rovimix	10 g	Mineral mixture	3.0 kg
		Vitamins supplement	10 g

Source: Dr. S.K. Ranjhan and co-workers

*Note: Expected growth rate with the above rations (about 0.5 kg) per head per day with a feed efficiency of 4 kg of meal to 1 kg wt gain.

Finishing feed will represent approximately 45–50% of the feed usage on a farrow-to-finish operation, so decisions to change or modify finishing diets must be made based on economics. Finishing pigs are more subjected to changes which affect feed intake, therefore, feeding programs which include summer vs winter diets, and (or) split-sex feeding can be economically justified.

Composition of some finisher rations as suggested by Professor S.K. Ranjhan:

Table 14.5 Finisher Rations

Ingredient		Ingredient	
Maize	40.0 kg	Maize	20.0 kg
Ground nut cake	12.0 kg	Wheat bran	20.0 kg
Wheat bran	30.0 kg	Groundnut cake	12.0 kg
Til cake	10.0 kg	Milo	20.0 kg
Fish meal	5.5 kg	Rice polish	10.0 kg
Mineral mixture	2.5 kg	Til cake	10.0 kg
Salt	0.5 kg	Fish meal	2.5 kg
		Mineral mix	5.5 kg
		Rovimix A+B_2+D_3	10 g

14.3.3 Gestation ration

An excellent, well balanced diet is very important during gestation. Gilts have greater requirements than mature sows because their diet will have to take care of their growth as well as that of the developing foetus. During gestation, the recommended feeding method for gilts and sows is a limited feeding program. However, it should be emphasized that a limited feeding program is limiting only the energy intake and not other nutrients, such as protein, minerals, and vitamins. The energy is limited in order to keep sows from becoming too fat. Excessive feeding of gilts and sows leads to increased costs and interferes with the potential to maximize reproductive efficiency.

Sows that are overfed immediately after breeding or throughout gestation, often suffer from high embryonic mortality and producing smaller litters than sows fed proper amounts. Sows that become too fat have a tendency to have more farrowing difficulties and crush more pigs. This is especially true during the summer, when the sows are subject to heat stress.

Diets for the pregnant female must meet her daily requirements for all essential nutrients. During normal (spring/fall) weather conditions, about 6000 kcal of metabolizable energy per head per day will keep sows in good condition. However, this energy intake may need to be adjusted up or down depending on the condition of the sow and as the weather changes. This is usually accomplished by increasing or decreasing the amount of feed given to the sows daily.

For sows and gilts in confinement, under ideal environmental conditions, 5000 kcal of metabolizable energy per head per day may be sufficient. During the winter, the sow should have about 7500 kcal metabolizable energy per head per day. Sow condition is a critical indicator of performance, thus high producing sows may require higher feeding rates to maintain adequate body condition.

The daily allowance for protein is 250 g, lysine 9g Ca 16 g and P 14.5 g. This allowance can be met by feeding 2 kg of 14% crude protein diet per day. During the summer, feed intake may be reduced to about 1.6 kg per head per day. In this case, the protein in the diet must be increased to about 16% to meet the 225 g per head per day requirement, assuming amino acid levels are adequate. Feeding levels lower than 2 kg will also require an increase in the levels of minerals and vitamins to maintain proper amounts on a daily basis. Suggested gestation diets are listed in Table 14.6.

The success of limited fed gilts and sows depends upon controlling the intake of each female. Care must be taken to see that each one gets her share. Individual sow feeding stalls are an effective device for controlling boss sows. If sows are group fed, it is imperative that the grain be spread across a larger area to reduce the amount of fighting and to ensure that all animals get the calculated energy requirement.

Interval feeding during gestation is a possible alternative to limit feeding. Interval feeding is accomplished by feeding the sows every other or every third day. Of course, the amount fed is adjusted accordingly. For an example, instead of feeding 2 kg each day during gestation, 4 kg is fed every 2 days. With interval feeding, it is necessary to have sufficient feeder space. Research results have shown that a minimum of 2 to 6 hr out of every 72 hr is an adequate feeding time. Interval feedings is not recommended for gilts.

A ration for gestating females may be relatively coarse in structure. The following composition may be followed:

Table 14.6 Gestation Ration

For preparation of 45 kg ration	
Maize	50%
Groundnut cake	20%
Molasses	5%
Wheat bran	18%
Fish meal	5%
Mineral mixture	1.5%
Salt	0.5%

14.3.4 Farrowing ration

It is considered good practice to feed highly and with bulky feeds from 4-5 days before and after farrowing by substituting wheat bran, oats, ground legume hay or dehydrated lucerne meal. At farrowing, about one-third of the ration may be made up of these bulky feeds.

Adding bulk to the ration at farrowing may help prevent constipation and reduce problems with mastitis-metritis-agalactia (MMA) syndrome. It has to be making sure that there is a good supply of freshwater at farrowing time.

14.3.5 Lactation ration

The feed requirements of the sow during lactation are considerably greater than during gestation. This is because the increase of nutrients required by the sow for milk production is greater than for producing young. Sows during lactation should be full fed in order to obtain maximum milk production. A sow will normally consume 4 to 7 kg per day. This intake will depend upon a diet's composition, sow's condition, previous gestation diet, and environmental temperature of the farrowing facilities. For maximum milk production, it is recommended that the sow be maintained in an environment of 16–22 °C. At higher temperatures, a reduction in feed intake will be evident.

Feed ingredients with a high fiber content such as beet pulp, oats, and wheat bran, may be used as laxatives to keep sows from becoming constipated. However, they also reduce the energy density of the diet and limit sow energy intake. Chemical laxatives, such as magnesium, potassium, or sodium sulfate, may be a preferred method of controlling constipation problems. The recommended level of magnesium sulphate is 5 to 10 kg per tonne or top dressing about 1 to 2 table spoons per feeding. Suggested lactation diets are listed in Table 14.7.

In smaller swine operations, it may not be practical to use two different diets for the sow herd. Therefore, the lactation diet, if properly formulated, can be fed during gestation at the rate of 2 to 3 kg per sow per day. Feed cost will be higher if the lactation diet is fed during gestation.

Table 14.7 Lactation Ration

For preparation of 45 kg ration	
Corn	33.25 kg
Soybean meal, 48%	9.15 kg
Fat	1.05 kg
Deflourinated phosphate	1.15 kg
Salt	0.23 kg
Vit-TM mix	0.15 kg
Lysine HCl	0.02 kg

14.3.6 Feeding replacement stock

Replacement gilts are either commonly reared in a grower finisher facility along with market hogs, or purchased at market weight from a seed stock supplier. It is becoming more common to purchase replacement gilts at 18 to 27 kg and then to isolate and allow them to acclimate to herd conditions for several months. By five to six months of age, gilts should be introduced into a gilt replacement pool and fed a diet with a higher nutrient content (Table 14.8). During the following two to three month period, they should be acclimated to the different housing conditions, exposed to sow herd diseases, have fence line or direct contact with boars, and be monitored daily for estrus activity. Gilts should not be bred until their second or third estrus.

During the period prior to breeding, two feeding strategies have emerged in the United States for replacement gilts.

The first strategy, used largely with maternal lines that are genetically lean, is to increase the gilts; body fat content during the pre-breeding period. By feeding a lower protein diet, the rate of muscle growth will be slightly reduced, but there will be an increasing body fat content. Producers with 'high-lean' gilts use this method because of the importance of body fat on later lactation and rebreeding performance.

The second strategy, used largely with gilts with less lean potential, invokes the feeding of a lower quantity of a diet during the pre-breeding period. This method results in a lower body weight at breeding. Moderate increases in body fatness occur prior to the gilts initial breeding while maintaining maximum lean. This strategy is used for gilts that have a lower mature body weight than the higher producing genotypes. If feed is not restricted, many of these gilt lines often get too fat and heavy, which later results in poorer lactation feed intake, less milk production, and lower litter weaning weights. Vitamins and minerals should be formulated at a higher concentration in order to meet the daily quantitative needs of the replacement gilt.

When gilts of either strategy are thin at breeding the provision of a high quantity of feed for 11 to 14 days pre-breeding (*i.e.* flushing) is recommended. Flushing will result in an increased ovulation rate and litter size.

The decision of when to breed gilts is unfortunately often based on the need to fill farrowing groups. Research has shown that genotypes with a high lean and/ or high producing capacity should be bred at heavier body weights than industry average gilts. Breeding gilts after they attain physical and body compositional maturity will help to ensure that they will have sufficient body nutrient stores to meet the metabolic challenges of reproduction. Back fat thickness at the 10th rib may be slightly less than that collected at the last rib.

14.3.7 Feeding of boars

Boars can be fed grain soybean meal diet fortified similarly to a gestation diet. The daily feeding rate has to be changed to reflect differences of season, condition, and workload of the boar. Boars under heavy use should be fed 6 pounds per head per day.

Muscle is the largest body component in the growing boar, growth of muscle mass generally parallels the growth of the whole animals. Improvements in daily gain, feed efficiency, loin eye area and lower back fats occur when boars are provided the dietary protein concentration that meets their requirements.

Although bone structure is an inherited trait, adequate dietary macro and micro mineral concentration must be provided for proper bone development. Boars have bones of a greater length and diameter than either gilts or barrows. An increased dietary calcium and phosphorus concentration is therefore necessary for the higher bone mass and bone mineral content both of which are essential in withstanding breeding stresses on the legs. An adequate dietary vitamin in D concentration will ensure optimum absorption and utilization of calcium and phosphorus. However, an excessive amount of vitamin D may cause calcification of connective tissue and decalcification of bone tissue.

When feed is restricted to the boar prior to puberty, both growth rate and sexual maturity will be delayed without permanently damaging the testes. The seminiferous tubules in the tastes, which are the origin of the sperm cells, however, will be reduced in diameter and size.

14.3.8 Flushing

It is the method of feeding sows and gilts before breeding. A good grower ration fed to sows and gilts seven to ten days before breeding helps in increasing ovulation rates in them. After breeding sows and gilts should be fed a limited but well balanced ration until the last six weeks of pregnancy and then full feeding should be resumed.

Table 14.8 Nutrient Recommendations for Gestation (as fed basis)

Item	Parity 1		Parity 2 and Later	
	Industry average	High Producing	Industry average	High producing
Energy, Mcal ME/kg	3	3	3	3
Protein %	14	15	12	13
Amino acids[a]				
Lysine %	0.65	0.75	0.55	0.60
Tryptophan %	0.10	0.11	0.08	0.09
Threonine %	0.42	0.48	0.31	0.36
Methionine + cystine %	0.39	0.45	0.32	0.35
Macro minerals[b]				
Calcium %	0.90	0.90	0.90	1.00
Phosphorus (total) %	0.70	0.70	0.70	0.80
Phosphorus (available) %	0.42	0.42	0.42	0.45
Sodium %	0.20	0.20	0.20	0.20
Chloride %	0.16	0.16	0.16	0.16
Salt %	0.50	0.50	0.50	0.50
Trace minerals[c]				
Copper, ppm	15	15	15	15
Iodine, ppm	0.15	0.15	0.15	0.15
Iron ppm	100	100	100	100
Manganese ppm	10	10	10	10
Selenium ppm	0.3	0.3	0.3	0.3
Zinc ppm	150	150	150	150
Vitamins[c]				
Vitamin A, IU/kg	4000	4000	4000	4000
Vitamin D, IU/kg	400	400	400	400
Vitamin E, IU/kg	60	60	60	60
Vitamin K, mg/kg	1.00	1.00	1.00	1.00
Riboflavin mg/kg	4	4	4	4
Pantothenic acid mg/kg	16	16	16	16
Niacin mg/kg	12	12	12	12
Vitamin B12, mg/kg	16	16	16	16
Biotin mg/kg	0.20	0.20	0.20	0.20
Choline mg/lb	0.50	0.50	0.50	0.125

a. Total amino acid recommendations reflect a diet composed of a corn-soybean meal.

b. Values reflect total dietary concentrations unless noted otherwise.

c. Values reflect the supplemental level to be added to the diet.

Source: Tri-state Swine Nutrition Guide Bulletin 869–98, The Ohio State University.

Table 14.9 Nutrient Recommendations for Lactation (as fed basis)

Item	Parity 1		Parity 2 and Later	
	Industry average	High producing	Industry average	High producing
Energy, Mcal ME/kg	3	3	3	3
Protein %	15	18	14	16
Amino acids (total) [a]				
Lysine %	0.75	0.90	0.70	0.80
Tryptophan %	0.15	0.18	0.03	0.15
Threonine %	0.50	0.55	0.47	0.53
Methionine + cystine %	0.45	0.47	0.40	0.45
Valine %	0.75	0.90	0.70	0.80
Macro minerals[b]				
Calcium %	0.90	1.00	0.90	1.00
Phosphorus (total) %	0.70	0.80	0.70	0.80
Phosphorus (available) %	0.42	0.45	0.42	0.45
Sodium %	0.20	0.20	0.20	0.20
Chloride %	0.16	0.16	0.16	0.16
Salt %	0.50	0.50	0.50	0.50
Trace minerals[c]				
Copper, ppm	15	15	15	15
Iodine, ppm	0.15	0.15	0.15	0.15
Iron ppm	100	100	100	100
Manganese ppm	10	10	10	10
Selenium ppm	0.3	0.3	0.3	0.3
Zinc ppm	150	150	150	150
Vitamins[c]				
Vitamin A, IU/kg	4000	4000	4000	4000
Vitamin D, IU/kg	400	400	400	400
Vitamin E, IU/kg	60	60	60	60
Vitamin K, mg/kg	1.00	1.00	1.00	1.00
Riboflavin mg/kg	4	4	4	4
Pantothenic acid mg/kg	16	16	16	16
Niacin mg/kg	12	12	12	12
Vitamin B12, mg/kg	16	16	16	16
Biotin mg/kg	0.20	0.20	0.20	0.20
Choline mgkg	0.50	0.50	0.50	0.50
Folic acid, mg/kg	1.50	1.50	1.50	1.50

a. Total amino acid recommendations reflect a diet composed of a corn-soybean meal.
b. Values reflect total dietary concentrations unless noted otherwise.
c. Values reflect the supplemental level to be added to the diet.
Source: Tri-state Swine Nutrition Guide Bulletin 869–98, The Ohio State University.

Table 14.10 Modified Nutrient Recommendations for Replacement Gilt Development

Item	Weight range (kg)			
Macro-minerals[a]	25–50	50–75	75–100	100–125
Calcium (total) %	0.85	0.80	0.75	0.75
Phosphorus (total) %	0.75	0.70	0.65	0.65
Phosphorus (available) %	0.49	0.45	0.40	0.40
Trace minerals[b]				

Table 14.10 (*Contd...*)

Item	Weight range (kg)			
Macro-minerals[a]	25–50	50–75	75–100	100–125
Copper ppm	15	15	15	15
Zinc ppm	150	150	150	150
Selenium ppm	0.3	0.3	0.3	0.3
Vitamin E IU/kg[c]	40	40	40	40

a. These nutrients are considered as modifications for replacement gilts. Other nutrient requirements are similar to those in tables.

b. Values are total dietary levels unless denoted otherwise.

c. Values are supplemental levels.

Source: Tri-state Swine Nutrition Guide Bulletin 869–98, The Ohio State University.

Table 14.11 Dietary Nutrient Recommendations for Replacement Gilts

Item	High producing >120 kg	Industry average > 120 kg
Protein %[a]	13 to 14	14 to 16
Amino acids[a]		
Lysine %	0.7	0.80
Lysine g/day	19.10	18.00
Tryptophan %	0.13	0.12
Threonine %	0.46	0.50
Methionine + cystine %	0.42	0.46
Macro minerals[a]		
Calcium %	0.75	0.75
Phosphorus (total) %	0.65	0.65
Phosphorus (available) %	0.40	0.40
Sodium %	0.20	0.20
Chloride %	0.16	0.16
Salt %	0.50	0.50
Trace minerals[b]		
Copper pm	15	15
Iron ppm	100	100
Zinc ppm	150	150
Manganese ppm	10	10
Iodine pm	0.15	0.15
Selenium ppm	0.30	0.30
Vitamins[b]		
Vitamin A, IU/kg	5000	5000
Vitamin D3, IU/kg	500	500
Vitamin E, IU/kg	60	60
Vitamin K, mg/kg	1	1
Riboflavin mg/kg	4	4
Pantothenic acid mg/kg	15	15
Niacin mg/kg	12	12
Vitamin B12, mg/kg	16	16
Biotin mg/kg	200	200
Choline mg/kg	350	350
Folic acid, mg/kg	1.50	1.50

a. Values are total dietary levels unless denoted otherwise

b. Values are supplemental levels

Source: Tri-state Swine Nutrition Guide Bulletin 869–98, The Ohio State University

Table 14.12 Nutrient Recommendations for Boars (as fed basis)

Item	Development phase			
	Early[a]	Middle[a]	Late[b]	Mature[b]
Body weight kg	25–60	60–100	100–150	150–300
Protein %	22	20	18	16
Amino acids[c]				
Lysine %	1.2	1.1	1.0	0.85
Tryptophan %	0.24	0.22	0.19	0.17
Methionine + cystine %	0.72	0.66	0.63	0.54
Macro minerals[c]				
Calcium %	0.95	0.85	0.80	0.90
Phosphorus (total) %	0.75	0.65	0.75	0.80
Phosphorus (available) %	0.75	0.65	0.75	0.80
Sodium %	0.12	0.12	0.20	0.20
Chloride %	0.08	0.08	0.16	0.16
Salt %	0.25	0.25	0.50	0.50
Trace minerals[d]				
Copper pm	15	15	15	25
Iron ppm	100	75	75	100
Zinc ppm	150	100	100	150
Manganese ppm	10	10	10	20
Iodine pm	0.15	0.15	0.15	0.15
Selenium ppm	0.3	0.3	0.3	0.3
Vitamins[d]				
Vitamin A, IU/kg	5000	4000	4000	5000
Vitamin D, IU/kg	500	400	400	500
Vitamin E, IU/kg	60	60	60	60
Vitamin K, mg/kg	1.2	1.2	1.2	2.0
Riboflavin mg/kg	12	10	10	12
Pantothenic acid mg/kg	20	15	15	20
Niacin mg/kg	30	25	25	30
Vitamin B_{12}, mg/kg	30	30	30	40
Biotin mg/kg	0.20	0.20	0.20	0.20
Choline mg/kg	0.50	0.50	0.50	0.125

a. Assumes *ad libitum* feeding.

b. Assumes limit feeding.

c. Values reflect total dietary concentrations unless noted otherwise.

d. Values reflect the supplemental level to be added to the diet.

Source: Tri-state Swine Nutrition Guide Bulletin 869–98, The Ohio State University.

CHAPTER 15

HOUSING OF PIGS

15.0 Housing of Pigs

15.1 Housing practises in India

In most of the developing world, pigs are raised by the farmers in their backyard like poultry. Up to 2–3 sows are generally kept for their own requirement and to meet out the part of the produce for neighbors.

In the recent past, the government has initiated various poverty alleviation programme for livestock development in which piggery development has been a focus of attention. In this case, BPL families are given various incentives to enhance their income through piggery production. One of the first programs launched by the government was to provide 5 sows and one boar free of cost and Rs 1000 for their housing. Under the programme, floor area of 50 sq. ft was laid in brick with a manger and water trough covered by thatch roof on bamboo and/or wrought iron poles to give protection from hot sun/rain and winter cold. This has led to income generation to the farmers, which has helped them to increase the number sows up to 10 and two boars and 15 sows and 3 boars.

In late sixty's government also developed piggery development programme on the Danish Model wherein bacon factory was established along with a large pig farm of 300 to 500 exotic sows on scientific lines to produce the raw materials for the factory and to provide exotic males for cross breeding to small and marginal farmers. The crossbreds so produced, could be processed and farmers would get reasonable price for the produce. In some of the bacon factories, in addition, healthcare and feed facilities were also provided to the farmers so that integrated piggery development could take place. The project enabled the farmers and other entrepreneurs to setup piggery farm and enterprises of 50 sow and even larger units.

Among the resource poor farmers, pig keeping is a major livelihood option as the pigs survive and produce on kitchen waste and scavenging for food in the by-lanes of the neighborhood. The housing requirement under these systems is minimal and generally back yard is having a thatch roof which is invariably an extension of the dowelling unit.

These pig farmers generally construct their pig sty with locally available materials like bamboo and woods (as they are cheap), located in road side slope area with a raised platform above 2–3 feet from the ground (to make them reptile, rat or small wild predator proof, to make cleaning easy and to prevent dampening of floor due to rain.

The floor space per adult was usually found to be inadequate (average 12 sq ft) in majority of the farms. The farm equipments which are used in housing included mainly iron vessel (Kerahi) for boiling feeds, empty mustard oil tin (modified form) or cut piece of wood or bamboo, tyres as feeding trough. Further it was recorded that supply of water mostly dependent to share with household nearby streams. Separate water storage facility for pigs and electricity were absent in most of the farms.

15.1.1 Basic principles of pig housing for commercial pig units

The improvement of housing has not kept pace with developments in the field of swine nutrition and breeding in this country which is largely due to socio-economic condition of the class of people involved in pig keeping. Accommodation for pigs and equipments used in the housing complexes are chosen so as to suit the type of management system adopted. However, there are certain similar principles and practises in most systems. These originate from the fact that most pig units will contain pigs of different ages and classes which need different types of accommodation.

General considerations

Housing requirement for pigs vary with its category. A breeder may, therefore, plan to have houses for weaner, grower, pregnant, lactating and dry sows and boars. The houses should be so arranged that shifting the animals from one house to the other becomes easy. Like next to weaner house, grower, boar, pregnant and farrowing houses should be constructed so that the animals could be shifted in a rotational manner.

Two basic considerations in providing proper housing to pigs are needs of pigs and needs of the pig farmer. Pig requires fresh air, protection from weather, and scope for free movement and exercise. Both the habits and characteristics of the pig provide clues to basic needs for pig housing as given below:

(i) Pigs being hairless have less protective mechanism against heat and cold, and so are highly susceptible to sudden changes of temperature and extremes of environmental variables. Prolonged exposure to chill, cold winds, damp cold can cause rheumatism and unthriftiness.

(ii) Pigs have poor development of sweat glands and so they find it difficult to keep cool in hot weather, which is prevalent in most parts of this country and so shade is needed.

(iii) Pigs are, by nature, clean animals and therefore generally does not urinate and dung in sleeping place. When pigs are seen dirty, it is primarily due to faulty system of management and housing or carelessness on part of the pig keeper in charge of he pigs.

(iv) In natural conditions, the pig obtains much of its food from rooting in ground for which it is endowed with strong jaws and powerful snout. It is therefore imperative that any enclosure or building structure should be soundly constructed and gaps avoided so that pigs may not be able to apply any leverage and cause damage to the structure.

The requirements of the farmer are primarily determined by his capacity of investment and the profits likely to be obtained.

Two systems of housing are generally built depending depending on climatic conditions and topography. In temperate climate, closed housing is required. In tropics loose housing which is also called open housing, is recommended.

In close housing system the climatic requirements are described below.

Climatic requirements

Pigs will grow most economically and maintain the best health only if the climatic conditions in their house are favourable for production. The values that can be recorded on the farm without difficulty include temperature, humidity and wind velocity. It may be possible on occasions to measure the air change or ventilation rate in a building well. The important principles to be kept in view for providing physical requirement of pigs are as follows:

(a) Temperature

Heat generated within the piggery will vary with the number of pigs. Loads will occur in the ventilation and through the structure. There is no doubt that temperature data are of the most immediate practical use. For this, a continuous recording instrument, such as a thermograph, is of most use to the farmer. This is essentially a bimetallic strip that contracts and expands in direct relation to the air temperature. To this is geared a pen which leaves an inked continuous record on a revolving chart worked by clock which may revolve once every 24 hr or preferably every 7 days. At the end of this period the, a new chart replaces the completed one and the pen is refilled with special slow drying ink. That is almost all the attention it needs apart from an occasional recalibration of the thermograph against an ordinary mercury thermometer with the National Physical Laboratory mark on it.

The sow require minimum of 20 °C whereas once her feed intake is increased to 4–5kg per day a minimum of 15–16 °C will be adequate.

Piglet needs a temperature of 30–33 °C for the first 4 hr. The temperature requirement reduces rapidly as the piglet grows so that by two weeks of age it will be comfortable at 24–25 °C.

Good control of both the farrowing house and creep area temperature will help improve piglet survival leading to higher weaning weight and reduced energy costs.

(b) Humidity

Some investigators found that a warm day environment was preferable to cold damp one. Humidity, however, has little bearing on well being within the range of desirable temperature.

(c) Light

For providing natural lighting for the pig house/farm, windows along the sides and ends are required. A common rule is to provide 1 sq.ft window space for each 20–30 sq.ft. floor space. In order to admit natural light to both sides of the house during the day, windows are placed in the side walls and the long axis of the pig house is usually placed north and south.

The importance of the length of the lighting period for breeding females was studied. The result showed that gilts given an 18 hr light period per day exhibited a stronger, longer and more regular oestrus than gilts exposed to a 6 hr lighting period. It was also found that gilts exposed to the longer periods produced 0.8 to 2.7 more piglets in the first litter than the controls.

Where the sow is concerned, she does not appear to be affected by daylight length as do some other breeding animals, though it has been suggested that natural day light may be an important factor in the breeding cycle patterns of sows and gilts confined in stalls.

The effect of temperature

Pig has a better mechanism for retaining heat, especially due to well developed subcutaneous fat cover. The pig possesses sweat glands only on the snout and it is unable to dissipate large amount of heat by sweating.

At lower temperatures the pig requires to divert food energy to increase heat production in order to maintain body temperature. The lower critical temperature will vary between pigs according to a number of factors, for instance (i) how fat or thin the pig is, (ii) how much food it is eating and therefore how much fat it is growing, (iii) whether it has bedding to help prevent heat loss, (iv) whether it can huddle with pen mates and (v) whether it can make postural changes to minimize heat losses. Eventually, with decreasing ambient temperature, the pig can no longer maintain its body temperatures in spite of high heat production and hypothermic condition can arise.

When environmental temperature approaches body temperature, the pig will attempt to increase evaporative heat loss by sweating (through its limited sweat glands) making postural and positional changes and wallowing in water and mud. In addition, it will reduce its energy output of the feed. A small concrete platform or step near the water bowl will enable the young pigs to reach the water. All the bowls are fixed with the lip 18 cm above the floor level.

15.2 Insulation System

In any building which maintains a temperature higher than that of outside, there will be a transfer of heat from inside to outside. The reverse will be the case when the outside temperature is higher than that of inside. The materials which comprise the walls and roof will offer some resistance to this transfer, but will not entirely prevent such heat movement. The purpose of thermal insulation is to reduce heat transfer. Choice, however, should always be made with knowledge of the insulation value of the composite construction. Walls should have a 'U' value not exceeding 0.33 and for roofs a 'U' value not exceeding 0.1.

Insulation of the roof, walls and floor is necessary in order to conserve the heat produced by the pigs' body within the building. It has been estimated that 10 pigs of 90 kg live weight will produce as much heat as 2 kwt electric fire.

It is generally agreed that a constant temperature of between 16–21 °C with a day atmosphere is preferable for fattening pigs.

15.2.1 Features of insulation

The principal features required for all pig accommodation is to provide the correct environmental conditions as cheaply and economically as possible.

The control of climatic conditions is completely dependent on three factors.

(i) The standard of insulation
(ii) The standard of ventilation and measure of control
(iii) The number of pigs housed.

It is well known that very hot weather has an adverse effect on pigs. Growth rate can suffer as markedly under very hot conditions as under cold.

The most interesting fact that emerges is that it is the roof through which most heat is lost, and indeed, this loss, together with that through ventilation, accounts for 80% of the total heat loss. It should be noted that this does not include heat loss by conduction from the pig to any surface with which it has contact.

Good thermal insulation not only serves to retain the heat in winter, it also keeps the building cool in summer. It helps to prevent condensation and dampness, keeps any heating costs down and enables the farmer to maintain uniform and near constant conditions in the house. The effects on stock are economically vital by helping to maintain an optimum environment, food costs are kept to a minimum and growth and good health are promoted.

Before dealing with the strictly practical aspects of insulation, we should have some knowledge of the way one can assess the respective insulation values of different materials or forms of construction. This will help us considerably in choosing our material.

First of all, attached to every building material has a thermal conductivity or 'K' value. This figure is the measure of a material's ability to conduct heat. It is the amount of heat in watts through a sq.mt of the material when a temperature difference of 1 °C is maintained between opposite surface of a metre thickness.

In this way one can grade different materials according to their insulating qualities and it goes some way to answer the question as to which are the best insulation.

In fact, 'K' values are of limited use because surfaces of pig houses are generally composite structures. *e.g.*, an insulated roof might consist of an outer cladding of corrugated asbestos sheets and an inner lining of mineral wool and fiber board and also an air space. What we really want to know is the rate of heat loss (or heat gain during very hot summer weather) through the whole structure rather than just the individual materials.

This value takes us much beyond the 'K' value and is known as the 'U' value. By definition, this is the amount of heat in watts that is transmitted through one square metre of the construction from the air inside to the air outside when there is a 1 °C difference in temperature between inside and outside. It is possible to build up the 'U' value of a complete wall or roof structure if one has the 'K' values of the individual materials (plus 1 or 2 other figures). It is economical in most piggeries now to aim to have a 'U' value of the roof of 0.40 or less. For the walls, a figure between 1.0 and 1.5 is acceptable; with the floor, however, we should aim to be as near as 0.40.

Insulating materials can be divided into three broad classifications.

(i) Rigid materials capable of resisting structural forces, e.g. no fine concrete blocks, building blocks constructed from foamed slag, clinker, pumice, etc.
(ii) Board materials, *e.g.* asbestos insulating board, compressed straw board, insulated fibre board, rigid glass wool boards, wood slabs etc. Some of these may only be used as structural members for roof coverings or as panel infilling in frame construction
(iii) Flexible materials and loose fill *e.g.* granulated cork, glass wool (loose or quilt form), exfoliated vermiculate.

Materials from the farm that can be used, where permanency is not important, are straw chaff, flax, chives etc. Flexible materials need to be supported and are, therefore, often draped over joists or bearers, or can, as is necessary, with loose fill materials, supported by being laid over a board lining or ceiling.

The total area of walls and roof per pig plays a large part in determining climatic conditions. *e.g.*, should a construction be chosen with the best (lowest) 'U' value, its effectiveness will be lost if the building is larger than needed, and conversely a moderate 'U' value can give reasonable results if the building is reduced to a minimum in size.

(a) Roof insulation

Most building materials are porous to air and this is equally true of most insulating materials. The amount of water vapour that the air can carry related to the temperature of the air, the warmer the air the more water vapour it can carry. Consider then the action that takes place in a piggery where the insulation materials in the roof are porous. The air in the piggery is warm and therefore carries a quantity of water vapour. This slowly percolates through the insulation and eventually reaches the asbestos sheet covering the roof. As this sheet will probably of the same temperature as the air outside, the moisture-laden air is cooled and so cannot carry as much water vapour, which is deposited as water on the underside of the asbestos. If this process continues, the insulation becomes wet and immediately losses its properties as an insulator and internally rot and fungi are encouraged to grow, thereby leading to rapid deterioration. Such a process is the cause of buckled, stained and wet board linings so commonly seen on piggery ceilings. This can be prevented by the incorporation of vapour barriers in all insulation work. Where insulation fibre board or ordinary hardboard is used as internal linings two coats of oil paint on the piggery side of the board will give useful barriers, but under high humidity conditions it would only be efficient for a short period of time and would require repairing to maintain efficiency. The use of fully 'compressed' flat asbestos sheet as a lining to hold up the insulating material and at the same time provides an efficient barrier. It does not require any maintenance.

Further, alternative vapour barriers are polyvinyl sheet laid between the supporting boards and the insulating material, or bitumen backed aluminum foil. The later material, having a highly polished surface, can also reflect radiant heat back into the piggery. It can only do so, however, when its polished surface is not in direct contract with other materials. In other words, it must be used with shiny surface next to an air cavity. As all the vapour barriers must be placed on the warm side of the insulating materials, the position of aluminum foil in roof construction is usually the immediate lining on the piggery side of the roof. This can make the roof construction difficult and the more use of foil in walls constructed of timber framing where a cavity is readily formed. It should be noted that foil is just as efficient in reflecting radiant heat from the pigs in summer as in winter. In winter this heat gain is welcome, but in hot summer it could be an additional embarrassment in keeping temperatures down and may necessitate increased ventilation.

(b) Floor insulation

Floor insulation is essential to prevent continued loss of heat from the pig into the ground when it is lying down. Insulating the floor will bring its surface temperature within a few degrees of air temperature and although some heat transfer from the pig will still take place, the insulation will allow fairly rapid heating and thereafter heat loss will almost cease. On all sites, a damp proof layer, immediately underneath the insulation, is recommended and on wet sites it is essential. In floors constructed of concrete or porous materials, it is best placed between the ground and the concrete.

The damp proof layer can be chosen from several materials, such as two coats of hot tar or bitumen, bituminous felt or 500 gauge polyvinyl sheeting, which can be obtained in long lengths and various widths. The wooden flat finish is best.

In connection with thermal insulation, it is worth stressing the merits of reducing the air space in house to reasonable proportions.

15.3 Ventilation System

Ventilation is the renewal of foul, moisture laden air and replacing it with clean fresh air. There is no doubt that pigs do better when they are housed in comfortable airy conditions, kept at the correct temperature.

A relative humidity of 70% will provide an atmosphere which feels dry and will prevent condensation in a well insulated building.

There are three main systems of ventilation: (i) natural ventilation (ii) forced ventilation and (iii) pressurized ventilation.

15.3.1 Natural ventilation

In this method air is extracted through a chimney type construction fixed in the roof apex and to allow fresh air into the building through hopper type windows or baffled inlets.

15.3.1.1 *Air outlet*

The outlet area should be approximately 32 cm per 45 kg live weight or 64 cm for every 100 kg bacon pigs and extraction of 100×64 cm^2 air may be needed. This would be achieved with a ventilation shaft measuring approximate by 80×80 cm.

15.3.1.2 *Air inlet*

The air inlet should be about three times the outlet area, thus we reckon approximately 100 cm^2 per 45 kg pig or 200 cm^2 per baconer. The inlet should be fixed in the side walls, at least 1 m above floor level and not less than 0.3 m below the eaves.

The disadvantages of natural ventilation are that, the system cannot be controlled automatically and therefore labour must be available to alter the inlets according to the outside weather conditions.

15.3.1.3 *Forced ventilation*

Forced ventilation entails the use of an electrically operated extractor fan fixed in such a position as to draw out foul air without causing drafts in the building. Usually the fans are fixed in or near the dunging passage, in order to extract the foul air from as near the source as possible. Fresh air is drawn in from a roof inlet.

The pig requires a minimum of 0.3–0.28 m^3 of air (m^3/h/kg) per kilogram live weight during the cold winter months, and 0.8–2.02 (m^3/h/kg) live weight in the summer months.

One of the disadvantages of extractor fans is that during the winter months, when only a small air movement is required, the houses may suffer from a drop in temperature if the fans are run fully. To overcome this problem it is necessary to connect the fan to an electric thermostat, which will stop the fan operating, if the temperature drops too low.

Pressurized systems

A recent development in ventilation is to draw fresh air into the piggery through a central position in the ceiling by means of an impeller fan. The air within the building will become pressurized and therefore as the pressure increases, foul air will be forced out through side vents. The main advantage of this system is that incoming draughts are virtually excluded because of the air pressure within the building. The impeller fan is connected to a thermostat so that the inside temperature is easily regulated.

It must be remembered that whichever system is used, the aim should be to provide warm, airy, draught free conditions, with a low humidity. When you have fed your pigs, always check that they are lying down comfortably in the sleeping quarters. Check that there are no draughts or strong smells inside the building.

15.4 Housing System

15.4.1 The Site

The site for setting up a pig unit should be selected keeping in view the topography of the land. It should be at a higher level so that the rain water does not accumulate and there are no chances of water logging in the area which may affect the health of pigs adversely and there may be possibility of worm infection. Further, low lying areas may make the management difficult in rainy season. A well drained site should be chosen for setting up of permanent structure. Pig houses should be simple, open sided structures as maximum ventilation is needed. A building for open confinement is, therefore, essentially a roof supported by poles. The roof supporting poles are placed in the corners of the sties where they will cause least inconvenience. A free span trussed roof design would be an advantage but is more expensive.

In some circumstances it may be preferable to have solid gable ends and one closed side to give protection from wind or low temperatures, at least for part of the year. If such walls are needed, they can often be temporary and be removed during hot weather to allow maximum ventilation. Permanent walls must be provided with large openings to ensure sufficient air circulation in hot weather. If there is not sufficient wind to create a draught in hot weather, ceiling fans can considerably improve the environment.

The following points should be kept in mind while selecting a site for pig housing.

1. The site should provide plenty of fresh air, sunlight and shelter from winds.
2. It should be away from human habitation but not too far away from attendant's quarters.
3. Accommodation for animal is best built in an open, well drained site.
4. The topography should be of higher elevation than the surrounding grounds to offer a good slope for rainwater and drainage of the wastes of the piggery to avoid stagnation within.
5. The site should be such that the structure could be oriented east to west.
6. Availability of cheap labour in the neighbourhood.
7. Availability of medicines and vaccine in the nearby market.
8. Availability of telephone facilities, school for children of workers, post office, bank, shopping centre, cinema hall etc.
9. Cheap availability of feed ingredients.
10. Availability of electricity, drinking water.
11. Available space for expansion of the farm.
12. Farms should be located nearer to town, if possible.
13. Trees acts as wind breakers and natural shades.
14. During erection of a house, care must be taken so that it must provide:
    - (i) Comfort to both animal and labour
    - (ii) Proper sanitation facilities.
    - (iii) Protection to the animal against extreme weather and predators.
    - (iv) House should be durable.
    - (v) Economy of construction and management is also desirable.

15.4.2 Choice of housing system

The principal factors determining the choice of a housing system for pigs are size, permanency of enterprise, type of pig to be produced (breeding stock, weaners and fatteners) and land and crops available. If keeping of pigs is on a large scale, specialized system of housing providing different type of houses for each class of stock is required and if the pig enterprise is on a small scale, or as a sideline, the conversion of an existing building, or construction of a new one of simpler type are more suitable for adaptation to other purposes, if required, may be preferred.

There are mainly 3 types of housing systems:

(i) Open air system, (ii) Indoor system, (iii) Mixed system,

Each system has its advantages and disadvantages. There are many variations of these systems and a pig may spend part of its life in one and part in another.

15.4.2.1 *Open air system*

Wild pigs live amongst bushes and the roots of tress. When pigs are kept with access to a warm, low area to lie and sleep in, as they would in the wild, the pigs do better.

Pigs can be kept in a field where they can feed on grasses and plants. If pigs are kept this way, the field must be surrounded by either a strong fence or a wall. Pigs will push their way out of a field if the fence is not strong enough. The animals are given shelters called pig arks to sleep in (Fig. 15.1). These can be made of wood or metal sheets and should contain bedding. The arks can be moved to fresh ground when necessary.

Fig. 15.1. Shelter for piglets in the field

Open air system is suitable for mild climatic conditions when pig enterprises are located on well drained land. In this system pigs are kept on big open enclosures with small simple building for shelter and sleep. It is suitable for young pigs and breeding stock due to plenty of fresh air, exercise and sunlight to provide good start to young pigs to get strong body frame in preparation for fattening or breeding. The system provides healthy environment and also minimizes risk of anemia in young piglets due to access to minerals in soil. Temporary buildings on farm and portable building can also be provided.

A growing interest has been shown in alternative pig production systems because of the low capital cost of outdoor systems, which varies from 40 to 70% of the cost for conventional indoor systems (Thornton, 1988). Concerns for animal welfare and awareness of niche marketing opportunities have increased interest in the production of free-range animals (McGlone, 2001). Outdoor housing on pasture or dirt pens accounts for less than 5% of the pigs finished in the United States; an additional 9% are housed in an open building with outside access (NAHMS, 2001). Success of outdoor pig finishing systems may depend on the details of the housing design, management, and location, including soil type and climatic conditions (Edwards and Turner, 1999).

15.4.2.2 *Indoor system*

For large scale pig enterprise and in extreme climatic conditions indoor housing is necessary. Yarding is suitable for all type of pigs, though more common with fatteners, so that they can be kept together in larger numbers than breeding stock. Permanent, specialized fattening house is considered essential for pig enterprisers who undertake fattening throughout the year. Choice of such housing will be influenced by relative cost, climate of the area and amount of straw available for bedding. Farrowing houses are always required at a pig breeding farm or breeding unit of a medium or large sized building. This consists normally of a series of pens arranged along a feeding passage and equipped with guard rails and creep for young pigs.

Because of the cost of a concrete floor, there is a tendency to reduce the floor area allowed per animal. However, too high stocking densities will contribute to retarding performance, increasing mortality, health and fertility problems and a high frequency of abnormal behaviour thus endangering the welfare of the animals. Increasing the stocking density must be accompanied by an increased standard of management and efficiency of ventilation and cooling. In particular, to aid in cooling, finishing pigs kept in a warm tropical climate should be allowed more space in their resting area than is normally recommended for pigs in temperate climates. Tables 15.1 and 15.2 lists the recommended space allowance per animal at various stocking densities.

15.4.2.3 *Mixed system*

This is more common system which comprises both outdoor and indoor systems In this system pigs are kept in the open for some time when the climatic condition is favourable and in the closed area during night and during unfavourable climate. In this system the pigs may be kept in a small group.

The figures listed for high stocking density should only be used in design of pig units in cool areas and where the management level is expected to be above average. The dimensions of a pen for fattening pigs are largely given by the minimum trough length required per pig at the end of the pen. However, the width of a pen with low stocking density can be larger than the required trough length.

Furthermore, the flexibility in the use of the pen will increase and the extra trough space allows additional animals to be accommodated temporarily or when the level of management improves. Sometimes finishing pens are deliberately overstocked. The motive for this is that all pigs in the pen will not reach marketable weight at the same time and the space left by those pigs sent for slaughter can be utilized by the remainder. Such over-stocking should only be practiced in very well managed finishing units.

15.4.3 Design, layout and management of buildings

The design of buildings should adhere to the basic dimensions to ensure optimum ventilation regulation. The following factors should also be kept in mind:

- Use of economical materials;
- Use of good quality concrete;
- Applying damp-proofing to the floors and insulate the floors with no-fines concrete, especially in wet areas.
- Insulating the roof where high temperatures can be expected.
- Buildings must be spaced at least 18m apart to ensure effective air movement between the buildings and also to combat the spread of disease;
- There should be no obstructions in the way of warm winds;
- If the land falls in the direction of prevailing warm winds, smaller spaces between the buildings may be considered;
- Obstruction to cold wind, however, are advisable.

15.4.3.1 *Constructional details of the sty*

Generally, farmers prefer to construct the sties under trees to provide shade to pigs during hot season. The pig farmers select a sloppy area for constructing pig sty so that the excreta is directly dropped from the sty to the sloppy ground and carried downwards away from the sty. Pigs can be kept alone or in small groups in a pig sty, a concrete or solid floored pen with a low shelter. When building a sty we should choose an area which is never flooded in the rainy season. It should not be too near to houses so that smells and flies which become a nuisance are avoided. The floor should be concrete and sloping away from the sleeping area so that urine flows out and away. The concrete floor should be laid on a good foundation and will need to be 5–6 cm thick. If the concrete is too thin and cracks, the pigs will soon start to dig it up. An earthen floor cannot be kept clean and will lead to problems with parasites and other diseases. The walls of the sty need to be fairly smooth so that they can be kept clean. Cracks in the walls will allow dirt and germs to accumulate.

The animals should be given plenty of bedding in the shelter. Pigs will always dung away from their sleeping and feeding areas. The dung can be removed every day allowing the pen to be kept clean and avoiding the build up of waste and smells.

Floor

The concrete floor should be laid on a good foundation and will need to be 5–6 cm thick. If the concrete is too thin and cracked, the pigs will soon start to dig it up. For all types of confinement housing a properly constructed easily cleaned concrete floor is required. 80 to 100 mm of concrete on a consolidated gravel base is sufficient to provide a good floor. A stiff mix of 1:2:4 or 1:3:5 concrete finished with a wood float will give a durable non-slip floor. The sty floors should slope 2 to 3% toward the manure alley and the floor in the manure alley 3 to 5% towards the drains. However, some part of the floor may be left bare so as to permit rooting. The floor may be of wood, concrete, bricks or slabs. Generally, the floor of sty is made up of wooden planks with a gap of 1–2 inch in between them so that the excreta directly fall down on the ground and not accumulated under sty preventing health hazard to pigs. Usually, the floor of the sty in front side is kept at least one foot high from the ground so that feeding and other management is easy.

Tables 15.1 and 15.2 indicate the floor space requirement of different categories of pigs.

Table 15.1 Floor Space Requirement for Different Categories of Pigs

Sl. No.	Class of animals	Covered area (Sq feet)*	Open area (Sq feet)
1.	Weaner	10–15	15–20
2.	Grower	12–20	20–30
3.	Boar	35–50	50–70
4.	Lactating sow	70–100	70–100
5.	Dry sow	20–30	30–50

One sq meter = 10.76 sq feet *

Table 15.2 Floor Space Requirement as per ISI Standards

Sl. no	Type of animal	Floor space requirement (Sq. Mt. per animal)		Maximum number of animals per pen
		Covered area	Open paddock	
1.	Boar	6.0–7.0	8.8–12.0	Individual pens
2.	Farrowing pen	7.0–9.0	8.8–12.0	Individual pens
3.	Fattener (3–5 m old)	0.9–1.2	0.9–01.2	30
4.	Fattener (above 5 m old)	1.3–1.8	1.3–01.8	30
5.	Dry sow/gilt	1.8–2.7	1.4–01.8	3–10

Roof

Roof should be waterproof and should not be bad conductor of heat. Keeping this in mind, a roof of thatch is excellent in hot climates, particularly in non-confined systems, but cannot always be used because of fire hazard and because it is attractive to birds and rodents, not durable and may harbour insects. Economically, asbestos sheet can be used. Since it is not suitable in sunny days, gunny bags can be put on this roof and water can be sprinkled over them. Height should be above 10–12 feet.

In climates where a clear sky predominates, a high building of 3 m, or more, under the eaves, gives more efficient shade than a low building. A wide roof overhang is necessary to ensure shade and to protect the animals from rain. If made from a hard material, the roof can be painted white to reduce the intensity of solar radiation. Some materials such as aluminium reflect heat well as long as they are not too oxidized. A layer of thatch (5 cm) attached by wire netting beneath a galvanized steel roof will improve the microclimate in the pens.

Walls

The height of the wall should be 4 feet above the floor. Brick and concrete can be used up to height of 3 ft. from floor and 1 ft. can be made up of wood (or) the railing of G.I. pipe. The walls should be smooth otherwise it may injure the animal.

Wooden or bamboo walls are cheaper, but less durable. In this case, the pillars are made up of wooden logs or cement. A farmer can choose any combination that suits him depending on the requirement and capital availability. In side walls, the bamboo or wood is fixed in such a way that enough gap exists between them to allow sufficient ventilation.

Doors

Doors have to be tight fitting and any other openings in the lower part of the wall surrounding the building should be avoided to exclude rats. Apart from stealing feed and spreading disease, large rats can kill piglets. They should be fitted without any gap to the floor up to the height of 2–3 ft.

Windows

The size of the window should be such that it can provide cross ventilation and sun light to the sties.

Guard rails

It should be made up of galvanized iron pipes (2 inch diameter) which may be fitted about 8–10 inches away from the walls of the farrowing pan in order to prevent crushing of piglets.

Wallowing tank

Pigs are more sensitive to high temperature due to absence of sweat glands and are unable to dissipate excess heat. Hence shades and wallowing tank should be used during hot weather.

Feeding trough

There should not be any wastage of feed so the trough should be made of concrete and with though walls. A trough space of 2.5 feet length for each pig is sufficient for proper feeding without scrambling and fighting. Galvanized sheet feeding troughs are also available in market.

Water supply

The water is required for cleaning and drinking purpose. Wholesome and clean drinking water should be provided to the pig and for this, water trough can be made of concrete or galvanized sheet, but some times the feeding trough can be used for the watering.

A necessary permanent fitting in the piggery is the automatic water bowl. As self-filling bowls are generally used and there is always some spillage, it is most satisfactory to place them in the dunging area; where this is impossible, they should be at least be situated at the lowest point in the pen adjoining the dunging area.

The bowl is best placed well within the passage so the pig has its whole body in the passage when drinking. To satisfy this requirement the bowl may either be placed on the dung passage door, connected by flexible piping, or recessed into the dividing wall between pen and passage way. The bowl lip should be 150 mm above floor level, but where young pigs before weaning are using it is good practice to lace a step up to it. This keeps it cleaner and less likely to be fouled. Allow one water bowl per 10–10 feeders.

The nozzle drinker has recent years achieved a large measure of popularity. The water flows when the pigs depress a valve on the end of a brass nozzle projecting from the wall, gate or pen division. The system is cheap, hygienic and should give little mechanical trouble.

Where restricted water is given rather than ad lib to save physical handling of the water, the bottom rail over the trough is, in effect, a water pipe. This pipe is individually controlled by a valve to each pen and is punctured on the base by a series of small apertures (3 mm dia) at 230 mm centres.

Table 15.3 depicts feeding/watering space requirements for swine

Table 15.3 Feeding/Watering Space Requirement for Swine (ISI standard)

1	2	3	4	5	6	7
Type	**Space/ pig (cm)**	**Total manger in a pen for 100 pigs (cm)**	**Total water through in a open for 100 pigs**	**Width of manger water through (cm)**	**Depth of manger water through (cm)**	**Height of inner wall of manger/water through (cm)**
Adult pigs	60–75	6000–7500	600–700	50	20	25
Growing pigs	25–35	2500–3500	250–350	30	15	20

j. Drainage facility

No elaborate drainage system is necessary in piggeries where pigs are kept in deep litter system as all the urine is expected to be absorbed in the litter. Surplus water may, however, be carried away through drains. But in all other piggeries, there should be good and suitable drainage system for disposal of urine and washings. Every 3–4 mt., the gradation of the slope should be 2 cm.

15.4.3.2 *Housing for piglets*

Breeding sows and their litters can be kept in sties or using the open field system. Plenty of bedding should be given to help keep the young animals warm and it must be changed frequently. If a litter is raised in a sty, the sty should be thoroughly cleaned and scrubbed out after the litter has been weaned and moved elsewhere. If a litter is raised in the field, the shelter should be moved to a new site for the next litter to avoid disease problems, especially from parasitic worms' development.

Whatever the housing method used, piglets should have access to a warm area which the sow cannot reach. This is called a creep and piglets can be given feed here and can lie down without the risk of the mother lying on top of them. The sow is prevented from entering the creep by placing a temporary wall of boards or strong rails across part of the shelter. The bottom rail is about 30 cm from the ground allowing the small piglets to pass under it.

Fig. 15.2. A creep

15.4.3.3 *Housing for dry sows and gilts*

Dry sows and gilts do not require any special purpose building. They should be housed away from the breeding boars. Fully or partially covered yards por lose boxes may be used for dry sows and gilts.

15.4.3.4 *Weaning and fattening pens*

The weaners, whether they come from a farrowing pen or a weaner pen, will be 12 to 15 weeks of age and be sufficiently hardened to go to a growing/finishing pen. Finishing can be accomplished either in one stage in a growing/finishing pen from 25 kg to 90 kg or in two stages so that the pigs are kept in a smaller growing pen until they weigh 50 to 60 kg and are then moved to a larger finishing pen where they remain until they reach marketable weight. In large scale production, the pigs are arranged into groups of equal size and sex when moved into the growing/finishing pen. Although finishing pigs are sometimes kept in groups of 30 or more, pigs in a group of 9 to 12, or even less, show better growth performance in intensive systems. An alternative, where growing and finishing are carried out in the same facility, is to start about 12 pigs in the pen and later, during the finishing period, reduce the number to 9 by taking out the biggest or smallest pigs from each pen.

Pigs should not be allowed to wander free around the community. This results in the spread of disease among the animals and also between them and people.

Fig. 15.3. Housing and pens for pigs

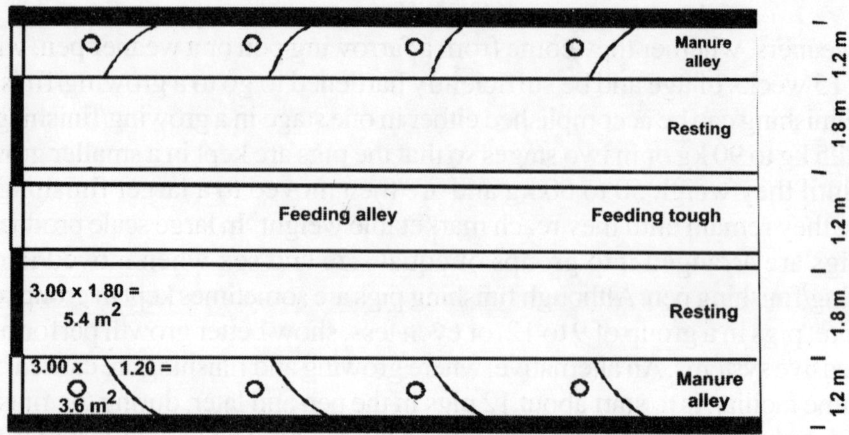

Fig. 15.4. Housing plan for growing/finishing pigs

15.4.3.5 *Replacement pens*

In intensive systems a sow will, on average, produce 3 to 6 litters before she is culled because of infertility, low productivity or age. Young breeding stock should be separated from the rest of the litter at about 3 months of age, since they should be less intensively fed than the fattening pigs. Gilts are first covered when they are 7 to 9 months of age or weigh 105 to 120 kg. After mating they can either be kept in the same pen up to 1 week before farrowing, or kept in the gestating sow accommodation, but in a separate group. Boars in the tropics are usually quiet if run with other boars or with pregnant sows but may develop vicious habits if shut up alone.

15.4.3.6 *Pig hatcheries*

A pig hatchery is actually a huge maternity ward where many sows are kept and their offspring marketed soon after weaning. Under these conditions there is no problem of feeding and fattening pigs, the operator can devote his full time to proper care, feeding and an advanced breeding program so that they can produce large numbers of weaned pigs efficiently. In addition, by proper culling, breeding and testing of sows, it should be possible to put out a more or less standardized quality product, namely that pigs offered for sale are uniform in size, type, weight and placed on the market castrated, vaccinated and most important of all, capable of making efficient gains being certain of a dependable market for the pigs. Unless conditions appear extremely favourable, it would be wise to start in a modest way, perhaps with only 5 or 10 sows, as the operation can quickly be expanded as one gains experience and a firm market is established.

15.4.3.7 *Farrowing pens*

For farrowing and rearing the litters, up to the stage of weaning, the sow should have separate accommodation where she can find her piglets all to herself. A farrowing pen should have an area of 5.5 to 7.5 sq mt. and should be fitted with automatic drinking bowl or a water trough. Attached to each farrowing pen there should be a small exercise paddock with a wallow of about 15 cm in depth especially in hot regions of our country. The pen should be warm and dry.

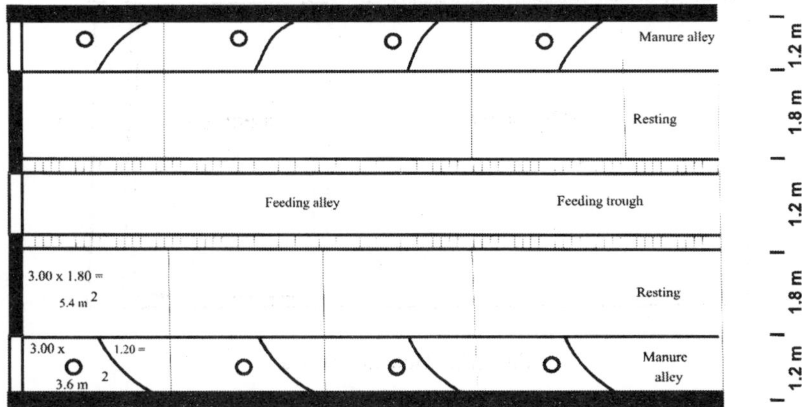

Fig. 15.5. Plan of a farrowing pen.

The fig. 15.6 gives the outline of a mixed system of housing in which different groups of pigs can be kept under one shed.

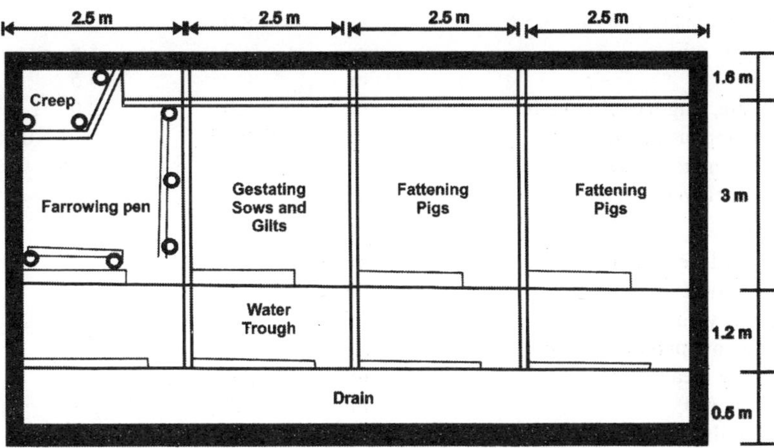

Fig 15.6. Outline of combined housing plan

15.4.3.8 *Housing for boars*

The boar should be housed individually away from the dry sow unit. A boar should not be kept beside a paddock of dry sow with a wire fence in between as when the boar moves up and down the fence it loses much of its energy. A boar house should be strongly built with a large open air paddock enclosed up to a height of not less than 1.5 m.

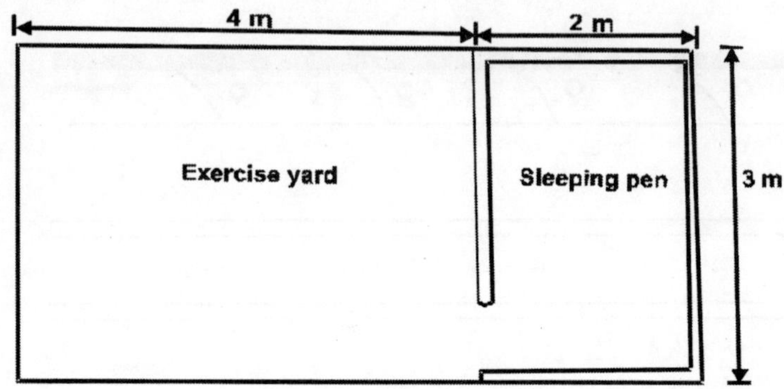

Fig. 15.7. Housing of boar

Table 15.4 lists the dimension and area of various types of pig housing.

Table 15.4 Dimensions and Area of Various Types of Pig Pens

	Units	Stocking density		
		Low	Medium	High
A. Farrowing/sucking pen				
Resting area, if weaner pens are not used	m²	10.0	7.5	6.0
Resting area, if weaner pens are used	m²	8.0	6.0	5.0
Manure alley width	m	1.7	1.5	1.3
Farrowing pen	m²	–	4.5	4.0
Farrowing crate, length excl. trough	m	2.0	2.0	2.0
Width depending on size of sow	m	0.65–0.75	0.6–0.7	0.55–0.65
Free space behind the crate	m	0.4	0.35	0.3
Piglet creep (incl. in resting area)	m²	2.0	1.5	1.0
B. Boar pen				
1. Pen with yard Resting area (shaded)	m²	6	5	4.5
Yard area (paved)	m²	12	10	8
2. Pen without yard	m²	9	8	7
C. Gestating sow pens in groups of 5–10 sows				
1. Loose resting area (shaded)	m²	2.0	1.5	1.1
Yard area (paved)	m²	3.5	3.0	2.5
Feeding stalls, depth × width	m	2.0 × 0.6	1.8 × 0.55	1.7 × 0.5
2. Individual stalls with access to manure alley, length of stalls excl. trough	m	2.2	2.1	2.0

Table 15.4 *Contd...*

	Units	Stocking density		
		Low	**Medium**	**High**
width of stalls	m	0.65–0.75	0.60–0.70	0.55–0.65
width of manure alley	m	1.5	1.4	1.3
3. Confined in individual stalls (L × W)	m	2.2 × 0.70	2.1 × 0.65	2.0 × 0.60
D. Weaner pen (to 25 kg or 12 wks)				
Resting area excluding trough	m^2/pig	0.35	0.30	0.25
Manure alley width	m	1.0	1.0	1.0
E. Growing pen (to 40 kg or 17 wks)				
Resting are excluding trough	m^2/pig	0.5	0.45	0.40
Manure alley width	m	1.1	1.1	1.1
F. Finishing pen, resting area excl. trough				
For porkers to 60 kg or 21 wks	m^2/pig	0.70	0.60	0.50
For baconers to 90 kg or 27 wks	m^2/pig	0.90	0.75	0.60
For heavy hog to 120 kg or 33 wks	m^2/pig	1.0	0.85	0.70
Manure alley width	m	1.2–1.4	1.2–1.3	1.2

15.5 Importance of Hygiene

The premises of any pig enterprise should have hygienic environment. The pig enterprise having indoor housing, greater care has to be taken to maintain proper hygienic conditions for which regular hygienic practises have to be adopted. Routine for regular cleaning of houses, cleaning of drains and feedings trough is needed. In case proper measures are not adapted there is always risk of having disease problems involving risk of mortality.

15.5.1 Sanitation, cleaning and disposal of wastes

In swine housing, provision of proper drainage system has gained importance. Use of dry bedding in farrowing pens and for young piglets, proper removal of soiled bedding from pig houses everyday and their disposal are essential. Pigs generally eliminate wastes away from feeding and sleeping areas. Dunging alley is cleaned and washed everyday. The drainage channel is generally 14 inches wide and 10 to 12 inches deep and may be kept covered, with removable cast iron lids. From time to time lime is sprinkled on these drains.

a. Removal of muck and cleansing

This is probably the most important part of the operation. All muck should be taken out and placed as far away from the premises as possible. This is very important. The subsequent cleaning may then be carried out in several ways. Some of the alternatives are:

(a) By water sprayed under pressure this is an effective and popular way in husbandry, being cheap and practicable.

(b) Steam cleansing this is effective both for cleaning and disinfection, using a suitable steam generator. However, the equipment is expensive, and operation laborious.

(c) Soak and scrub with hot water containing a detergent of 4% washing soda. Soaking may be done in cold water followed by scrubbing but this represents the most laborious method of all.

b. Disinfection of drains

A drain under normal circumstances need not be disinfected or deodorised, if they are not faulty and need reconstruction. So far as the actual disinfection of drain is concerned, this is very seldom done and is practically impossible to carry out satisfactorily. Pouring liquid disinfectant down the drain is quite useless owing to the great dilution and the rate of passage through the drain. If the disinfection of drain becomes essential then plug the drain or pipe and fill them to their utmost capacity and the disinfectant must be left *in situ* for a sufficient length of time. Sulphate of iron is recommended at 500 g to 5 litres of fluid or use of lime. A weekly cleaning of buckets and a thorough flushing with water may be required for cleaning of containers etc.

c. Manure disposal

Manure may be handled as solid or as liquid. Used bedding absorbs liquids. The material may be scraped by hand into a gutter or directly into a storage pit.

In case of liquid manure, handling system involves the use of water under pressure to remove the manure to a storage tank which should be big enough to hold 3 months accumulation. About 10 to 15 litres of liquid manure will be produced per hog per day depending on amount of water used in cleaning the pens.

Manure pits of brick and concrete flooring be constructed about 3 ft long 3 ft wide and 5 ft deep. Provision must be made to empty and clean the pits periodically.

Table 15.5 Approximate Daily Manure Production of Pigs

Age (weeks)	Live weight kg	Volume of solid and liquid manure in litre (gal)
8–12	14–24	1.5–2.0
13–15	24–37	2.0–3.0
16–20	37–54	3.0–4.5
21–24	54–72	4.5–7.0
25–28	72–90	7.0–8.0
Sow with litter	–	14.0

15.5.2 Hygienic measures for prevention of diseases

Prevention of the spread of infectious diseases is one of the most important and difficult duties. Each case must be treated according to its own requirements. There are however, certain methods of preventing the spread of infection that are common to all diseases and a consideration of these, forms the basis of all preventive medicine. The great resistance of some infective agents, the insidious nature of many infectious diseases for which the animal may be an active carrier without giving any indication of the fact until the disease has become widespread. The very nature of microbes or infective agents favours the spread. They find the resting places and by all sorts of means they are, in turn, passed from place to place and animal to animal.

15.5.2.1 *Infection transmission*

Infection is transmitted from the diseased to the healthy animals either by direct contact or by indirect way.

Any material that has been in contact with an infected animal may carry the contagion. An infective material which has been in contact with an infective animal may pass the contagion onto other material which in turn may transmit it to a receptive animal. Disease is carried from the diseased to healthy animal through other animals acting as passive carriers. Man may act as passive carrier by conveying the infective material on their hand, clothes and boots, vermin, birds, flies and other insects are usual modes of transmission. Food, water and air are also common transmitters of infection. The contagion of disease may enter the body by inhalation, ingestion, inoculation or by absorption.

15.5.2.2 *Preventive measures*

These measures include the following:

(i) Isolation of infected material and animal
(ii) Notification of the infection
(iii) Disinfection of all materials likely to hold or carry infective material
(iv) General prophylactic steps

(i) Isolation

The most important active measures, is the complete isolation of the sick or suspected animals. Partial or indifferent isolation is very dangerous as it tends to promote a false feeling of security. Not only animals but all other material belonging to animal must be completely isolated from contact either directly or indirectly with healthy animals. The attendant of the patient must be regarded as equally infective as the sick animal.

It is better to have separate persons to attend healthy and sick animals but if it is not possible then the sick animals be attended in the last and the attendant must make due precaution to clean himself, before passing among the non-infected stock. The period of isolation must extend beyond the recovery of the animal, and not lifted until all possibilities of infection have passed away.

Quarantine

The object of quarantine is to give time to the disease that may be latent to become active. During this period measures are taken to disinfect material that may be infective.

(ii) Notification

It is very necessary to control and eradicate the diseases that are considered dangerous. Some diseases are not easily diagnosed as might be thought from their text book description. This is why it is important to notify any condition of mass ailments or deaths.

(iii) Prophylaxis

Prophylactic measures taken to prevent appearance of diseases as far as possible, while the term is generally applied in connection with infectious diseases. The steps taken to prevent the onset of any preventable disease are prophylactic in character.

15.6 Common Disinfectants and their Application in Sty

The environment of the pig shed is favourable for the growth of microorganisms if the unhygienic condition prevails in the sty created by the dung, urine, split of milk, uterine and nasal discharge, feed residue, etc. Pigs housed in such environment are liable to suffer from infectious diseases and as a result there is high mortality, loss in production, economic losses due to treatment etc. Therefore, proper disinfection of the shed must be carried out in order to reduce the microbial infection and bring optimum profit to the pig rearers. Disinfection is the process of eliminating all pathogenic micro-organisms from the shed. The process of disinfection can be divided into two broad categories:

(a) Natural, and (b) Artificial.

15.6.1 Natural disinfection

15.6.1.1 *Sunlight*

It has most potent and enormous power of destruction of microorganisms due to its ultra violet range of spectrum having highest wavelength of 1800–2400 Angstrom. One draw back of this ray is that it cannot penetrate through glass, translucent roofing material, cloud and industrial haze. So, the animal sheds should be constructed in such a manner that the direct sun light should enter the house. Though it is very much effective in open housing system but not of much help in case of intensive system of housing.

15.6.1.2 *Heat*

High temperature accelerates the destruction of exposed micro-organisms. Presence of organic matter hinders disinfection by heat. Heat is applied in four forms.

(i) Dry heat

It is applied with the flame thrower. It is comparatively less active than moist heat. Most of the bacteria can not withstand more than a few minutes of dry heating at 80 °C with some exception like *Clostridium* and *Bacillus* groups. Therefore, the transitory heating from a flame gun must be at a high temperature to achieve disinfections.

(ii) Moist heat

It is relatively more effective than dry heat. It is applied in the form of steam. It is most effective in case of equipments as a disinfectant but its efficacy to disinfect building is reduced where microbes may be protected in cracks and crevices. Incorporating with a detergent or chemical disinfectant can increase its efficacy. It is used under 3 kg pressure.

(iii) Hot water

Equipments can be disinfected by boiling for 4–5 minutes. It is not effective in floor as it loses its heat soon.

(iv) Fire

It is the best method to dispose infected material and carcass.

15.6.2 Artificial disinfection

It is done by using chemical disinfectant, radiation and filtration and aerosol fumigation. Out of these, Chemical and Fumigation methods are widely used.

15.6.2.1 *Chemical disinfectant*

These are number of chemical agents. These are able to cause disinfection by coagulation, hydrolysis, oxidation, precipitation, or otherwise through penetration of protein and in particular the essential microbial enzymes and disruption of cell wall.

Factors affecting chemical disinfectant activity

There are three basic phenomena for disinfections by chemical means.

1. Absorption of the compound by the cell wall of the micro-organism.
2. Penetration into the cell cytoplasm.
3. Reaction of the compound with one or more of the cell constituents.

The absorbability and penetrability of a chemical disinfectant depends upon the chemical constituent of the compound and immediate environment, which affect surface tension and other physicochemical properties. The nature of the solvent is also very much important for the efficacy. Chemical disinfectant can most readily attack cell via aqueous phase neutral to the organism.

Any solvent that reduces the concentration of the disinfectant in the aqueous phase has the consequent effect of reducing the activity. Conversely a germicide of high oil/water solubility is easily absorbed by the liquid fraction of the cell and thus the germicide may be expected to be more effective against organism of high fat content. Presence of inorganic salts also increases the activity of the disinfectant by their action on surface tension and osmosis. Microorganism is highly vulnerable to changes in tonicity. Reduction of surface tension usually increases the activity of a disinfectant. Disinfectants have a selective action on microbes. As cell wall of micro-organisms differ from each other in respect to their lipoprotein composition and as every protein has its isoelectric point, each responds individually and is influenced by acidity and alkalinity of the disinfectant.

Disinfection takes place gradually. Although many more microorganisms are killed at the beginning of the process than at the end, there is an initial lag phase before the activity commences. The destruction of micro-organisms is very fast after the lag phase and then tends to slow up. The concentration and temperature of disinfectant greatly influence the rate of death of the micro-organisms. An increased concentration of disinfectant increases the death rate of pathogens. The activity of most of the disinfectant increases with the increase of temperature.

Organic matter always interferes with the action of disinfectant in the following ways.

1. The organic matters form a coating on the cell and thus prevent the ready access of the disinfectant.
2. Some disinfectants react chemically with the organic matter, giving rise to non-germicidal reaction product.
3. The reaction between organic matter and disinfectant may sometime results in an insoluble compound and reduces its potentiality.
4. The organic matter sometimes remains as particulate and colloidal state and absorb the antibacterial agents and results in reduction of disinfection potency.

(a) Phenols and related compounds

Phenols are bactericidal and fungicidal, but has no action against spores and virus. Small concentration change of phenols gives rise to marked difference in their killing rate. They are more active and effective in acid solution and rise of temperature. Phenols are also more active in saline solution and additions of certain proportions of metallic salts to the solution increase their effectiveness.

(i) Cresols

These are slightly soluble in water and usually emulsified in soap. These are effective against a wide range of bacteria but are not very effective against spores. Excessive quantities of soap reduces their effectiveness. It should be used in 2% saponated solution.

(ii) Lysol

It is used at a concentration of 2% for general use and 5% for killing spores.

(iii) Synthetic phenolic disinfectant

These types of disinfectants are non-toxic, non-irritant and have a pleasant colour. One of the important disinfectants of this group is chloroxylenol. A solution of chloroxylenol BP is prepared by mixing of 5% chloroxylenol, 10% terpeinol, 20%

alcohol and 7.5% caster oil. It is mainly used in the form of aerosol and can be safely used for air disinfection.

(b) Alcohols

These are bactericidal against vegetative organisms but are ineffective against spores. The ethyl alcohol, benzyl alcohol, ethylene glycols, propylene glycols are being used now-a-days. The glycols are mainly active in aerosol forms and as viricidal agents. The ethyl alcohols are effective in reducing the bacterial flora of the skin at a concentration of over 70%. When 1% of a mineral acid or caustic alkali or 10% of amyl-m-cresol is added to the alcohol, it is able to kill most of the resistant spores with in 4 hr.

(c) Halogens

i) Chlorine compounds

Chlorine is available for sterilizing farm utensils from 2 sources.

Sodium hypochlorite (NaOCl) and calcium hypochlorite ($CaOCl_2$)

- 5% Sodium hypochlorite (NaOCl) solution is applied @ 0.5l/m^2 to inactivate Foot and Mouth disease (FMD) virus. It can be used for irrigation of wound after diluting 10 times. It dissolves necrotic tissue and blood clots. It should not be used in a hot solution.
- 5% Calcium hypochlorite ($CaOCl_2$) solution is recommended @ 0.5l/ m^2 in Anthrax, Tetanus and Tuberculosis *etc.* It can be used in floors and gutters by dusting.

Organic chlorine compounds:-Chloramine-T, Dichlorodimethylhydantoin and sodium or potassium salt of dichloro or trichloro-isocyanuric acid. The organic compounds are blended with a detergent to produce a stable powder combining detergent and sterilizer.

Chlorine compounds are bactericidal in nature and it's efficiency and effectiveness is greatly impaired by the presence of organic matter and also depends on (a) concentration, (b) temperature, (c) contact time and (d) pH value

(d) Iodine

This is a powerful germicide and effective against vegetative organisms, spores, bacteria, virus, fungus, etc. Its efficiency is greatly impaired by presence of organic matter. It is non irritant and odorless. It is used at a dilution of around 25 parts of available iodine per million. *e.g.* iodine trichloride, iodinum compounds like katiodin and Iodophor preparation like Iodicide.

As iodine is most effective as a bactericide in an acid solution the phosphoric acid is usually incorporated with iodophor for farm use.

(e) Sodium carbonates and sodium hydroxide

Sodium carbonate is used as a disinfectant against FMD virus, fowl pox virus, etc. Although its efficiency increases when used as hot 4% solution, it is mostly used to prepare a site before applying an approved disinfectant.

Sodium hydroxide is more effective against viruses and Gram +ve bacteria. It is better to use at 2% level for normal use in animal buildings and 5% for highly infected surface. It is caustic in nature and dangerous to use. Therefore rubber gloves, goggles and protective clothing should be worn during handling. This is the only disinfectant which is less active when warm.

(f) Ammonia

This is the most effective agent for the destruction of coccidial oocyst in 10% aqueous solution.

(g) Quaternary ammonium compound (QAC)

These compounds are surface active, colourless, odourless, non-toxic, non-corrosive, stable and compatible with most alkaline detergents. It has bacteriostatic and bactericidal activity against a wide range of micro-organisms and should be applied as 1% solution for washing/wiping of udder and hand.

(h) Quick lime (CaO)

This is not very effective. This is occasionally used to disinfect ground @ 2 tons of quick lime per acre and covering of carcass, disposed off by burial method.

(i) Slaked lime ($CaOH_2$)

This is used as a white wash on the wall of animal house. This should be mixed with other disinfectant such as 1% phenol and used after proper cleaning of the wall.

15.6.2.2 *Gaseous and aerial fumigation*

This method is cheap and usually harmless to the materials used in the construction of the house. This may be applied at or slightly above normal temperature. This spray may be toxic to humans. Therefore care must be taken in their use. The compounds in use are:

Formaldehyde

The formaldehyde gas is detrimental to bacteria, spores even in presence of organic matter. It can be used in number of ways.

- Mixture of formalin and potassium permanganate in the ratio of 3:2. The container should have sides high enough to prevent the mixture from bubbling out due to a risk of fire. All the combustible materials including litter and wooden parts should be kept out of range. Operator should wear a respirator.
- Formalin vapour: This is achieved by dispersion of a mixture of formalin and water as an aerosol of small particle size.
- Cetyl Fumigation Lamp: This lamp is a self contained unit containing candle heat generator and the fumigant *i.e.*solid paraformaldehyde.The formaldehyde gas is released after heating of paraformaldehyde. It is useful for small building up to 60 cubic meter.

Precaution

Humidity plays an important role in the efficiency of formaldehyde and 60–80% humidity is considered to be optimum. The mixture should have a temperature of 22º–23 ºC. The space after fumigation should be kept closed for at least 12 hr.

The disadvantage of formaldehyde is that it gets absorbed on the exposed surface as a film of polymerized formaldehyde and difficult to be removed after its use. This problem can be dealt with by sprinkling a dilute solution of ammonia in the shed. Double fumigation should be done in case of out break.

15.6.3 Procedure for disinfection of animal building and equipments

1. Procedure of regular disinfection without any disease in the herd

- All removable, detachable equipments and fittings should be dismantled. It should be taken out and soaked in a bath of disinfectant, power sprayed or steam sterilized after thorough cleaning.

- Roof and other structure of the house should be dusted and cleaned properly preferably with the help of the vacuum cleaner.
- The lower portion of the wall, floor should be soaked and scrubbed with a detergent disinfectant.
- Earth floor should be soaked in a solution of one-pint formalin to 12–gallon water.
- After cleaning and disinfections of house and equipment, the house should be sprayed with aerosol formulation.

2. Procedure after the disease

- The building should be closed to the visitors and should be prohibited.
- The litter, all area and materials, which have been in intimate contact with the stock, should be sprayed with a strong disinfectant.
- The litter should then be removed and burnt or buried.
- Portable equipments and fittings should be dismantled and disinfected properly.
- Lower part of the wall, floor should be cleaned and scrubbed with a disinfecting detergent solution.
- The roof, other structures, fittings should be cleaned and dusted properly. In case of earthen floor it should be covered with polythene or tarred paper before new litter to be put down.
- Removing out of top few inches of soil from a heavily infected area should be advisable.

Footbaths and foot dips with a disinfectant should be provided at the entrance of the building.

CHAPTER 16

MANAGEMENT SYSTEMS

16.0 Management

Successful management of pigs in any part of the world depends primarily on intelligent planning that is based on knowledge of the biology of the pig.

In most tropical countries in the past, the indigenous producers did not attempt to obtain maximum productivity from their pigs but managed them primarily as scavengers. Although pigs are still used as scavengers, in most countries there is also an ever expanding commercial pig industry. Methods of management in this new commercial section should not necessarily be based on those now practised in the temperature zone.

16.1 Adaptive Physiology

The pig is essentially a non-sweating species and is very sensitive to changes in the climatic environment. While discussing the origin of our present major breeds it was suggested that the majority of the pigs managed in the tropical world today are derived from the wild species that was adapted to a warm, shaded, humid environment. These facts probably explain why temperate type breeds of pigs, unlike temperate type breeds of cattle, thrive in the humid tropics under suitable managerial and feeding conditions.

The following facts have been established with regard to the effect of ambient temperature on pigs.

The baby piglet at birth does not appear to possess a very efficient temperature regulating mechanism. It is incapable of protecting itself against either excessive heat or cold. Newland *et al.* (1952) have shown that the body temperature of the baby pig of typical American breeding falls 1.7° to 7.2 °C during the first 30 minutes of life and then slowly returns to normal during the next 48 hr. The body temperature falls most rapidly in small piglets that weigh under 0.9 kg and takes a longer time to recover to normal if the air temperature is low. These workers suggested that cold air temperatures contribute to an increase in the mortality of piglets during the first 2 to 3 days of life, particularly as chilled piglets stand and shiver, become sluggish in their movements and are likely to be more easily 'laid on' by their mother. Later work has confirmed these suggestions and in practise during the first 2 days of life the ambient temperature for piglets should exceed 32.2 °C and be gradually lowered as the piglets age. It is now normal practise in temperate zone countries to use infra-red lamps to warm the piglets immediately after birth, so that they do not get chilled. This managerial practise unquestionably reduces piglet mortality. In the tropics where mean annual air temperatures vary around 26.7 °C, the problem is not as acute as it is in the temperate climatic zone. However, it has been found that even in tropical climate piglet mortality due to overlaying may be reduced by the use of an additional heat source for the baby piglets during the first few days of life.

As pigs age and grow, the optimal ambient temperature for maximum live weight gain and efficiency of food conversion changes. Heitman and Hughes (1949) raised pigs in a controlled climatic chamber for periods averaging 7 days in air temperatures ranging from 4.4° to 46 °C at a comparatively constant relative humidity and airflow. They found that live weight gain and efficiency of food conversion was at a maximum at approximately 24 °C for pigs weighing 32 to 65 kg and at approximately 15.6 °C for pigs weighing 75 to 118 kg. They also noted that if the air temperature rose, the respiration rate of pigs also rose very rapidly. Other American workers have shown that at temperatures of 32.2 °C and above, respiration rates of 150 to 200 per minute are common in pigs. Under these circumstances the pigs stop eating and lose weight, and if forced to exercise may even die of heat exhaustion. The general observations of Heitman and Hughes (1949) have been confirmed by other investigators. Recently, however, Verstegen *et al.* (1973) have stated that although energy retention in the pig depends upon ambient temperature and feeding level, nitrogen retention is not influenced by ambient temperature.

During the daytime, particularly during the hottest months, tropical ambient temperatures are usually well above 24 °C so that tropical pigs weighing 32 to 65 kg are probably being reared under almost optimal environmental conditions, but as pigs grow older and heavier, normal tropical temperatures would be too high for maximum productivity. Thus, in the tropics the aim should be to raise a porker weighing approximately 54 to 64 kg and the larger fatteners, as well as the gilts, sows and boars may require some amelioration of the climatic stress if they are to produce at a maximum. Under such conditions relief from the adverse climatic conditions can be obtained by the provision of adequate shade and find water sprays or wallows. For adequate shade the roof of the pig pen should not be too high for maximum productivity. Thus, in the tropics the aim should be to raise a porker weighing approximately 54 to 65 kg and not bacon or lard pigs weighing 91 to 109 kg.

For adequate shade, the roof of the pig pen should preferably be constructed of thatch (unless vermin are a major nuisance) or asbestos sheet or tile. If it is necessary to use corrugated iron then the roof should be painted black on the underside with aluminum paint on the top surface. The most suitable site for water sprays is in the dunging passage, if the buildings are provided with such a facility. Wallows should be approximately 25 cm (10 inch) deep with a surface area of approximately 1.5 m^2 (16 ft^2) per sow and should preferably be covered with a roof. Wallows should be constructed so that the water in them can easily be changed as they rapidly become very dirty. If energy is relatively cheap and engineering skill available, an alternate cooling device that has been advocated is a forced cool air draught. This can be particularly useful for cooling the sow in a farrowing pen when the young piglets require a relatively high ambient temperature, whereas the sow, if it is required to milk adequately, requires a lower ambient temperature. Under the restricted conditions of a farrowing pen a forced cool air draught can be directed on to the head of the sow and some relief can be provided for her without radically reducing the overall ambient temperature within the pen.

There is limited information on the effect of high temperature on carcass characteristics. In two experiments Holmes (1971) compared the carcass characteristics of a group of large white and landrace pigs raised at 31° to 32 °C and 32° to 33 °C with a group raised at 21° to 24 °C and 22° to 26 °C. Carcass length (one experiment) and backfat thickness (both experiments) were significantly greater in the heat stressed group and the weight of the liver was significantly less (one experiment). If future investigations confirm that heat stress increase back fat thickness this will be an additional reason for adopting practical methods of ameliorating heat stress in fattening pigs.

It may be thought that pigs in the tropics can be properly managed either in or outdoors and that it might be less expensive to provide adequate shade and wallows outdoors. Unfortunately, there is one major difficulty experienced in managing pigs outdoor is the very high incidence of certain internal parasites. This is particularly so in the humid tropics which provide an almost perfect environment for many parasites. The most dangerous of these is the kidney worm (*Stephanurus dentatus*), and in many regions the population of this parasite is so high that pigs can only be managed properly on floors that can be cleaned daily. Even a very strict rotation of pigs around a series of outdoor paddocks is an inadequate precaution. Pig breeds do vary in their tolerance of a high incidence of kidney worm infection and some indigenous Southeast Asia breeds appear to be considerably more tolerant than the breeds originating from the temperate zone.

16.2 System of Management

These may be conveniently classified in to those suitable for the peasant or village producer and those that can only be practised by large scale commercial producers.

16.2.1 The peasant or village producer

In most villages, in regions where pigs are kept as domestic livestock, they are free to roam where they will. They are useful as scavengers, sometimes cleaning up human and domestic animal faeces and always picking up offals where they can.

Quite simple arrangements could be made to improve the productivity of these scavenging village pigs. Some of these are as follows:

1. The feeding of supplementary feeds, either once or twice a day. In an area adjacent to the house the pigs could be fed waste feed, such as rice bran and the peelings of root crops. If the householder is willing to cook the waste feeds, it is much better. This is a system that is widely practised by the Dayak people of Sarawak, who boil roots and green leaves and pour the hot mixture over rice bran spread on the bottom of wooden troughs. The greatest difficulty encountered is that of keeping neighboring pigs away from the feed.
2. Where land is plentiful, the pigs can be managed in simply fenced paddocks adjacent to the household in which some root crops are grown and in to which all household offals are thrown and where the cooked feeds can be fed secured from neighboring pigs. The fences might be made of netting wire, if this is available at an economic price, or they can be made of platted bamboo, paling wood or a closely planted live-fence species. The paddock should be sub-divided in to four to six smaller areas so that the pigs can be moved from one enclosure to another at 10 day to 2-week intervals, thus reducing the incidence of parasitic infection. Water and shade would have to be available within the paddock. Pigs raised in this manner might not be very much more productive than scavenger pigs, but females could be bred to selected sires so that the stock could be slowly improved.
3. A further improvement would be to construct simple pens in which pigs could be confined. Productivity would of course, only be improved if

there was sufficient food available from village resources to feed the confined pigs and if they were regularly fed and watered. Several types of simple pen are constructed in different regions of the tropics. Some suitable types are as follows:

(a) A simple type of deep litter pen. This could be constructed of rough timber with a thatched roof and an earth floor. Coarse hay, straw, rice hulls, reed, etc., can be thrown continuously into the pen in order to create a suitable type of litter. It would be necessary to construct such a pen on a well drained sift.

(b) A conventional pen with a concrete floor that can be washed or cleaned in some other manner. The pen could be constructed of rough timber with a thatched roof. If it was built close to a stream that was not otherwise used by humans, water could be diverted to run through it. This running water could be used both for drinking purposes and for cleaning the pens.

(c) A timber pen with a thatched roof could be built with a slatted bamboo floor, either over a fish pond or over a drainage channel. This is a type of simple and practical pig pen that is often used in Southeast Asia.

4. Still further improvements could be effected by the distribution of improved sires for upgrading purposes and by the provision of high protein and mineral feed supplements. In any upgrading programme great care must be taken not to upgrade too quickly or too far. The possible use of high protein and mineral supplements depends upon whether the farmer has a sufficiently high income to afford to purchase supplements and/or the availability of such supplements.

5. There is of course no reasons or circumstances for permitting the village pig keeper to use some of the more advanced managerial practises described below. The suggested managerial improvements described above are only considered as useful first steps in the raising of the general level of management of pigs in the village.

16.2.2 The large scale producer

The managerial methods used will depend upon what labour and feed supplies are available and at what cost, and on the incidence of disease and parasites.

The total number of large scale pig farms in the tropics has been increasing rapidly during the last decade, particularly in Southeast Asian countries like Philippines, Singapore, Malaysia and Thailand. Despite these developments, it is likely that the majority of pigs in the tropics are still managed under village conditions, although the proportion of the total pigs population managed by large scale producers is likely to continue to increase.

Accompanying this increase in large scale operations, there has been increasing specialization and development of the use of ever increasing quantities of commercially prepared feeds. Nevertheless, the pig industry produced feeds with a considerable proportion of by-product feeds being incorporated into commercial feed mixes. This is a highly desirable development as available by-product feeds will, in general, be used more economically.

Large scale pig production is therefore likely to develop in those regions of the tropics where ample supplies of by-product feeds are available and East Asia is one such region. Not only very large quantities of rice milling by-products are available there, wheat by-products are also now available from new mills at the ports and maize by-products from new processing plants. In addition, a variety of high protein meals are produced, such as coconut, sesame, peanut, cottonseed and oil palm, and there are expanding fishmeal and abattoir by-product industries.

Despite increasing specialization within the industry, the large scale breeding of purebred lines and/or hybrid pigs for use by the commercial sector has not yet developed in most tropical countries, and the government is often the only source of supply of breeding stock and crossbred pigs. It is therefore usually necessary for the large scale pig farmer to raise the majority of his own breeding stock. In order to do this he will need accommodation for farrowing, creep feeding and fattening and for gilts and sows, boars and young breeding stock. He will also probably require feed milling and mixing equipment, storage for straight and mixed feeds, weighing facilities, loading places, a piped water supply and facilities for the removal of manure.

16.2.3 Intensive systems

In this system all pigs should be raised on concrete floors or on some other form of flooring, such as one made of slats, that can be cleaned daily. This should ensure that internal parasites can be adequately controlled and that labour costs are reduced to a minimum. Concrete floors should not be too smooth or the pigs may skid on them, nor should they be too rough. Litter may or may not be used according to circumstances. If a slatted floor is favoured the slats may extend over the dunging passage or cover the entire area of the pen. The latter is more expensive but preferable. The slats maybe made of wood, concrete, steel and/or aluminum and should be spaced sufficiently close so that the pigs do not get their feet trapped. The slat width should be 10 to 13 cm (4 to 5 inch) and the space between slats should be 2.5 cm (1 inch). If the slats extend over the whole pen there is no need to provide a dunging passage. The space below the slats should slope towards a drainage outlet so that dung can be flushed off the slats with water. Slats should not normally be used for the floor of farrowing pens. If they are used, the slats should be covered with a grating before the sow farrows.

One of the most suitable and cheapest pens is one that is half covered by a roof so that the pigs can shelter if necessary. The roof should be 2.4 to 3 m (8 to 10 ft) at the highest point and 1.8 to 2.1 m (6 to 7 ft) in height at the eaves. It can be made of thatch (coconut frond, nipa, reed, grass, etc.) or of a conventional material such as galvanized iron. A layer of thatch of 5 cm (2 inch) is attached by wire netting beneath a galvanized iron sheet. The galvanized iron can be painted black on the underside and with aluminium paint on the topside, or aluminium roofing material can be used that is painted black on the underside. The pen can be constructed of any suitable material, but perforated are superior to solid internal walls. Due consideration must be given to both the free circulation of air and the provision of shelter from cold and rain.

A simple and very flexible system for the smaller farm is a series of pens that can be adapted for farrowing, fattening or breeding stock, according to the dictates of farm policy. Some difference in the size of pens is desirable as this increases the flexibility of the system. Farrowing pens should be equipped with farrowing rails or a farrowing crate and with creep feeding facilities. A 2.4 m × 4 m (8 × 13 ft) pen will accommodate a sow and her litter, up to twelve porker pigs, eight bacon pigs or three breeding sows.

Fig 16.1. Semi pucca housing of pigs

The larger pig farmers will want to build more specialized housing with some general details on housing requirements.

Feeding troughs should be designed in such a way so that the minimum of labour is used for feeding and so as to prevent feed from wastage. They can be fixed or movable and should be made of materials such as sealed concrete, glazed pipe or galvanized iron so that they can easily be cleaned and are not pitted. Feeding troughs finished only in raw concrete will soon be pitted by food acids, particularly if skim-milk is fed. Concrete troughs should therefore be finished inside with a substance that gives a smooth, glazed and permanent finish. There are several such proprietary compounds on the market. If self-feeders are provided, one self feeder hole will provide feeding space for four pigs under 15 weeks and three pigs over 15 weeks of age.

Water should be available in all pens for drinking purposes and in all feed alleys for cleaning purposes. The feeding troughs can also be used as water troughs, but pigs tend to lie in them and automatic water cups are preferable if they are available at a reasonable price. One automatic water cup is required in each pen of 20 to 25 pigs. Water should also be available for sprinklers and/or wallows. Drinking water should be as cool as it is possible to provide and water pipes should not be exposed to the hot sun if other arrangements are practicable.

Bedding may be provided on concrete floors, but is not essential in most tropical environments.

Tree shade over the piggery building is usually desirable if it can be provided with the exception of buildings in the hurricane zones.

In specialized piggeries, all kinds of labour saving devices can be introduced, including automatic feeders. When planning a piggery, it should be ensured that all feeding stuffs and manure are carried only down hill. This can be arranged by setting the feed mixing and/or storage shed at the highest level and the midden or manure collecting area at the lowest. It should be possible to site the piggery so that manure can be removed with minimum effort.

Pig manure may be sun dried and sold as a fertilizer. In some Southeast Asian countries sun-dried manure is a very profitable by-product of the industry. It can also be used for the production of methane gas or for the culture of chlorella. Details of these processes should be obtained from a local extension officer or an NGO specilizing in animal agriculture.

In some areas of Southeast Asia, pig farming is associated with fish pond culture. Effluent from the piggeries is run in to fish ponds as it is believed that it improves the growth of micro organisms and plants on which the fish feed. This practise is controversial as often only phosphatic fertilizers are needed in the ponds, nitrogen being fixed very effectively by blue green algae and potassium being very rarely in short supply. Effluent nitrogen can in fact be counter productive as it may inhibit the production of blue green algae. Also, the organic materials in the effluent may produce deoxygenation in the pond water as they contain carbohydrates that have to be broken down by bacteria which use oxygen dissolved in the water. Nevertheless, large quantities of fish are produced in ponds in to which pig effluents flow particularly in Southeast Asia.

Pig effluent may also be channeled in to irrigation canals in order to fertilize fruit or other crops or it may be collected in a sump, filtered and the liquid fraction pumped in to an overhead spray irrigation system.

In most temperate zone countries the disposal of effluents from large scale piggeries has become a major problem because of stringent environmental regulations with regard to disposal methods, but in most tropical countries no such regulations have yet been enacted. Farmers should consult their local extension office for information on regulations concerned with the disposal of effluents.

16.2.4 Semi-intensive system

There are many variations of the semi-intensive system. Unfortunately, this system can only be practised in those regions of the tropics where the kidney worm and other internal parasites can be adequately controlled. As the kidney worm takes at least one year to grow to maturity within the pig and produces eggs that are voided in the pig's urine, some authorities advocate the management of breeding stock on pasture in regions where there is a low intensity of kidney worm infestation by only retaining gilts to produce three or four litters.

Usually breeding pigs are raised outside on grass and fattening pigs are raised intensively in buildings. The most common system is to allow the gilts and the in pig sows to graze with or without the boars. They must be rotationally grazed around a series of paddocks. These should be located on well drained soils, low lying marshy areas being fenced off, provided with adequate shade and a water supply and be well fenced, preferably with pig netting. Mud wallows inevitably become centres of parasite infection and if they are used they should be frequently cleaned and dried out in the sun. Sows that rot should be nose ringed.

Sows with litters, housed in portable sheds, can also be rotated across grazing. The portable shed can be fenced with portable mesh or an electric fence, or alternatively the sow can be tethered. This system is labour intensive as feed and water have to be carried to the pigs, but in regions free of the kidney worm the young pigs are usually very healthy.

Breeding pigs or fatteners can be run in semi covered yards, fresh litter being thrown in to the yard daily. This is a form of deep litter management.

16.2.5 Extensive system

All pigs can be put out on grazing or in semi covered yards. Rotation is essential on grazing and labour costs are high. It is doubtful whether this is a very suitable managerial method in the tropics. One reason is that it needs more supervision and skilled labour than intensive methods, and both are in short supply in most tropical countries. Another reason is the possible presence of kidney worm.

16.3 Accommodation for Gilt and Weaned Sows Dry Quarters

Dry quarters where the maiden gilt, whether brought in as home bred or weaned sow, can find shelter at all times form rain, wind, cold or even hot sun is all that is required. A concrete bottom to the yard is essential for thorough cleaning to avoid the buildup of parasites.

Space requirement for sow yards

Space requirement for sow yards should be 30–40 sq ft per sow. Newly weaned sows or gilts should be housed next to the boar's living quarters as sight, sound and smell will encourage the newly weaned sow to come in heat. An open pen gate or open gates need to be fairly robust and frames with 25 mm square section will do this job nicely. Yards with individual feeders are ideal, but floor feeding with large nuts would be acceptable. Fresh air and exercise are important at this stage and they should be housed partially on concrete to firm up legs and action.

16.4 Accommodation for Dry Sows

16.4.1 The fully-covered yard

An excellent way, perhaps the best way from the pig's point of view of housing, of dry and in pig sows, is to keep them in a completely covered yard. Such a yard can provide complete protection from the weather and under these circumstances the amount of straw used need not be prohibitive.

Whilst the system enables the sows to be kept under the healthiest and most invigorating type of conditions, it keeps them well protected from bad weather, allows plenty of exercise and makes the provision of good stockmanship easy. Individual feeders can be provided. The system is based on a line of feeders down one side of the yard which are raised above the general level of the yard itself. The further the raised area the greater the build up. A drop of 750 mm will allow a build-up of approximately 3–4 months.

A suitable yard on these lines can be provided by having a span of 9 m and dividing it into bays of 4.5 m along its length. This gives total area of 42 m^2 per bay, which is suitable for eight sows. This is an ideally small number to keep together as the likelihood of fighting and bullying rises as the number kept together increase. Along one side the individual feeders have a length of 2.1 m including the trough, so that the actual lying and exercising area is just over 3.7 m^2 per sow, which can be considered a satisfactory and generous one.

16.4.2 The partly-covered yard

Good as the covered yard is, it does represent a relatively expensive way of housing breeders. A cheaper way of dealing with this matter, and probably little inferior in practise, is to have only part of the yard strawed and covered and the remainder composed of a concreted area partly for exercising and partly for the feeding, containing the usual individual feeders. The simplest layout would be similar to that in the totally covered yard, with the lying area at the back in the form of small kennels, allowing 0.93 m^2 of lying area per sow, and a reasonably generous concreted area, in front of 2.8 m^2 area per sow. At the far side of the unit will be the individual feeders served by a concrete apron.

16.4.3 Sow stall

Stalls are similar in size to a farrowing crate and allow the sow only to stand up and lie down; she can neither turn nor exercise, with collar and chain round the neck, a trough in front and with timber or metal partitioning. The stalls are 750 mm wide, 900 mm high and 1.97 m long, with two retaining chains behind.

It can be economical form of housing as the stalls may be placed in simple narrow buildings some 4.2 m wide, though it is more usual to have them in two or more rows. The system has also the advantage that it ensures each sow a fair share of food, freedom from fighting and bullying throughout pregnancy, and uniformly equable conditions.

Sow kept in stalls should be able to see their fellows and if two rows are used they should housed face-to-face rather than back-to-back as in the latter case the sows strain to see what is going on behind them. The sow-stall house must be very well insulated and ventilated, bearing in mind that often the sow will have no bedding, she will be fed frugally under modern techniques and there will be no heat generated from exercising. Extreme of temperature and dampness will therefore be potentially harmful.

Two interesting approaches have been evolved which keep sows in small groups. Sow cubicles, consist of a group of three or four 2.1 m long × 600 mm wide free choice cubicles for feeding and lying with communal dunging area behind them of 1.8 m × 1.8 m. The unit would normally be placed under a covered yard and with gates between each dunging area could be mechanically cleaned. Cubicles may also be placed in an outside hut with separate or communal dunging areas, one design has a door that can be opened and closed by the sow. These are all useful approaches with the same aim as all dry sow husbandry to give a stress free environment at minimum cost and with easy management.

16.4.4 Rearing pens

The fullest protection for the piglet is required for the first ten days to two weeks of life, when both crushing and chilling are at their commonest. In a house with a central feeding and service passage, pens are placed on either side, with dimensions of 3m × 2.4 m, the longer side running along the front. The creep is placed along the form of the pen, measuring 1.5 m × 910 mm minimum and adjoining it, is the sow's trough and water bowl and a 600 mm wide gate. Only a dwarf wall 600 mm high is needed adjoining the creep enough to prevent draught but allowing ease of access.

Using a central service passage, drainage can be by open gulleys on either side of this so that there is no contact between pens. The fall in the floor towards the end and corner taking the drainage away should be a good one, of the order of 100 mm from back to front and 50 mm from side to side.

16.4.5 Multiple sucking pens

One approach is the 'farrow to finish' pen where pigs are taken from birth to finishing in the same pen. Another is to mix three to five sows and litters together, usually at about three weeks of age and thereby form a 'weaner pool' which includes the dams. After two or three weeks the sows are removed and the group of 30 to 50 weaners are left for a further period almost always ad lib feed until ready for the finishing stages of fattening.

At 3 weeks of age the piglets' resistance to infection is about the lowest in their life, since they have lost most of the 'passive' immunity they had from their mother either in utero or via the colostrums and yet have not developed the 'active' immunity which will be produced consistently from now on. Under ordinary farm conditions the piglets will develop this 'active' immunity to disease satisfactorily and gradually if they go into clean quarters without too close a contact with piglets already excreting organisms which are capable of producing disease. The need to 'health' groups of piglets through an early weaning unit is therefore essential and helps to ensure that the groups are not too large.

The second essential for piglets weaned at three weeks is that they are reared on as clean a floor surface as possible with a system that makes sure they have absolute freedom from the pollution of their own dung and urine. This means that the lying area must be absolutely clean, and the dunging area, if separate, should be either perforated or if solid, should be so frequently cleaned that the chances of pollution are minimal. Provided these essentials are there, systems using a completely perforated floor or partly solid and perforated, or all solid flooring with or without bedding, can be utilized.

The 'micro-environment' can be placed as the third essential. The temperature at the start should be about 27 °C; the more important issue is to keep it uniform within this range. In practise the question of the air movement is almost as vital as that of the temperature, as the piglet, with no coat to speak of, is extremely vulnerable to draughts.

16.4.6 Cage rearing

A pig farmer rear piglets away from the sow from 7 to 10 days onwards in cages. Piglets are grouped by weight, nine at a time in cages, three, four or five tiers high. Each cage measures 1200 mm long × 600 mm wide × 390 mm high and has a floor of 12 mm × 12 m × 12 gauge wire mesh. These cage batteries are placed in housing kept at 27 °C, and in subdued light. The piglets remain here until about 7 kg weight after which they are usually moved to flat deck cages.

Another interesting method is the 'Chediston' two tier system in which the heat form the older pigs in the bottom tier warm the newly weaned pigs in the top tier. The top tier is designed to take the newly weaned piglets from about 3 weeks of age and the follow onwards. It is a further useful way of conserving heat and reducing costs, but there is an increased disease risk unless each section in periodically depopulated in the usual way.

16.4.7 Fattening accommodation

Whilst there are several fundamentally different forms of piggery for the fattening stages, probably the most popular is the totally enclosed piggery where environment is under complete control and all attendance to the pig is under cover. In its traditional form it involves a central feeding passage 1.2 m wide, side dunging passages 1.05 m and a pen of 3 m × 1.8 m, to hold ten pigs to bacon weight.

The design of fattening accomodation present a large open air space in which the dunging passage, pen and feeding passage were separated only by a 1.05m high wall. Current practice is to screen off the dunging passage from the pen, leaving only a pop hole between the pen and the passage. In this case the environmental control is very much improved, as the pigs in reality lie in a building within a building. In such a design making use of good insulation and mechanical ventilation taking in fresh air from the ridge, temperature within the range of 16–21°C may be obtained without difficulty. Only the central part of the house needs complete insulation. It is always best to place a number of cross partitions and perhaps aim to have not more than 100 pigs within a common air space and 300 to 400 pigs in one building.

A popular practise is still to have a length of pen, from 3–4.5 m, taking 10 to 15 pigs, and a depth behind the trough of 1.5 m or 1.7 m. The pen itself should be raised 50–100 mm above the dung passage. The height to eaves need be no more than 1.8–2 m and a low pitch on the roof helps to conserve heat.

16.4.8 Pen size

A fundamental question in the design of fattening quarters is how many pigs should be penned together and how we should grade the pens to make maximum use of the area. The general consensus of opinion is that groups of fatteners are best in lots of not more than 15–20, with 10 perhaps the ideal. It must, nevertheless, be stressed that some farmers can rear up to 40 together with apparent success with a high and unique standard of stock management.

It is essential that when the pigs are lying down in the pen, they more or less cover the entire floor, otherwise dirty habits will develop and muck will be deposited in the pen. The problem arises as to how one can ensure this when a weaner will occupy only about 0.18 m^2 of floor space when recumbent, where as a baconer occupies some 0.46 m^2 and a heavy pig 0.50–0.55 m^2 .

Another solution for baconers and 'heavies' is to have pens of two sizes, one for the growing stage from weaning to say, 16 weeks and the finishing pens form 16 weeks (45 kg) to finishing. If it is desired to have 15 pigs to a pen, the area of the grower pen would be 2.4 m × 1.8 m and the finishing pens could be 3.6 m × 1.8 m. This arrangement envisages ab lib or floor feeding in the grower stage and floor feeding in the finishing stage. If troughs were inserted, 12 to 13 pigs only could be penned under this arrangement. For bacon production two finishing pens would be needed for every grower pen.

Yet another arrangement is to have a weaner pool at eight weeks in which young pigs are placed in fairly large pens, 20 to 30 to a unit. They can be allowed 0.18–0.27 m^2 of lying area and kept there under 45–54 kg. At this stage, the best 10 are taken off to the finishing pens, leaving the remaining number for a few more days when they can be divided off into well balanced groups in the finishing pens. It is likely that the mixing of several litters at weaning creates a 'stress' from which the pigs may take some time to recover under intensive conditions and it is for this reason that the deeply bedded yard with warm kennel lying area is more popular, allowing up to 0.74 m^2 area per pig. The same system can be used if multiple sucking is practiced, but without the severe weaning stress of several changes at once.

16.5 Farrowing Policy (merits and demerits)

Depending on requirement of pigs for breeding, market or for meat processing units, when the demand may fluctuate as it is more during winters and less during summer, the farrowing policy, has to be drawn out on that basis. It can be either multiple farrowing *i.e.* farrowing every month throughout the year or seasonal farrowing so that sows farrow only during two periods of the year. There are merits and demerits of both and should be considered while deciding on farrowing policy for any pig enterprise.

Generally, depending on the size of breeding farm, efforts to make equal number of sows farrow each month be made so that availability of piglets throughout the year is arranged. The objective should be to have two farrowing per breeding sow in a year. Sow has to be watched for onset of heat and then mated. Sows generally exhibit heat within 3 to 5 days after weaning. Production programme can be well planned, when farrowing are planned for every months. In case of seasonal farrowing, the sows have to be mated at such time, so that the sows farrow only two times in a year. This system results in problem of farrowing accommodation over working of boars, as sows have to be covered in particular period and employment policy also have to be modeled as more labour may be required when more farrowing take place. It also creates problem of marketing as lower prices are available due to seasonal run.

In general, therefore, the farrowing policy of having farrowings every month throughout the year is preferred, as there is economical use of capital involved of breeding stock, equipment and accommodation. This also provides for even distribution of labour, reduces the uncertainty of speculation.

16.5.1 Farrowing accommodation

The pig farmer should be ever mindful that there is an appalling mortality in piglets before weaning and average in excess of 20% of those born alive. He should also know that the majority of these losses are due, not to disease, but to bad management, housing playing an important, if not principal, part in this. Surveys have shown that about 50% of the piglets that die, perish due to chilling and crushing, also much of the disease that occurs may be induced by the stress of rearing them under unsatisfactory conditions. The yardsticks for farrowing quarters may be said to be protection, warmth and hygiene.

Some form of protective crate for the sow can be considered, therefore, essential to prevent her clumsy movements crushing the piglets. At the same time, the piglets must be encouraged to spend their resting period away from the immediate proximity of the sow and this is best done by providing nests close to, or as part of the crate itself.

It must be borne in mind that the pigman has to keep a watchful eye on the sow and piglets over the farrowing period without worrying or unnecessarily interfering in any way. In other words, it is as important to give him good facilities as it is for the piglets. All too often this is quite forgotten. We can say that the piglets need a draught proof nest temperature of 21°–27 °C and the sow 10 °C minimum.

Sows and piglets need hygienic surroundings, which means that an essential requirement is periodic depopulation, fumigation and disinfection of the building. Also, sows need quietness and they get this much better in the smaller building. We aim, therefore, to have a unit of a maximum of 16–20 pens within the building, and preferably less. But the liming factor is that it must be small enough to be emptied of all stock regularly and there should be absolutely no compromise of this.

16.5.2 Farrowing crates

There are numerous makes and designs of farrowing crates in the market. Some are portable, semi portable or permanent. A farrowing crate is a device for confining a sow and her litter in such a way that the sow may farrow normally, without hazard to the pigs, and the pigs will have space of their own on either side.Sow's compartment should be long and wide enough to permit them to lie down and get up comfortably but not to turn around. They can be set permanently or can be made a self contained unit that can be removed and it becomes desirable to put the space to other use. Elevated, slotted floor for farrowing crates, designed to greatly reduce the labour involved in cleaning, can be constructed over any concrete floor.

16.5.3 Farrowing crate unit

In its simplest conception a farrowing crate consists of a pair of three parallel rails. The top rails are set 530 mm apart, as also are the centre rails. The bottom rails are 750–800 mm apart, depending on the type of sow and to allow adequate room when she lies down. The bottom row is a minimum of 250 mm from the floor and the second row is 300 mm above the bottom row and 300 mm below the top. Thus the crate has a total height of 850 mm from the floor. It can be constructed of 25 mm bore tubing which may be fixed into concrete blocks at the front and back of the crate. There are escape nests on each side, a minimum of 530 mm wide. Thus the total width of the crate is 1.6–1.65 m or with dividing walls approximately 1.8 m. The wall at the outside of the nests and the front of the crate can be solid, and there is a gate at the back for access by the sow and the attendant. On the inside of this gate there should be a semi circular metal bar 250 mm from the base extending up to 230 mm inside the gate. This will prevent the sow backing right up against the gate and crushing the piglets.

It is desirable to have a cover of plywood, hardboard or asbestos sheet over each nest, and on each side there should be a heat source usually an electric infra red lamp. Great care must be taken with the falls in the floor to make sure that any water or urine runs towards the back and away from the crate and creep; this may seen obvious but it is surprising how frequently the falls are incorrectly made and lead to muck accumulating within the crate. The top on the nest is important to reduce floor draught due to high speed convection currents induced by the infra red lamp.

16.5.4 Indoor farrowing

Farrowing house should contain a soundly constructed creep to provide warmth, food and protection for the piglets, a comfortable sleeping area for the sow and a dunging and feeding area where a clean supply of water is always available. The partitions walls should be at least 1 m high, but preferable built to ceiling hight, as this will isolate the pens and help to prevent the spread of airborne disease. A clear floor area of 2 m will accommodate a first litter sow, whereas a mature sow for safety should have a $^{2}/_{2}$ or $^{2}/_{2.5}$ m pen exclusive of space occupied by guard rails or slope of the roof.

16.5.5 Guard rails

Installation of guard rails on three sides of the box type pen and on the back wall will be made of metal pipe, pieces of 5 to 8 cm straight pipes fastened rigidly to the wall about 23 cm from the floor in such a manner that they project out at right angles from the wall a distance of 15 to 20 cm.

The creep is an essential part of any farrowing accommodation. Simultaneously low power bulb raise the temperature of the creep to around 21°– 27 °C and will serve the purpose and the light will attract the newly born pigs away from the sow. The piglets return to the sow only when they are hungry. Artificial heating of the creep is essential in obtaining the temperature required. The use of infrared lamps for this purpose is standard whilst electric floor heating will provide desired temperature.

Where possible, build the creep adjoining the feeding passage, so that the pig has access for feeding and inspection with our entering the sow's pen, allowing at least 0.13 m^2 per pig. Build the creep square if possible, 1.2 m × 1.2 m are satisfactory measurements. The square, rather than the long creep, will encourage a more even temperature inside the pen. The side walls must be strong. Solid walls are preferred to rails. A roof over the creep is essential as this will keep the heat in and help to prevent draught at ground level.

16.5.6 Creep area

A creep area where the temperature requirement of the piglet can be obtained and a feeding that is free from perishability, are commonly made in the creep design. An electric pig breeder equipped with a 100 or 150 watt eclectic lamp is very satisfactory and can be made at very little expense. The breeder should be placed in the corner of the farrowing pen for the first week or two of the pigs' life.

16.5.7 Alternative crates

Indeed, many of the most successful units consist of a row of crates, single or double with a feeding passage in the front and a small passage for cleaning and movement of the pigs at the back. To ensure easy access to the creep, to attend to the piglets, the walls adjoining this portion need be only 530–600 mm high to retain the piglets. A trough and, if desired, a separate water bowl or nozzle drinker may be placed in front of the crate. With this design it is best to use raise in the front, to assist feeding, or it may be fitted quite successfully, but more expensively, with a swinging trough front.

With these units the cross sectional size of the building can be 4.8 m with a single sided unit (cute and trough 2.7 m, passages each 1 m, or with a double sided unit 8.4 m) both being standard sizes in the building industry.

16.5.8 Bunker design

Some farmers believe that it is important to provide exercise for the sow. The crate can still be used in this form. Here the crates are placed alternatively head to tail and there is an exercise area behind each extending area over the width of two crates. The exercise area is 1.5 m wide and 3.6 m long; with the feed and water trough, and the sow will probably be allowed out to feed and exercise twice a day for around half an hr.

This design is rather more expensive than the one previously described. A single row of crates can only be accommodated in a building of 5.4 m wide. It also means 1.5 m wide doors along the exercise and feeding area, and a common passage way for cleaning. Whether the extra expense is really worth the luxury of allowing the sow to exercise for the short period during which they are in the crate, is arguable, but taken all in all, this may be said to be a most satisfactory system for the large enterprise, provided it is divided into smaller units.

16.5.9 Slatted and slotted floor farrowing pens

There has recently been a considerable interest in farrowing pens with either part slatted or entirely slatted floors. In the case of the latter, a suitable slat has been either 65 or 100 mm concrete one with a normal gap or 10–12 mm but with an enlarged gap of 20–28 mm in a 0.18 m^2 area behind the sow. For the first week after farrowing, the area behind the sow is covered with expanded metal sheet to prevent the baby pigs catching their feet in the gap.

16.5.9.1 *Slat floor*

Most of the piggeries are with solid floored, side dunging passages is to clean out with shovel and barrow and provide trapped drains to take off excess liquid. An alternative is to have a solid floor dunging passage of 75–100 mm, omit the drains, but have a virtually flat floor so that the passage can be cleaned out either with a mechanical scraper or with a squeegee that is made to exactly fit the width of the passage. The simplest arrangement for dung disposal is the slatted floor with automatic drain of sludge to pit underneath.

Slatted floors for pigs can be constructed of several different materials, concrete, metal or welded wire mesh. Reinforced concrete is probably the most popular and widely successful material, whilst wire mesh is the cheapest.

The following measurements are those commonly used. For concrete slats, a width of 50–75 mm at the top tapering to 38–50 mm at the base, a depth of 60–75 mm and gap of 21 mm between the slats.

Welded mesh is usually 75 mm × 12 mm at 10 gauge or 75 mm ×15 mm at 5 gauge. Perforated steel panels, 14 gauge, are made 1 mm × 0.2 mm or 1.2 mm × 0.2 mm and holes are punched to give about a 50% void. If they are not dip galvanized, their life is much prolonged and they have the advantage that they are suited to all ages of pigs, probably the only perforated flooring that comes into this category.

Other metal performed floors are made of flattened expanded metal 17 mm, 10 gauge or steel straps 30 to 37 mm wide and 5 to 7 mm thick with a space between of 9 mm for farrowing and 17 mm apart for nursery units.

In many respects, concrete is the material of choice for slats as, if it is properly made, it is almost indestructible. It is also load bearing and self supporting. It is usual to use concrete slats of 50 to 75 mm width, but narrower widths are more susceptible to lodging of faeces than wider slats, for example of 100 mm, although narrower gaps must be used in the narrower slat in order to ensure comfort. Also, with the narrower slat the sides have to be perpendicular to the face in order to provide adequate cover to the steel reinforcement. However, slats bounded by parallel concrete faces 75 mm deep are subject to bridging of the manure.

In calculating the volume required for sludge, one should allow for approximately 0.08 m^3 of sludge per pig per week with whey feeding (which produces the maximum), down to 0.04–0.05 m^3 per pig per week with meal feeding. If lengthy storage facilities are needed, the cost of the sludge tank will be considerable and one of the main advantages of this system.

16.5.10 Housing the boar

The boar is half the herd from the genetical stand point is true enough and this emphasizes his importance. The correct form of the housing is also important because it prolongs his existence and use in the herd and aids his fertility. Boars are apt to 'go off their legs' and suffer from a number of mechanical troubles of the legs and limbs which may be less likely with good housing. From the health and vigour point of view, probably the best way of keeping the boar is outside in a paddock with the simple protection from heat and rain. Where more confined accommodation is required, a well bedded place adjacent to the service area is essential with at least 7.2 m^2 or 9.3 m^2 if the boar pen is combined with the service area. Preferably a separate service pen usually works better, away from his own quarter. A work routine can then be planned which allows the stockman to bring sows to the service pen without interference from the boar. Safety is an important factor in boar pen design and layout. An open pen front or open gates need to be fairly robust and frames with 25 mm square sections will do this job nicely. A square service pen, at lest 3 m × 3 m, without projections and with a non-slip floor, make ideal operating conditions for boar and stockman. On no account should a boar be kept in cold, damp conditions, right away from other stock or he may develop leg troubles and become vicious.

16.5.10.1 *Boar sty*

Boar sty should not be more than 24 pens under one roof and one pen shall accommodate not more than one boar

Simple hygienic layout

The simplicity of this design will be apparent. Also, it can form a desirably hygienic layout. There is no contact between pens as the drainage can run outside in small open channel at the side of the passage way. A good fall on the floor will take the urine and water to the corner of the doorway and out to the channel. The open channel may seem to be an desirable feature, but in reality it is probably a much more hygienic arrangement than a closed drain running under the passage. Drainage to the back of the pen and trap to the outside will give good hygienic drainage, but this is an expensive arrangement.

16.6 Densities and Numbers

A critical factor in the successful rearing of pigs from three weeks onwards is to have correct balance between the numbers in the group and the space they are allowed. Many trials have been conducted on pigs of all ages. The results of the trial have shown that (a) overstocking retards growth, (b) small groups do better than large ones, and (c) the litter group is the ideal.

For example, 50 piglets are not uncommon in one pen, but the troubles they can produce range from uneven growth, poor food conservation and vices such as tail biting, scour, pneumonia and rhinitis. In general, litter group is ideal and often obtainable with flat decks, up to 25 seems perfectly acceptable with the absolute necessity where larger numbers are used of providing rather more space to compensate, particularly so far as the lying area in kennels is concerned. The optimum total floor area should approximately be 0.09 m^2 for 9 kg live weight of pig. If the flooring is partly solid and partly perforated, then the lying area may be reduced by about 25%, but an extra area for the dunging is added on top of this.

Demerits

They found that group size affected level of performance and that the best group size was governed by the size and weight of the pig and by environment. A lower rate of gain took place at the higher stocking densities, which was due to lower food consumption due to heat stress. The feed conversion was not affected by the stocking density.

About 12–20 pigs per pen is probably the ideal and is unlikely to be an unwise choice; extremely dense stocking may, however, retard the growth of pigs unless the environment conditions are carefully maintained at the optimum.

16.7 Weaning

Weaning involves removal of young piglets after birth from access to milk provided by the mother after sometime. In natural course, the piglets normally become accustomed to foods other than milk by process of exploratory behaviour which is due to decline in milk available from mother with increasing appetite of developing pig. Normally the piglets are weaned from dam at 8 weeks of age when the milk yield declines in sows. By earlier weaning, sow is removed from the responsibility of providing piglet with nutrition from the time of weaning and the weaned piglets have to be provided with substituents for its mothers milk. Early weaning will therefore depend on balance between value of sow feed and cost of extra milk substitute required by early weaned pigs. If the cost of weaning food is not too expensive relative to sow food and there is marked increase in number of weaners per sow per year due to early weaning, then the practise of early weaning should be adopted.

When piglets reach 4 weeks of age they have better immunity and more mature digestive system to minimize both the extent and effect of post weaning stress and is also not so demanding in terms of environmental temperature. This 4 weeks age is the optimum age for weaning as it is more than compensated by increased litter size and low mortality after weaning as compared to earlier weaning at 2 to 3 weeks. Relative availability and price of ingredients suitable for including in the diet of early weaned piglets, should be kept in a view for deciding about the age of weaning. The objective of weaning process is to obtain steady and uninterrupted growth with no check and live-weight gain accelerated steadily over the period. Mortality from birth to weaning should be minimal *i.e.* less than 5%. Piglets should not show any enteric problem after consuming reasonable quantities of creep and post weaning diet.

From 1 week to 4 week age, growth rate should be aimed to have gain in weight of 214 g per week and in the 5th week about 286 g and in 6th week 350 g and this should progressively increase in subsequent weeks. Weaning weight of 20 kg can be achieved by end of 8 weeks.

16.7.1 Minimizing stress at weaning

Social and psychological upset is caused to piglets when they are suddenly deprived of mothers' presence and further accentuated by disturbed behaviour of litter mates.

It looses main sources of nutrient (mothers milk). The weaning process is so adapted so as to minimize the stress and then only growth curve can be maintained smoothly.

16.7.2 Climatic environment

Piglets perform best when temperature is maintained between lower and upper critical temperature. In cold conditions pigs have to be kept warmer. For this, well insulated floors, keeping of pigs in groups, prevention of draught by using curtains be provided. At the same time under hot conditions during summer, piglets be kept in cool condition. For this, adequate floor space is needed so that pigs are allowed to spread out as they avoid huddling.

16.7.3 Advantage of early weaning

Advantages of weaning at birth are discussed here:

(a) More litters could be obtained from the sow, as the sow comes in heat early after weaning and can then be served;

(b) The sow does not loose condition as a result of prolonged lactation and there is saving in food;

(c) Sow can be returned back to breeding herd early and so more can be kept;

(d) Piglets are not in contact with parasite and other infections;

(e) Better control of anemia can be carried out as milk substitutes can be fortified with iron etc.,

16.7.4 Pigs born and weaned

The most critical period in pig production is at farrowing time, during the suckling period, at weaning time and during the period from weaning until pigs are marketed. If due care is not taken the losses are great. Even if proper farrowing takes place and large litters are produced, there is still chance that pigs will not be saved. About 25% or so pigs farrowed are lost in weaning stage and losses occur about 80% or so within 3 or 4 days after farrowing. It should be appreciated that loss of each pig affects the economy of the enterprise as the investment in feeding of sow from breeding to weaning is lost proportionately and thus affects the potential profit. The economy of the enterprise is affected depending on average number of farrowing per sow per year, average litter size, total number of pigs weaned per litter. Generally in India, the average litter size is of 9 pigs per litter and average number weaned per litter is 7. Profit or loss from any swine breeding herd is dependent upon number of pigs weaned and marketed per sow. Feeding and management of herd during gestation influences number of pigs born and weaned.

If management is proper more pigs will farrow per sow producing larger and healthier pigs at birth and producing fewer dead pigs, runts and abnormal pigs per litter and better production of milk per sow and more heavier pigs are weaned per litter.

16.7.5 Rearing of orphan piglet

Health of the sow (dam) due to illness, absence of milk in dam or otherwise soon after farrowing require artificial rearing of the orphan piglets. If the dam is alive and milk is available piglet be fed necessarily for three days to obtain colostrum as it is useful for providing immune bodies from dam to piglet. Then milk substitutes, primarily designed for dry feeding, can be reconstituted with cow milk or water to produce required food for feeding piglets either through shallow trough, or bottle. Fresh drinking water should always be provided. Piglets are fed in such cases four times daily for first 10 days and later reduced to three times daily. Long night intervals should be avoided when the litter is weaned from sow.

Artificial rearing of piglets without colostrum is not desirable and following aspects to be considered:

(i) as an attempt to save a litter when the sow dies during parturition;

(ii) as a method of creating a nucleus of breeding animals from a valuable herd of pigs with one or more endemic diseases;

(iii) as a measure of rearing, infection free animals may be given *E. coli* antiserum at birth, may supply some of the essential factors which the piglets receive through colostrum and an injection of vitamin-A and D is advisable.

Use of heater and infrared lamp is essential for rearing orphan piglets for keeping them warm and kept in a place free from drought and sufficient warm dry clean bedding.

16.7.6 Birth and weaning weight measurement

At the time of farrowing, each individual piglet of the litter is weighed and litter weight is also taken and recorded. Efforts should be made for gaining maximum weaning weight whenever the weaning is carried out between 2 to 8 weeks. Live weight from the average birth weight of 1.5 kg till end of 4th week is restricted to abut 210 g per week but later in 5th week, the growth is fast, about 285 g, in 6th week it is about 350 g, in 7th week it is 425 g, in 8th week it is 570 g and weaning weight is about 20 kg. The growth curve is more steep from 5th to 8th week. In India birth weight of 1 kg to 1.2 kg and weaning weight of 15 kg and around is generally achieved and this weaning weight can be improved by better quality of creep ration. The objective is that the young piglet after weaning should have a strong, well developed skeleton and high proportion of good quality lean meat and less fat. System of feeding with highly digestible diets for very early weaning leads to a greater degree of contentment. This practise also ensures more equitable food intake for all pigs within a pen as compared to restricted feeding. When early weaning is practiced, rationing are desirable. Once these weaned pigs become used to, they gradually utilize the high quality starter feed well. They are provided cheaper ingredients depending on age and stage of growth. High intake of creep feed is achieved prior to weaning. Diets of pigs in crucial period after weaning should be properly evaluated keeping in view daily gain in weight, feed efficiency, piglet losses, securing consistency of results from week to week and ease of management. Generally the high quality diet is more cost effective. The reduced cost per kg of live weight gain is more important than low cost of tonnes of diet.

The practise of keeping pigs of uniform weight into one pen at weaning is important, as they can grow well in this group and smaller and more vulnerable pigs together, so that they can get special treatment. Development of 'runts' takes place when in sucking stage, weak piglets, which is not able to get milk, gets weaker and does not thrive and when in weaning stage, it is unable to feed itself due to many factors such as less space available at feeding trough, higher weight pigs kept with the weak ones etc. So, it is advisable in such cases to keep small once separately and provide special diet etc. for their better weight gain.

If pigs remain healthy and have high feed intake, it shows that most weaning diet and factors like group size, floor space and feeding space are adequate. Live weight gain is the best indicator of good weaning system being practised. If pigs of uniform weight are put in a pen at weaning and they grow well, this system is considered fairly sound.

16.8 Management of Growing and Fattening Pig

When pig reaches about 15 to 20 kg live weight after weaning within about 8 to 9 weeks of age, management system is considered to be good. Its digestive system by this time is capable of effectively dealing with wide range of ingredients having energy and protein concentrate.

In case of growers, balanced diet with high energy content should be provided. If diets having less than the average energy content and high fibre content is provided, then the pigs above 20 kg live weight fail to maintain same energy intake. The faster the growth and shorter the time to reach slaughter weight, the producer will get the return more quickly on investment. Similarly, the faster the growth, the more the pigs can be produced in the same building and the overhead cost per pig will be reduced. Similar will be the reduction on labour cost. Some consider that restricted feeding helps to achieve superior feed conversion efficiency by checking feed wastage which is associated with *ad lib* feeding.

Average daily food intake increased from 2 to 2.5 kg per day in pigs with live weight range from 20 to 90 kg and this progressive increase in feed intake produce fatter pigs. In order to produce pigs with higher proportion of lean to fat the practise of providing ad-lib feeding up to a gain of 50 kg live weight and later restricted feeding till it gains 85 to 90 kg weight will reduce deposit of fatty tissue. This system is useful.

Management should be so carried out in growers and fatteners that the objective of good health, minimizing losses and development of good meat carcass quality, good food conversion efficiency, low food cost per unit of live weight gain, low total cost per unit of live weight gain and good return on investment and capital.

It is necessary that efficient genetic stock are available for growing and finishing pigs in the enterprise so that all expensive inputs such as those on feed, labour and housing will yield a good financial return. Healthy stock can also exploit these expensive inputs more effectively than the unhealthy stock. If growers about 20 to 30 kg live weight are purchased from outside, every precaution to purchase them from healthy source should only be taken and the disease status of the sources should be ascertained before purchase. Diets for growers between 20 to 50 kg and fattiness between 50 to 90 kg will vary and diet should be so formulated, so that nutrient and digestive requirements for various groups is cost effective. It may prove useful to increase the protein quality of diet to achieve lean tissue growth rate in superior genotypes for development of better grade carcasses. Two different diets, one for pigs between 20 to 50 kg and the other above this weight range *i.e.* between 50 to 90 kg slaughter weight be formulated and the cost effectiveness of these two diets be made and changes effected in those traits on the basis of trial and monitoring their result on live weight gain as well as lean tissue growth rate and back fat thickness. A combination of *ad lib* and restricted feeding system keeps a check on carcass quality and less back fat. It will be good that dry feed mixture is given in feeding troughs but it may be made wet before consumption by pigs to avoid wastage of dry feed. Sufficient feeding space for all pigs fed at one time in a pen should be provided. Shoulder width measurement be treated as space required for feeding for one pig. It has been estimated that the wet feed is superior to the extent of 5% in respect of live weight gain as well as fat conversion efficiency as compared to dry feed. Grouping of these growers and fatteners should not be more than 30 in a pen, this is the upper limit as there is belief that pigs do not recognize more than 30 of their fellows in a pen, and it is good developing a stable social order. Their should be adequate feeding and watering space order depending on number of pigs in a pen. Adequate floor space requirement is the space taken by a pig in fully recumbent position (lying full) stretched on its side.

16.9 Care and Management of Pregnant Animals

The gestation period of sow varies from 109–120 days with an average of 114 days. Pregnant animals should be housed in groups in separate enclosures and should not be mixed with new animals to avoid fighting which at times may result in abortion. It would also be advisable to house pregnant gilts and sows in separate groups during gestation. About 3 m^2 of dry housing should be available for each sow. The pregnant animals should be allowed to move about every day in the morning on a free range or a pasture if available. A pasture area is presumed to be clean if a cultivated crop was raised.

16.10 Management of Boars and Gilts

There is obvious differences in effectively managing castrates, boars and gilts.

Boars and gilts have higher heat production at maintenance levels of feed intake than castrates. In case of boars it is 10% higher and in gilts about 15% higher than of castrates, it is due to higher lean content. Sex hormones like oestrogen in gilts and androgens in boar are responsible for lower appetite level in boars and gilt as compared to castrates. There is definite superiority in live-weight gain on entire male in relation to gilt during the period of growth between 50 to 90 kg of live weight. Gilts also deposit fatty tissue at faster rate between 50 to 90 kg than male and castrates deposit fatty tissue even faster than gilts at this stage. The entire males use dietary protein more efficiently than gilts and they are more efficient than castrates. It has been worked out after study that for producing carcasses of same weight and similar lean to fat ratio between 20 to 90 kg live weight, feed intake reduced by 16% in castrates and 8% in gilts as compared to boar.

Two sexes, boars and gilts may be managed separately and instead of separating them at a later stage, which often results infighting amongst pigs, it is useful to segregate immediately after weaning. Gilts will have to be slaughtered at lighter weight than boars and castrates.

Gilt should be bred when they are about 10 months to 12 months of age and should be well grown. In pig sows should be flushed *i.e.* given extra feed during last fortnight before farrowing time. The extra quantity required will depend on age and condition of the sow and the quantity of feed mash.

Boars should be in good, thrifty condition and well managed. If vitality of boars is too low or if he is used heavily, this will weaken male reproductive cells (spermatozoa) to such an extent that it will not be able to fertilize all the eggs produced by the female and may affect litter size. If condition of boar is poor, he should be put to some extra quality feed but should not be fat. About 50 to 60 gilts during a breeding season can be mated to a mature boar. It is always better to bring boar to gilt rather leaving boar to run with sows. Not more than 3 sows can be mated to a boar in one day. Due to excess of fat some boars become inactive. For these non-breeders or those showing low 'sex libido', veterinarian should be consulted. Exercise be given and feed intake can be reduced to restricted feeding if it is too fat.

16.11 Castration

Pigs should be castrated at an early age of one or two weeks. If they are castrated later, problems may arise. It should be performed at least at two week interval from weaning or after deworming and should be carried out in warm weather, pen should be provided with clean, dry bedding. Two persons are required for castration, one holds the pig while the other operates. Pig is held by a front and hind leg on opposite sides with its back on the floor or by the hind legs with pigs head and shoulder between the assistant's knee. Castration is carried out by open method. The scrotum is washed with soap and water and mild antiseptic solution is applied. Using a sharp knife the incision is made over each testicle or in between the testicles parallel to middle line of body. Incision should pass through the skin from the top of testicle and covering of the testicle. The testicles are then slowly pulled through the incision and the attachments are separated which may cause little bleeding. The second testicle is removed in similar manner. Incision should be fairly long so as to provide proper drainage to the wound. Remove as much chord as possible. Some disinfectant may be applied to the wound. In case of 'rigs' *i.e.* male pigs which show only one testicle, the other may be in the body of the pig. Such pigs be handled by qualified veterinarian

16.12 Removal of Needle Teeth

Piglets are born with four pairs of sharp teeth, with two pairs on each jaw. They are of no practical value to the piglets and they may irritate the sow's udder during nursing or cause injury to other piglets. Clipping of these teeth shortly after birth will prevent the injury of the udder caused by the needle teeth.

16.13 Hints on Selection and Culling of Animals

For proper selection of breeding stock, physical selection, (phenotype) should be seen along with breed characteristics and for genotypic selection, the performance of breeding herd and pedigree should be considered. Performance in relation to average litter size, average farrowings per year, piglets weaned are considered.

Birth and weaning weight, feed conversion efficiency and mortality, carcass quality and genetic defects in progeny, if any, are important. Boars having good progeny testing record should be considered.

Culling of female from breeding stock be made at the earliest age, at about 50 kg liveweight. If they are unfit for breeding they should be fattened and marketed for meat. Preliminary selection for both male/female can be made at earlier age *i.e.* about 90 days and those males which are considered unfit for breeding castrated and fattened for meat purpose. In case of gilts, 2nd selection, when they gain 50 kg live weight, be made. Gilts which are progeny of problem mother should be culled.

16.14 Determination of the Number of Pens and Stalls Required in a Pig Unit

One objective in planning a pig unit is to balance the accommodation between the various ages and numbers of pigs. Ideally, each pen should be fully occupied at all times, allowing only for a cleaning and sanitation period of about 7 days between successive groups.

In the following example the number of different pens required in a 14–sow herd, where 8 week weaning is practised, will be determined.

1. Determining the farrowing interval and number of farrowings per year

Average weaning to conception interval	20 days
Gestation	114 days
Suckling period (7 × 8 weeks)	56 days
Farrowing interval	190 days

Number of Farrowings per sow and year 365/190 = 1.9

2. Determining the number of farrowing pens

The piglets remain in the Farrowing pen until 12 weeks of age.

Before Farrowing	7 days
Suckling period	56 days
Rearing of weaners	28 days
Cleaning and sanitation of pen	7 days
Occupation per cycle	98 days

Thus one Farrowing pen can be used for: 365/98 = 3.7 Farrowings per year.
A 14 sow herd with an average of 1.9 Farrowings per sow and year requires (14 × 19)/3.7 = 7 Farrowing pens.

3. Determining the number of servicing/gestating pens

Average weaning to conception interval	20 days
Gestation period less 7 days in Farrowing pen	107 days
Cleaning and sanitation of pen	7 days
Occupancy per cycle	134 days

Thus one place in the servicing/gestation accommodation can be used for: 365/134 = 2.7 Farrowings per year.
With a total of 27 farrowings a year 27/2.7 = 10 places would be required.

4. Determining the number of places for replacement stock

Presume the sows on average get 5 litters, then 20% of all litters will be from gilts.

Rearing of breeding stock (12 to 35 weeks)	168 days
Gestation less 7 days in farrowing pen	107 days
Cleaning and sanitation of pen	7 days
Occupancy per cycle	282 days

About 30% more animals are separated than the required number of gilts thus the required number of places in the 14 sow herd will be

$(14 \times 1.9 \times 0.2 \times 1.3 \times 282)/365 = 6$ places

5. Determining the number of places in the growing/ finishing accommodation

One stage finishing:

Fattening of pigs 12 to 27 weeks of age, (25–90 kg)	105 days
Extra period for last pig in the pen to reach marketable weight	21 days
Cleaning and sanitation of pen	7 days
Occupancy per cycle	133 days

Assuming that 8 pigs per litter will survive to 12 weeks of age the number of places required in the finishing accommodation will be:

$(14 \times 1.9 \times 8 \times 133)/ 65 = 78$

That is 8 pens with 10 pigs in each or 10 pens if each litter should be kept together.

Two stage growing/ finishing unit

Growing pigs 12 to 20 weeks of age will occupy a growing pen for 63 days including 7 days for cleaning.

$$(14 \times 1.9 \times 8 \times 63)/365 = 37 \text{ places is required in the unit.}$$

Finishing pigs 20 to 27 weeks of age will occupy a finishing pen for 70 days including 14 days emptying period and 7 days for cleaning. The emptying period will be shorter if the pigs are sorted for size while being transferred from the growing to the finishing pens.

$$(14 \times 19 \times 8 \times 70)/365 = 41 \text{ places is required in the unit}$$

From the above example it will be appreciated that the number of pens of various kinds required in a pig unit is based on a number of factors. It is, therefore, not possible to lay down hard and fast rules about the relative number of pens and stalls.

16.15 Manure Management

Farmers who grow pigs typically collect manure in lagoons outside the barns in which pigs are raised. In the lagoons, manure is degraded by anaerobic bacteria; carbon-containing compounds decompose and become carbon dioxide and methane; organic nitrogen is converted to ammonia. Farmers usually use lagoon liquid as fertilizer, applying it to fields. Using the liquid in this way takes advantage of the nutrients in the manure. The liquid from lagoons is a nutrient resource, and proper management and use of this resource can reduce the need for commercial fertilizers.

Livestock manure represents a valuable resource, which if used appropriately with minimal loss, can replace significant amounts of mineral fertilizer in areas with intensive livestock production. On the other hand, large volumes of animal manure are not only a source of valuable plant nutrients but also a source of air pollution and a threat to aquifers and surface water.

In India, manure may be discharged into waterways and liquid manure is leaching to groundwater that may be used by water abstraction plants. This poses a great risk to human and animal health, as livestock manure contains numerous pathogens (bacteria, viruses, parasites). It is also likely that pathogens, *e.g.* foot-and-mouth disease virus, may be transported with animal effluent into waterways. Thus, an infected farm may cause infection of farms downstream.

Livestock production units emit gases that contribute to global warming, putrefaction and bad odours. European emission inventories showed that livestock production constitutes 70–80% of the total ammonia emissions. The atmospheric concentration of methane (CH_4) a greenhouse gas (GHG), has increased with 45% since 1850 and livestock manure is estimated to contribute 5% to the total emission of CH_4 in the 1990s. Nitrous oxide (N_2O) which is a very potent GHG, emission has increased from 11 Tg year-1 in 1850 to 18 Tg year-1 in mid 1990s, mainly due to increase in agricultural sources and manure have contributed significantly to this increase.

The environmental hazards described above can be mitigated by the environmental friendly recycling of animal manure. Recycling will contribute with plant nutrients to crop fertilization, thereby reducing the need for nitrogen, phosphorus, potassium and micro-nutrients applied to the fields in mineral fertilizers.

In India, manure on all sizes of farm holdings is mainly separated manually in solid and liquid fractions inside the animal house. Urine from the housing diluted with washing water, *i.e.* liquid manure, is transported through open channels to outside the animal houses. On smallholder farms, pigs, dairy cows and buffalo may be raised under confinement on floors bedded with rice straw, thereby producing manure with a high carbon content, which may contribute to improve soils poor in organic matter. Poultry manure is collected in a solid dry form.

Solid manure on most livestock farms is composted in quantities so small that only a limited temperature increase is achieved. On a few farms the manure is covered with plastic or composted in-doors. The advantages of composting are a reduced risk of spreading pathogens and weed seeds, a reduction of volume, stabilization of the manure and the reduction of hatching of insects. It has been observed that hatching insects can be a problem even after the application of manure. Solid manure is in some regions a commodity sold to farmers producing high value crops such as vegetables, coffee, or to fish producers.

Liquid manure is spread to fields manually, a practice that is demanding much manpower and is unpleasant. Problem aggravates when the manure is transported through villages to the fields. Some medium and large-scale farmers are discharging the liquid manure into canals or rivers after anaerobic treatment in lagoons, coupled in series.

Nutrient flows in manure handling systems

For the purpose of assessing the amount of plant nutrients available in manure, information is needed about the excretion of N and P by animals, and of fractionation of plant nutrients in solid and liquid manure. The N and P concentration may well be affected by water used to clean animal houses and to cool the animals, thus it is estimated that between 30 and 50 litre water is used per pig. There is no information about leaching losses of N and P from solid manure and from liquid manure stored in lagoons and surface runoffs. Solid as well as liquid manure is frequently used for crop production. Presently, composted manures may be applied for either fruit orchards or vegetables, whereas for wetland rice. To some extent the manure may be applied directly onto the field without composting. Normally mixing with human waste is not recommended, but in some villages this practice may be carried out, as toilet or latrine waste is diverted into manure biogas digesters or waste ponds.

Farms are using the liquid manure as fertilizers for crops although it is the impression that most liquid manure is discharged to fish ponds, where it contributes as feed for the fish and as fertiliser for the plants that is eaten by the herbivorous fish. It is the impression that there is little recognition of the risk for transmission of pathogens with animal manure and little is done to reduce the risk of diseases, to farmers, that is caused by inappropriate management of manure.

The present lack of technologies for transport and application of animal manure is a barrier for an efficient use of plant nutrients in animal wastes and therefore become an increasing risk for the environment. These structural changes should be followed up by regulations. The regulations should be based on research, and policies focusing on the development of manure handling systems, that support a harmonious relationship between plant nutrients present in manure and crop/ fish pond requirements. Further, there may be a need to prohibit direct discharge of untreated effluents into rivers and lakes. It appears that manure processing with the objective of reducing the risk of pathogen transmission is only carried out if there is no imminent need of the manure as a fertilizer. Otherwise, manure is spread directly without storage or pre-treatment. Thus, the research and development of appropriate manure handling technologies should focus on manpower requirements, plant nutrient availability and pathogen reduction. Outcome should be guidelines for efficient and sanitary safe use of animal manure, which include recommendations for introducing technologies to reduce nitrogen losses and improve the utilisation of organic nitrogen from manure. Furthermore, farmers should be familiar with the capacity of manure as a fertilizer. If manure is used efficiently this would increase farmer incomes.

Solid manure

Swine manure was historically handled as a solid, either deposited directly by grazing animals, or collected in bedding placed on solid shelter floors to absorb the urine. Pastured animals spread the manure over the land as they grazed. Manure deposited on solid floors is typically stored where it falls, with more bedding added as needed to maintain a dry floor. Liquid drains away from the manure dropped on an outside lot and must be collected in storage, leaving the solid manure behind. The manure composts in place somewhat and is removed every few months. Fertilizer value is recovered by spreading on cropland and to complete the nutrient cycle. Solid manure is normally surface applied, but in some cases may be incorporated into the soil with a farm tillage operation shortly after spreading. Composting is another option for solid manure management.

Lot runoff

Manure is typically scraped from outside lots every week or two and stacked until it can be hauled to cropland. It is important to keep an outside lot relatively free of manure to control odour and so that rainfall runoff stays mostly free of manure. This facilitates storage of relatively clean runoff for irrigation onto cropland. It is even possible to divert runoff from small operations directly to pasture or to a vegetated filter strip where it can infiltrate. It must be prevented from entering waterways. Clean upslope water and roof water should be diverted away from the open lot to minimize the amount of wastewater that must be handled as a manure.

Liquid manure

Manure typically falls through a slotted floor (with the size of slot depending on the size and age of animal) into either a gutter or a concrete storage pit. Storage pits provides from 3–12 months storage of the manure. This pit may either be located directly under the slotted floor and may be from 4' to 10' deep. In some operations, the manure falls into a shallow pit or gutter which is periodically pumped, flushed or drained to a large outside storage. The outside storage may either be constructed in the earth or commercial steel or concrete storage purchased and erected onsite. This avoids the need to apply manure during the crop growing season and when weather conditions are unsuitable.

Lagoons

Lagoons are different from liquid manure storage because they are operated to encourage anaerobic digestion of organic material while it is being stored. This reduces odor when the treated manure is land applied. A properly designed and operated treatment lagoon is much larger and more expensive than liquid manure storage with the same storage time, and the organic solids are much less concentrated in the liquid.

An equal part of relatively clean water must be added for each part of manure. Furthermore, manure must be added slowly and uniformly to the lagoon, to avoid an upset (and subsequent release of odors) to the biological treatment system. One common method of doing this is to utilize shallow pits or gutters under slotted floors and drain or flush manure to the lagoon on a frequent basis, usually every three days to three weeks. This is done by simply pulling a plug in the bottom of the pit, called gravity drain, use of a scraper system running in the underfloor gutter, through a process called a "hairpen" gutter or by recirculating a volume of relatively clean effluent from the lagoon to flush manure out of the building and into the lagoon. Recirculation involves either a flushing action that takes place several times a day or a "pit recharge" system that works basically like a toilet that is flushed every few days.

Fig.16.2. Schematic diagram of a lagoon

A portion of the lagoon contents or "minimum design volume" must be left in the lagoon after its contents are pumped to the land to provide a large number of microbial organisms to treat the new manure entering the system. In spite of proper operation, there is an "over turning" of the lagoon contents that occurs in the fall of the year for a couple of weeks, as ambient temperature drops and cools the top layer of liquid in the lagoon. As its density increases, it "overturns" or drops to the bottom of the lagoon, forcing the bottom layer, containing partially digested manure solids, to the top. This phenomenon results in higher odor levels for a week or two around the lagoon. Multiple lagoons in series normally emit fewer odors than single cell lagoons.

Lagoon contents are normally applied to cropland by spray irrigation systems. If the lagoon is properly designed and operated, spray irrigation should not release much odor because most of the organic solids should have been biologically degraded. In a well-operated lagoon, typical effluent should have only about 20% as much nitrogen (N) and about 30% to 40% as much phosphorous (P) and potassium (K) as the raw manure, because of treatment and sedimentation of solids to the bottom of the lagoon. Note that the P and K "lost" actually accumulate in the sludge and must be utilized properly when removed. These solids, or sludge, must be removed every few years and the operation should plan to handle them as a part of their nutrient management plan. Because this material is more concentrated, it may be practical to haul the sludge off site to more distant cropland that can better utilize the nutrients contained in the sludge. Because of the nuisance potential of this partially stabilized material, it should be incorporated as liquid manure if possible.

Manure treatment and utilization

Liquid manure and other organic waste are a byproduct of agricultural production with a high content of nutrition and organic matter. Manure contains pathogens and weeds, which are considered to be a greater problem when manure is applied to the soil.

Manure treatment and utilization technology approach is based on the extensive research and technological work, which is based on the concept that all manure and other organic waste should be treated and utilized without increasing the potential for pollution problems.

Wastewater and sludge treatment technology and processes are well known and can be adapted for liquid manure treatment. The scale of an operation is a significant factor in the selection of manure treatment process. There are many factors to consider in selection of manure treatment technology but most importantly it needs considerable technical expertise for proper design.

Solid-liquid separation approach

The technology includes conventional equipment for gravity separation device, pumps and piping. The removed solids are going to composting treatment following land application and liquid pumped to irrigation water storage.

The ability to design and accommodate complete flexibility makes the solid-liquid separation technology adaptable to a wide range of productivity and requires minimum farmer's attention and can be run with unskilled personnel. This approach can be used as an improvement step on existing farms. Most important manure should not be stored near the barn.

Thermo-treatment approach

The main consideration in manure processing is of getting a product which is easy to utilize and environmentally acceptable. For this reason the thermo-treatment of liquid manure brings complete sterilization with simultaneous preparation for dehydration and avoids the possibility of weed germination.

CHAPTER 17

HANDLING AND CARE OF SWINE

17.1 Handling and Care of Swine

The pigs are generally nervous individual and inclined to be mischievous and destructive in groups. Nevertheless, if approached quietly and firmly, pigs will soon respond to handling and become docile. Cruel and harsh methods never pay in handling pigs as that usually results in excitement and injury to pigs. A good hog raiser avoids sudden moves and actions which startle the hogs and cause them to break and run. Necessity of handling is for drenching, wound dressing, vaccination and castration.

17.2 Handling and Catching

17.2.1 Handling of piglets

Piglets may be easily caught by grasping either of the hind legs just above the hock joint and lifting them off the floor. Heavier pigs should be caught by grasping them behind the shoulders, using your outstretched hands. In this way pigs up to about 50 kg may be handled with little difficulty. Heavier pigs should be run into a cage or restricted by using a rope.

17.2.2 Handling the older pigs

Pigs will naturally head for a gap (or opening) when you approach them or try to catch them. You can use this practise to make the pig to go where you want. If two pig boards (wooden board of 0.8 sq mt) are placed on either side of the pig's head, it will move forward in the direction the handler wants it to go. As the animal gets older it can be trained to move under the control of one handler who uses a board and a wooden bat of about 1 m long.

17.2.3 Restraining of pigs

Restraining can be done by two methods; physical restraining and chemical restraining.

17.2.3.1 *Physical restraining*

It can be done by the use of bull nose ring and self piercing copper wire ring. Bull nose ring is used for boars and sows whereas copper ring is used for growing stock. Physical restraining can also be done by using snout rope, harness rope, using snout snares and by using tongs.

Several suitable types of ropes are available which are slipped over the upper jaw of the hog and provide the person with sufficient leverage to hold the hog readily when it pulls backward.

For holding large sows and boars, use a piece of stronger rope about one-half inch in diameter and several feet long. Make a loop in one end and slip it over the upper jaw of the hog. As the hog pulls backward, the noose is tightened. The other end of the rope may be quickly snubbed around a post. The rope stays tight because the natural tendency of the hog is to pull backward when held in this manner. For ringing a large number of hogs, use a special ringing crate.

On a pig farm it is desirable to have a "crush" to hold pig for inoculation and other operations. It can be built with wooden posts fixed on either side getting narrower and narrower at one end so that there is space for only one pig. By slipping a wooden partition behind and allowing just sufficient space for pig to pass out its head through the front gate, the animal can be controlled. The pig can be released by opening the front gate, which should then be closed and the rear partition removed so as to allow the next pig to enter after which it should be replaced to secure the pig.

Hold and carry a larger pig by grasping both hind legs. A suitable position for castrating is achieved by holding the pigs with its back toward the holder's legs and by gripping its head between the knees of the holder. For weighing a pig at 56 days, provide a heavy sash cord about two feet long with the ends tied together; loop around one hind leg above the hock, and suspend the pig from a spring balance on a tripod.

Casting

There are three types of casting: (i) By using snout rope and one hind leg; (ii) holding both legs from one side using two ropes and (iii) by using hog shackle and snout snares.

Pig catcher

Instrument to restrain adult pig made of an iron bar, 1 to 2 cm thick and 1 m long with handle at one end and a ring of 10 cm in diameter at the other.

Nose rings

By instinct most hogs do some rooting, but it is likely to be especially damaging to pastures. When rooting starts, the herd should be "ringed"; and this is applied to all hogs past weaning age. Older animals can be restrained by a rope or snare placed around the snout, whereas young pigs can be held.

Two types of rings

Self piercing copper wire rings, which is most suited for growing stock and the small "bull ring" which is recommended for boars and sows.

For restraining, an assistant is required to catch hold the pig in sitting position and the mouth is held firmly closed. Rings are usually placed in the snout just back of cartilage but away from the bone; though some producers prefer the use of ring that is placed through the septum (the partition of the nose). Nose ring prevent from rooting up pastures.

The 'noose' is tightened firmly and then the end tied to a post. The pig will squeal and put backward and in so doing further tighten the cord. The self piercing bull ring, which is approx 40 mm in diameter, is fixed between the inside of the two nostrils. A small screw is then used to hold the sides of the ring together.

17.2.3.2 ***Chemical restraining***

Used for major surgery. Different anesthetic drugs are used for this purpose. Acepromazine @ 0.22 mg/kg body weight through deep intramuscular route Ketamine @ 10.20 mg/kg body weight through intramuscular route. Pelazon (Zolapoza/Telatamine) @ 2.4 mg/kg bodt weight through IM or IV route.

17.3 Clipping the Boar's Tusks

It is never safe to allow the boar to have long tusks. With these they may inflict injury upon other boars or even prove hazardous to the caretaker. Above all, such tusks should be removed well in advance of the breeding season, at which time it is necessary to handle the boar a great deal. The common procedure in preparation for removing the tusks consists of drawing a strong rope over the upper jaw and tying the other end securely to a post or other object. As the animal pulls back and the mouth opens, the tusks may be cut with a bolt clipper.

17.4 Removing Needle Teeth

New born pigs have eight small, tusk like teeth (so-called needle or black teeth), two on each side of both the upper and lower jaws. As these are of no benefit to the pig most swine producers prefer to cut them off soon after birth. This operation may be done with a small pair of wire cutters or with forceps made especially for the purpose. In removing the teeth, care should be taken to avoid injury to the jaw or gums, for this reason only the tips of needle teeth should be clipped about $^{2}/_{3}{}^{rd}$ of each tooth.

17.5 Tail Docking

Tail docking seems to be the best method of preventing, or at least reducing, tail biting. Cutting of tails may be advisable wherever pigs are to be raised in total confinement.

The tail should be clipped to about 0.75 to 1.0" (19 to 25 mm) from the bone of the tail. Either sterilized wire cutter or an electric cauterizing blade can be used. A protectant spray or dip may be applied to the tail stump. Instead of cutting, a tight rubber cord which will result in slaughtering of the tail within a few days, may also be applied.

17.6 Medication

After birth a pig receives his first injection. Proper injection involves the right size needle and the best site for the injection. For piglets, 0.5 to 1" (13 to 25 mm) 20 gauge needle works for thin liquids where as an 18 gauge needle is best for thick liquids. There are three ways by which a pig can be injected: (a) sub-cutaneously, which means under loose fold of skin, (b) intramuscular, which is to inject directly into the muscle, and (c) intravenously.

(a) Sub-cutaneous injection

The most suitable site for injecting under the skin is at the base of the ear, where loose folds of skin are clearly available. This site also has the advantage of being in a clean area of the pig's body. An alternative site is in the groin region.

(b) Intramuscular injection

The intra muscular injection is usually made at the base of the neck with adult pigs and the fleshy part of the hind legs in young pigs. For intramuscular injection a 30 mm needle is used. Clean the site with surgical spirit and then inject deep into the muscle.

(c) Intravenous injection

The injection is usually made in the ear vein

17.7 Transportation

The stress which confront the pig during transportation are:

(1) The handling at loading and unloading time
(2) The new surroundings
(3) With strange pigs
(4) Physical discomfort of the journey
(5) Heat stress (under tropical conditions)

Measures to be taken to minimize these stress are:

(1) Transportation by truck ensure that the loading ramp is properly designed with solid walls and is at a correct height of the cart, truck or trailer.
(2) Handle the pigs quietly and gently at all times. Avoid the use of sticks and prodders.
(3) Do not feed pig for 12 hr before loading.
(4) Spray the pigs with cold water before loading and again in the truck.
(5) Provide a cover on the truck, good ventilation, adequate bedding and ensure that the floors are not slippery. Ensure that the sides of the truck are high enough to prevent the pigs from jumping out.
(6) Never mix the pigs of different weights.
(7) Do not stop enroute to slaughter house.
(8) Carry only 20–25 pigs in a truck.

17.8 Identification

Not only identify each animal, it should be able to prove identification to the satisfaction of others. It avoids disputes with lost or stolen stock.

Each pig must be identified in a mode to keep a record for it. In pigs, four methods of identification such as fire branding, tattooing, ear notching, and ear tagging are used.

17.8.1 Fire branding

Fire branding has the advantage that (i) it is a method of marking live animals as well as carcasses; (ii) this method is effective if carefully applied with a suitable brand which is not overheated or held too long or pressed too deeply on the pig, as it results in a clear and legible skin and body mark. Suitable copper firebrands last longer than iron brands. Size of the marking should not exceed 1.75" by 1.25"; (iii) when properly applied, firebrands on pigs will be legible for about two months. Apply mustard oil mixed with zinc oxide to encourage healing.

17.8.2 Body tattoo marking

Tattooing is the best and most practical way of marking for the identification.

Procedure

Area on which the tattooing is to be applied should be clean and free from accumulation of muck. The tattoo needles are dipped in the paste or ink is then firmly pressed with marker. The best position on the body for the tattoo mark is on the shoulder. The efficiency of tattooing as a means of identification depends on:

1. The effective use of the tattooing instrument.
2. The use of an instrument of a reliable type with strong sharp needles.
3. Taking time to do the job properly; and
4. The use of a reliable brand of ink, paste, or paint.

17.8.3 Ear marking

No system of identification is perfect, but for the identification of live animals both earmarking and ear tattooing are practical and readily applied. The earliest age at which an identification mark becomes necessary in pig is between one and two months.

17.8.4 Ear tattooing

Tattooing the ears is another method of marking pig belonging to light coloured ears. The method is to punch several small holes with a dye (meant for this purpose) in the form of numbers or letters through the skin on the inside of the ear and then fill them with tattoo ink. If done correctly, this is a permanent mark. The usual practice is to start the marking with '001' and continue the same up to '999'. Its disadvantage is that animal must be caught and the inside of the ear cleaned to be able to read the identifying marks.

Equipment

Tattooing forceps and numbers, antiseptic tattoo paste, surgical spirit and swab.

Procedure of ear tattooing

(i) Place the tattoo letters and numbers in the applicators, (ii) Check the letters are in correct position by piercing a piece of card board before tattooing the pig (iii) Pick up the pig and swab the outside of the ear with surgical swabs (iv) Tattoo the ear by piercing the back of the right ear with the tattoo set, (v) Put the block of the antiseptic tattoo paste into the markings with your thumb or an old tooth brush.

Precaution

Ensure that: (i) it is indelible, (ii) ears and instrument are perfectly clean before the operation is performed, otherwise septic trouble may result and a fibrous wart growth set up around the mark. (iii) next to cleanliness, it is important that the needle blocks be firmly placed in the jaw of the pliers. The area to be punctured should first be cleaned by wiping over with a cloth soaked in methylated spirits, these remove grease, then the marking ink or paste should be rubbed on in to perforations made by the needs.

Disadvantages

It has the disadvantage that when pigs fight or tear their ears on wire fences or where the ears are damaged in dehairing machines at the factory, this identification marks may be marred or destroyed.

17.8.5 Ear notches

A satisfactory system of earmarking pigs by notches in the ears. Normally 'V' shaped notch along with the border of the ear is done. The litter number is notched in the pig's right ear and the individual pig number in the pig's left ear. Leave at least $^1/_2$" between the notches. For this system of ear marking unit numbers 1 to 9 are placed in right or off ear, and tens (10 to 90) in the left or near ear. Notches should be completely healed in about 1 week to 10 days. As the position of the notch on the ear determines its value, it is important that positions 1 and 10 and 4 and 40 be kept well towards the bottom and the tip of the ear respectively, to prevent confusion with the positions 2 and 20 in the middle of the ear. Care should be taken to avoid ear vein and use of clean equipment for ear notching.

17.8.6 Ear tags or buttons

All tags are subject to being pulled or torn out or to be crushed, mutilated, or disfigured to such an extent as to be unreliable as a means of identification. Liability to be pulled or torn out when the animal fights or rubs against wire netting, and consequent unreliability as a permanent identification. If not properly inserted, the ear tag may disfigure the ear. The method of applying the ear tag by use of combination pliers is that one portion of the instrument is used to punch a hole in the ear into which the tag fits. The other potion of the pliers is to seal the tag.

17.8.7 Hair clip marking

Marking pigs by clipping away the hair on any particular portion of the body. Very strong objection to their use lies in the fact that an unscrupulous person could readily disfigure the mark and thus cause confusion and annoyance.

17.8.8 Uses of identification

By making identity of each animal, we should be able to prove the identification to the satisfaction of others. It alleviate dispute like (a) cost, (b) stolen, (c) or strayed stock.

17.9 Dentition

There are three varieties of teeth–incisors, canines, and molars. Incisors are cutting teeth and are situated in the front of both upper and lower jaws, usually with sharp edges and single fangs. A deciduous incisor is a temporary or milk tooth and is later replaced by a permanent incisor.

A canine tooth, sometimes called dog-tooth or tusk, is one of the four sharp pointed teeth lying between the incisors and the molars. In the boar they develop into two pairs of prominent tusks. Those in the upper jaw are sometimes called eye teeth. Molars are the back teeth usually with a blunt, grinding surface and more than one fang.

1. At birth the young pig has 8 teeth 4 temporary incisors and four temporary tusks.
2. At one month four incisors are out, two in the upper and two in the lower jaw.
3. About the sixth week the temporary molars are visible
4. At 3 months 3 more are added to each jaw and at this period all the temporary or milk teeth are in position.
5. At 6 months, in the majority of pigs, a small tooth comes up on each side of the lower jaw behind the temporary tusks, between them and the molars, and in the upper jaw directly in front of the molars.
6. At 6 months the fourth molar appears through the gums.
7. At nine months, the corner incisors fall out and the permanent incisors make their appearance. The permanent canines are also cut at this period and the fifth molar on each side of both jaws cuts its way through the gums.
8. At one year the middle incisors are replaced by permanent ones, and the temporary incisors and molars are all replaced by permanent ones.
9. In its mature form the pig has 44 teeth. The upper jaw 6 incisors, 2 tusks, 2 pre-molars, 12 molars. The lower jaw-6 incisors, 2 tusks, 2 pre-molars, 12 molars.

The dental formulae of pig is given below:

Temporary dentition $^3/_3$, $^1/_1$, $^3/_3$, $^0/_0$ total 20 teeth.

Permanent dentition $^3/_3$, $^1/_1$, $^4/_4$, $^3/_3$ total 44 teeth.

The mouth of a pig is large and lip fissures extend far back. The upper lip is blended into the snout. The canine teeth are large, thick, dense and well developed especially in the boar. The lower canine or tusks are long an curved outwards and upwards; the upper ones pass downwards and outwards. They continue to grow during the lifetime of the animal. The tusk or canine are more prominent in male than in female.

CHAPTER 18

IMPORTANT DISEASES OF PIGS AND HEALTH MANAGEMENT

18.0 Introduction

Present day concept of disease for any profitable pig enterprise is the herd performance data in regard to food conversion efficiency and daily live weight gain which are sensitive indicator of status of diseases, specially for sub-clinical form of diseases. Moreover, health problems in pigs have shown that various diseases occur due to many factors rather than one factor. These different factors may have synergistic effect or may lead to disease condition while effecting in sequence. Disease problem causes un-profitability of pig production. Disease affects appetite, feed-efficiency, growth and their ability to raise healthy piglets.

The herds though initially healthy, may be gradually contaminated with infectious organisms. Efforts should always be made by the pig keepers in any pig enterprise to constantly attend to management; hygienic practises; preventive medicine be given preference in respect of preventive measures to check incidence of disease and timely vaccination against important diseases.

18.1 Signs of Normal Health

A normal healthy pig shows good bloom, moist snout, warm ears, curl in tail and is always alert. The mucous membranes of eyes and vulva have pinkish colour. Pigs which are ailing, loose this picture of health and refuse their food, eyes sunken, ears cold, tail hangs and is dull and depressed. These early signs of ill health are generally noticed by experienced attendants and pig keepers.

18.2 Microbial Diseases of Pigs

Domestic pigs are susceptible to a number of viral and bacterial infections. This chapter deals with some of the common microbial infections in pigs.

18.2.1 Viral

18.2.1.1 *Swine fever*

Syn. Hog cholera, Pig typhoid. This is a highly contagious viral disease of pigs, which occurs nearly all over the world. The severity of this disease varies with the strain of the virus, the age of the pig and the immune status of the herd and range from heavy mortality, mild illness and in some cases reproductive problems and birth of weak piglets. Swine fever occurs in much of Asia, some Caribbean islands, the African countries of Madagascar and Mauritius and much of South and Central America.

Cause

The cause of swine fever is a virus which is associated with certain secondary invaders, *e.g. Salmonella suipstifer* or *B. choleraesuis*, *Pasteurella suipstifer* and *Actionomyces necropherus*.

Clinical symptoms

The disease has an incubation period of 5–10 days (range 2–30 days) after which the following symptoms develop:

Hyper acute type

One or two pigs may be found dead and others may show signs of acute type when examined. Disease terminates fatally in 4–8 days.

Acute type

Affected animals are dull, lethargic, anorexic and show fever of 40.5°–41.5 °C at first. These symptoms are followed by conjunctivitis in which eyelids may be stuck together by exudates and constipation followed by diarrohea, usually of greenish or brownish colour and always possessed a fowl odour with occasional vomiting. The animals often huddle in the bedding in piles and walk reluctantly with a swaying of the hind quarters. Cyanotic discoloration of abdomen, inner thighs, ears and tail may develop. Dynproea occur and may be present until death. Nervous symptoms, in particular, convulsions occur early in the disease and are followed by circling, in-coordination and atexia. Death occurs within 9–19 days.

Chronic type

In this form pigs are dull, do not come readily to the trough, burrow in the bedding, uncurl their tails, and occasionally vomit. To a casual observer the general health of the herd may appear to be good. When temperature is taken of a few pigs, a surprising number registers 2 to 3 or more degrees of fever, while one or two will be found to have a vary high temperature, perhaps 41 °C or more. Other symptoms may include cough, unsteady gaits and conjunctivitis. Abortion or early loss of litter may occur. The sow is not a carriers, since after the birth of her piglet the virus, having crossed the placental barrier, no longer remains within her body. A period of 56 days may elapse between the last death on a farm and a recurrence of the disease.

Low virulence virus

This may give rise to transient pyrexia and inappetence with no other clinical symptom. Post ataxia has been recorded four months after birth in pigs born and sows infected with low virulent virus. There may be reproductive symptoms such as abortion and the birth of mummified or stillborn piglet.

Diagnosis

The clinical symptoms accompanied by the findings of high fever and some deaths in large herd of pigs should raise the suspicion of swine fever, particularly if the herd affected is well fed. Typical 'button ulcers' in intestines are important lesion seen on postmortem. Classical swine fever can be diagnosed by detecting the viral antigens by direct immunofluorescence assays and ELISAs. The virus can also be isolated in cell cultures like PK-15 cells and identified by direct immunofluorescence or by immunoperoxidase staining. Reverse transcriptase polymerase chain reaction (RT-PCR) can be used to detect and diagnose SF. Antibodies develop after 2 to 3 weeks and persist lifelong.

Treatment

The best results are derived from hyper-immune serum. Sulpha drugs and antibiotics are not effective against the virus but combat secondary invaders.

Prevention and control

Vaccination with live attenuated lapinised vaccine can be given either before colostrums is taken by new born piglets or at 30–50 days of age. Vaccination should be repeated at 9 months interval. Vaccine virus may be present in the semen of recently vaccinated boars and cross the placenta.

Control

Swine fever is controlled by a slaughter policy in which all in contact pigs with confirmed cases are slaughtered and buried or burned on the affected farms. In India, all stock should be vaccinated regularly. Pens are disinfected and left depopulated for a period of 60 days. Pig movement in the affected area is strictly controlled. All fed to pigs is heat treated. Strict hygienic measures are adapted and suspected cases isolated.

18.2.1.2 *Foot-and-mouth disease*

Foot-and-mouth disease in pigs is an acute and very contagious viral disease characterized by fever, formation of vesicles on the coronary band but less frequently on the lips and tongue. Morbidity is high but mortality is low except in young pigs.

Foot-and-mouth disease is caused by aphthovirus of which seven types are now recognized including the three known as A, O and C. An important feature of the disease in relation to its spread is the excretion of virus before symptoms become evident to the owner of the animal. There is no cross immunity between types and only partial immunity between sub-types with the existing types.

Mode of infection

Infection takes place by inhalation and ingestion and can also result from infection of abrasions on the skin and other body surfaces. Virus is shed even before the appearance of the lesions, particularly from the upper respiratory tract. Infection results from direct or indirect contact with affected animals, carcasses or animal products. Transmission may be by the aerosol route, by contact with other species of affected livestock, by mites, and from infected material of animal origin or incompletely cooked swill. Following infection, pigs do not remain carrier.

Incubation period

The incubation period is 2–7days.

Clinical symptoms

The sudden onset of severe lameness is the commonest finding in affected pigs, the feet of which are obviously painful. The back may be arched, reluctant to move or movement may be accompanied by squealing. Vesicles appear as raised white areas 0.5–1 cm in diameter on the dorsum of the tongue and on the snout and may rupture readily to leave small ulcers. Frothy saliva may be present. Vesicles appear in the coronet or in the inter digital space, and on the supernumerary digits. Fever up to 41 °C accompanies in the earliest stages of the disease. Affected animals are depressed, anorexic and loss of condition. Mortality rarely exceed 5% though it may reach 50% in piglets.

Prophylaxis and control

Early recognition, followed by slaughter, disinfection and strict control on the movement of pigs. Vaccination should be undertaken to cordon off disease. Inactivated vaccines are made from virus grown in cell culture. Pigs are not readily immunized and require high concentration of virus, although immunity has been demonstrated for up to 9 months using adjuvant vaccine. These cause granulomation reactions and should be given into the pinna of the ear or intra peritoneally. One or more serotypes may be included in preparation of vaccine. Protection may take 7–20 days to develop and revaccination is required every 6–8 months.

18.2.1.3 *Swine pox*

A mild infectious disease caused by a pox virus in which red, circular pox lesions appear on the skin of the belly, maxillae, face and teat in young pigs.

Incidence

Widely spread but rarely reported.

Incubation period

3–6 days but may be up to 14 days and the lesions persist for 1–3 weeks.

Clinical symptoms

Slight pyrexia 40 °C may accompany the appearance of the lesions. Red 1 cm papules appear on the ventral abdomen and rapidly form circular red brown scabs, which is rapidly blacken. In young piglet the bursting of the vesicular stage of the face may lead to wetting scab formation and conjunctivitis. Slight inappetence may accompany with development of the lesions.

A large percentage of successive litters of suckling pigs may be affected, but the disease is rare in adults. Mortality is rare although infection presumed to be transplacental, may cause deaths in new born pigs. Lesions on the edge of the tongue, lips and sometimes elsewhere may be seen.

Diagnosis

The size and colour of the lesions is characteristic.

Treatment and control

Treatment or control are rarely attempted, because of the mildness of the disease, but the use of insecticides to eliminate lice and thorough disinfection of the pens in which outbreaks of the disease have occurred helps to reduce the incidence.

18.2.1.4 *Swine influenza*

Swine influenza is a respiratory tract infection with influenza type A virus resulting in coughing, dysponea and prostration. It is caused by influenza virus an orthomyxovirus 80–120 milli micron (mμ) in diameter. The virus has variant in the form of ABC virus.

Pathogeneses

In classic form of swine influenza, the virus enters the respiratory tract and multiply rapidly to give rise to the clinical symptoms and affected animals rapidly recover. Viraemia can occur and transplacental infection has been recorded up to 40 days before parturition and may result in failure of the lungs to develop. Secondary bacterial infections may complicate disease condition like pasteurella multocida, E. coli and Salmonella typhimurium.

Clinical symptoms

In typical out breaks there is rapid, virtually 100% involvement of all susceptible animals, which exhibit severe prostration, anorexia, fever (41.8 °C), dyspnoea, conjunctival discharges, cough and loss of condition.

18.2.1.5 *Porcine reproductive and respiratory syndrome (PRRS)*

Porcine reproductive and respiratory syndrome is caused by the PRRS virus, an RNA virus coming under genus *Arterivirus* of family *Arteriviridae*. The syndrome is characterized by reproductive failure of sows and respiratory disease in pigs. The reproductive syndrome is recognized by abortions in late gestation, early or delayed farrowings that contain dead and mummified fetuses, stillborn pigs, weak-born pigs and an increase in repeat breeding. The respiratory syndrome is recognized by dyspnoea, fever, anorexia, and listlessness. The virus is primarily transmitted directly via infected pigs and also by faeces, urine and semen.

18.2.1.6 *Rabies*

Rabies is caused by a *Lyssavirus* of the family *Rhabdoviridae*. Pigs show variable signs. Some animals show excitement and a tendency to attack or dullness and incoordination. There may be twitching of the snout, rapid chewing movements, excessive salivation, paralysis and death. The condition can be diagnosed by FAT done on brain tissue, microscopical examination to demonstrate Negri bodies or by viral nucleic acid detection.

Treatment and control

No treatment must be attempted after the onset of clinical signs. Immediately after exposure the wound should be irrigated with soap solution and water. Post exposure vaccination then can be given, depending on the nature and extent of risk involved. Euthanasia must be avoided and suspected animals must be kept under close observation. Control is effected by the destruction of wild fauna in and around animal holdings and the vaccination of all domestic cats and dogs being maintained at the premises. Live and inactivated vaccines both of chick embryo origin and tissue culture origin are available and can be used.

18.2.1.7 *Rotavirus infection*

This infection can occur in suckling piglets of 2 weeks of age. There will be watery diarrhea which lasts for 3–5 days. Mortality is rare. The condition can be diagnosed by detection of the virus by electron microscopy, virus isolation and ELISA.

18.2.1.8 *Aujeszky's disease (pseudorabies)*

Aujeszky's disease is caused by Aujeszky's disease virus, a member of the family Herpesviridae it is primarily a disease of pigs, fatal in piglets below 2 weeks of age and the mortality rate decreases in older animals. The virus is latent in older pigs which recover from the infection. The disease occurs in parts of Europe, Southeast Asia, and Central and South America. Aujeszky's disease should be suspected in pig herds with high mortality and CNS symptoms in young piglets, and lower mortality and respiratory signs in older animals. Aujeszky's disease virus is usually transmitted between pigs by the respiratory or oral routes, indirectly via fomites and through carcasses. Venereal and transplacental transmission is possible. The incubation period is usually 3 to 6 days in weaned or adult pigs and lesser in suckling piglets.

In pigs, the clinical signs vary with the age of the animal. In piglets less than a week old, fever, listlessness, anorexia, tremors, paddling, seizures and paralysis of hind limbs are seen. Once neurologic signs develop, the animal usually dies within a couple of days. Older piglets also show similar signs, but the mortality rate is lower. In weaned pigs, the disease is mainly a respiratory illness (sometimes complicated by secondary bacterial infection), with symptoms of fever, anorexia, weight loss, coughing, sneezing, conjunctivitis and dyspnoea. Weaned pigs tend to recover after 5 to 10 days. In adults, the infection is usually mild or in apparent, with respiratory symptoms predominating. Pregnant sows may abort or give birth to weak piglets.

On necropsy, many pigs have rhinitis, pulmonary edema, congestion or consolidation and secondary bacterial pneumonia. Affected pigs may also have necrotic tonsillitis or pharyngitis, congested meninges or necrotic placentitis.

Aujeszky's disease can be diagnosed by virus isolation (from nasal swabs, oropharyngeal fluid, tonsils and brain), detection of viral DNA or antigens and serology. Control of the disease is by keeping domesticated herds well separated from feral swine, thorough disinfection and vaccination.

18.2.1.9 *Swine vesicular disease*

Swine vesicular disease is caused by swine vesicular disease virus, a member of the genus *Enterovirus* in the family *Picornaviridae*. The virus affects only pigs and the resulting illness is manifested by the formation of vesicles and erosions. It resembles foot and mouth disease as far as clinical signs are concerned. The virus spreads by direct contact with infected animals or via environmental contamination. The virus enters the body through broken skin or mucous membranes, and by ingestion. Contaminated undercooked pork or other scraps can act as source of infection. The incubation period is usually 2 to 7 days. The symptoms are usually more severe in young animals. Most pigs recover completely within 2–3 weeks.

The disease is characterized by the development of vesicles and erosions on the legs and around the mouth. Vesicles then appear around the coronary bands, in the interdigital spaces and on the skin of the lower legs. The vesicles soon rupture, leaving shallow erosions. Vesicles are also seen occasionally on the snout, lips, tongue and teats; they are relatively rare in the oral cavity. Swine vesicular disease may be subclinical, mild or severe, depending on the virulence of the strain and the husbandry practises. The morbidity rate can reach 100%; but death is not seen.

Swine vesicular disease can be diagnosed by detecting viral antigens by ELISA or immunohistochemistry. RT-PCR may be resorted to for detection of the viral nucleic acid. Serological tests used for diagnosis include ELISA, virus neutralization, immunodiffusion, radial immunodiffusion and counter-immunoelectrophoresis. Preventative measures include screening imported pigs, restricting the importation of pork products that may contain virus and restricting garbage feeding to pigs. Outbreaks are controlled by quarantining infected farms and regions, tracing pigs that may have been exposed, culling all infected and in-contact pigs, and cleaning and disinfecting the affected premises. No vaccine is commercially available against the disease.

18.2.2 Bacterial

18.2.2.1 *Leptospirosis*

Infection with *Leptospira* spp. may be inapparent or may cause fever, icterus and death in piglets and abortion and still birth in sows.

Cause

Leptospira ictero haemorrhagica and *L.canicola* occurs throughout the world in pigs.

Clinical symptoms

Three main clinical symptoms are associated with leptosprial infection in pigs

(i) Sub-clinical

This form of infection is only identified by serological tests.

(ii) Acute or sub-acute infection

A high fever of 40 °C for 3 to 5 days has been recorded. Affected pigs become dull, anorexic, show diarrohea and icterus with haemoglobinuria and heavy mortality may occur. Rarely, some nervous symptoms may be seen *e.g.* weakness of the hind quarters or tremors if leptospira enters nervous tissue.

Abortion, still birth and neonatal mortality accompanied by fever, loss of milk and jaundice in sows are common consequences of leptospiral infection in breeding herds. In sows which are infected later, weak piglets are born. Mummified and macerated fetuses also occur commonly amongst the litters. Loptospirotic abortion in the last trimester of pregnancy is common.

Diagnosis

Abortion in the last trimester of pregnancy, the birth of week piglets and weaners.

Along with the clinical symptoms, postmortem findings and serological tests, isolation of organisms from kidney, seminal vesicles, testes or fallopian tubes.

Treatment

The parenteral administration of penicillin, sreptomycin and tiamutin. Abortions may be prevented and renal carriers eliminated by parenteral treatment with streptomycin 25 mg/kg body weight as a single dose for 3–5 days. Medication with tetracycline may also be used.

Control

The disease may be eliminated by hygiene, vaccination, treatment or slaughter of carriers or a total slaughter policy combined with disinfection, elimination of rodents and restocking.

18.2.2.2 *Tuberculosis*

Tuberculosis in pigs may be due either to the bovine, avian or human strain of mycobacterium organism. The avian strain is usually associated with lesions localized to the sub-maxillary and mesenteric glands, but the infection may also be generalized.

The method of infection is by ingestion, the organism entering the body either from the pharynx, nostrils or from small intestine.

General tuberculosis

This is associated with loss of weight, local discharging sinuses and chronic pheumonia is usually caused by infection with mammalian strain. Lesions occur in the liver, kidney and in many lymphnodes.

At postmortem, encapsulations and calcifications of the lesions is seen. Cases of generalized tuberculosis have been recorded in pigs under two months old. Other distinctive feature of porcine tuberculosis include the frequency with which the sub maxillary glands are affected.

Localized tuberculosis

Usually occurs only in the pharyngeal, cervical and mesenteric lymph nodes and is usually associated with a vain tubercle bacilli. The lesions vary between small yellowish white calcareous lesions to generalized enlargement of the node. Granulomata are fleshy and are rarely classified.

Diagnosis

Usually based on finding at slaughter and a semen's stained by Zichl Neelson's method, but chronic pneumonia may lead to suspicion of tuberculosis.

Control

Eradication and control centres round the identification of carriers and their disposal and disinfection of the premises.

18.2.2.3 *Anthrax*

Anthrax is caused by *Bacillus anthracis*. The disease is characterised by high fever, swelling of throats in some animals and by the presence of the causal organism in the blood stream. It may attack all the domestic animals and man, while numerous wild animals are also liable to suffer from it.

Incidence

The disease is wide spread, occurring in all parts of the world, particularly in the tropical and sub tropical areas.

Method of infection

The commonest method infection in animals is by ingestion. Spores are unharmed by the gastric juice and pass on to the intestine, where they set up infection.

Clinical symptoms

Septicemia, pharyngeal and intestinal forms of the disease occur in pigs, most commonly as a result of entry of the organism. Oedema and swelling of the neck region, associated with dysponea, usually a high fever up to (41.5 °C) are the commonest symptoms noted. Depression, vomiting and inappetence may occur. Death usually follows within a day of the onset of cervical oederma.

Recovered animals may remain as carriers and pig to pig transmission may occur. Anthrax may affect and kill man.

Treatment

Anthrax responds to penicillin treatment.

Prevention

For prevention of the spread of infection following points should be considered:

(a) The disposal of carcass be efficient and safe.
(b) The careful observation of animals that have been in contact with diseased ones and their isolation on exhibition of rise of temperature.
(c) Strict supervision of the carcass until such time as it can be disposed of, with efficient methods of sterilization of any blood or discharges that have spilled.
(d) Use serum for passive immunization of in contact animals.

18.2.2.4 *Salmonellosis*

Salmonellosis usually occurs as out breaks of septicaemia, acute enteritis or chronic enteritis and wasting in weaned pigs of 10–16 weeks of age. Morbidity and mortality may be high in affected group.

Cause

Salmonella choleraesnis, S. typhimurium and S. Dublin

Clinical symptoms

Pigs of all ages can be affected although outbreaks are commonest in pigs aged between weaning and 3 to 4 months. The septicaemie form is commonest among younger pigs. Animals may be found dead (mortality almost 100% in this form), depression, dullness and weakness or even nervous symptoms. Affected animals often burry themselves in straw and show manure red cyanosis of the ears, limbs and the centre of back. Affected pigs have a temperature of (40.6–41.7 ºC) and die within 24–48 hr. The acute enteric form also occurs in younger pigs. Affected animals pass thin watery yellowish diarrohea and may be dull and fevered (40.6–41.7 ºC). Pneumonic signs, weakness and nervous symptoms such as paralysis and tremors may occur. In severely affected cases skin discoloration is present.

Recovered pigs may slough affected ears and tails. In the chronic enteric form affected pigs appear severely emaciated and may have intermittent fever. There is persistent diarrohea which may contain greenish shreds of necrotic epithelium it rarely contains blood.

Epidemiology

S. cholera suis infection may be enzootic on some forms and than appears only in weaned pigs 12–16 weeks of age. Infection is usually introduced in a herd by the purchase of carrier pigs or animals infected by contact at markets. Outbreaks of salmonellosis due to other *Salmonella* sp occur when contaminated feed stiffs are fed.

Diagnosis

(a) Acute and chronic enteritis

Must be distinguished from swine dysentery. The presence of fever and small intestinal involvement with haemorrhagic mesenteric lymph nodes at postmortem is usually sufficient.

(a) Bacteriological examination

Isolation of *S. cholerone* suis or other *Salmonella* is significant in the absence of other pathogens.

Treatment

Affected animals may be treated individually by the daily injection of a suitable anti-bacterial agent. Tetracyclines, streptomycin, aprimycin, neomycin, ampicillin amoxicillin, streptomycin, trimethoprium, sulphonamide and chloramphenicol are all effective.

Control

Where disease is due to *S. cholerae* suis and disease in enzootic in the herd, vaccination may be practised. Pigs should be vaccinated at least 14 days before the risk period. Where pigs are brought in batches and from different origins should not be forwarded for laboratory examination to determine the causative organism and its antibiotic sensitivity.

18.2.2.5 *Pasteurellosis*

Pasteurellosis caused by *Pasteurella multocida* manifests as a bronchopneumonia which may be accompanied by pleuritis and pericarditis. Usually seen in pigs of more than one year of age. The disease can become chronic with polyarthrirtis and thoracic lesions. The condition can be diagnosed from the thoracic lesions during post mortem and concurrent recovery of the organism in culture.

18.2.2.6 *Staphylococcosis*

Staphylococcus aureus is responsible for botryomycosis, a chronic suppurative granulomatous condition of mammary glands in pigs. *S. hyicus* causes exudative epidermitis (Greasy Pig Disease) and arthritis.

Diagnosis

To confirm the diagnosis, the organism must be isolated from the meninges of clilnically affected pigs and identified in a laboratory. The disease must be differentiated from joint infections, flssers disease, generalised septicaemia, salt poisoning, aujeszky's disease and hypoglycaemia.

Treatment

Exudative epidermitis is a generalized dermatitis that occurs in 5 to 60 day old pigs and is characterized by sudden onset, with morbidity of 10–90% and mortality of 5–90%. The acute form usually affects sucking piglets, whereas a chronic form is more commonly seen in weaner pigs. It has been reported from most swine producing areas of the world.Successful treatment requires that the antimicrobial be given in high dosages early in the disease and for a period of 7–10 days. Success is greatest when antimicrobial therapy is combined with daily applications of antiseptics to the entire body surface. Treatment is less effective in very young pigs and ineffective in advanced cases. In severe outbreaks, in contact pigs should also be given antibiotics for several days.

18.2.2.7 *Streptococcosis*

Streptococcus suis infection in pigs is associated with meningitis, arthritis, septicaemia and bronchopneumonia in pigs of all ages. Outbreaks are common in intensively reared pigs when subjected to stressful conditions. The litter can be affected and neonatal death can occur. Meningitis usually leads to nervous signs and death usually follows. *S. porcinus* causes submandibular lymphadenitis.

Diagnosis

The organism must be isolated from the meninges of clinically affected pigs and identified in a laboratory.

Treatment

Streptococcus suis is usually sensitive to penicillin, synthetic penicillins or sulphonamides. Injections of penicillin should be given twice daily. Good nursing is equally important because the condition is very painful. Remove the piglet during the first 3 to 6 hr from the litter to a warm environment and carefully supplement it with milk via a stomach tube.

18.2.2.8 *Actinobacillosis*

Actinobacillus pleuropneumoniae can cause pleuropneumonia in pigs of all ages. Highly contagious disease primarily affecting pigs of less than 6 months of age in intensive rearing conditions. In acute cases, pigs are found dead. Some show respiratory signs and pyrexia. There may be blood stained froth from the nostrils and mouth and some pigs show cyanosis. Post mortem lesions are consolidation and necrosis in lungs with fibrinous pleurisy.

Diagnosis is by isolation and identification of the organism, FAT or PCR. *A. suis* infection can occur in pigs less than 3 months of age. Disease characterized by septicaemia and death. Clinically fever and respiratory signs are seen. Mortality may be up to 50% in some litter.

18.2.2.9 *Brucellosis*

Brucellosis in pigs is mainly caused by the Gram negative bacterium *Brucella suis*, and rarely by *B. abortus* and *B. melitensis*. Pigs usually get infected by ingestion of materials containing the bacteria such as fetuses and membranes. Venereal transmission and spread through fomites, feed and water can also occur. The organism can persist in the environment for months especially in cool and humid conditions and can withstand drying. The incubation period is variable.

Symptoms

The most common symptoms are abortion, at any time during gestation, and birth of weak or stillborn piglets. Some sows develop metritis. *B. suis* can also cause epididymitis and orchitis in boars which may lead to sterility. Abscesses and swelling are sometimes seen, and the testes may be atrophied during the final stages of disease. The lesions are often unilateral. Swollen joints and tendon sheaths, lameness, incoordination, posterior paralysis, spondylitis, metritis, arthritis, bursitis and osteomyelitis and abscess formation in various organs may also be observed in some animals. Some animals remain asymptomatic. When the disease is introduced into a herd, reproductive problems like abortions, still births etc. increase; in endemic herds, this disease may be manifested as non-specific infertility. To demonstrate the organism, the smears must be stained by modified method.

Diagnosis

In swine, serology is generally considered to be more reliable for identifying infected herds than individual pigs. Serological tests like ELISA, plate agglutination test and complement fixation can be used to detect infection. Isolates of *B. suis* can be identified to the species and biovar level by phage typing and cultural, biochemical and serological characterization. Polymerase chain reaction (PCR) techniques are available in some laboratories.

Control

Brucella species are readily killed by most commonly available disinfectants like hypochlorite solutions, 70% ethanol, isopropanol, iodophores, phenolic disinfectants and formaldehyde. To disinfectant contaminated surfaces 2.5% sodium hypochlorite, 2–3% caustic soda, 20% freshly slaked lime suspension or 2% formaldehyde solution can be used. No vaccines are currently available for *B. suis*. *B. suis* can infect human beings especially laboratory workers, abattoir workers, farmers, herders and veterinarians.

18.2.2.10 *Clostridial infections*

Pigs are moderately susceptible to tetanus toxin secreted by *Clostridium tetani*. They are also susceptible to the toxin of *C. botulinum*. *C. perfringens* Type A causes necrotizing enterocolitis in pigs and Type C can cause haemorrhagic enteritis in piglets. Diagnosis of clostridial infections is usually by demonstration of the bacilli in Gram stained smears from the lesion and by toxin neutralization tests. Treatment is done by administration of hyper immune serum, penicillin and other drugs depending on the clinical signs. Vaccines are available to prevent most of the clostridial infections.

18.2.2.11 *Escherichia coli* infections

Enterotoxigenic *E. coli* and sometimes verotoxigenic *E. coli* are associated with clinical illness in swine. Both can cause “post weaning diarrhoeic disease” in pigs.

Disease occurs within a week or two after weaning often following changes in feeding regimens or in management. Clinically there is diarrhoea and purplish discolouration of areas of the skin. Some animals die suddenly. "Oedema disease" of pigs is caused by verotoxigenic *E. coli* strains. Usually occurs 1–2 weeks after weaning, characterized by posterior paresis, muscular tremors and oedema of eyelids, face, larynx (which leads to a hoarse squeal). Death occurs within 36 hr of onset of clinical signs. A flaccid paralysis may be seen prior to death.

18.2.2.12 *Glasser's disease*

The causative agent is *Haemophilus parasuis* a Gram negative bacterium. The disease is basically a polyserositis and leptomeningitis seen in pigs from weaning up to 3 months of age. The incubation period is 1–5 days. Anorexia, pyrexia, lameness, recumbency and convulsions are seen. Cyanosis and thickening of the ears may also be present. In young growing pigs meningitis or middle ear infections are common together with pneumonia, heart sac infection, peritonitis and pleurisy. Sudden death in good sucking piglets is not uncommon in herds with a problem and in particualr when immunity in gilt litters is low.

Chronic form

Sucking piglets ae often pale and poorly growing and 10–15% may be affected in a litter. Such pigs then continue into the growing period with poor growth. When long standing pericarditis is a feture sudden deaths occur.

Diagnosis

This is confirmed by clinical observations, post-mortem examinations and isolation of the organism in the laboratory. Post-mortem and bacteriological examinations are required to differentiate.

Treatment

Haemophilus parasuis has a wide antibiotic sensitivity including amoxycilliln, ampicillin, streptomycin. Treatment must be given early, particularly if cases of meningitis are occuring and this can only be done by isolating the respective organisms from the brain. Identify the onset of disease in sucking pigs and inject 3 to 4 days prior to this prevent disease with long-acting penicillin.

Control and prevention

Where the disease is a problem in sucking pigs, the sows feed can be top dressed daily 7 days before and 7 days after farrowing with phenoxymethyl penicillin. The lactating and creep rations can be medicated with 200–300 g of phenoxymethyl penicillin.

18.2.2.13 *Atrophic rhinitis*

A complex clinical syndrome, in which sneezing occurs in young pigs, followed by atrophy of the turbinate bones and distortion of the nasal septum, sometimes accompanied by shortening and twisting of the upper jaw. Depression of the rate of weight gain may occur.

Etiology

The two most commonly associated microtus are *Bordetella bronchiseptica* and *Pasteurella multocida*.

Symptoms

The acute form is to be found in 2 or 3 weeks old piglets, where there is no deformity of the snout to be seen and not always an overflow of tears. Sneezing is perhaps the most common symptom. The eyelids may be outlay and sometimes the piglets have a copious discharge from its nose and breathes through its mouth accompanied with the blockage of lachrymal ducts or even epistaxis.

Diagnosis

The characteristic clinical symptoms and pathological lesions are normally a sufficient basis for diagnosis in late case. In early cases or there in which turbinate atrophy is present without deflection of snout, x-ray examination of snouts may show atrophy of the dorsal turbinate and distortion of the septum.

Treatment and control

The regular clinical inspection and the regular postmortem examination of piglets found dead are more effective. Endoscopy or assessment of mal-occlusion of the incisors may give an objective indication of the severity of lesions at the time when they are occurring.

Vaccination of sows and piglet with combined vaccine of *B. bronchiseptica* and P. multocida are of more value as far as protection is concerned. Medicated early weaning can be used for the foundation of atrophic rhinitis-free herds.

18.2.2.14 *Swine erysipelas*

Swine erysipelas is an infectious disease of pigs caused by erysipelothrix rhusiopathiae. The acute form is characterized by septicaemia, skin discoloration and gastroenteritis, where as the chronic form is recognized by general unthriftiness and sometimes lameness due to arthritis.

Source of infection

The transmission of the disease is usually through the alimentary canal by the ingestion of contaminated food or water. Occasionally the infection may occur through a wound. It seems probable that 'caviers' play an important part in introducing the disease in a herd.

1. Clinical symptoms

The disease occurs in hyper acute or chronic forms.

(i) Hyper acute form

Sudden death may occur or affected pig collapses with a high temperature (41.1°–42.2 °C) and often show a reddish purple patchy or diffuse discolouration of the skin. This form is common in adults.

(ii) Acute form

In younger pigs such as gilts, fattening pigs and young boars, anorexia, and high fever (40.6–41.7 °C) is common. There may be reddish purple, patchy or diffuse discoloration of skin and ears. Affected animals may die within 12–48 hr with cyanosis of the body and dysponea. In older pigs anorexia and thirst are most commonly noted and a high temperature is found. The pathogonomic diamond skin lesions appear within 24–48 hr of the onset of clinical symptoms.

(iii) Chronic form

Affected animals may recover completely, but the skin lesions may become necrotic, turn black and slough. The tips of the ears may also be lost. Affected joints become hot and painful to the touch but become swollen and fresh after 2–3 weeks and may stiffen. Affected pigs become lame and may lose condition.

Diagnosis

Erysipelas should be considered if a high fever of 106 °F or more is found in any adult pig, which has gone off feed and shows no respiratory symptom. The development of pathogenic skin lesions, lameness and fever together is sufficient to confirm the diagnosis of erysipelas.

Treatment

Penicillin is the drug and the response to treatment is normally rapid. It is advisable to give 2 to 3 daily injections to prevent relapse or the persistence of the organism to give chronic infection. Hyper immune serum is commercially available for treatment.

Control

It is advisable to clean out and disinfect the pens of affected animals. Control is normally exercised by prophylaxis and this takes the form of preventing the spread of out break to other susceptible animals by prophylactic injections of penicillin or hyper immune serum. Vaccination is commonly employed to prevent clinical swine erysipelas in breeding stock and young fattening pigs.

18.2.3 Parasitic infection

Endo-parasites

18.2.3.1 ***Ascariasis in pig***

The main importance of ascariasis is its impaction on economics of the farm as the infestation leads to condemnation of livers and plucks and poor weight gain. Pig also die due to pneumonia caused by the migratory larvae.

1.Muscular form of ascariasis

(a) Clinical symptoms

In mild form pigs show irregular appetite, thirst, vomiting, mild colicky pain, tenderness and swelling of abdomen. In severe form, obstruction of masses of ascarids or from perforation of the intestinal wall followed by peritonitis. If there is obstruction in the bile duct, jaundice arises from bite retention. There may also be intoxication from secretions of ascrids, causing high temperature, mechanical obstruction, and lacerations.

Control

Control is difficult because of the adhesive nature of eggs and its longevity on the ground (up to 5 years). In premises where ascariasis is a continuous problem, thorough cleaning of farrowing and fattening pens with detergent or hot washing soda be combined with anathematic treatment of the show stock. The use of horticultural flame gun has given good results in reducing the number of ascarid ova in concrete floor.

2. Respiratory form of ascariasis

Ascarid pneumonia is caused by the invasion of the large alveoli of the lungs by large number of larvae of ascaris sum during migration.

Clinical features

In severe form of ascarid, pneumonia develops as an acute non febrile disease. There is depression, anorexia and frequent heavy coughing with a serous discharge which sometimes contain blood. Death may occur a few days after the onset of symptoms. In chronic form, dry, repeated painful cough, accompanied by marked dysponea, spasmodic breathing with accentuated expirations, susceptible to secondary viral and bacterial infections. At the beginning of such infection, there is marked rise in temperature. The lesions are a serio-haemorrhagic alveolitis, and many patechiae or areas of echymosis with oedematious infiltration, rich in poly nuclear leucocyte and eosinophils in sub-acute inflammatory areas; there are eosinophilic granuloma also.

Diagnosis

The disease may be suspected when a number of young pigs show the breathing symptoms. Presence of ecteric pig with no fever and postmortem or slaughter findings of milk spot liver, and ascarids in the intestines are indications for treatment of the herd.

3. Verminous bronchitis in pigs

This is caused by the presence and development in the bronchi and bronchioles of metastrongylus elongata and *M. pendendolectus*. These worms are 2–6 cm long × 300–400 μ filamentous and white.

Clinical features

The pigs kept out doors and youngs are most susceptible hosts. The first symptoms appear 10–15 days after pigs have access to earth worms containing infective larvae. There are digestive disturbances usually diarrhoea due to the passage of larvae through intestinal wall. Later, respiratory disease develops in which both bronchi and bronchile obstructions are present. Affected pigs have dry, repeated painful cough, which increases when the animal move. There is mucus discharge from the nose, which increases in amount with coughing, often contain masses of worms. There is marked dysponea with critical paroxysoms of coughing and suffocation may results.

As a rule there is no rise in temperature unless complications with other types occur. Pigs with verminous bronchitis may be chronically affected. Coughing at all times of the year and remaining in poor condition are some of the definitive symptoms.

Diagnosis

Demonstration of larvae and presence of embryo or egg in faecal specimen.

Treatment

Carbon tetra chloride 33% in oil (@ 0.3 mg 1 kg) subcutaneously.

18.2.3.2 *Flatworm infection (Fascioliasis)*

Fasciola hepatica (liver fluke) is found in bile ducts of pigs in endemic areas. It is found in pigs which have access to areas of low lying pastures. Infection produces anemia, emaciation and digestive disturbance. S/c injection of carbon tetrachloride is effective.

18.2.3.3 *Parasitic encephalitis or cerebral compression*

This affection in pigs is caused by the presence of cysticercus cellulose and the larvae of Taenia solium in the brain, while the normal location of the cellulose is striated muscle. Similarly during the dissemination of embryos of *Taenin solium* by blood some may find their way to nerve centres in brain, where they become encysted larvae.

Clinical features

Pigs become infected by the ingestion of infected human faeces or contaminated feed. The cysts cause localized inflammatory reaction in the nerve substance causing cerebral compression, which is non febrile in character. Young pigs are most affected and show symptoms such as in coordination of movement difficulty in rising, attacks of giddiness, abnormal position of head and heck. The severity of different symptoms vary according to the location of cysts.

Diagnosis

Impossible during life.

18.2.3.4 *Echinococcus granulosae*

The larvae stages produces cysts in the liver and lungs mostly. Cysts can be seen on the under surface of the tongue, in the region of eye, anus and vulva in the living animal. In carcases, cysticercosis cyst are recognised by their piseferm shape and the characteristic hooks, particularly in the caseathed or calcified form which have been immersed in 5% hydrochloric acid for an hr.

18.2.3.5 *Cocciodiosis*

Coccidiosis in pigs is simply caused by small protozoa called coccidia that actually live and multiply inside the host cells, typically located in the intestinal tract. There happen to be three types which are Eimeria, Isospora and Cryptosporidia. The disease is pretty common and widespread typically in piglets in the suckling period and occasionally it does occur in pigs up to 15 weeks of age.

Typically tiny-egg like infected structures, often called oocysts, are passed out in the faeces into the environment where they can develop or sporulate. Typically this takes place within 12–24 hr at temperatures ranging from 26–36 °C.The oocytes can survive outside of the pig's body for many months and are often very difficult to kill because they are resistant to most disinfectants. The only disinfectant that is able to be effective against the oocytes is Oo-cide (Antec).

The oocysts are eaten by the pigs and then it basically undergoes three complex developments in the wall of the small intestine to essentially complete the cycle and during this period the damage occurs. It is important to remove sow faeces from the farrowing houses daily.

Symptoms

In piglets, it can cause diarrhoea due to the damage caused to the wall of the small intestine. Like many other diseases, this one is typically followed by secondary bacterial infections. The faeces may vary in color as well as consistency from yellow to grey green or bloody depending on the severity of the condition. Dehydration is pretty common in pigs with coccidiosis. Sometimes coccidiosis may be seen in young boars and gilts that are housed in permanently populated pens that are floor fed. The secondary infections causeed by bacteria and viruses can result in high mortality but the mortality due to coccidiosis is rather low on its own.

Treatment

For the treatment to be effective in the pigs, it must be given before the invasion of the intestinal wall because once the clinical signs have appeared the damage in the intestines has already been done. It is better to treat them before clinical signs, is to medicate the sow feed with amprolium premix 1kg/tonne, monensin sodium 100g/tonne or sulphadimidine 100g/tonne. Typically, this can be fed from the time the sow enters the farrowing house and throughout lactation. Each litter can be injected with a long-acting sulphonamide when they have reached six days of age. Also small amounts of milk powder with a coccidiostat such as amprolium or salinomycin can be given. Give small amounts of this daily to the piglets when they are three days of age along with top dressed on the creep feed.

To control the disease 1 or 2 doses of toltrazuril at a level of 6.25 mg/kg can be given. It is basically prepared by mixing 250 ml of glycerol, 125 ml of water and 125 ml of Baycox together. From then a 2 ml dose may be given once at 4, 5, or 6 days of age. The response determines the exact time and is repeated again at ten days of age. However, if there is no response, then it is very unlikely that coccidiosis is the problem in swine herd. Also you should specifically discuss this kind of method of treatment of your swine herd with your veterinarian who may also prepare it for you.

Prevention

One of the most important things that can be done to prevent the swine herd from getting coccidiosis is to control the insects and the hygiene of pigs essentially by removing the sow and piglet faeces daily. You should also make sure that the slurry channels are completely emptied between farrowing and also make sure to prevent any sort of movement of faeces from one pen to another. Make sure to disinfect the farrowing houses and keep the pens as dry as possible, especially areas of the floor where piglets often defecate. One method that seems to be effective is to actually cover the wet areas with some shavings and just simply remove them each day.

For some that may still traditionally have outdoor pigs, it is more harder to control in the herd. One thing that can be done for sure is to burn the sows bedding after farrowing and always move the farrowing grounds to new areas. If you use boards in the farrowing areas, again you must disinfect these with antiseptic solution. If you have wallows, they can really be an ideal focus of infection, especially during the lactation period. Always remember to place wallows far away form the food source as it can get contaminated.

18.2.3.6 *Kidney worm (Stephanurus dentatus)*

Stephanurus dentatus are stout-bodies worms (2–4.5 cm long) found encysted in pairs along the ureters from the kidney to the bladder. The kidney worm is found worldwide, particularly in taropical and subtropical areas. Infection is by skin penetration or ingestion of the infective larvae (earthworms may serve as paratenic hosts). The larvae migrate to the liver, where they migrate extensively for 3–9 months. Larvae then penetrate the capsule and migrate through the peritoneal cavity to the perirenal area. Occasionally, some larvae errantly migrate to other tissues and organs and to developing fetuses. Infections usually become patent in 9–16 months but may be found as early as 6 months.

Clinical findings and diagnosis

When present in large numbers, kidney worms may adversely affect growth. The principal econmic loss results from condemnation of organs and tissues affected by migrating larvae. Kidney and lung damage are also possible.

When worms are in the kidney or in cysts that open into the ureter, eggs may be recovered in the urine. Prepatent infections ae difficult to diagnose, and a definitive diagnosis depends on demonstration of the worms or lesions at necropsy.

Control

Good control practises are indicated in areas when the worm is known to occur. More commonly, anthelmintics and sanitation (rearing on concrete or in confinement) ae used to control kidney worm. Ivermectin (in-feed for 7 days at 1.8 g/ton) and fenbendazole (in-feed for 3–12 days at 9 mg/kg/day) are effectuve agausbt Stephanurus so. Levamisole (in-feed at 0.36 g/ton) is also approved for use against this worm.

18.2.4 Ecto parasites

18.2.4.1 *Ring worm in pigs*

Ring worm is an infectious disease of pigs. The skin, is affected by trichophylon or nicrosprum. Ring worm lesions often occur behind the ear as expanding rings or circles of reddish or light brown scabs which may beraised above the surrounding skin, but are not common or the belly. Prurites occurs in trichophyton mentagrophyte infections.

Diagnosis

The presence of the fungus in skin scrapings or it isolation from them confirms a diagnosis based on clinical symptoms.

Treatment

If found topical treatments with sodium bendazol, undecyclinic acid preparations, copper salts may be adequate. Disinfection with formaldehyde, hypo chlorites or detergents should be carried out, steam and phenolic disinfectant have also been used.

18.2.4.2 *Mange*

The most common form of mange in pigs is that caused by *Sarcoptus scabei var suis*, which burrows in the skin and causes intense purities resulting in loss of condition and trauma to the skin, particularly in the region of ears.

Life cycle

S. scabei is a small burrowing mite 0.5 mm in length, which lives in galleries, in the hairy skin. They are circular and may be identified by the pattern of sucker and bristle in the legs. The normal life cycle is usually 14–15 days. Multiplication can only occur on the host, although the mites may survive for up to 2–3 weeks in moist place in piggeries etc.

Pahogensis and clinical symptoms

Lesions appear at least 3–4 weeks after initial infestation, as the host must become sensitized. The burrowing and feeding activities of the mites cause intense purities, which causes scratching which in turn results in the liberation of fluid from small vesicles near the burrows of mites. Affected pigs scratch continuously and may lose condition. The first lesions appears as small red pappules and general erythrema around the eyes, around snout, on the concave surface of the auricles of the external ears, in the axilla and on the front of the lock where the skin is thin. Scratching results in the excoriations of these affected areas and the formation of borrwish acabson the damaged skin. Subsequently the skin becomes wrinkled, caused with crusty lesions and thickened.

Epidemiology

Spread of infection is via pig to pig in contact. Infection of weaners usually occurs from the dam.

Diagnosis

Deep skin scrapings should be examined for the presence of mites, finding of which is pathogonomic. The most likely site in which to demonstrate the mite is from the anterior portion of the inside of ears.

Treatment and control

Mange can be treated by washing or spraying gently with a snap sack or pressure washer. The whole pig must be thoroughly treated with commercial mange dressings such as bromocyclin Alugan concentrate (Hoechst) and BHC (carpers mange dressing) or liazine (fison likeand mange wash). Treatment may have to be repeated within 10 days or as advised by manufacturer. It is particularly important to treat the ears.

Control

Depends upon treatment being applied to sows before farrowing in order to reduce the infection of piglets and also upon the provision of rested accommodation (empty for 3 weeks for treated pigs).

18.2.4.3 *Lice*

Hog lice, by their blood sucking habits, cause economic loss to swine producers, and they may be responsible for the spread of infection. The louse is about six millimeters long and greyish brown in color. During the winter it may be found in the ears, in folds of skin around the neck, and around the tail.

The female lays several eggs a day during the winter. These eggs are attached to the hair, and hatch in two to three weeks. They mature in another two weeks.

Prevention

The pigs should be dusted with an insecticide, such as DDT powder or sprayed with a 0.5% DDT solution. The pigs should be treated as often as necessary.

18.2.5 Non-specific diseases

18.2.5.1 *Mastitis in sows*

Mastitis in sows may be localized to a single gland or may involve more than one to cause fever, depression and death. Loss due to pig mortality may be considerable. It is of two types.

(i) Coli form mastitis, (ii) Chronic mastitis.

(i) Coli form mastitis

Coli form mastitis, may be caused by any entero-bacterial organism. Infection may be localized to a single gland or may involve more than one gland to cause fever, depression and death. Agalactia is common and pig mortality may be considerable.

Clinical symptoms

Acutely affected sows are usually depressed, inappetent, and pyrexia (40.5°–42.0 ºC). The udder is usually swollen and oedematous, often with massive congestion. Any secretion that may be obtained is purulent, pain in the udder may lead to restlessness in the sows.

Treatment and control

Acutely ill sows may be saved by parenterial treatment with antibioties like neomycin, tetracycline, ampicillin, anoxycillin, or by the use of trimethoprin. Sulphonamide injection for 2–4 days. Control depends upon the use of hygiene, bedding other than saw dust, clipping ofpiglet teeth, early treatment.

(ii) Chronic mastitis or sporadic mastitis

This type of mastitis affects single gland particularly in older sows resulting in abscess formation and loss of individual gland often with few systemic clinical symptoms.

Cause

C. pyogenes, streptococci, staphytococci, *Bacterioide* spp. Clostridia do not usually cause the systemic reaction. Bedding, unclipped teeth, floor, any factor causing damage to teat or udder.

Clinical symptoms

Often noted only when an affected gland fails to return to normal after weaning. It may be noted as local inflammation and pain over that portion of udder during lactation and pus may be expressed.

Diagnosis

Clinical symptoms, confirmation by isolation of organisms from the discharge of the teats.

Treatment and control

Parenteral antimicrobial including penicillin when staphylococci or streptococci are involved.

Control can be established by improving husbandry. Affected sows should only be retained if they, have sufficient functioning teats.

18.2.5.2 *Pneumonia*

More commonly it affects young piglets particularly up to 4 months of age. It is clinically manifested as a sequence of infection of upper respiratory tract. It is caused due to bacteria mycoplasma, *Chlamydia* and virus either alone or in combinations. Sudden onset of pneumonia is noticed and if one piglet is affected the entire litter shows symptoms, as they are infected by aerosols produced by infected pigs or by direct contact. Clinical symptoms of respiratory distress coughing and sneezing and nervous disorders may also be occasionally exhibited.

Bacterial pneumonia caused due to *Pasteurella, Salmondella Mycoplasma* require treatment with sulphadimidine 33.3% solution by injection, and treatment continued for about 5 to 6 days. Broad spectrum antibiotics are effective. In pneumonia due to slamonella I/m injection of chloramphenical is good andwhen it is caused by mycoplasma, then tylosin and sulphonamide are useful.

Separation of affected and susceptible animals in different units will prevent transmission. Cleanliness, disinfection, and proper ventilation too are essential. Exposure to inclement weather be avoided. Fresh water, and nourishing food be provided. Antiseptic inhalations prove effective if large number of animals are affected.

18.2.5.3 *Enteritis*

This disease condition is caused in pigs by parasites, bacteria, or virus. Escherichia coli is the main source of infection. Some predisposing factors such as unhygienic conditions, change of diet and cold result in this condition. Dehydration due to excessive diarohea takes place and may even lead to death in some cases. In case of enteritis due to clostridium perfingens, fatal necrotic and haemorrhagic enteritis in piglets of about 7 days of age is caused. Sulphonamides or these in combination with streptomycine is good if it is due to E. coli and the chloramphemical is good if cause is salmonella. For control, food be reduced and new changed diet be provided.

18.2.5.4 *Foot lesions in pigs*

Foot lesions are an important cause of debility in pigs and occur in a number of systemic and nutritional conditions. In addition foot lesions of the baby pigs occur in response to the quality of the flooring and in the adult, poor quality of slatted floors cause a posture leading to compression of the spinal nerves and eventual paralysis.

1. Foot lesions in baby pigs

Erosion of the sole, brusing of the heel and of the accessory digits and erosion of the knees can all occur in suckling piglets. Once the skin has been eroded, infection enter and can give rise to septic arthritis of the pedal joints or to synovitis and permanent lameness. Such conditions have a marked effect on growth rate and may impair it permanently. Prevention depends on proper antiseptic dressing regularly.

2. Foot lesions in older pigs

(a) Foot rot

The term used to describe a variety of septic conditions affecting the claws of pigs of all ages. The primary lesion in sows in the form of defect or penetration of wall or bearing surface of hoof, which provides point of entry for secondary bacterial invasion. These include erosions of sole, head and toe, fissures in the wall (sand crack) and separation of wall from sole at the white line. Infection spreads in the hoof in 3 possible ways, a deep necrotic ulcer may develop involving laminae or coronary band; necrotic may reach the coronary band and form ulcers, or infection may penetrate deeply and involve the deep digital flexor tendons, or phalangeal bones of the joints. When abscesses burst at coronets the condition is known as '*bush foot*'. Abrasive and chemical effects of newly laid concrete contribute to the production of hoof defects. Wet, unhygienic conditions and poor bedding also contribute to what is often a herd problem. Bacterial such as *F.neorophorus, C. pyogenes* and spirochaetes may infect the lesions.

Clinical symptoms

Lameness depends on the number of feet involved. Affected animals tend to walk on tip toe with paddling of 'goose sleeping' gait, and are reluctant to rise and move and may sit on their launches. Lateral claws especially of hind feet are most commonly affected. The affected claws are warm, painful and the primary lesion is usually apparent. Severe pain occurs when abscess develops at the coronary band and the leg is often held off the ground. *Cellulites* may occur in the limbs and reach the corpus or the hock. The heel and coronary band become swollen and blue black in colour, with multiple sinus formation. Septicemia and bacteraemia can occur and secondary abscess as may occur elsewhere *e.g.* in brain, spine and liver.

Treatment

Improve hygiene and management, in particular. Ensure that slats have plain edges and are at least 100 mm wide. Pigs should be run through foot baths containing 5–10% formalin (2–3 times a week) during septic hoof lesions, poulticing and bandaging help but rarely economic. Inject with antibiotics such as tetracyclines, ampicillin and use a spray *e.g.* tetracycline on local lesion.

(b) Laminitis

Occur mainly in boars and heavily pregnant or recently farrowed sows. Sometimes associated with post parturient fever, the signs of which can mask the laminitis.

Clinical symptoms

Stiffness, reluctance to move; affected animals walk on the front of carpie and are recumbent for long periods. Claws are warm and tenders. Pulsation of digital arteries is prominent.

Treatment

Corticosteroids, antihistamines, reduced feed and if there is fever, give a broad spectrum antibiotic parenlerally.

(c) Over growth and deformity of claws

Claws are not identical in size or shape. The lateral claws are slightly larger and broader. This is specially evident in the hind feet of pigs over 6 months. The toe is curved in the lateral and pointed in the medial claw. Overgrowth is mainly seen in pigs over 1 year. The commonest causes are keeping pigs in muddy ground or on very deep litter with insufficient exercise. Boars, especially those kept in small pens, are often affected. Generally it only affects the main claw, but sometimes the accessory claws may also be affected. When a single claw is over grown, it often results from a change in weight bearing, secondary to some painful lesion. Gross deformity is less common and is usually congenital and may occur in one or a number of claws in the same pig.

18.2.5.5 *Agalactia*

Syn. post parturient fever for sows. It occurs within 12 hr to 3 days after a normal farrowing. The animal goes off feed, slight feverish and apt to resent suckling by her piglets. The udder is hard. The hardness beginning at the rear and extending forward. A watery or white discharge from the vagina is not invariably present. The uterus may not be involved at all.

Cause

Following farrowing, may be due to prior feeding with excessive quantities of fodder, due to the inflammation of the uterus or due to endocrine failure, wet cold floors and cold. Post pasturent fever, draughty premise, bacterial invasion of mamary gland and subsequent production of endotoxin which are absorbed to give systemic symptoms.

Clinical symptoms and treatment

Affected sows should be given frequent small doses of oxytocin and if fever is present, a broad spectrum antibiotic such as ampicillin, tetracycline or trimethoprin, sulphonamide, B. Methsone has been suggested. Animals may be given a saline purgative. The litter should be given glucose or sow milk suppliment.

Control

Hygiene and exercise for the sow prior to farrowing and during the early stage of lactation may help. Restriction of the feed for the last 2 or 3 days before farrowing and replacing part of it with bran, may reduce the incidence

18.2.5.6 *Transmissible gastroenteritis (TGE)*

Transmissible gastroenteritis is an enteric disease of pigs caused by transmissible gastroenteritis virus (TGEV), a member of family *Coronaviridae*. The virus multiplies and damages the enterocytes lining the small intestine, producing villous atrophy and enteritis. Diarrhoea and vomiting occur in pigs of all ages; mortality due to dehydration is highest in piglets less than 7 days of age.

The symptoms are profuse watery greenish grey diarrhoea. Vomiting is also seen. The disease may be tentatively diagnosed from the clinical signs and pathology of the intestinal tract. Fluorescent antibody technique may be employed in mucosal impression smears. Virus isolation and serology can also be employed in diagnosis.

18.2.5.7 *Vomiting and wasting disease*

This is the name given to the condition caused by the haemagglutinating encephalomyelitis virus which is a coronavirus. Piglets of 4 days to 3 weeks of age are susceptible. There may be vomiting, anorexia, depression, emaciation and development of nervous signs. Most infections are inapparant and no gross pathology is seen. Diagnosis is by virus isolation and serology.

18.2.5.8 *Heat stroke*

As environmental temperatures increase form 22 °C to 40–41 °C pigs seek shade, water, mud or slurry in order to a wallow and cool themselves. Animals in transit, stressed and unable to cool to themselves in sow stalls, farrowing crates or in full sun, become dyspnoeic, salivate, breath with difficulty and become restless. They may later become frenzied, cyanotic and die. A body temperature of 41 °C to 43 °C may be recorded.

On autopsy, blood stained foam in the nostials and trachea, congestion and haemorrhage in the carcass and some oedema, particularly in lungs may be seen.

Diagnosis

History of high temperature and absence of other causes of death.

Treatment

Involves tranquilizers. Cooling the limbs and belly with water and injection of corticosteroids.

Prevention

Increased ventilation and cold running water in each pen.

18.2.6 Mycotic diseases

18.2.6.1 *Mycoplasma infections*

Enzootic pneumonia caused by *Mycoplasma hyopneumoniae* occurs in intensively reared pigs. Conditions that induce stress like poor ventilation, overcrowding etc.

may lead to outbreaks. Pigs of all ages are susceptible. Coughing and poor growth rate are the clinical signs seen. On necropsy, consolidation of the lungs is the important change seen. The condition can be diagnosed by isolation and identification of the agent, FAT, CFT or ELISA. Drugs which can be used for treatment include tylosin, lincomycin or tiamulin. Inactivated and adjuvant vaccines are available but their use is not widespread.

M. hyorhinis infection is seen in pigs of up to 10 weeks of age. It is a progressive polyserositis and is characteised by fever, laboured breathing, lamness and swollen joints. On post mortem examination, serofibrinous pleurisy, pericarditis and peritonitis may be seen.

M. hypsynoviae can cause a polyarthritis in pigs of 10–30 weeks of age. The disease produces a temporary lameness and is self limiting.

18.2.6.2 *Dermatophytosis*

This condition is uncommon in pigs. The dermatophyte associated is *Microsporum nanum*. If infection is seen, lesions can occur anywhere on the body as thick brown crusts.

18.2.7 Vitamin deficiency

18.2.7.1 *Vitamin A*

The absence of vitamin-A from the diet, its presence at low levels or poor absorption of the diet causes infertility and the birth of weak or non-viable piglet with congenital defects to sows with poor body reserves of vitamin. Nervous, skin changes and reduction in the rate of bone growth may occur.

Clinical symptoms

Nervous symptoms such as in-coordination, tilting of the head and eventual paralysis of the hind limbs occuring in growing pigs in which skin changes especially splitting of the tips of the bristles may occur.

Sows in particular and gilts may produce stillborn or moribund piglet at full term in the most severe form of the disease. Stillborn piglets may have ascities and moribund animals are listless, weak and lie on their sides around the sows.

Diagnosis

Based on history and postmortem finding and liver biopsy.

Treatment

Commercial rations normally contain adequate vitamin A, but home mixed feed should be examined for vitamin A content. If condition is diagnosed, affected animals or those at risk can be treated parenterally or by oral supplement of vitamin A.

18.2.7.2 *Vitamin-B*

1. Thiamine (vitamin B-1) deficiency

This is likely to occur in pigs fed on rations containing large amounts of heat treated carbohydrates *e.g.* bakery waste in swill to which no balancer has been added. Requirement of thiamine may be increased by high ambient temperature. Clinical symptoms include inappetence, poor growth rate, emaciation, a fall in temperature and respiratory rate resulting in the death within 56 weeks from congestive cardiac failure. There is cyanosis of the skin and mucus membranes.

2. Riboflavin (vitamin B-12) deficiency

Clinical symptoms of riboflavin deficiency consist of slow growth, frequent diarrohea, conjuctivities, skin changes such as alopecia and matting of the hair with sebaceous secretions. Irregular oestrus, delayed oestrus and eventually anoestrus have been reported in this deficiency. Inclusion of 2–3 g of riboflavin pertonne of feed is preventive.

18.2.7.3 *Vitamin-D*

If dietary calcium and phosphors levels are adequate no ricket condition develop of pigs. Treatment with parenteral ADE preparations may be used as an aid to recovery. This condition is not prevalent in India.

18.2.7.4 *Vitamin-E*

Low dietary levels of Vitamin E (X-tocopherol) is associated with a number of syndromes.

Clinical symptoms

Sudden death is most commonly seen. Piglets affected by muscular dystrophy may be dyspnoeic with normal body temperature and no symptoms of enteritis. Clinical symptoms may appear in a few piglets ina litter at about 2 weeks of age.

Diagnosis

Clinical symptoms and postmortem findings.

18.2.8 Mineral deficiency

18.2.8.1 C*opper*

If iron therapy alone does not cure anemia copper supplementation may be necessary. This is unusual occurrence and is rarely necessary.

Few clinical symptoms, other than anemia and leg weakness have been directly related to copper deficiency but the inclusion of this element usually as copper sulphate is common in growing pig rations at a level 50–200 ppm. Copper, iron metabolisms are linked and high levels of zinc may reduce liver copper as may iron. Copper toxicity is evidenced by inappetence and jaundice.

18.2.8.2 *Piglet anemia*

A hypochronic-microcytic anemia of rapidly growing piglets housed on concrete, which results in poor performance and in the death of severely affected animals. It is caused by primary deficiency of iron in the diet.

This is a rare condition under modern husbandry practises but may occur in individual litters or where oral iron dosing is accompanied by diarrhoea.

The piglet is born with limited iron and copper reserve and blood hemoglobin level of 12 mg/100ml occurs but the correct level is subsequently regained in normal healthy piglets with adequate available dietary iron.

Where animals are farrowed and reared on concrete or outside on iron free soil the requirement of iron outstrips the supply, thus the intake of iron in creep feed or otherwise becomes essential to supplement the iron requirement for proper growth.

Clinical symptoms

The clinical symptoms appear mostly when a severe anemia may have developed, particularly in pigs in good condition. Affected pigs may be plump, although their growth rate is less than that of normal pigs. They appear pale, the ears, belly and the mucosae may be yellowish in colour. There may be oedema of the head and forequarters. Lean, and pale pigs are more commonly seen. Diarrohea is common but the faeces is normal in colour. Affected pigs are lethargic and when disturbed show dysponea and a prominent apex beat. Severely affected piglets may die suddenly. Not all piglets in an iron deficient litter may show clinical symptoms of iron deficiency.

Treatment and control

Supplementation of the sows diet is normally ineffective in preventing the condition and the piglets must be provided with 15 mg iron per day to prevent the occurrence of the condition by one or more of the following methods, as the case may be, to ensure the availability of required quantity of iron to the piglet.

1. Intra muscular injection of iron compounds for prevention. Piglets be injected in the hind limb at 3 days of age with at least 200 mg of elemental iron.
2. Pastes of granules of iron preparations as may be convenient. Often iron alone does not cure the condition and copper supplementation be necessary. This is why crude ferri sulph powder is often used as paint on udders to provide adequate quantity of iron in conjunction with copper. Incidentally crude ferrus sulph also provide traces of cobalt ions which plays a significant role in haemopoiesis.

18.2.8.3 *Iodine*

As a non metallic element, iodine is required by the body for the formation of thyroxin, the hormone produced by the thyroid gland. Intake below the optimum level affects the health and fertility in pigs.

Clinical symptoms

The rate of development is lower and the pigs are hairless.

Treatment

The remedy is to provide salt licks or mineral mixtures containing traces of iodine.

18.2.9 Zoonotic diseases

18.2.9.1 *Sarcocytosis*

Cause

Sarcocystosis of pigs is caused by three species of sarcocystis, a sporozoan parasite.

They are *Sarcocystis miescherian, S. suihominis,* and *S. porcifelis.* It is characterized mainly by a cystic invasion of most tissues of the body, especially skeletal muscle and nervous tissue. The overall prevalence of sarcocysts in pigs appears to be relatively low and the incidence appears to be decreasing largely due to methods of husbandry where pigs are being reared indoors. S. *suihominis* uses humans and non human primates as definitive hosts. It is important as a zoonotic agent.

Life cycle

Sarcocystis sp. develop in two host cycles, the definitive host being dogs, cats and wild carnivores, while farm livestock, including pigs are intermediate hosts. Humans can also act as intermediate hosts. Only the intermediate hosts are harmed, the organisms rarely causing disease in definitive hosts. Once ingested, sporozoites will be released from sporocysts in the small intestine of intermediate host by the action of digestive enzymes. The sporozoites then divide by multiple fission into first generation schizonts and numerous merozoites will be liberated from each schizont. These merozoites then enter the blood stream and develop into second generation schizonts which further liberate merozoites. Merozoites from second and third generations penetrate muscle cells and neurons and glial cells in the brains where they encyst, produce the infective cysts about 2–3 months after initial ingestion.

Mode of infection

Humans acquire infection by ingestion of uncooked pork containing viable cysts of sarcocystis. Pigs become infected through eating food stuffs and pasture contaminated with the sporozoites from dogs, foxes, cats etc., which acquire infection from eating infected meat and offal. The affected pigs exhibit weight loss, purpura of the skin especially of the legs and buttocks, dyspnea and muscle tremors. Post mortem examination of affected carcasses will reveal distribution of schizonts throughout the body especially in striated muscles, heart, brain, liver, lungs and kidneys.

Diagnosis

Diagnosis can be done by histopathological examination of organs and muscles, wherein we can demonstrate the cysts lie within or between individual muscle fibers with characteristic cigar shaped. The larger cysts may lie loose in the perimuscular connective tissue and are globular, oval or bean shaped. In heavy infestation the carcass in condemned. In moderate to light infestations the lesions are removed and the carcass passed.

Zoonotic transmission of sarcocystosis is rare but has been reported in humans. To prevent infections, meat should be cooked before human consumption. Infection in pigs can be prevented by avoiding the feeding of human faeces or effluent to pigs. Obviously, this practise is not a component of modern hog production systems.

18.2.9.2 *Taeniasis*

Cause

Taeniasis is an infection of the small intestine of man with the adult stage of the pork tapeworm (*Taenia solium*). Cysticercosis is the tissue infection that involves larval cysts (*Cysticercus cellulosae*) of *Taenia solium*. Taeniasis and cysticercosis are of great public health and economic importance. The adult tapeworm exists in humans while the metacestodes exist in domestic and wild pigs as well as humans. Humans infected with *T. solium* excrete eggs into the environment as long as the worm is active in the intestine. This disease exists under conditions where pigs have access to tapeworm eggs and segments.

Mode of infection

The ingestion of food and water contaminated with human faeces containing proglottids by scavenging pigs is the most frequent way of transmission of cysticerci to swine. Larvae hatch from eggs in the pig intestine and they further migrate to muscle tissue, brain, liver and other organs. The use of inadequately treated human excrements as fertilizer is the other cause of larva (measly pork) lead to infection in humans. Additionally, autoinfection may occur in humans by means of fecal-oral contact wherein direct transfer of *T. solium* ova from the faeces of an individual harbouring an adult worm take place.

The larvae become infective after 2–3 months of infection which is characterized by the appearance of invaginated head as a white spot in the encysted site. The predilection sites of *Cysticercus cellulosae* in pigs are heart, diaphragm, masseter muscles, tongue, neck, shoulder, intercostals and abdominal muscles.

Cysticercosis in pigs is usually asymptomatic but heavy infections can produce muscular stiffness and loss of conditions.

Symptoms

In humans, clinical symptoms associated with adult worm in the small intestine can include abdominal pain, digestive disturbances, diarrhea/constipation and loss of weight. In case man acts as an intermediate host *i.e.* if harbor the larvae, the common site of severe symptomatic infection is the central nervous system (neurocysticercosis) and is manifested with headache, dizziness, hydrocephalus, loss of vision and nausea. In the brain parenchyma, cysticerci form a thin capsule of fibrous tissue that thickens with time.

Diagnosis

Post mortem examination of the exposed muscular surfaces, especially those of the diaphragm, abdomen, thigh and shoulder will reveal cysts. One of the usual site in post mortem examination is triceps barchii muscle, wherein an incision is made about 2–2.5 cm above the elbow joint. Heavy infestation with cysticercus cellulosae calls for carcass condemnation. In case or moderate infection, the carcass may be conditionally approved pending heat or freezing treatment.

Zoonotic transmission of taenia solium occurs due to scavenging nature of pigs in countries where pigs have access to human faeces. The infection is found only in free range animals and not sty raised ones. Prevention of animal to human transmission can be achieved by thorough meat inspection and adequate cooking/ freezing of contaminated meat.

18.2.9.3 *Trichonellosis*

Cause

Trichinollosis, a disease of great zoonotic importance, is caused by *Trichinella spiralis,* a nematode parasite which occurs in pigs, rats, mice, and many mammals. The life cycle of *Trichinella is* unusual because the worm could undergo complete development from larva to adult to larva in the body of a single host. Infective larvae are found encapsulated within cysts in the muscle.

Mode of infection

Infection in humans results from the consumption of raw or undercooked flesh of pigs containing viable encysted larvae. Pigs can acquire infection from the flesh of another pig via raw or unboiled garbage. The rats and mice also could obtain infection from similar source and via cannibalism. Ingestion of meat with viable larvae leads to digestion of the cyst by the host's digestive enzymes thus releasing the larvae. The larvae invade the lining of the small intestine where it matures into an adult. After mating with a male, the female worm release larvae. The newborn larvae pass to the striated muscles by the lymphatics and the blood stream. In the muscle they grow, curl up in a spiral coil and are encysted, the entire process may complete in one month time from the infection. The predilection sites for these larvae are voluntary skeletal muscles especially those poor in glycogen *viz*. the tongue, diaphragm, eye, masticatory and intercostals muscles. In the muscles, larvae may persist for a long period of time or they may die and become mineralized.

Symptoms

The affected pigs during ante mortem examination exhibit fever, stiffness, muscle pain, dyspnoea and facial oedema due to the presence of larvae in muscles.

Diagnosis

Post mortem diagnosis can be done by visual examination of pork carcass and trichinoscopic examination of pigs' meat wherein we could demonstrate the calcified cysts in the muscles, but the calcification take place only after a year of getting the infection. Therefore, immunological tests like ELISA and complement fixation tests are recommended for perfection. If trichinellosis is confirmed then the whole carcass must be condemned.

Prevention of zoonotic transmission in hogs can be achieved by strict adherence of garbage feeding regulations, rodent control, preventing the exposure of pigs to dead pigs and other animal carcasses, prompt disposal of dead pigs and other animal carcasses and provision of barriers between pigs and other animals. Trichinellosis is not a major problem in tropical countries including India, where the ambient temperature is high.

18.2.10 Hygienic measures for prevention of diseases

Prevention of the spread of infectious diseases is one of the most important and difficult duties. Each case must be treated according to its own requirements. There are however, certain methods of preventing the spread of infection that are common to all diseases and a consideration of these forms the basis of all preventive medicine. The great resistance of some infective agents, the insidious nature of many infectious diseases for which the animal may be an active carrier without giving any indication of the fact until the disease has become widespread. The very nature of microbes or infective agents favours the spread. They find the resting places and by all sorts of means they are in turn passed from place to place and animal to animals.

18.2.10.1 *Infection transmission*

Infection is transmitted from the diseased to the healthy animals either by direct contact or by indirect way.

Any material that has been in contact with an infected animal may carry the contagion. An infective material which has been in contact with an infective animal may pass the contagion onto other material which in turn may transmit it to a receptive animal. Disease is carried from diseased to healthy animal through other animals acting as passive carriers. Man may act as passive carrier by conveying the infective material on their hand, clothes and boots, vermin, birds, flies and other insects are usual modes of transmission. Food, water and air are also common transmitters of infection. The contagion of disease may enter the body by inhalation, ingestion, inoculation or by absorption.

18.2.10.2 *Preventive measures*

These measures include the following:

(i) Isolation of infected material and animal;
(ii) Notification of the infection;
(iii) Disinfection of all materials likely to hold or carry infective material;
(iv) General prophylactic steps;

(i) Isolation

The most important active measures, is the complete isolation of the sick or suspected animals. Partial or indifferent isolation is very dangerous as it tends to promote a false feeling of security. Not only animals but all other material belonging to animal must be completely isolated from contact either directly or indirectly with healthy animals. The attendant of the patient must be regarded as equally infective as the sick animal.

It is better to have separate persons to attend healthy and sick animals but if it is not possible then the sick animals be attended in the last and the attendant must make due precaution to clean himself, before passing among the non-infected stock. The period of isolation must extend beyond the recovery of the animal and not lifted until all possibilities of infection have passed away.

(ii) Quarantine

The object of quarantine is to give time to the disease that may be latent to become active. During this period measures are taken to disinfect material that may be infective.

(iii) Notification

It is very necessary to control and eradicate the diseases that are considered dangerous. Some diseases are not easily diagnosed as might be thought from their text book description. This is why it is important to notify any condition of mass ailments or deaths.

(iv) Prophylaxis

Prophylactic measures taken to prevent appearance of diseases as far as possible while the term is generally applied in connection with infectious diseases. The steps taken to prevent the onset of any preventable disease are prophylactic in character.

18.3 Health Schedule and Calendar of Operations for Control of Diseases

In order to control mortality at pig breeding farms and units effectively due to diseases, it is advisable to adapt a health schedule and follow calendar of operations given as a guideline which can be amended as per requirement, as this has an important bearing on economics of any pig enterprises.

For checking incidence of anemia in piglets, which is often noticed at some farms and declares due to iron deficiency largely, the udder of the dam is painted with iron preparation, and when the piglets suckle this is ingested along with milk.

For prevention against swine fever, which is one of the very fatal diseases of pigs, and unanimity has to be imparted to pigs, by carrying out swine fever vaccination in the third month to piglets to prevent occurrence of swine fever. This vaccination has to be repeated every year.

For checking parasitic infection, piglets be dewormed after weaning and subsequently also if worm load continues.

Vaccination against foot-and-mouth disease periodically.

Preventive measures will be required in case, there is incidence of tuberculosis, brucellosis and leptospirosis.

18.3.1 Protection from infection

Following considerations be made to protect any pig farm from infection:

- Stock should be purchased from clean sources and also transported under hygienic condition and keep the herd closed as far as possible.
- Maintain high standards of hygiene.
- All fresh stocks should be properly guarantied for about 3 weeks and minimum contact between groups of pigs should be avoided.
- Follow a regular health schedule to keep control on diseases and prevent occurrence of diseases by regular deworming and preventive vaccination.
- Avoid stressful condition in pigs as they are predisposing factors for occurrence of some important diseases.
- Avoid factors such as dust, extreme of temperature, ill ventilation, over crowing, under feeding, dietary deficiencies to mitigate effects of diseases.
- Provide suitable 'sick pens' accommodation for treatment.

Table 18.1 Vaccination Schedule against important diseases of Pigs

Name of disease	Age of first vaccination	Booster doses	Type of vaccine	Remarks
Anthrax	One year	Before monsoon every year	Spore	Generelly not done unless epidemic
Swine fever	Immediately after weaning 6–8 weeks	Every 6 months	Lapinised	Compulsory
Foot-and-mouth disease	Immediately after weaning 6–8 weeks	Every 6–9 months	Tissue culture	Compulsory
Swine erysipelas	3–4 weeks of age	3–6 weeks after first and later every 6–9 months	Alum treated	To be practised in endemic areas
Hoemorrhagic septicaemia	6–8 weeks (immediately after weaning)	Immediately before moonsoon	Mono/poly valant killed	

All the vaccins should be given through S/C route

CHAPTER 19

MAINTENANCE OF RECORDS

19.1 Need and Importance of Records

For any pig enterprise essential records should be maintained as they are an invaluable aid to management and planning. They are required to be maintained for the following economic traits which are of great value:

1. Litter size and birth weights
2. Weaning % and weaning weights
3. Growth rate
4. Feed consumed and feed conversion ratio
5. Mortality rate in piglets, weaners or fattening stock and adult breeding pigs
6. Financial records-these include profit and loss account, balance sheet, daily expenditure and Annual budget
7. Inventory of building and farm assets
8. Repair and maintenance register

For getting information on the above characters individual records have to be maintained for breeding, production, pedigree and herd records, feed records, labour records, marketing and complete enterprise records. Unless the pig breeder maintains proper records in all these aspects, he cannot analyse the efficiency of the enterprise and compare with standards laid down for various activities.

For profitable pig enterprise, production of large litters and finisher pigs to be produced along with economical use of feed and labour, as well as most judicious use of buildings and equipment and timely control of diseases are some of the essential components.

Where pig enterprises require financing from financial institutions and utilize Government subsidies, which is necessary due to confined feeding, controlled housing and high costs involved in labour, feed and medication, besides payment of interest, it has made it imperative for owners of pig enterprises to maintain proper and essential records to support their request for credit facilities from banking and financial institutions including government's ante-poverty programmes. Since large sized pig units are being established through credit financing, the record maintenance play an important role.

These records are also essential for the pig farmer to determine the status of his enterprise, its functioning on profitable lines, helping in taking proper management decisions, to pinpoint the practises which contribute to the efficiency or cause losses. Owner is also able to maintain careful watch on cash flows and income and expenses of each component of the enterprises.

Following are some of the basic records to be maintained. From this basic data, periodical analysis of performance or determination in achievement of goals, which critically affect the profitability of any pig enterprise running on commercial lines can be made.

19.2 Type of Records

Following type of records are required to be maintained but those will vary according to the production programme, objective of establishing the pig enterprise, size of the pig farm or unit and the agency financing the enterprise from whom credit facility is taken.

(i) Breeding records
(ii) Pedigree and herd records
(iii) Production records
(iv) Feed records
(v) Labour records
(vi) Complete enterprise records including financial records

1. Breeding register

As the entire profitability is primarily dependent on production of piglets in an enterprise which is a pig production unit, breeding stock with proper pedigree record are essential to be maintained which must indicate:

a) Number of sow
b) Date each sow or gilt is bred, boar no. to whom mated and number of services required
c) Date of confirmation of pregnancy (by checking each sow mated about 18 to 21 days after service)
d) Anticipated and actual date of farrowing
e) Date of weaning
f) Date of subsequent mating

This record will help to carry out breeding programme and implement farrowing policy and programme, so as to achieve the goal of farrowing per sow per year. This will help in identifying problem of infertility in the herd. This record will also be useful for identifying pedigree of the animal as well as in selection and culling of breeding stock. For maintaining proper breeding records identification of animal is a must either by ear tagging or notching of ears. This record will also help in judging performance of each boar. It will also be useful to separate the pregnant sow from the herd just 14 days prior to farrowing to the farrowing pen for preparation for farrowing.

2. Pedigree and herd records

For this purpose records are required to be maintained in respect of the following:

(a) Farrowing performance record

About 8 live piglets must have been farrowed per farrowing and the sire, dam and the litter piglets sold should be free from rigs, inverted teats, and other inherited defects.

(b) Growth register

Growth of litter piglets be recorded by noting down weight of piglets at the time of birth and a month there after and at weaning time.

(c) Feeding register

Gaining ability of meat type of breeding stock should be efficient. Rate of gain and feed requirements records are essential (this indicates feed conversions efficiency of the litter and thereby its dam).

(d) Slaughter record/register

The carcass quality tests indicate the performance, especially regarding carcass length, back fat thickness, area of eye muscle, dressing percentage etc.

The above records will help in determining the pedigree and herd performance. These records also help in progeny testing and sire evaluation, This information will be helpful in selection and culling of breeding stock as well as planning herd improvement programme *i.e.* selection of best boars and best sows and replacement stock. This will affect profitability of the enterprise.

3. Production records

It has been noticed that certain breeds and out of them certain lines make more rapid gains and make more efficient use of feed and produce more desirable carcasses. Some of them show better prolificacy than other.

On the basis of this consideration especially in a breeding farm, inferior lines can be weeded out. In this connection following Registers / records have to be maintained.

(a) Size of litter farrowed by each sow and litter born dead (Farrowing Register)

(b) Number of piglets weaned and the weaning weight, weaning loss can be worked out at 8 weeks of age.

(c) Weight of piglets at 21, 35 and 56 days indicates the productions of each breeding sow (Growth Register)

(d) Regular weight of growers/fatteners will help in ascertaining growth rate and feed conversion efficiency and will help in management decision about replacement of ingredients and change in ration formula. This information can be maintained breed wise and year wise which will prove useful in selection and other assessments (Conversion Efficiency Register).

4. Feed register/records

A record of the food used in the swine enterprise is of considerable value in checking the efficiency of swine production. If we know how much feed is consumed in the enterprise during the year and determine the quintal of live marketable hogs produced, we are in a position to compute the feed required per quintal weight. Since feed costs comprise about 80% of the total cost of producing pigs, a fairly accurate estimate may be made of the total costs if the cost of feed is known.

It may prove useful to maintain feed records in respect of date, amount and price of each kind of feed. Information about the quality of feed is ascertained by knowing and recording information on carcass quality regarding, (a) % of lean cut, (b) back fat thickness in inches, (c) carcass length in inches, (d) loin eye area (squire ft).

5. Labour records

The expenditure on labour is an important component in cost of production per pig. For this, time spent on feeding and caring for the herd during each month, and keeping record of houses, feed mixing, castration, vaccination etc. Labour expenditure should be restricted to 10 to 12% of total cost. There is no sense in maintaining any detailed record in this respect as it will be more time consuming.

6. Health and mortality records/register

Proper records pertaining to disease problem, treatment, vaccination and castration, mortality in different group: piglets, weaners, growers, fatteners and adults be maintained. Cause of death on the basis of post mortem reports should also be maintained. Tabulation of incidence of various diseases and mortality due to them in different age groups should be maintained. The disease status of herd is important for disposal of stock.

7. Post mortem and disposal register

It is important to know the cause of death of each animal died in the herd / farm. Post mortem examination will confirm the cause of death and will help in taking precautionary measures to avoid repetition as far as practicable. Post mortem record is a must for swine enterprises.

8. Financial records and registers

To maintain and study complete financial performance of the enterprise, proper financial records should be maintained regarding expenditure on components like, seed, labour, piglets produced, piglet death, total pork produced, price received per 100 kg, total return per 100 kg produced, Profit and loss accounts, Balance Sheet, Daily Expenditure Record, Provision for interest on loan and capital/working capital involved, Schedule of repayment of loan etc. Budgeting is also an important requirement for any production unit.

These records will help in analyzing performance, maintenance of pedigree of the herd, working out profit and loss of the enterprise as well as cost of production and weaknesses of the enterprise and pin pointing where losses are involved.

9. Keeping a simple journal or diary

In raising pigs, it is desirable to have a small notebook or diary to record various happenings and items of importance which may be forgotten unless they are written down as they happen. In this diary, recording of purchases of feed, labour performed by others, information on breeding and farrowing, if not recorded elsewhere, and similar information should be done. As we apply certain improved practises, it should be written down together with the date the practices have been applied. Losses, sales, changes in rations, weights, etc. may be included in the diary. The items directly and immediately entered in other records need not be noted in the diary.

19.3 Analyzing and using of Records

The records kept of the pig enterprise are valuable only if one analyze them carefully and use these records to determine (a) how efficient he is as a pig raiser (b) the extent to which the enterprise is improving from year to year (c) the strong and weak places in the enterprise and (d) what practices should be emphasized to improve the enterprise. A careful study of the records should be made in making effective use of them.

19.4 List of Records and Registers to be Maintained

1. Breeding register
2. Pedigree and herd records
    (a) Farrowing performance record
    (b) Growth register
    (c) Feeding register
    (d) Slaughter register
3. Production register
4. Feed register
5. Labour record
6. Health and mortality register
7. Post mortem and disposal register
8. Financial records and register
9. Simple journal or diary for day to day activity record

Table 19.1 Proforma for Maintenance of Breeding/Production Record

Sow record		
Sow number	Farrowing date	
Sow litter i, ii, iii	Total farrowed	male female
Date of birth of sow		
Weight of sow	Farrowing live	male female
Weight of sow on 180 days	Total weaned	male female
Date of weaning		
Weight of litter on		
i. 21st day		
ii. 35th day		
iii. 56th day		

Table 19.2 Record regarding litter

Name and number breed			**Boar number breed**		
Piglet no/ ear marking	Sex	Birth wt/ weaning wt (kg)	Age in days milch wt actual wt (kg) on 154 days	Carcass quality back fat/carcass wt (kg)	Remarks

(a) Avg. 154 days wt of litter

(b) Feed required 50 kg of body wt gain

CHAPTER 20

PROCESSING OF PIGS FOR MARKET

20.1 Introduction

World pork production is 104 million tonnes (2005) and forms 38.9% of total meat production. Pork is a major meat in a number of countries such as China, Denmark, USA, UK and Canada. There are 13.5 million pigs in India comprising of desi, cross breds and pure bred exotic pigs. The Uttar Pradesh ranks first among the states, having 2.3 million pigs and possess 17% of total pigs in the country. Pig meat production in India is supported by a large number of indigenous breeds of pigs, crossbreeds (constituting about 17% of total pigs) and exotic breeds including Yorkshire, Landrace and Hampshire.

During 2007–08 meat of swine (fresh, chilled or frozen) export was 1710.09 MT valued at Rs 2463.69 lakh. Pig production remained largely a scavenging activity with a very little input costs and primarily an activity of weaker section of people. For these people it is not only a source of income and livelihood but also a choice of meat for consumption. Availability of quality pork for a variety of consumers is a scarce item. A major programme was taken up for production, processing and marketing of pigs/pork products by setting up 8 bacon factories in different parts of the country during 1960s. Bacon factory is a composite structure with pig rearing unit, slaughter and products processing and marketing facilities. However, bacon factories were not a success story mainly due to low capacity utilization and managemental inadequacies.

Meat is an important commodity in the diet of people not only of developed countries but also of the developing countries in view of the preference for meat in the diet and the complimentary role of meat in balancing the diet with rich source of aminoacids, vitamins and minerals particularly iron and zinc. Meat is particularly important in the diets of young children and pregnant woman because of its rich protein and iron content.

Pork continues to occupy an important position as a food resource in developed countries as well as in developing countries. The pig has been a scavenger and in early domestication it was raised as a means of utilizing human food wastes. In many countries pig still performs this function as a 'backyard' inhabitant. Raw garbage need to be essentially cooked for use in feeding.

The income elasticity of demand for meat, dairy products and eggs are generally closely united. Incomes have a substantial effect on the demand. The expected increase in incomes in the developing countries are believed to cause a shift towards more meat, milk and eggs in the human diet. However, in the developed countries with current high consumption levels, future per capita meat consumption may not change or even decrease if incomes rise further. In Indian situation with liberalization of trade in the post WTO period a dynamic change in the food consumption aspects is likely to occur not only due to increase in incomes but also due to increase in tourists arrival and a greater number of Indians getting exposure to meat consumption in the foreign countries and expecting such products in India. Quality Pork and pork products demand would increase but there are no modern plants available to meet such demand. Imported products would have the disadvantage of higher price and would not be in reasonably fresh form. Thus the prospects of plants for production of high quality hygienic and processed pork products would be enormous and be a successful venture.

Pig genetics and breeding in the recent years has resulted in modern pigs of lower backfat thickness and lower fat percent and higher protein percent in retail carcass weight, resulting in lower kilocalories of energy per 100 g of retail carcass to meet the consumer preferences. Consumers would be desiring for a healthy, nutritious, tasty and safe pork product that satisfies their life style desire. Modern pork processing plant need to produce pork products of high quality and safety in addition to producing products of consumer choice.

Current status of slaughter operations

Slaughter of animals is a state subject and the State legislative have the exclusive power to legislate as per entry 15 of list II in the 7th Schedule in the Constitution of India. States have to follow Directive Principles while making laws and these legislative powers are regulated by State Animal Prevention Acts. Slaughter houses are regulated by local bodies and private slaughter houses are to be authorized by local bodies. Meat hygiene is regulated as per the by-laws of the local body and meat as food was earlier covered under prevention of Food Adulteration Act, 1954 and now under Food Safety and Standards Act, 2006.

Reasons behind poor hygienic status of pork in India

(a) Production System

Difficultly in applying hygienic practises in a herd of few pigs, say 10 or less which is very common in our production system. Further, our production systems are characterized by 'production by masses compared to the mass production by a few' in the European countries where the pig production practises.

(b) Socio-religious-cultural-political obstacles

The taboos related to pig rearing and further pork processing.

(c) Lack of producer awareness and low economic status

Majority of the pig farmers in our country belong to the below poverty line strata.

(d) Lack of consumer awareness

Most of our pork consumers are either not aware of or not worried about the consequences of eating unhygienicaily produced pork *viz.* zoonosis, microbiological and even physico-chemical aspects.

20.2 General Considerations for Constructing Pig Abattoirs

20.2.1 Selection of site: factors which need to be considered

a. Land
    i. Level: select high level areas to prevent water stagnation
    ii. Quality: sandy/red soil are the most suited
    iii. Size: enough space for future expansion
    iv. Cost: need to be economical
b. Electricity: requires adequate power supply
c. Sewage facility: sufficient sewage facility to remove the treated effluents
d. Transport facility: enough access to main roads

e. Distance: best site will be in the outskirts of city. Need to be at least 10 km away from airports and 1000 m away from residential areas.

f. Wind direction: best if the orientation is opposite to the wind direction (leeward)

g. Area requirements

   (i) Small abattoir *i.e.* <30000 LSU*/year : 1–2 acres

   (ii) Medium abattoir *i.e.* 50000–100000 LSU/year : 2–4 acres

   (iii) Large abattoir *i.e.* >100000 LSU/year : 4–6 acres

   *LSU=Livestock units

   1LSU=2 pigs

20.2.2 Water supply

Site should have the provision for adequate amount of potable water supply. In addition to this there should be facility for hot water at 80–85 °C for sanitation purpose.

a. Water requirement for dressing operations: on an average 454 liters/pig/day or about 10000 liters of water/tonne of pork produced.

b. Floor washing requires pressure in the range of 200–300 Pa and that for carcass washing may be in the range of 1000–1700 Pa.

20.2.3 Civil construction

(a) Antemortem pen

A crush of suitable size shall be there for conducting antemortem examination.

(b) Lairage

Large pigs require a pen size of 0.7 m^2/pig and small ones require 0.6 m^2/pig. The pens shall have adequate facilities for providing drinking water and feed.

(c) Abattoir building

i. Floor: shall be non-absorbent, non-slippery with sufficient gradient.

ii. Coves: radius at the junction of wall and floor shall be more than 8 cm and round in shape to facilitate easy cleaning and to maintain the plant hygiene.

iii. Interior walls: shall be smooth and flat with washable surface, preferably with acid proof tiles, up to a height of 3 m from the floor and rest apoxy painted.

iv. Ceilings: shall be at a height of 5 or more meter height from floor with smooth and flat surface.

v. Windows: shall be about 1.2 m from the floor level with ledges slope at 45°.

vi. Doors: preferably self closing types made of rust resistant materials with 1.5 m width and 3 m height.

vii. Screens: shall be provided for windows and doorways to control insects and flies.

viii. Rodent proofing: screen with a mesh size of not more than 1.25 cm shall be provided.

ix. Opening for chutes: Floor openings for chutes shall have curbs of not less than 30 cm height to prevent floor drainage getting access into the chutes.

x. Lighting: Roof can be provided with transparent glass covering of about $1/4^{th}$ of the floor area for better lighting inside the plant. Further, enough light source shall be provided to ensure more than 500 lux at a height of 1.5 m from floor in the inspection area and 200 lux and 100 lux at a height of 1 m from the floor level in work places and other areas respectively.

xi. Drainage: one drain per 40–45 m^2 area shall be provided with a slope of 1:50. However, the drainage for blood can have a slope of 1:20–25. The drainage shall have minimum of 15 cm width.

xii. Ventilation: Sufficient ventilation shall be provided to prevent accumulation of odour and dust inside the plant. Care must be taken to ensure sufficient fly proofing for these ventilators.

Equipment installation: minimum of 1–1.5 feet distance shall be provided between the equipment and wall to ensure better accessibility for cleaning, repairing etc. Similarly, they may be fixed at a minimum height of one foot from the floor level.

20.3 Pig Supply for Abattoir

Abattoir should receive good quality pigs from the known sources of production either from attached own farm as integrated facility or from adopted farms (contract farming) with provision of input services and marketing arrangement with quality supervision or a combination of both own farm supply and adapted farms. The abattoir could run for slaughter and marketing of pork carcasses or could lease out to entrepreneurs on annual charges or could operate as a slaughter facility and charge for the slaughter services.

Transport of pigs should be done at night or during the cooler part of the day. Stocking density in the truck should be appropriate. The roof of the truck should be insulate to reduce radiant heat gain. Pigs could be given cold shower on arrival at the abattoir for reducing transport stress.

20.4 Pig Receiving and Holding in Lairage

Pigs are received and unloaded in a holding area and are examined for health and only healthy animals are allowed to enter lairage. Transport, holding and handling of pigs for slaughter is the most controversial aspect of pre-slaughter care. Stress factors emerge when they are penned before slaughter and natural aggressive instincts are amplified in slaughterhouse conditions. Overcrowding in the lairage should be avoided. Insulated floors help to reduce stress and if they are to be kept overnight, clean, dry straw helps them to settle down. The straw must be removed and the pens cleared out before another batch is held in the same pen. Every effort should be made to handle them without excitement. Water should be provided. Pigs are generally fasted for a day to reduce the amount of intestinal contents.

Animals are given rest prior to slaughter for producing hygienic and better quality pork. Animals that are excited, exhausted and badly handled would result poor quality pork. Pigs which are stressed produce some level of pale, soft, watery muscle which is low in quality. Stress increases the possibility of 'blood splash'. Tepid water sprinkled from overhead in the restraining passages helps to calm pigs down. Pigs enjoy water sprays and they quite down for slaughter.

20.5 Antemortem Inspection

Animals presented for slaughter should be clean, healthy, fasted, free from blemishes, unstressed, easy to handle and well muscled and not over fat. Dirty stock is hygienic risk and extra handling of dirty animals results in stress which adversely affects meat quality.

Pigs are examined during holding as well as in lairage and sick animals are isolated and appropriately dealt after detailed examination. Animals should be examined individually at rest and motion for general behaviour, level of nutrition, cleanliness and clinical signs of diseases recorded. Abnormalities in respiration, behaviour, gait, posture, structure and conformation and abnormal discharges and protrusions from body openings, abnormal colour and abnormal odour should be examined. Animals are broadly categorized as normal animals and abnormal animals. Normal animals are passed for slaughter and abnormal animals are moved to separate pens for thorough examination.

20.6 Post mortem Inspection

Post mortem inspection is essential for wholesome pork production, because many diseases and abnormal conditions are not detectable during ante-mortem examination. It shall be done as soon as possible after dressing operations as setting of carcass may render it difficult to examine the lymph nodes. Care shall be taken to avoid unnecessary mutilations in the carcass.

Post mortem inspection involves visual examination and palpation of organs and incisions where necessary and laboratory tests wherever confirmation is required. Examination of 'meat lymph nodes' is important as these glands drain different parts of the body and the condition of specific lymph node can give an indication of condition of the part of body drained by it. Swelling or discolouration of lymph nodes indicates pathological condition. Meat lymph nodes in pigs include prefemoral and popliteal lymph nodes. Viscera and head shall remain identifiable with the carcass until the inspection is completed.

Stages of post mortem inspection

a. Examination of carcass

Inspect for bruising, haemorrhage, discolourations, local or generalized oedema, swelling of joints etc. Diaphragm shall be lifted and examined for tuberculosis. Muscles near shoulder shall be incised to detect cysticercosis. Diaphragm, abdominal and intecostal muscles shall be examined for trichinosis.

b. Examination of head

Surface of tongue can be examined for lesions of FMD and a portion of tongue can be inspected under microscope for trichinellosis. Masseter muscle can be incise and examine for *Cysticercus cellulosae* infection.

c. Examination of viscera

All viscera shall be inspected immediately after their removal. Every organ and their associated lymph nodes shall be examined by visual, palpation and if necessary by incisions. Lungs shall be examined for pleurisy, pneumonia and tuberculosis. Similarly, heart, especially pericardium shall be examined for tuberculosis and pericarditis.

Judgments

a. Passed for human consumption.
b. Total condemnation *e.g.* FMD, swine fever, generalized *Cysticercus cellulosae* infection and trichinellosis.
c. Partial condemnation *e.g.* localized tuberculosis.
d. Conditionally passed *e.g.* localized *Cysticercus cellulosae* infection.

20.7 Live Pig Weighing

When weighing of live pigs is desired, gently drive them onto the platform of the weigh balance. Allow the needle to come to rest or for digital scales, allow the digital readout to equilibrate and record the live pig weight.

20.8 Slaughter of Pig

Pre-slaughter care

Pigs should be kept off their food for about 12 hr before slaughter and allowed all the water they will drink. A well rested and fasted animal will give a better carcass, as the muscle is in good condition and the blood stream will not be gorged with nutrient substances from the digestive system. In ordinary circumstances, most of the Slaughter. The pigs selected for slaughter should be free from disease and in a healthy condition, gaining and not losing weight contamination that takes place at slaughter is of intestinal origin, and for this reason, the intestinal content should be reduced to minimum.

The chain method of slaughtering is used in killing and dressing pigs. In this method, the following steps are carried out in rapid succession

Stunning

The pigs are rendered insensible by use of captive bolt stunner, gunshot, electric current or carbon dioxide.

Shackling and hoisting

The pigs ae shackled just above the hoof on the hind leg and are the hoisted to an overhead rail.

Sticking

The man doing the sticking takes a position scuarely in front of the pig, holds down the snout and opens the skin for a distance of about three inched in front of the breast bone. He then inserts the knife, edge upwards, taking a line with the base of the tail, for about four or five inches, lowers the wrist, which brings the point of the knife upwards and withdraws the knife. The animal is allowed to bleed for 5 to 6 minutes. Blood provides an ideal medium for the growth and multiplication of putrefactive organisms and it supplies a vehicle for their distribution throughout the animal. Thorough bleeding, therefore, has a profound influence on the keeping quality of carcass.

Scalding

The animal is placed in a scalding vat for about 4 minutes. The carcass should be kept moving, so that all parts get a uniform scald and a clean white skin is produced. The temperature of the water is about (66 ºC). The scalding process loosens the hair and scruff. A slow scald is better and much safer then a quick scald.

Dehairing

After the pig has been lifted from the scald and placed on a bench or table, The scraping must be done as quickly as possible, as the hair will again adhere if allowed to cool.

Returning to overhead tracks

As the animals are discharged from the dehairing machine, the gam cords of the hind legs are exposed; and gambrel sticks are inserted in the cords. Then the carcass is again hung from the rail.

Singeing

Singe the carcass for the removal of reminiscent hairs using a blow lamp. Scrape the scruff and wash the carcass thoroughly with cold water.

Removing the head

The head should be removed before opening the carcass to permit complete drainage of blood from the carcass. A cut should be made just above the ears at he first point of the backbone and across the back of the neck. The gullet and windpipe should be severed to permit he head the drop. The cut should be continued around the ears to the eyes and to the point of the jaw bone, thus permitting the head to be removed while leaving the jowls on the carcass.

Evisceration

Stand at the back of the carcass and grasp the tail then cut around the pelvic arch to loosen the bung, care being taken to keep the point of the knife against the pelvic bones.

With the belly facing the operator, make a shallow cut from between the back legs to the throat; cut deep to the bone between the back legs and open the abdominal cavity, care being taken not to picture the bladder. Insert the left hand and keep back the intestines and stomach and continue the cut through to the breast bone; the knife may be pointed downwards and inserted into the chest cavity in order to continue the cut through the centre of the breast bone and thought. Now pull the bung through the pelvic cavity and ease the intestines down by severing the attachments to the backbone; cut around the skirt or diaphragm and pull out the lungs and heart; cut through just below the gullet.

The carcass is then thoroughly washed both inside and out. The kidneys and leaf fat are also removed and the carcass left neat and trim, then allowed to cool thoroughly.

Splitting

Split the carcass by sawing down the midline through the centre of the backbone.

Removing the leaf fat

While the carcass is still warm, loosen the leaf fat which is found on the inside of each half of the carcass.

Washing

Washing the carcass and then keeping it in the chilled room over night, where the temperature is held from around 1 °C.

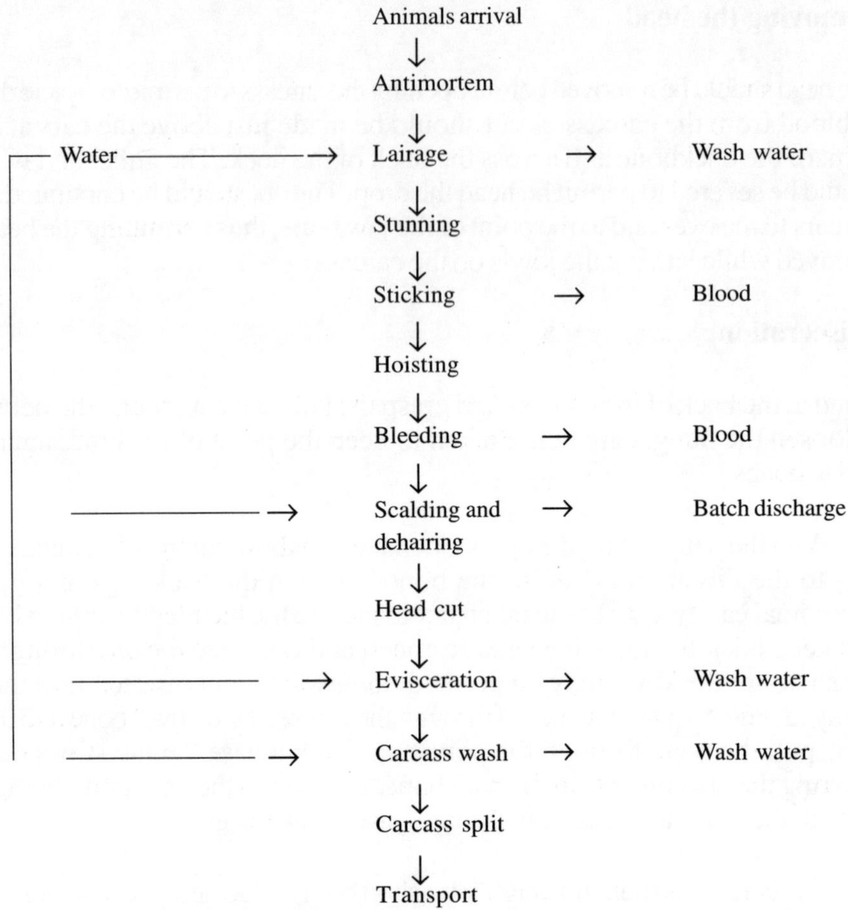

Fig. 20.1. Processing flow chart for pig slaughtering

20.8.1 Quality of carcass

Because consumer preference is such an important item in the production of pork, it is essential that the producer, the packer, and the meat retailer be familiar with these qualities, which are summarized as follows:

Quality

The quality of the lean is based on firmness, texture, marbling, and color.

Firmness

Pork muscle should be firm so as to display attractively. Firmness is affected by the kind and amount of fat. *e.g.* pigs that are fed liberally on peanuts produce soft pork. Also, pork with small quantities of fat will contain more moisture and tend to be soft.

Texture

Pork lean that has a fine-grained texture and porous, pinkish bones is preferred. Coarse textured lean is generally indicative of greater animal maturity and less tender meat.

Marbling

This characteristic contributes to buyer appeal, feathering (flecks of fat) between the fins and within the muscles is indicative of marbling.

Colour

Most consumers prefer pork with a white fat on the exterior and a grayish pink lean marbled with flecks of fat.

Maximum muscling, moderate fate

Maximum thickness of muscling influences materially the acceptability by the consumer. Also, consumers prefer a uniform cover of not exceeding ¼ inch (6 mm) of firm, white fat on the exterior.

Repeatability

The consumer wants to be able to secure a standardized product meat of the same tenderness and other eating qualities as the previous purchase.

The clean dressed carcasses are now ready for transfer to the refrigeration chamber for taking off the animal heat and allow the carcass to set properly and mature. The clean halves of the carcass after weighment on rails are transferred to the cold storage, which are maintained at 1 °C, so that the temperature of carcass near the bone is brought down to 2–3 °C and relative humidity of carcasses kept between 90%. Proper air circulation is also maintained and air exchange in cold storage room be changed about 5 to 6 times for each day.

20.8.2 Cutting of carcasses

After setting of the carcasses in cold storage overnight when they are properly set, the carcasses are brought to cutting room after weighment, where the halves of the carcass are cut into ham, bacon, loin, belly or streaky bacon. For cutting of ham the symphysis of pork is located and cut at a point not farther than 2.5 inches from it and feet cut just above hock joint, a litter above or below depending on requirement. Fat from the face of the ham is removed and all jugged skin trimmed off to give proper shape to ham.

The bacon side is separated from the shoulder at a point between 2nd and 3rd rib or between 3rd and 4th rib. The bony portion is sawed out and fleshy portion cut off with the help of knife. The part left after removal of the shoulder pieces and leg, pork is called side bacon. If loin pieces are required for fresh meat, then ribs are not removed. A cut is made starting from shoulder and just below the vertebral column with the help of a saw and if back bacon or steamy bacon is to be made, ribs are first removed and the cartilaginous portion of ribs are moved with a knife on same lines as loin cutting. Portion from upper half is called back bacon and portion of belly part is called steak bacon. Shoulder piece is cleansed of blood spots if any and irregular portions are trimmed off and shaped properly. It is used for sale as fresh meat. The front foot is cut off at knee joint, spare rib and cervical vertebrae (neck bone) are removed. The shoulder piece also de-boned and meat can be used for various types of sausages. The joint is separated.

Various type of cutting of pork joint or pieces is prevalent according to demand of trade.

20.8.2.1 *Fresh pork cuts*

A range of fresh pork cuts would be produced depending on the market demand and ability to promote marketing. Different countries follow different methods of preparing retail cuts from whole sale cuts and market them. A large variation is observed in the production and marketing of consumer cuts. Marketing could be done as primal cuts or bulk or retail consumer cuts based on market requirements.

Efficient marketability at higher returns is the prime consideration in processing fresh pork cuts.

20.8.3 Processed (cooked) pork products

A very large range of processed or cooked pork products could be produced depending on the market demand and sale prices available. Smoked/cooked ham, bacon, loin and picnic shoulder and a range of sausage products are commonly produced to utilize different cuts in a profitable manner. Utilization of edible by products such as skin, stomach, intestines, liver and fat trimmings is essential for profitable production of processed products. Production of newer products such as restructured products utilizing tough and lower value cuts to produce better quality products of higher sale value is an important consideration for better marketability of pork products. Pork nuggets, bites, slices etc. are important products that may find high acceptability and variety among Indian and foreign consumers as well.

Effect of cooking on heat sensitive nutrients in pork

Time and temperature of cooking are the major factors influencing the eating quality of pork and pork products. On exposure to heat, muscle proteins undergo denaturation followed by coagulation, thus results in loss of solubility. Disintegration of z-disk occurs at higher temperature. In well cooked meat increased rigidity *i.e.* protein hardening occurs above 65 °C. Sarcoplasmic proteins denature more quickly. Among myofibrillar proteins alpha-actinin is most heat labile. Actin becomes insoluble at about 80 °C, whereas tropomyosin are denatured at about 85 °C. Collagen undergoes shrinkage at a temperature above 60 °C followed by increase in solubility. In presence of moisture, collagen is hydrolyzed to gelatin *i.e.* collagen becomes more tender on heating whereas myofibrillar proteins become tougher on heating. However, at temperatures of about 70 °C rapid shrinkage of collagen occurs followed by protein hardening and toughening. Above 80 °C oxidation of '-SH groups' results in increased tenderness.

pH increases by about 0.2–0.3 units on heating and consequently the pH of the meat will be shifted to higher side resulting in no-apparent change in the water holding capacity. The increase in pH can be attributed to unfolding of actomyosin complex during the cooking process which exposes the basic groups to outside. Cooking results in considerable decrease in weight of pork either due to drip loss or evaporation. Change in weight ultimately alters the percent of protein, fat and ash content in the cooked pork. This loss of moisture further results in decrease in juiciness in the cooked pork. Cooking by moist heat results in more nutrient loss compared to that by dry heat. B-vitamins are mainly affected as they are water soluble in nature. Thiamine is the mot sensitive vitamin to heating. Riboflavin, niacin and Vitamin B_6 can also be partially destroyed by cooking.

Further, cooking results in change in colour of pork surface to brown due to surface dehydration and maillard reaction, which occurs only at a temperature of above 90 °C. Brown colour of cooked pork can also be attributed to increase amounts of metmyoglobin and globin-hemimyochromogen. Certain volatile compounds *viz.* sulfydril and nitrogen compounds, ketones etc. are driven off during cooking and impart characteristic flavour to the cooked pork. Dry heating mainly imparts flavour at the exposed surfaces where the temperature is high, whereas moist heating result in flavour development in the deep tissues. Thus, palatability of pork is influenced by time temperature combinations and the method of cooking practised.

Full exploitation of Indian spices and condiments in association with other culinary practises would facilitate production of a range of products for beneficial utilization of different pork cuts and their efficient marketability.

Table 20.1 Primal and Retail Cuts of Fresh Pork

Wholesale/primal cut	Retail cuts	Cooking methods
Fresh pork Ham	Fresh ham roast	Roast
	Fresh ham butt	Roast
	Center cut roast	Roast
	Rolled ham roast	Roast
	Fresh ham shank	Cook in water
	Center cut steaks	Braise
Loin	Center cut loin roast	Roast
	Ham end loin roast	Roast
	Shoulder end loin roast	Roast
	Loin pork chops	Braise
	Rib chops	Braise
	Tenderloin	Braise, roast
Spareribs	Spareribs	Braise, roast

Table 20.2 Whole sale/Primal Cut and Retail Cuts

Spare ribs	Spare ribs	Roast, braise cook in water
Shoulder butt	Boston butt	Roast
	Boneless butt	Roast
	Blade steaks	Braise
Picnic shoulder	Fresh picnic roast	Roast
	Arm steaks	Braise
	Cushion style picnic	Roast
	Rolled picnic	Roast
	Hock	Cook in water

Table 20.3 Processed Pork Products

Wholesale cut	Retail cut	Cooking methods
Smoked pork ham	Whole ham	Bake (roast)
	Half ham	Bake
	Ham slice	Broil, pan broil, braise
Loin	Canadian style bacon blices	Roast (in piece) broil, pan broil (in slice)
Bacon	Breakfast bacon	
	Sliced bacon	Broil, pan broil
Cottage roll	Cottage roll	Bake
Picnic shoulder	Smoked picnic	Bake

20.9 Preservation and Manufacture of Meat Products

The ancient people by experience found out that salting of meat helps checking putrefaction which is caused due to micro-organisms they produce certain acids. Salt was used by them which prevents acid formation or delay it, besides this is also antiseptic and improves taste of meat.

20.9.1 Curing

The preservation of meat through process of curing by use of dry salt is practised. Use of salt-peter is associated with process of curing. When salt is sprinkled on meat block or piece, it enters into deeper tissues by process of osmosis and meat juices mix with salt and form brine and reaches all parts. Salt peter when used gives pleasing and stable pink colour to meat. With the development of refrigerator, curing can be done throughout the year and salt and saltpeter in combinations is used for dry curing of meat to give long life. Desired temperature for curing has to be maintained, if temperature gets higher, meat can get sour. Those producers who do not have refrigerator facilities, curing is restricted during winter when temperature are favorable for this. Temperature of cold storage should be maintained between 3°–5 °C. Both dry and wet curing methods are sued. Besides proper cold storage, vats or cement concrete tanks, equipment for brine making and storing, scales for weighing ingredients, brine pumping sets, saltpeter (KNO_3) or Sodium Nitrite ($Na\ NO_3$), or Sodium Nitrate ($Na\ NO_2$) and sugar are required. The brine made for pumping in the ham/bacon is known as pump pickle and those used in vats and tanks for immersing them is called cover pickle. Quality of salt should be good uniform in brand and free from impurities so as to maintain uniformity of products.

Wet curing

For this, pickle solution is required to be prepared. First pickle solution is stored in cool place at 3°–5 °C as at higher temperature it sours. Hygienic conditions should be maintained in handling and curing otherwise it will introduce harmful effect causing souring and off colour.

Proper colour development

Development of pink red colour in meat by curing with salt and saltpeter is important. The pigment of meat muscle myoglobine is red in colour which looses its colour when cooled due to coagulation of myoglobin, while in case of cured meat it does not loose its pink colour. The pork meat which cures well is slightly acidic pH 5.4 to 6.0.

Factors affecting colour fixation are:

(a) Ratio of lean fat
(b) Temperature of curing
(c) Duration of curing period
(d) Curing ingredients and formula used

Ṣaltpeter also inhabits growth of an crops in meat which are putrefactive. Proper limits of nitrates in cured meat as prescribed by Indian Standard Institution (ISI) have to be maintained as them its harmless. The nitrate content is final processed material should not exceed 0.02% by weight (or 200 parts per million) sugar is added to pickle solution to soften salts freshness. Regular testing of pickle for nitrate and nitrate content and bacterial flora be made to keep check on quality.

Curing of ham

Hams are graded according to weight, smaller hams are kept on top in the tank as they are cured earlier. Pump pickle is first injected into hams, one shot in smaller pieces and two in large pieces. These injected pieces are than kept in cover pickle in vats and about 20–25 litter of cover pickle is required for 45 kg of meat block. Hams are overhauled thrice on 5th, 15th and 30th day of its curing periods. After curing hams are placed on wooden racks to drain for about 48 to 78 hr and then they are sow bed in water to remove excess salt and hung on to air dry and then sent for smoking.

Curing of bacon

Same process is adapted as in ham except that strength of cover pickle is kept at 70º salinometer. About 60 ml of pickle in one stroke is injected and kept in cover pickle to overhaul twice on 3rd and 5th day and cured at rate of 1.5 day per 500 g. It is then drained of and dried in air and then sent to smoke room. These beacon pieces can also be directly kept in cover pickle and cured at the rate of 2.5 days per 500 g and overhaul on 3rd 7th and 15th day during curing.

Dry curing

This is also practised for ham and bacon at many places, specially by small process or where adequate refrigeration facilities do not exist. Application of salt and covering ingredients directly on the surface of meat block in a highly concentrated form gives protection to meat and hastens during at somewhat higher temperature although proper temperature of 3–5 ºC give uniform cured product with minimum changes of spoilage. In case of ham 50 ml of mixture is rubbed thoroughly per

500 g of weight of ham. About 3 rubbing at a interval of 3 to 5 days is made. In places 7 days per inch of thickness of hams is taken in curing. These pieces are piled one up on the other and covered to exclude air.

20.9.2 Smoking

To smoke cured meat is a common practise. This is resorted to impart a flavour to smoked products which is liked and relished by people. There are many types of smoke house, besides modern type of smoking cabinet. Smoke house have pits below the floor covered with grates where saw dust or special quality of wood is burnt for imparting smoking flavour. Meat pieces are hung at about 8 ft height on racks. The wood fire is allowed to smolder slowly so that smoke is kept forming. Hung pieces should not overlap and space allowed between pieces. Bacin is smoked for 8–12 hr depending upon weight of pieces. Ham is smoked for 12 to 20 hr. Colour of the meat blocks get established when interval temperature is reached to 52 °C. Temperatures of smoke houses have to be controlled and air circulated inside should be uniform. After smoking pieces are removed and allowed to cool in atmospheric temperature and then kept in cold storage. In smoking cabinet, the temperature and humidity is automatically controlled, besides the inflow of smoke to the smoking chamber and circulation is also automatically controlled. This is easy to operate and uniform products are produced and more hygienic to handle. These are being used in number of meat plants recently built in India.

Effect of smoke

It helps to bring out colour of cured meat and improved keeping quality of meat block. It imparts antioxidant properties to the fat and antiseptic and germicidal properties to the surface of meat block. It tenderizes the meat and imparts fine gloss and finish to surface of smoked products which is pleasing to sight.

20.9.3 Processing of sausages

The trimmed and chopping of meat, which cannot be retailed profitably, is minced and spiced and filled in casings, guts internal lining and offered for sale as sausages. With the coming up of organized factories in this country for meat processing, variety of sausages are being produced as per consumer appeal and demand. The trimmings, unsolvable portions of bacon, ham and fat are available from different sections of factory, besides regular cuts from carcasses are also used for manufacture of sausages. They are classified into a fresh, smoked and dry.

Fresh sausages

Two varieties of fresh sausages, breakfast and cocktail, are made of lean meat.

This meat is put in bowl chopped and spices and binder are added and consistency watched, after which it is filled in hog casings.

The ingredients used in different fresh sausages vary *viz* pork meat, fat, binder, dextrose or sugar, salt and spices. Cereal flour 'rice', soybean milk powder is used as binder, which holds fast the meat fibres and absorb extra water. The equipment required for sausages mainly is meat mincer, blow chopper and sausage filler with linking device. Fresh sausages are highly perishable products, they should, therefore, be cured immediately after processing and stored in cold storage having 2 to 5 °C temperature.

Smoked sausages

There are varieties of smoked sausages originated in Europe where veal and other meat is mixed, while this is not used here. In this country, lean pork meat is selected, tongue, tire, liver and heart are also used. This meat is then cured using a mixture of sodium nitrate, sodium nitrate dextrose and this mixture is sprinkled uniformly on meat to be cured and thoroughly mixed. This is then placed in cold storage for 24 hr having temperature of 3 ° to 5 °C for curing. The cured meat is then mixed in mixing machine and then chopped in blow-chopper and spices and binder are then uniformly mixed. When proper binding has taken place, which takes near about 5 minutes, then it is transferred to sausages filling machine and filled out in the type of casings distinctive for variety of sausage desired to be made. These sausages are then smoked, cooled and stored in cold storage. Bologna sausages, frankfurter sausages, ham sausages, liver sausages etc are made. The spices, variety and proportion of pig meat and fat varies. Their smoking and cooking time also varies.

20.9.4 Canning

Varieties of canned pork products are also manufactured. They are packed in 500 g or 1 kg can which are lacquered insides, so that the tine plate does not affect quality of meat. These cans are properly cleaned and then the products filled in these cans leaving some air space and lids are clinked and put in a exhaust box and steamed under pressure to expel the air from cans they are then steamed in double steamer and then put in to big autoclave where sufficient heat to all the meat in cans is applied depending on product filled, pH, bacterial load etc. for predetermined length of time so as to ensure proper sterilization and product should be commercially sterile, otherwise tins are spoiled due to micro-organisms and get puffed up and becomes unsafe for human consumption. After which the tins are cooled, labeled and packed and then product can be kept for about 1 year for use.

Cooled ham is prepared by keeping the ham, which is properly cured, in stainless steel frames and then cooked in cooking cabinet, which has automatic temperature control or it can be cooked in pans.

20.9.5 Labeling, packing and transport

All the goods whether fresh, processed or canned have to be properly labeled. Under meat control order for proper labeling, name of the product, date of manufacture, name of ingredients, weight and price have to be clearly indicated. All the packing materials for fresh goods used should be of standard and special quality and checked from time to time. Different size of packing should be properly sealed so as to avoid any damage to the product. In case of canned products they are cleaned with saw dust so as to remove any grease and then labels are pasted. Date of manufacture and expiry date of product is embossed on tin lids. These tins are then packed in wooden boxes or cartons, depending on means of transport to be used and distance to which they are to be dispatched. In India ISI and army has laid down standards for labeling and packing of goods in regard to quality of wood, size, type of boxes and marking and labeling of products as well as boxes and then boxes are properly secured with iron hoops etc.

For meat and products, packaging shall provide the wholesalers, retailers and most importantly the consumers with optimally sized product in a safe and wholesome condition. Packaging of meat, despite its highly perishable nature, been a matter of less importance to our retailers, inspection agencies and most importantly consumes, especially in the rural areas. Selling the fresh cut in unwrapped conditions is still popular in our wholesale and retail meat trade. Packaging, thus, is not simply surrounding the meat with some materials, but shall address all the specific demands of marketing *viz.* containment, protection, preservation, convenience, communication etc.

The common packaging technique used for selling fresh pork and pork products in India *i.e.* wrapping, belongs to non-preservative packaging types wherein the packaging contains and protects the product from contamination and water loss without creating in-pack conditions very different from outside *i.e.* these types of packaging have no/little effect on extending the shelf life of pork. The product life can be achieved by different packaging methods *viz.* vacuum packaging, modified atmospheric packaging, controlled atmospheric packaging, active packaging etc. wherein an inhibitory environment is created and maintained in the in-pack conditions. A detailed explanation on the characteristics of these packaging materials is out of scope for this book and the interested readers may refer Sharma, B.D. and Sharma, N. (2000) for further information.

The large uneven cuts like bacon, ham, loin etc. can be wrapped neatly by shrink packaging. The materials for shrink packaging shall have high structural strength and the commonly used ones are polypropylene, polystyrene and poly vinyl chloride. These films are manufactured by streching the film under controlled temperature so that they are mono or biaxially oriented to stay stretched at ambient temperature and then locking the film in this stretched condition by cooling.

Due to the increased concern over the 'non-biodegradable' nature of plastic packaging films, the concept of using biodegradable packaging materials is picking up momentum in the recent past. Any material claiming to be biodegradable shall completely decompose into carbon dioxide and water within 6 months period. Biodegradable films are mainly vegetable in origin and the most common ones are derived from polysaccharides *viz.* starch, cellulose, gelatin, gum, gluten etc. Such films have good mechanical and optical properties but have poor water vapour barrier property which makes them unsuitable for packaging fresh pork.

The fresh and smoked goods have to be transported after proper packing either through refrigerated bans under cool conditions, if they are to be sent to nearly markets, or in ice packing, in which case they are put in tin boxes which are closed and then ice is placed over this box and again kept in a wooden box and insulation provided and sent through rail or trucks to distant place which may take 24 to 48 hr some parties use thermocole insulated box, so the fresh products reach them in safe conditions. The railways are also providing this facility in this country on long routes. Insulated railway wagons for transportation on special rates for quick delivery, provide sufficient load of goods are available for dispatch regularly.

20.10 Utilization of By-Products

In western countries maximum utilization of slaughter house and meat factories waste and by-products is made, which has enabled them to improve their economic return from such units, as they are able to sell their finished products at a much cheaper rate. In India most of these materials are generally wasted and full benefits are not derived from them. Proper utilization of these products can substantially contribute towards improving the economy of these units provided care in collection, preservation and facilities for their proper utilization are made available. Wastes and by-products are of following varieties:

(a) Blood
(b) Bone
(c) Meat, condemned part and organs
(d) Fat

(e) Viscera
(f) Lung, liver, kidney, ears, head
(g) Trotters and hooves

20.10.1 Utilization of the wash and by-products

(a) Blood

In Indian conditions, due to lack of facilities for collection at most of the places, blood is wasted. About 70% of this is not utilized in this country. In some places it is used for human food, or for pharmaceutical purposes. Dried blood is good source for fertilizer, and it contains nitrogen which is required for growth of plants. It is also used as manure in tea gardens, coffee and rubber plantation and agriculture farm. It contains 13% nitrogen. Fresh blood, if properly collected, can be converted into blood meal by dry rendering or blood dryer. Blood meal is produced in ratio of 6:1.

(b) Bone

Bone meal is made out of skeletal bones, head bones, feet, ribs etc., from which meat had been scraped and bone meal is produced in dry rendering mill or bone digester. From bones about 70% of bone meal is produced. While producing bone meal, some technical fat is also produced (about 10%). It is used in livestock/ poultry feed.

(c) Meat

Meat cuttings and condemned meat etc., after steaming and drying, is converted into meat meal. It is mostly used as a supplement for the livestock feed. The drying rate is 4:1.

(d) Fat

Fat available from slaughtered animal is rendered and converted into good quality edible lard and canned and sold at good price. The other inferior quality fat after rendering is utilized by soap manufacturers in soap industry.

(e) Casings and gut

After stripping of intestine of all food material and then washing and cleaning, they are processed in automatic gut making machine for making casing which is utilized for sausage making. Roughly 0.4 rings of grade A per animal can be produced.

(f) Viscera

Viscera can be utilized for animal feed after cleaning.

(g) Glands

Glands like pancreas, pituitary, and ovaries are collected and used for manufacture of pharmaceuticals. It requires proper collection and preservation in proper manner under hygienic conditions.

(h) Trotters

Trotters can be sold as such as they are used for making infector soaps.

20.11 Sanitation Practise of Slaughter Houses And Meat Factory

Plant sanitation and hygiene in slaughter houses and meat factories can not be overlooked. It is essential for production of quality product. For ensuring effective sanitary control, it is essential to observe proper procedure for cleaning and maintenance of equipment, personal hygienic, environmental sanitation, sanitary designing of slaughter houses and meat plants and proper plant sanitation *i.e.* water, floor, fixtures etc., and disposal of wastes and handling of meat till it reaches consumer.

In slaughter houses as well as in meat factories, the edible and inedible department should be kept totally separate and this should be kept in view while planning unit in future. This helps in avoiding objectionable conditions affecting preparation and handling of edible products.

The water supply is very important requirement for maintaining proper sanitary condition in any unit. Water should be sufficient and potable. Proper chlorination be carried out and this should be regularly tested from time to time and sufficient cold water for washing carcasses and floors made available. Besides this, steam for sterilization of knives and other equipment and cooking is needed. Warm water for washing is also needed for slaughter house and meat factories.

Proper drainage be provided for the unit. Drainage from toilet should be separate on the factory and abattoir drain from which liquid and solid contents be separated. Affluent and waste treatment plant for environmental hygiene is essential.

20.12 Guidelines for Establishing Pork Processing Plant

20.12.1 GMP requirements

Food safety management systems are based on prerequisite programmes like good manufacturing practise (GMPs) and good hygiene practise (GHPs). Good Manufacturing Practices provide the basis for the regulations and plans required. GMPs are designed to assure that the foods are produced under hygienic conditions and that microbiological, chemical and physical hazards are prevented. These are the control factors relating to the entire manufacturing operation, not just processes used, and include programmes for facilities and grounds, equipment and utensils, pest control, receiving and storage, process control, product recall and personnel training. Prior to development and implementation of Hazard analysis at critical control points (HACCP) plans, a company should first review existing programmes to verify that all GMPs are in place and are effective. GMPs are similar to any policy programme of a firm and require a written programme, an appropriate training programme and schedule, a maintenance schedule and management commitment. The written programme applies to all areas covered under the GMPs. It will include who, what, where, when, why and how actions or procedures are put into place. The written programme explains the scope of the GMPs, responsible individuals, parameters, monitoring activities and records, corrective actions and records of those and verification activities. Programmes should be written by teams of employees from various areas of the company that can bring technical and operational expertise to the table. Within the GMP programme, cleaning and hygiene are given their own subsection referred to as GHPs. This may be defined as those operations involved in providing a clean sanitary environment for the preparation, handling and storage of meat. In other words, the GHPs define what has to be done in relation to cleaning and hygiene, when it has to be done and by whom. Areas covered by the GHP programme include cleaning of plant and equipment, staff health in relation to food handling and staff cleanliness, the cleanliness of the raw materials including live animals, ensuring all detergents, sanitizers and other non-food chemicals are properly packaged, labelled, comply with their specifications and are stored correctly. The following guidelines for developing Good Manufacturing Practises for meat plant operations are recommended for voluntary consideration and use in developing plant-specific procedures. These GMPs are not designed to control specific hazards, but are intended to provide guidelines to help processors' produce safe and wholesome products.

Basic requirements and criteria for GMPs in meat plants

A . Meat plant structure

All buildings and surroundings should be designed, constructed and maintained in a manner so as to minimize contamination of meat and meat products. Management should have a programme in place to monitor and control all structural elements and maintain appropriate records.

1. Meat plant premises

All buildings on the premises should be designed to permit proper cleaning and sanitation. The general area outside the factory should have a perimeter fence. The yard area should be free from all obstructions and accumulations of refuse. Buildings should be of sound construction and maintained in good repair and must not present chemical, microbiological or physical hazards to the carcasses.

2. Production line

All floors, walls, coving, doors, windows, ceilings/overhead fixtures and stairs in the production area should be constructed of material that is durable and easy to clean. All production floors should be sufficiently sloped for liquids to drain into grated trapped outlets. Where specified risk material is removed, drain traps must have a screen apparatus or mesh, of no more than 6 mm in size, fitted. All walls should be light coloured and have coving at the bottom. All windows should be equipped with close fitting screens. All doors should have a smooth, non-absorbent surface and, where appropriate, be self-closing. All stairs and overhead structures should be designed and installed in a manner that prevents the contamination of the product. All light fixtures that are suspended over the production area of the plant and all other areas should be enclosed in shatterproof diffusers, to prevent the contamination of products in the case of breakage. All parts of the plant should also be adequately lighted. The ventilation system should eliminate, as far as possible, the build-up of condensation and remove contaminated air. The ventilation openings should be equipped with close fitting screens and should be located in areas that prevent the intake of contaminated air. The sewage system should be designed and constructed so that there is no cross-connection between the effluent of toilet waste and any other waste that comes from the production process. The equipment used to decontaminate hand-held tools (knives, hooks and saws), commonly referred to as 'sterilizers', should contain water at a temperature of 82 ºC or higher and there should be sufficient numbers of 'sterilizers' correctly located near the operators' workstations . The production line should contain sufficient number of conveniently located workstation, wash hand basins with properly trapped waste pipes connected to drains. The workstation wash hand basins should be non-hand/arm operable. The workstation wash hand facilities should have a supply of premixed water at a suitable temperature and anti-bacterial soap.

3. Chills and frozen storage

Due to the perishable nature of the end product, temperatures in the chilling area and frozen storage area should be properly maintained at +12ºC and -12ºC

(-18 °C or lower is recommended), respectively, to reduce or prevent microbial growth. Chilling room should be large enough to hold the products until the internal temperature is reduced to no more than +7 °C (although +4 °C or lower is recommended). There should also be adequate air circulating around the carcasses. Consideration should be given to temperature rises that occur during loading, defrost cycles and in inactive chills.

4. Dry goods store

All packaging material and dry goods should be food grade, transported, stored and handled in a manner that prevents chemical, physical or microbiological contamination of the carcasses. Management should monitor and control this operation and maintain the appropriate records. Certification of incoming materials by letters of guarantee or other satisfactory means should be demanded from suppliers.

5. Sanitary facilities

There should be suitable and adequate changing room facilities including lockers, wash basins and showers. All the toilets should have self-closing doors, separate from and not leading directly into the meat plant, and should be correctly ventilated and maintained. It is recommended that there should be at least one toilet and wash hand basin for every fifteen male employees and one toilet and wash-hand basin for every ten female employees. All toilet areas should have hand-washing facilities with a supply of premixed water at a suitable temperature, anti-bacterial soap, disposable paper towels and a cleanable bin. There should be sufficient number of maintained sinks with properly trapped waste pipes connected to drains. In the hygiene lobby, there should be a boot-wash or equivalent for people to clean their boots on entering and /or leaving the abattoir. The hand washing facilities should be non-hand/arm operable. The hand washing facilities should also have a supply of premixed water at a suitable temperature, anti-bacterial soap, disposable paper towels and a cleanable bin.

B. Maintenance of meat plant equipments

This programme should outline procedures that ensure satisfactory conditions are maintained, areas to be inspected, tasks to be performed, person(s) responsible, inspection frequencies and records that should be kept.

1. Planned maintenance/calibration

Documented maintenance programme that lists all the equipment and utensils together with maintenance procedures should be available. The programme should specify the necessary servicing of the equipment (including frequency), minimum yearly calibration or as per manufacturers' recommendations, the replacement of parts, the person(s) responsible, methods of monitoring, verification activities and record keeping. All monitoring devices and any equipment that could impact food safety should be listed together with their intended use. All critical food safety measuring equipment should be calibrated to recognized national standards. Protocols and calibration methods should be established for equipment and monitoring devices. *e.g.* this equipment may include thermographs and refrigeration control units.

2. Equipment design and installation

The equipment should be maintained in a manner that prevents contamination of the carcasses. Management should have a programme in place to monitor and control the use of all equipment and maintain the appropriate records. All production equipment and utensils should be constructed of corrosion resistant material. All meat contact surfaces should be non-absorbent, non-toxic, smooth, free from cavities, unaffected by the product and capable of withstanding repeated cleaning and sanitation. All equipment and utensils should be installed and maintained in a clean and sanitary manner that prevents the contamination of the product. Adequate space must be provided within and around equipment for maintenance and inspection. All lubricants that are used in abattoir production equipment should be of food grade. The lubricant used should be tasteless, odourless and be resistant to bacterial growth and rancidity.

C. Sanitation of the meat plant

The management of the meat plant should maintain and document a full list of services. The list should include details for water treatment, water flow, the storage of chemicals, cleaning programme, pest control and waste disposal.

1. Water and steam quality programme

The water used in the meat plant must be potable and should be evaluated from a microbiological, chemical and physical perspective. Where applicable, this should also include the quality of the water used for the steam supply. If steam is used in the plant, it should be generated from potable water and meet operational requirements. Management should have procedures in place to deal with water that does not meet specified standards. All records of water potability tests and treatments applied should be maintained and filed. Some selected parameters for water quality used in carcass washing and meat processing plants are presented in Table 20.4.

2. Water supply

Potable water should be provided at pressures and in quantities sufficient for all production and cleaning requirements. Microbiological testing of water should be carried out. All records of water potability testing should be available upon request. When chlorination is carried out, two basic controls must be in place:

(a) A metering device for adding the correct concentration of chlorine that is designed to readily indicate malfunction

(b) Twice daily checks to determine total available chlorine or an automatic analyzer equipped with a recorder and an alarm.

There should never be cross-connections between potable and non-potable water supply systems. Non-potable water should never be used in the production process. It is permissible to use non-potable water for cleaning the lairage and animal transport vehicles. All hoses, taps and cross-connections should be equipped with anti-backflow devices. When hoses are not in use, they should be properly stored. A map of the water distribution system should be available for inspection and indicate the source, storage, treatment and the distribution of both potable and non-potable water within the factory.

3. Cleaning chemicals

All cleaning chemicals should be received and stored in a lockable, dry, well-ventilated chemical store, which is separate from the meat plant. There should be no possibility of cross contamination of the product from the cleaning chemicals. All cleaning chemicals should be mixed in clean, correctly labelled containers as specified under manufacturers' guidelines. The chemicals should be dispensed and handled only by authorized and properly trained cleaning personnel. All cleaning chemicals that are used in the abattoir should be of a food grade standard. Basic ingredients of cleaning and disinfecting agents that can be used in meat plants are presented in Tables 20.5.

4. Cleaning programme

The effectiveness of the cleaning programme depends on cleaning procedures, cleaning chemicals, competency of the cleaning staff and the structural standard of the premises. Verification of cleaning depends on microbiological monitoring. Production can commence only after a pre-production visible inspection of the premises has been carried out and all sanitation requirements are met. Records of monitoring, corrective actions and verification results should be made available upon request. Current guidelines recommend that there should be a documented cleaning procedure to ensure that the plant has been properly cleaned and disinfected before the commencement of production. The cleaning programme should specify:

- The areas, equipment to be cleaned, the frequency and the person(s) responsible
- Special instructions for cleaning food machinery and the person(s) responsible
- The cleaning equipment that is to be used along with the instructions for its proper operation (*e.g.* pressure and volume of water)
- The detergent/sanitizer to be used including commercial and generic names, dilution factor, water temperature
- The method of application of the solution, contact time, foam consistency, scrubbing if necessary, high/low pressure
- The rinsing instructions, water temperature
- The sanitizing instructions, commercial and generic names, dilution factor, water temperature, contact time
- The final rinsing instructions
- The safety instructions for the handling of all cleaning chemicals.

5. Microbial testing

All licensed meat plants to implement checks on the general hygiene and conditions of production in the establishment by means of microbiological checks. The purpose of microbiological testing at various points around the abattoir is to determine if surfaces are microbiologically acceptable. Microbiological testing should provide for the following:

- Procedures for sampling and the number of swabs to be taken daily
- The microbiological method for the examination of the samples
- Recording of test results
- Where results are unacceptable, corrective action should be applied, reviewing the process controls, thus ensuring that a reoccurrence of unacceptable results is prevented.

Sampling procedure

Only experienced persons with suitable aseptic precautions must undertake sampling for bacteriological purpose. The sample should be a true representative. Take care to protect sample from extraneous contamination. A sampling plan is the choice of particular sampling procedure and the decision criteria to be applied to a lot, based on the examination of prescribed number of sample units by defined methods. It should be administratively and economically feasible and should take into account the heterogeneity of distribution of microorganisms. The stringency of sampling plan should be based upon the hazard to the consumers from pathogenic or spoilage microorganisms. The choice of plan must therefore consider seriousness of hazard and future conditions to which the lot is exposed. A decision criterion includes the microbiological limits of the number of samples which should conform to these limits. Limits should be based on microbiological data appropriate to the food and to the kind of criterion on question. It should take into consideration the risk associated with the organisms likely to affect the acceptability of the food. Numerical limits should also take into account the distribution of microorganisms in the food and inherent variability of the analytical procedures.

6. Pest control

Establishing procedures for pest control is an important component of GMP. Management should have a properly documented pest control programme to monitor and regulate all elements of pest control. The pest control programme should include:

- The name of contact person(s) at the plant for pest control
- The name of the extermination company where applicable or the name of the person(s) responsible for the programme
- The list of chemicals and methods used
- The frequency of inspection
- Pest survey and control reports

The effectiveness of the pest control programme is verified by on-site inspection of areas for the presence of insect and rodent activity. Records of all monitoring results, recommendations and actions taken should be available on request.

7. Waste disposal

There should be facilities provided for the storage of all waste types prior to its removal from the premises. This area should be properly drained for any run-off that may occur and located away from the production area, preventing contamination of the end product. Containers for waste material should be clearly identified, leak proof and fitted with covers (if stored outside).

D. Operation of the meat plant

1. Training

This programme should provide, on an ongoing basis, in personal hygiene and food safety for all personnel working in the abattoir. Training should be updated annually. Training should also be evaluated to determine the needs of those involved, using management and operatives to provide feedback as to how to improve training. Management should monitor, control and maintain the appropriate records/ certificates, to prove that training has been carried out. All production personnel should be trained, to recognized standards, to produce products that are microbiologically acceptable. They should understand what the critical limits are, the importance of maintaining these limits and the action they must take if the limits are not adhered to correctly.

2. Communicable diseases/injuries

It is the management's responsibility to ensure that there is an annual renewal of medical certificates for all meat plant employees. No person, while known to be suffering from, or known to be a carrier of, a disease likely to be transmitted through food, or carrying an infected wound, skin infection, sores or suffering from a gastro enteric illness, is permitted to work in the meat plant. When returning from an illness, management must demand a medical certificate from that person indicating that they have no impediment to return to work. This is to prevent such a person contaminating the product with pathogenic microorganisms. A person with an open cut or abrasion should not handle the product unless the cut is completely covered with a coloured, waterproof covering.

3. Personal cleanliness

All personnel working in the plant should maintain their own personal cleanliness. Protective clothing includes light coloured overalls or a coat and trousers, chain mail gloves and aprons (where applicable), footwear and hair/snood coverings. These should be worn and maintained in a sanitary manner (*e.g.*, light coloured overalls, coat and trousers should be changed daily). All persons entering the plant should remove objects such as wristwatches and/or jewelry, the exception being a plain wedding ring/band, from their person, which may contaminate the carcasses. All personal belongings and clothing should be stored in an area away from the plant, in designated lockers. Smoking and eating and/or drinking are not permitted in the production area. Hand washing should be conducted on entering and leaving the plant, immediately after finishing any task that involved contact with intestinal contents/faecal material on the carcasses and after using the toilet facilities. Washing hands thoroughly with premixed water and anti-bacterial soap is necessary to remove microbial contamination.

4. Controlled access

The access of visitors should be controlled to prevent contamination. All necessary precautions should be taken to prevent cross-contamination, including the use of protective clothing, hair covering and footwear by all visitors.

As has been dealt with, GMPs concentrate on the broader areas of meat production and play a vital role in implementation of specific quality assurance programmes in any meat plant. They are not hazard specific, but are the guidelines to safe and wholesome meat production.

Table 20.4 Selected Parameters for Water Quality used in Carcass Washing and Meat Processing

Parameters	Limits (max)
Hardness	100.0 mg/L
Conductivity	2500.0 mS/cm
Chloride	250.0 mg/L
Sulphate	250.0 mg/L
Sodium	200.0 mg/L
Aluminium	0.2 mg/L
Nitrate	50.0 mg/L
Nitrite	0.1 mg/L
Ammonium	0.5 mg/L
Copper	2.0 mg/L
Fluoride	1.5 mg/L
Arsenic	0.01 mg/L
Cadmium	0.005 mg/L
Cyanide	0.05 mg/L
Chromium	0.05 mg/L
Iron	0.2 mg/L
Lead	0.01 mg/L
pH	6.5–9.0
Temperature	20 °C

Table 20.5 Basic Ingredients of Cleaning and Disinfecting Agents

Ingredients	Examples	Function	Concentration	Contact time
Acids	H_3PO_4	Removal of inorganic deposits	–	–
Alkali	NaOH	Removal of organic deposits	–	–
Hypochlorites	Sod.hypochlorite (14–30% available chlorine)	Disinfection	Meat plant 130–200 ppm carcass washing 100 ppm (max)	3–30 min
QAC*	Cetyl trimethyl ammonium chloride	Disinfection 50–500 ppm	Meat plant	1–30 min
Active Iodine	–	Disinfection	0.005–0.03%	–
Active Oxygen	–	Disinfection	0.03–0.5%	–

*QAC's are not suitable for CIP, as they often form foam vigorously.

20.12.2 Regulations

A number of regulations have to be followed in establishing and operating pork processing plant. The plant must be licenced under Meat Food Products Order,

1973 . The plant must observe different pollution control norms of the respective State where the plant is located as stipulated by the respective *State Pollution Control Board.* The plant must follow provisions under The *Prevention of Food Adulteration Act and Rules*, there under. It must also be permitted by the local body where it is proposed to be situated as per the bylaws of the local body. Meat export is regulated under *Export (Quality Control and Inspection) Act,* 1963 and the *Export (Quality Control and Inspection) Rules,* 1964. For export of processed products the plant must be registered under APEDA after inspection by the Meat Plant Registration Committee as per the *Processed Meat (Quality Control and Inspection) Rules,* 1995. Water (prevention and control of pollution) Act, 1974; Air (prevention and control of pollution) Act, 1981 and Environment (protection) Act, 1986 also stipulate requirements for meat processing plants.

20.12.3 Water

Water is used in the food industry as an ingredient, as a production process aid and for cleaning. Its use as an ingredient and as processing aid can give rise to potential microbial or chemical contamination problems, and so it is important to use water of a high microbiological and chemical quality (*i.e* of potable quality). Water used in hand washing facilities also poses a potential problem. Stagnant water is particularly hazardous as microbial levels can multiply under favourable conditions. Hardness of water must be considered since detergents are formulated in relation to the degree of water hardness. Potable and non-potable water should be in separate independent systems.

Table 20.6 Different Grades and Uses of Water in Food Processing Operations

Grade of water	Use
Treated potable water	Product
	Cleaning of product containers (high-risk)
	Cleaning of raw materials (high-risk)
	Cleaning of process machinery (high-risk)
	Boiler feed water
	CIP feed water
Potable water	Product
	Washing of containers
	Washing of raw materials
	Washing of machinery
	Washing of production areas
	Transporting product
	Processing product
	Washing facilities for staff and visitors
	Drinking
	Prewash and final rinse of materials and containers
	Heating and cooling

Table 20.6 *(Contd...)*

Grade of water	Use
Recycled treated or potable water	Secondary washing of materials and containers Secondary heating or cooling
Recovered water	Flushing of toilet Heating or cooling Washing of non-production areas Vehicle washing Fire fighting Garden irrigation

- Throughout the year, 95% of samples should not contain any coliform organisms or *Escherichia coli* in 100 ml
- No sample should contain more than 10 coliform organisms per 100 ml
- No sample should contain more than two cells of *E. coli* per 100 ml
- No sample should contain more than one or two cells of *E. coli* per 100 ml in conjunction with a total coliform count of three or more per 100 ml
- Coliform organisms should not be detectable in 100 ml of any two consecutive samples

Table 20.7 Selected Parameters of Water Quality (EU standards of potable quality)

Parameter	Units	Limit
Temperature	℃	20
pH	pH units	6.5–9.0
Conductivity	mS/cm	2500
Chloride	Mg/l	250
Sulphate	Mg/l	250
Sodium	Mg/l	200
Aluminium	Mg/l	0.2
Nitrate	Mg/l	50.0
Nitrite	Mg/l	0.1
Ammonium	Mg/l	0.5
Permanganate oxidation	Mg/l	5.0
Boron	Mg/l	1.0
Iron	Mg/l	0.2
Maganese	Mg/l	0.05
Copper	Mg/l	2.0
Fluoride	Mg/l	1.5
Arsenic	Mg/l	0.01
Cadmium	Mg/l	0.005
Cyanide	Mg/l	0.05
Chromium	Mg/l	0.05
Mercury	Mg/l	0.01
Nickel	Mg/l	0.02
Lead	Mg/l	0.01
Antimony	Mg/l	0.005
Selenium	Mg/l	0.01

Table 20.7 (*Contd...*)

Parameter	Units	Limit
Pesticides, individual	Mg/l	0.1
Pesticides, total	Mg/l	0.5
Benz 3,4 pyrene	µ/l	0.01
Trichloroethane	µ/l	14.0
Tetrachloroethane	µ/l	8.0
Total THMs	µ/l	100
Acrylamide	µ/l	1.0
Epichlorhydrin	µ/l	0.1
Aldrin	µ/l	0.03
Dieldrin	µ/l	0.03
Heptachlor	µ/l	0.03
Heptachlor epxide	µ/l	0.03
Benzene	µ/l	1.0
Bromate	µ/l	10.0
1,2 Dichlorethane	µ/l	3.0
Vinyl chloride	µ/l	0.5

20.12.4 Sanitation programme

The cleaning process consists essentially of two stages:

Cleaning

Removal of organic and inorganic deposits.

Disinfection

Sanitizing the equipment to kill pathogenic and spoilage bacteria.

Sanitation programmes are concerned with both the timing of cleaning and disinfection and the sequence in which equipment and environmental surfaces are cleaned and disinfected within the processing area. Sanitation programmes are so constructed as to be efficient with water and chemicals, to allow selected chemicals to be used under their optimum conditions, to be easily managed and to reduce manual labour. A sanitation sequence has to be established in a processing area to ensure that the applied sanitation programme is capable of meeting its objectives and that cleaning programmes are implemented on a routine basis. The following basic sanitation sequence has been demonstrated to be useful in controlling the proliferation of undesirable microorganisms.

1. Remove gross soil from production equipment
2. Remove gross soil from environmental surfaces
3. Rinse down environmental surfaces (usually to a minimum of 2 m in height for walls)

4. Rinse down equipment and flush to drain
5. Clean environmental surfaces, usually in order of drains, walls then floors.
6. Rinse environmental surfaces
7. Clean equipment
8. Rinse equipment
9. Disinfect equipment and rinse if required
10. Fog (if required)
11. Clean the cleaning equipment.

The sequence must be performed at a 'room' level such that all environmental surfaces and equipment in the area are cleaned at the same time. A key issue is timing of sanitation programmes. In practise, the content and timing of daily and periodic sanitation procedures will be a balance between the nature of production operations and an assessment of the hygienic quality of the processing environment. If the processing environment is not clean, a more frequent and rigorous sanitation programme may be required.

Managing sanitation programmes different job functions

- Selection of suitable chemical supplier
- Selection of sanitation chemicals, equipment and methodology
- Development of cleaning schedules
- Implementation of sanitation programme monitoring systems
- Representation of hygiene issues to senior management.

Table 20.8 Ingredients of Cleaning and Disinfecting Agents

Ingredient	Function	Concentration
Acid (*e.g.* nitric, phosphoric acid)	Removal of inorganic deposits	
Alkaline (*e.g.* sodium hydroxide)	Removal of organic deposits (Proteins, fat carbohydrates)	
Sequestrants (*e.g.* EDTA)	Removal of inorganic deposits	
(*e.g.* didecyldimethylammonium choloride, alkyldimethy-benzy lammouium chloride)	Quaternary ammonium compounds Disinfecting, removal of fat	0.05 – 2 %
Active chlorine (*e.g.* sodium	Disinfecting hypochlorite, chloramines T, sodium dichloroisocyanurate)	0.015–0.03% (active chlorine)
Active iodine (iodophor)	Disinfecting	0.005–0.01% (active chlorine)
Active oxygen (*e.g.* hydrogen peroxide with/without peracetic acid)	Disinfecting	0.03–0.5% (active chlorine)

20.12.5 Personnel hygiene

People are a large reservoir of microorganisms. Hence personnel hygiene is essential. If the staff contaminate the food because they are dirty, do not wear protective clothing or are liable to transmit diseases, all other controls will not ensure food safety. Specified people not being allowed to work in food handling areas include, people who are known or suspected to suffer from or be a carrier of a disease likely to be transmitted through food and also people with infected wounds, skin infections, sores or diarrhoea. The factory hygiene policy to include the following:

1. Protective clothing, foot wear and headgear issued by the company must be worn and must be changed regularly. Hair clips and grips should not be worn.
2. Protective clothing must not be worn off the site and must be kept in good condition.
3. Beards must be kept short and trimmed and a protective cover worn when considered appropriate by management.
4. Nail polish, false nails and make-up must not be worn in production areas. Strong after shave or perfumes must not beworn.
5. False eyelashes, wrist watches and jewelry (except the wedding ring or the national equivalent, and sleeper ear rings) must not be worn. Studs and ear rings if worn should be covered in appropriate dressings.
6. Hands must be washed regularly and kept clean at all times.
7. Personnel items must not be taken into production areas unless carried in inside overall pockets (handbags, shopping bags etc., must be left in the lockers provided).
8. Food and drink must not be taken into or consumed in areas other than the rest areas and the staff canteen/restaurants.
9. Sweets and chewing gum must not be consumed in production areas.
10. Smoking or taking snuff is forbidden in food production, warehouse and distribution areas where 'no smoking' notices are displayed.
11. Spitting is forbidden in all areas on the site.
12. Superficial injuries (*e.g.* cuts, grazes, boils, sores, and skin infections) must be reported to the medical department or the first aider on duty via the line supervisor and clearance obtained before the operative can enter production areas.
13. Dressings must be water proof, suitably coloured to differentiate them from product and contain a metal strip as approved by the medical department.

14. Infectious diseases (including stomach disorders, diarrhoea, skin conditions and discharge from eyes, nose or ears) must be reported to the medical department or first aider on duty via the line supervisor. This also applies to staff returning from foreign travel where there has been a risk of infection.
15. All staff must report to the medical department when returning from both certified and uncertified sickness.

20.13 Benchmarks for Slaughterhouse

A benchmark is a number that acts as a guide to the level of best practice that is achievable in a specific area, for example environmental performance. Often, suitable benchmarks are difficult to obtain and difficult to use. However, when they are available they can be useful in assessing the relative performance of a process or organization. Environmental indicators sometimes used by abattoirs to benchmark performance are water consumption, energy consumption and the organic load in effluent (COD or BOD), expressed as figures per unit of production. However, other indicators such as nitrogen and phosphorus loads in effluent have also been used. In some industries, environmental benchmarks are used extensively to gauge the performance and competitiveness of a manufacturing process. For the meat processing industry however, benchmarking of environmental performance is not common and it is difficult to find examples. The lack of environmental benchmarking is thought to be due to the considerable variation in production processes and scales of operation within the industry. The issue is further complicated by the fact that there is no widely recognized standard unit of production. Units used to describe production at abattoirs vary from country to country and even within a country. An additional problem is that existing benchmarks do not necessarily relate to specific types of processes. For example, in order to compare one process with another, or to compare a process with a specified benchmark, the scale, age, efficiency and type of process should be similar to enable sensible comparison. It is recommended that companies should first establish environmental benchmarks internally. It may then be possible to compare performance with other similar organizations within the same state or country. From there, the next step may be to compare performance with industries in other countries as long as the factors contributing to those countries' level of performance are understood.

Table 20.9 Benchmarks for Pig Abattoirs (90 kg pigs)

Technologies	Traditional	Average	Best available
Water L/animal	1400	700	300
Heat and electricity kW.h/ animal	125	50	30
BOD5 g/animal	2500	1000	500

1 COWI, 1999

Checklist of general housekeeping ideas[1]

- Keep work areas tidy and uncluttered to avoid accidents.
- Maintain good inventory control of consumables, such as cleaning chemicals, packaging materials, food additives etc., to avoid waste.
- Ensure that employees are aware of the environmental aspects of the company's operations and their personal responsibilities.
- Train staff in good cleaning practices.
- Schedule regular maintenance activities to avoid inefficiencies and breakdowns.

[1]UNEP cleaner production working group for the food industry, 1999.

Code of hygienic practise for meat (CAC/RCP 58–2005) developed by Codex Committee on meat hygiene

Scope and use of this code

1. The scope of this code covers hygiene provisions for raw meat, meat preparations and manufactured meat from the time of live animal production up to the point of retail sale. It further develops 'The Recommended International Code of Practise: General Principles of Food Hygiene' in respect of these products. Where appropriate, the Annex to that code (Hazard Analysis and Critical Control Point System and Guidelines for its Application) and the Principles for the Establishment and Application of Microbiological Criteria for Foods are further developed and applied in the specific context of meat hygiene.
2. The Code reflects contemporary developments, including risk-based hygienic measures,"farm to plate" approach, targets for hazard control measures that are necessary to achieve Appropriate Level of Protection (ALOP) and the changing role of different stake holders.
3. OIE is currently working on guidelines on application at national level addressing 'ante and post mortem activities in the production of meat to reduce hazards of public and animal health significance'.
4. Working principles for risk analysis for application in the framework of the codex alimentarius (Codex Procedural Manual, 14th Edition); Proposed draft working principles and guidelines for the conduct of Microbiological Risk Management (CX/FH 05/37/6).
5. For the purposes of this code, meat is that derived from domestic ungulates, domestic solipeds, domestic birds, lagomorphs, farmed game, farmed game birds (including ratites) and wild game. This Code of Practise may also be applied to other types of animals from which meat is derived, subject to any special hygienic measures required by the competent

authority. Further to general hygiene measures applying to all species of animal as described above, this code also presents specific measures that apply to different species and classes of animals, *e.g.* wild game killed in the field.

6. The hygiene measures that are applied to the products described in this code, should take into account any further measures and food handling practices that are likely to be applied by the consumer. It should be noted that some of the products described in this code may not be subjected to a heat or other biocidal process before consumption.
7. Meat hygiene is by nature a complex activity, and this code refers to standards, texts and other recommendations developed elsewhere in the Codex system where linkages are appropriate, *e.g.*, Principles for Food Import and Export Inspection and Certification (CAC/GL 20–1995). Proposed Draft Principles and Guidelines for the conduct of microbiological risk management (CX/FH 01/7 and ALINORM 03/13 paras 99–128).
8. General guidelines for use of the term "Halal" (CAC/GL 24–1997) and recommendations of the *Ad hoc* intergovernmental task force on animal feeding (ALINORM 01/38 and ALINORM 01/38A).
9. To provide information that will enhance consistency, linkages should also be made to the standards, guidelines and recommendations contained in the OIE terrestrial animal health code that relate to zoonoses.
10. Subsets of the general principles (Section 4) are provided in subsequent sections within 'double-line boxes'. Where guidelines are provided at the section level, those that are more prescriptive in nature are presented in 'single-line boxes'. This is to indicate that they are recommendations based on current knowledge and practice. They should be regarded as being flexible in nature and subject to alternative provisions so long as required outcomes in terms of the safety and suitability of meat are met.
11. Traditional practices may result in departures from some of the meat hygiene recommendations presented in this code when meat is produced for local trade.
12. The Code covers Code of Practice-general principles of food hygiene (LAC/RCPL-1969) and food risk managements guidelines (CAC/GL21-1997).

Definitions

Abattoir

Any establishment where specified animals are slaughtered and dressed for human consumption and that is approved, registered and/or listed by the competent authority for such purposes.

Ante-mortem inspection

Any procedure or test conducted by a competent person on live animals for the purpose of judgement of safety and suitability and disposition.

Carcass

The body of an animal after dressing.

Chemical residues

Residues of veterinary drugs and pesticides as described in the definitions for the purpose of the codex alimentarius.

Competent authority

The official authority charged by the government with the control of meat hygiene, including setting and enforcing regulatory meat hygiene requirements.

Competent body

A body officially recognized and overseen by the competent authority to undertake specified meat hygiene activities.

Competent person

A person who has the training, knowledge, skills and ability to perform an assigned task, and who is subject to requirements specified by the competent authority.

Condemned

Inspected and judged by a competent person, or otherwise determined by the competent authority, as being unsafe or unsuitable for human consumption and requiring appropriate disposal.

Contaminant

Any biological or chemical agent, foreign matter, or other substance not intentionally added to food that may compromise food safety or suitability.

Disease or defect

Any abnormality affecting safety and/or suitability.

Dressing

The progressive separation of the body of an animal into a carcass and other edible and inedible parts.

Equivalence

The capability of different meat hygiene systems to meet the same food safety and/or suitability objectives.

Establishment

A building or area used for performing meat hygiene activities that is approved, registered and/or listed by the competent authority for such purposes.

Establishment operator

The person in control of an establishment who is responsible for ensuring that the regulatory meat hygiene requirements are met.

Food safety objective (FSO)

The maximum frequency and/or concentration of a hazard in a food at the time of consumption that provides or contributes to the appropriate level of protection (ALOP).

Fresh meat

Meat that apart from refrigeration has not been treated for the purpose of reservation other than through protective packaging and which retains its natural characteristics.

Good hygienic practise (GHP)

All practices regarding the conditions and measures necessary to ensure the safety and suitability of food at all stages of the food chain

- These and other procedures and tests stipulated by the Competent Authority, may also be conducted, in particular for the purposes of animal health.
- Procedural Manual of the Codex Alimentarius Commission.
- The Competent Authority provides official assurances in international trade of meat.

- Recommended International Code of Practise: General Principles of Food Hygiene (CAC/RCP 1–1969, Rev 4–2003).

Hazard

A biological, chemical or physical agent in, or condition of food with the potential to cause an adverse health effect.

Inedible

Inspected and judged by a competent person, or otherwise determined by the competent authority to be unsuitable for human consumption.

Manufactured meat

Products resulting from the processing of raw meat or from the further processing of such processed products, so that when cut, the cut surface shows that the product no longer has the characteristics of fresh meat.

Meat

All parts of an animal that are intended for, or have been judged as safe and suitable for, human consumption.

Meat hygiene

All conditions and measures necessary to ensure the safety and suitability of meat at all stages of the food chain.

Meat preparation

Raw meat which has had foodstuffs, seasonings or additives added to it.

Mechanically separated meat (MSM)

Product obtained by removing meat from flesh-bearing bones after boning or from poultry carcasses, using mechanical means that result in the loss or modification of the muscle fibre structure.

Minced meat

Boneless meat which has been reduced into fragments.

Official inspector

A competent person who is appointed, accredited or otherwise recognized by the competent authority to perform official meat hygiene activities on behalf of, or under the supervision of the competent authority.

Organoleptic inspection

Using the senses of sight, touch, taste and smell for identification of diseases and defects.

Performance criterion

The effect in frequency and/or concentration of a hazard in a food that must be achieved by the application of one or more control measures to provide or contribute to a performance objective (PO) or a food safety objective (FSO).

Performance objective

The maximum frequency and/or concentration of a hazard in a food at a specified step in the food chain before the time of consumption that provides or contributes to a food safety objective (FSO) or appropriate level of protection (ALOP), as applicable.

Post-mortem inspection

Any procedure or test conducted by a competent person on all relevant parts of slaughtered/killed animals for the purpose of judgement of safety and suitability and disposition.

Primary production

All those steps in the food chain constituting animal production and transport of animals to the abattoir, or hunting and transporting wild game to a game depot.

Process control

All conditions and measures applied during the production process that are necessary to achieve safety and suitability of meat.

- WHO Teachers Handbook, 1999.
- Definitions for the Purpose of the Codex Alimentarius. Procedural Manual, 14th edition.

- 13 These and other procedures and tests stipulated by the Competent Authority, may also be conducted, in particular for the purposes of animal health.
- The "process" includes ante and post mortem inspection.

Process criterion

The physical process control parameters (*e.g.* time, temperature) at a specified step that can be applied to achieve a performance objective or performance criterion.

Quality assurance (QA)

All the planned and systematic activities implemented within the quality system and demonstrated as needed, to provide adequate confidence that an entity will fulfil requirements for quality.

Quality assurance (QA) system

The organisational structure, procedures, processes and resources needed to implement quality assurance.

Raw meat

Fresh meat, minced meat or mechanically separated meat.

Ready-to-Eat (RTE) products

Products that are intended to be consumed without any further biocidal steps.

Risk-based

Containing any performance objective, performance criterion or process criterion developed according to risk analysis principles.

Safe for human consumption

Safe for human consumption according to the following criteria:

- has been produced by applying all food safety requirements appropriate to its intended end-use;

- meets risk-based performance and process criteria for specified hazards; and
- does not contain hazards at levels that are harmful to human health.

Sanitation standard operating procedures (SSOPs)

A documented system for assuring that personnel, facilities, equipment and utensils are clean and where necessary, sanitised to specified levels prior to and during operations.

Suitable for human consumption

Suitable for human consumption according to the following criteria:

- has been produced under hygienic conditions as outlined in this code;
- is appropriate to its intended use
- meets outcome-based parameters for specified diseases or defects as established by the competent authority.

Validation

Obtaining evidence that the food hygiene control measure or measures selected to control a hazard in a food is capable of effectively and consistently controlling the hazard to the appropriate level.

Verification

Activities performed by the competent authority and/or competent body to determine compliance with regulatory requirements.

Verification (operator)

The continual review of process control systems by the operator, including corrective and preventative actions to ensure that regulatory and/or specified requirements are met.

Veterinary inspector

An official inspector who is professionally qualified as a veterinarian and carries out official meat hygiene activities as specified by the competent authority.

- This is an interim definition for the purpose of this Code.
- ISO 8402.
- This does not preclude interventions for the purpose of pathogen reduction.
- This is an interim definition for the purpose of this Code.
- See *e.g.* the General Guidelines for Use of the Term "Halal" (CAC/GL 24–1997).
- This is an interim definition for the purpose of this Code.
- These may include animal health objectives.

General principles of meat hygiene

i. Meat must be safe and suitable for human consumption and all interested parties including government, industry and consumers have a role in achieving this outcome.

ii. The competent authority should have the legal power to set and enforce regulatory meat hygiene requirements, and have final responsibility for verifying that regulatory meat hygiene requirements are met. It should be the responsibility of the establishment operator to produce meat that is safe and suitable in accordance with regulatory meat hygiene requirements. There should be a legal obligation on relevant parties to provide any information and assistance as may be required by the competent authority.

iii. Meat hygiene programmes should have as their primary goal the protection of public health and should be based on a scientific evaluation of meat-borne risks to human health and take into account all relevant food safety hazards, as identified by research, monitoring and other relevant activities.

iv. The principles of food safety risk analysis should be incorporated wherever possible and appropriate in the design and implementation of meat hygiene programmes.

v. Wherever possible and practical, competent authorities should formulate food safety objectives (FSOs) according to a risk-based approach so as to objectively express the level of hazard control that is required to meet public health goals.

vi. Meat hygiene requirements should control hazards to the greatest extent practicable throughout the entire food chain. Information available from primary production should be taken into account so as to tailor meat hygiene requirements to the spectrum and prevalence of hazards in the animal population from which the meat is sourced.

vii. The establishment operator should apply HACCP principles. To the greatest extent practicable, the HACCP principles should also be applied in the design and implementation of hygiene measures throughout the entire food chain.

viii. The competent authority should define the role of those personnel involved in meat hygiene activities where appropriate, including the specific role of the veterinary inspector.

ix. The range of activities involved in meat hygiene should be carried out by personnel with the appropriate training, knowledge, skills and ability as and where defined by the competent authority.

x. The competent authority should verify that the establishment operator has adequate systems in place to trace and withdraw meat from the food chain. Communication with consumers and other interested parties should be considered and undertaken where appropriate.

xi. As appropriate to the circumstances, the results of monitoring and surveillance of animal and human populations should be considered with subsequent review and/or modification of meat hygiene requirements whenever necessary.

xii. Competent authorities should recognise the equivalence of alternative hygiene measures where appropriate, and promulgate meat hygiene measures that achieve required outcomes in terms of safety and suitability and facilitate fair practices in the trading of meat.

Specific meat hygiene requirements should address biological, chemical and physical hazards; and pathophysiological and other characteristics associated with suitability for human consumption.

Working Principles for Risk Analysis for Application in the framework of the Codex Alimentarius, Procedural Manual, 14th edition; Codex Committee on Food Hygiene, proposed draft Principles and Guidelines for the Conduct of Microbiological Risk Management (CX/FH 05/37/6); Report of a Joint FAO/ WHO Consultation on

Principles and Guidelines for Incorporating Microbiological Risk Assessment in the Development of Food Safety Standards, Guidelines and Related Texts; Kiel, Germany, 18–22 March 2002.

Principles of meat hygiene applying to primary production

i. Primary production should be managed in a way that reduces the likelihood of introduction of hazards and appropriately contributes to meat being safe and suitable for human consumption.

ii. Whenever possible and practicable, systems should be established by the primary production sector and the competent authority, to collect, collate and make available, information on hazards and conditions that may be present in animal populations and that may affect the safety and suitability of meat.

iii. Primary production should include official or officially recognized programmes for the control and monitoring of zoonotic agents in animal populations and the environment as appropriate to the circumstances, and notifiable zoonotic diseases should be reported as required.

iv. Good hygienic practice (GHP) at the level of primary production should involve for example the health and hygiene of animals, records of treatments, feed and feed ingredients and relevant environmental factors, and should include application of HACCP principles to the greatest extent practicable.

v. Animal identification practises should allow trace-back to the place of origin to the extent practicable, to allow regulatory investigation where necessary.

Hygiene of slaughter animals

Both primary producers and the competent authority should work together to implement risk based meat hygiene programmes at the level of primary production that document the general health status of slaughter animals, and implement practices that maintain or improve that status, *e.g.*, zoonoses control programmes. QA programmes at the level of primary production should be encouraged and may include application of HACCP principles as appropriate to the circumstances. Such programmes should be taken into account by the competent authority in the overall design and implementation of risk-based meat hygiene programmes.

Working Principles for Risk Analysis for Application in the Framework of the Codex Alimentarius, Procedural Manual, 14th edition.

So as to facilitate the application of risk-based meat hygiene programmes:

- Primary producers should record relevant information to the extent possible on the health status of animals as it relates to the production of meat that is safe and suitable for human consumption. This information should be made available to the abattoir as appropriate to the circumstances.
- Systems should be in place for return from the abattoir to the primary producer, of information on the safety and suitability of slaughter animals and meat, in order to improve the hygiene on the farm and, where producer led QA-programmes are applied, to be incorporated into these programmes to improve their effectiveness.
- The competent authority should systematically analyse monitoring and surveillance information from primary production so that meat hygiene requirements may be modified if necessary.

The competent authority should administer an official programme for control of specified zoonotic agents, chemical hazards and contaminants. This should be co-ordinated to the greatest extent possible with other competent authorities that may have responsibilities in public and animal health. Official or officially-recognised programmes for specified zoonotic agents should include measures to:

- control and eradicate their presence in animal populations, or subsets of populations, *e.g.*, particular poultry flocks;
- prevent the introduction of new zoonotic agents;
- provide monitoring and surveillance systems that establish baseline data and guide a risk-based approach to control of such hazards in meat; and
- control movement of animals between primary production units, and to abattoirs, where populations are under quarantine restrictions.Official or officially-recognised programmes for chemical hazards and contaminants should include measures to:
- control the registration and use of veterinary drugs and pesticides so that residues do not occur in meat at levels that make the product unsafe25 for human consumption, and
- provide monitoring and surveillance systems that establish baseline data and guide a risk-based approach to control of such hazards in meat.

Animal identification systems, to the extent practicable, should be in place at primary production level so that the origin of meat can be traced back from the abattoir or establishment to the place of production of the animals.

Animals should not be loaded for transport to the abattoir when:

- the degree of contamination of the external surfaces of the animal is likely to compromise hygienic slaughter and dressing, and suitable interventions such as washing or shearing are not available,
- information is available to suggest that animals may compromise the production of meat that is safe and suitable for human consumption, *e.g.*, presence of specific disease conditions or recent administration of veterinary drugs. In some situations, transport may proceed if the animals have been specifically identified (*e.g.* as "suspects") and are to be slaughtered under special supervision; or

Guidelines for the Establishment of a Regulatory Programme for Control of Veterinary Drug Residues in Foods (CAC/GL 16-1993) *(under revision).*

- conditions causing animal stress may exist or arise that are likely to result in an adverse impact on the safety and suitability of meat.

CHAPTER 21

ECONOMICS OF PIG FARMING

21.1 Status of Piggery Development

In India pig keeping by and large has remained confined to socio-economically backward people. These people lacked resources as well as technical know how of pig production. Due to general apathy to this occupation, pig rearing remained neglected. Piggery development experienced many constrains such as shortage of high quality breeding stock, insufficient availability of low cost, balanced feed and poor market condition.

Pig farming received attention in last few Five Year Plans and about 100 pig breeding farms/units where nearly 30000 pigs including their progenies are being maintained in different states and union territories of India. At these farms about 5000 adult breeding stock of Large White Yorkshire, Middle White Yorkshire Hampshire, Landrace, Saddleback and Berkshire breeds are maintained, and they supply boars and sows to farmers for crossbreeding and upgrading of indigenous pigs. In last five years nearly 79000 piglets were distributed from these farms, to 28000 farmers. Through special livestock development programme economic status of the pig farms is also being improved by setting up pig production units, for which a mix of subsidy and loan is provided to small/marginal farmers and agricultural labourers. The scheme also envisages development of infrastructural facilities and training of farmers. Women, scheduled castes and scheduled tribes are to be encouraged for which minimum percentage has been fixed.

In Eighth Plan, modernization/improvement of existing slaughter houses, establishment of model slaughter houses for export of meat and to promote effective linkages between producers, processing industry for hygienic production of meat and to improve employment opportunities particularly for weaker section, who are generally involved in the field of meat production and processing and allied fields were started.

21.2 Importance of Pig Farming and its Contribution to National Economy

The pig population of the country is 13.5 million as per the 2003 livestock census and constitutes around 1.30% of the total world's population. During 2003 the production of pork and pork products were estimated to be 63000 MT with 3.03% growth rate in last decade. It comprised over 38% of the total world meat production. Indian share in piggery meat production moderately increased from 0.53% in 1981 to 0.63 in 2002. The contribution of pork products in terms of value works out to 0.80% of total livestock products and 4.32% of the meat and meat products. The contribution of pigs to Indian exports is very poor. About 1720 MT of pork and pork products were exported during 2007–08. The value of pork and pork products exported is Rs 24.6369 million.

Table 21.1 Swine Meat Production in India

Qty in 000 MT

Year	1985	1990	1995	2000	2003
Quantity	85	360	420	578	630

Source: FAO Production Year Book and FAOSTAT website.

Table 21.2 Export of Swine Meat from India 2005–06 to 2007–08

Qty in MT, Value in Lakhs

2005–06		2006–07		2007–08	
Quantity	Value	Quantity	Value	Quantity	Value
320.70	207.38	1523.47	865.30	1710.89	2463.69

Source: DGCIS Annual Data.

The pig farming is integral to the livelihood of rural poor belonging to the lowest socio-economic strata. They have no means to undertake scientific pig farming with improved foundation stock, proper housing, feeding and management. Therefore, suitable schemes to popularize the scientific pig breeding cum rearing of meat producing animals with adequate financial provisions are necessary to modernize the pig industry to improve the productivity of small sized rural pig units.

In view of the importance of pig farming in terms of its contribution to incomes of rural poor and possible potential for pig rearing as major employment option for the poor and socially backward classes, Government of India has initiated measures to promote the pig farming on scientific lines under it's five year plans. The first major step in this direction was to establishment of eight bacon factories and number of pig production units in rural areas in the catchment area of the bacon factories. In order to make available good foundation stock, regional pig breeding stations were also established for each bacon factory. Further expansion of pig breeding programmes paved the way for establishment of 115 pig breeding farms (1992–93) throughout the country. The location of bacon factories and pig breeding farms are given in table 21.3 and table 21.4 respectively.

Table 21.3 List of Bacon Factories

State	Capacity (No. of pigs/days)	Address
Uttar Pradesh	100	Bacon factory central dairy farm, Aligarh
West Bengal	20	Bacon factory, Harringhatta, Mohanpur, Nadia. West Bengal
Andhra Pradesh	100	Government bacon factory, Gannavaram, Krishna Dist.
Bihar	50	Government bacon factory, Kanke Ranchi
Maharashtra	100	MAFCO Bacon Factory, National Park, Borivali, Mumbai
Rajasthan	50	Meat Complex, Alwar
Kerala	50	Meat Products of India, Koothattukulam, Ernakulam
Punjab	20	Pork processing plant, Kharar

Table 21.4 Statewise Location of Pig Breeding Farm

State	Location of breeding farms
Andhra Pradesh	Gannavaram, Gopannapalem, Muktalya, Padavagi, Tirupathi, Vishakapatnam
Arunachal Pradesh	Karsingsa, Loiliang
Assam	Diphu, Haflong, Kaliapani, Khanapara (University), Khanapara (AICRP), Khanapara (Govt.), Khanikar, Marigoan
Bihar	Gaurikarma, Hotwar, Jamshedpur, Kanke
Dadra and Nagar Haveli	Port Silvasa
Goa	Curti Ponda, Ela
Haryana	Ambala City, Hisar
Karnataka	Hassarghatta, Koila, Kudige
Kerala	Ankamaly, Koothattukulam, Kunnamkulam, Mannuthy, Mundayal, Parasala, Thalayda Parambu
Madhya Pradesh	Bastar, Jabalpur, Sakalo
Manipur	None Tamenglong, Senepati North, Tarang, Torbumg
Meghalaya	Baghmora, Dalu, Jowai, Mairang, Mawryngkneg, Nongstoin,Pynursla, Rongjeng, Rongkhon
Mizoram	Kolasih, Lunglei, Selesih, Thenzawl
Nagaland	Alukute, Medziphema, Merang, Phek, Suthazu, Tijit, Tunesang
Orissa	Bhaminagar, Chiplima

Table 21.4 *(Contd...)*

State	Location of breeding farms
Punjab	Badal, Chhaju Majra, Gurdaspur, Jalandar, Ludhiana, Maltowara
Rajasthan	Alwar, Bharatpur
Sikkim	Gyalsing, Tadong
Tamil Nadu	Chettinad, Hosur, Pudukottai, Saidpet, Udagamandalam
Tripura	Amarpur, Birchandramanu, Gandhi Gram, Mendhihaor, Nabincherra, Nalkata
Uttar Pradesh	Aligarh, Arzilins, Barabanki, Basti, Dehradun, Izzatnagar, Lalitpur, Moradabad, Nilgaon
West Bengal	Bijanbari, Haringhatta, Pedong, Singruntaum, St. Mary's Hill Turki

21.3 Special Features of Pig Farming on Commercial Lines

Pig is a prolific breeder with a short generation interval. Not only this is a litter bearing animal which makes it most suited for commercial meat production, it has acquired a prominent position in meat production the world over, being the most efficient converter of feed into meat and fat with less requirement of labour. The quality of meat available per unit live weight is larger in pigs than in other livestock. Thus the returns over the investment are substantial and quick. In spite of these traits in favour of commercial pig rearing, we have not been able to establish it as an industry. One of the main reasons is that in a predominantly vegetarian population as well as social taboo has discouraged educated, economically sound people of socially advanced sections of the society to take up this enterprise, thus leaving this vocation confined principally in the hands of those, who provided scavenging as the only method of rearing pigs making it a nonviable commercial enterprise.

In order to establish pig husbandry in any form as a viable food production industry not only to cater to domestic demands, but for the export market, to earn foreign exchange on one hand and to play a significant role in improving the socio-economic status of the people engaged in pig rearing, it is essential to give proper importance to this enterprise. So to attain these objectives, an entirely new approach has to be adopted giving utmost care to each aspect of pig husbandry. Modern principles of business management are to be incorporated in pig based enterprises. Broadly the approach can be classified under the following heads:

1. Studying the existing status of the pig farming in the area
2. The various important linkages of material procurement–feed, livestock etc. and marketing and supply linkages.
3. Selection of the entrepreneur and work–During selection the interactive aptitude, educational and family background, outlook, desire to bear risk and to bear the hazards of the vocation.

4. Training of the personnel. Training in various diversified nature of jobs to be provided to each category of workers separately *viz.* rearing, processing, material management, marketing etc.
5. Organization of the enterprise
6. Selection of the site for the location of the enterprise
7. Profitability of the pig business
8. Financial support

21.4 Broad Approach to Start up Pig Enterprise

As it is customary to have a market and resources survey for the establishment of any enterprise, whether manufacturing or marketing, the same stands valid for the establishment of pig based enterprise be it rearing, breeding or marketing. It is, therefore, essential to undertake a bench mark survey of the area to judge the suitability of the area, willingness of the people, availability of resources *viz* man and material infrastructure and the scope for the exploitation of the existing resources. On the basis of this information collected during the survey, the following information may be summarized for consideration by the project formulation agency/ personnel:

1. Number of pig rearing families
2. Economic and educational status of the farmers
3. Number and kind of pigs reared per unit
4. Purpose for which the pigs are reared
5. Practises adopted by the farmers with respect to feeding, rearing, housing and marketing
6. Income pattern from the pig vocation
7. Profit/loss with causes
8. Status of pig disease in the area and mortality pattern
9. Availability of veterinary aid in the area

21.4.1 Selection and training of farmers and personnel

While selecting personnel it is essential that the following information about them are obtained to assess the suitability of the person for one of the many activities connected with the pig rearing or pig food production industry. It is an established fact that every person can not be expected to do well in all the spheres of activities related with the commercial pig industry. The information should be collected in a format to provide the detail information about the individual:

1. Why he wants to go in for pig rearing or any of the jobs connected with pig industry?

2. Back ground of the worker, under what specific sector of pig farming he has been brought up, if any. The extent of his exposure in the pig business.
3. What are his educational and vocational attainments and details of practical experience
4. Is he willing to continue in the present vocation? If he wishes to change, then why. Asses the difficulties he faced which ultimately led him to change the sphere of activity.
5. Information as to what have been his performance, his income, the constraints and change facilities for better returns from the vocation.
6. How he plans to arrange finances for his vocation and what will be his share in the investment.

Government of India and various state governments are organizing courses on swine husbandry and the Indian Council of Agriculture Research organizes special 9 months Post Graduate Diploma in Swine Husbandry and Pork processing which has proved to be useful. Now this course has been taken over by the Agriculture Universities.

21.4.2 Pre-planning for pig enterprises

Before commencement of any pig enterprise, pre-planning is needed regarding location, selection of site, layout of farm and types of building required along with the infrastructural facilities available in the hinter land. Detail of selection of stock, feed management practises and disease control along with marketing aspects have already been discussed in preceding chapters of this book and the entrepreneur should keep them in view while formulating the project and working out its economic viability to run it on commercial lines. Project formulation is important from the point of view of managing financial resources for undertaking the pig enterprise with the assistance of financial institutions and nationalized banks also under anti-poverty programme and socio-economic development programme of the government. Pig units of different sizes have been advocated. Financial institutions and banks also extend financial support for specialized large sized pig farms and meat processing units.

21.4.3 Economic feasibility of the enterprise

For formulating economic feasibility report of large size enterprises, detailed year wise expenditure and receipts, cash flow and forecast of profit and loss, break even points, pay back period of loan etc. is needed. It will be advisable to have performance budgeting which must correlate expenditure with production according to standard anticipated for different operations and activities. In preparation of profit and loss consideration for meeting burden of interest on loans and capital employed with, working capital as well as capital formation be considered. It will also take into account the burden of depreciation on plant, machinery, equipment and buildings etc. The direct expenses and overheads will also be provided for. The thumb rule for economic performance of any enterprise should be that after meeting the burden of interest, depreciation and taxes out of gross profit, it should leave net profits of about 20 to 25% on capital employed.

21.4.4 Financial assistance available from banks/NABARD for pig farming

NABARD is an apex institution for all matters relating to policy, planning, and operations in the field of agriculture credit. It serves as refinance agency for the ground level institutions/banks providing investment and production credit for various activities under agriculture and allied sectors for ensuring integrated rural development. It co-ordinates the development activities through a well organized Technical Services Department at the head office and Technical cells at each of the regional offices.

For undertaking the pig farming on scientific lines, loan from banks with refinance facility from NABARD is available. For obtaining bank loan, the farmers /entrepreneurs should apply to the nearest branch of a Commercial, Co-operative or Regional Rural Bank in the prescribed application form, which is available in the branches of financing bank. Necessary help or guidance can be obtained from the technical officer attached to or the manager of the bank in preparing the project report, which is a prerequisite for sanction of the loan.

For piggery development schemes with very large outlays, detailed project reports will have to be prepared for which specialized consultants are available. The items such as land development, construction of sheds and other civil structures, purchase of the breeding stock, equipment, feed cost up to the point of income generation are normally considered under bank loan. Other items of investment will be considered on need basis after providing the satisfactory information justifying the need for such items. The cost of land is not considered for loan. However, if land is purchased for setting up the piggery farm exclusively, it can be considered as beneficiaries' margin money.

21.4.5 Scheme formulation

In case of commercial piggery units, the banks are expected to submit a project for availing the refinance. The scheme normally should include information on land, livestock markets, availability of water, feeds, veterinary aid, breeding facilities, marketing aspects, training facilities, experience of the farmer and the type of assistance available from State Government's Regional Pig breeding centres.

The scheme should also include information on the number of and types of animals to be purchased, their breeds, production performance, cost and other relevant input and output costs with their description. Based on this, the total cost of the project, margin money to be provided by the beneficiary, requirement of bank loan, estimated annual expenditure, income, profit and loss statement, repayment period, etc. can be worked out and included in the project cost.

(A) Technical feasibility–this would briefly include

1. Nearness of the selected area to financing bank's branch.
2. Availability of good quality animals in nearby livestock markets/breeding farms.
3. Source and availability of training facilities.
4. Availability of concentrate feeds and kitchen/hotel/vegetable market waste and broken grains from Food Corporation's godowns.
5. Availability of medicines, vaccines and veterinary services etc.
6. Availability of breeding centres and marketing facilities near the scheme area.
7. Reasonability of various production and reproduction parameters.

(B) Economic viability–this would briefly include

1. Capital investment
2. Variable cost
3. Fixed cost
4. Income
5. Birth and death register
6. Bank loan required
7. Profitability
8. Cash flow statement

(C) Bankability

Repayment schedule (*i.e.* repayment of principal loan amount and interest). Other documents such as loan application forms, security aspects, margin money requirements etc. are also examined. A field visit to the scheme area is undertaken for conducting a techno-economic feasibility study for appraisal of the scheme. The economics of piggery units of different sizes are given in Table 21.5 and Table 21.6.

Sanction of bank loan and its disbursement

After ensuring technical feasibility and economic viability, the scheme is sanctioned by the bank. The loan is disbursed in stages against creation of specific assets such as construction of sheds, purchase of equipments and animals. The end use of the fund is verified and constant follow-up is done by the bank.

Lending terms-general

Unit cost

Each Regional Office (R.O.) of NABARD has constituted a State Level Unit Cost Committee under the chairmanship of RO-in-charge and with the members from developmental agencies, commercial banks and cooperative banks to review the unit cost of various investments once in six months. The same is circulated among the banks for their guidance.

Margin Money

NABARD has defined farmers into three different categories and where subsidy is not available, the minimum down payment, as shown below, is collected from the beneficiaries.

Sl. No.	Category of farmer	Beneficiary's contribution
(a)	Small farmers	5%
(b)	Medium farmers	10%
(c)	Large farmers	15%

Interest rate for ultimate borrower

Banks are free to decide the rate of interest within the overall RBI guidelines. However, for working out the financial viability and bankability of the model project we have assumed the rate of interest as 12% p.a.

Security

Security will be as per NABARD/RBI guidelines issued from time to time.

Repayment period of loan

Repayment period depends upon the gross surplus in the scheme. The loans will be repaid in suitable half yearly/annual installments usually within a period of about 5–6 years with a grace period of one year.

Insurance

The animals may be insured annually or on long term master policy, where ever it is applicable. The present premium rate for non IRDP schemes is 6% per annum.

Table 21.5 Financial Scheme for Pig Unit for 10 Sows and 1 Boar

Sl.No.	Item description	Amount in Rs
A. Capital Investment		
1.	Cost of the male Rs 2000 each	2000
2.	Cost of the female Rs 2000 each	20000
3.	Cost of housing.	
(i)	Covered area 1162 sq.ft. @ Rs 200/sq.ft.	232400
(ii)	Open area 1132 sq.ft. @ Rs 100/sq.ft.	113200
4.	Cost of equipment	4000
5.	Miscellaneous	2000
	Total Capital Investment	**373600**

B. Working cost (Variable cost)

Sl.No.	Details	I year	II year	III year	IV year	V year	VI year
1.	Green fodder	8469	14285	14501	14333	12917	9151
2.	Concentrates						
(i)	Feed for one boar	7373	7373	7373	7373	7373	7373
(ii)	Feed for 5 gilts (during breeding)	18188	13068	11654	13068	11654	6534
(iii)	Feed for sow during pregnancy	48904	53146	54560	53146	55314	28354
(iv)	Feed for sow during lactation	19512	19512	19512	19512	19512	19512
(v)	Feed for piglets (0–30)	3060	3060	3060	3060	3060	3060
(vi)	Feed for grower (31–90)	58536	90404	85202	78048	78048	78048
(vii)	Feed for Grower (91–150)	58536	117072	133656	137558	138534	117072
(viii)	Feed for Finisher (151–240)	103424	232704	232704	232704	232704	267608
3.	Labour 1500/month	36000	36000	36000	36000	36000	36000
4.	Veterinary aid	4800	4800	4800	4800	4800	4800
	Total Variable cost	**366802**	**591424**	**603022**	**599602**	**599916**	**577512**

C. Working cost (fixed cost)

Sl.No.	Item description	Amount in Rs
1.	Insurance on cost of sows @ 2.25%	495
2.	Depreciation @ 10% on (a) cost of sows. (b) cost of equipment's and (c) cost of shed	37160
	Total (fixed cost)	**37655**

Table 21.5 (*Contd...*)

D. Income

Sl.No.	Item description	Amount in Rs
1.	By sale of live piglets at the rate of Rs 80/kg.	512000
2.	By sale of manure 2 ton/year adult, 5 q/year piglet @ Rs 300/ton	19200
	Total Income	**531200**

E. Birth and death register

Sl.No.	Item description	I year	II year	III year	IV year	V year	VI year
1.	Male	1	1	1	1	1	1
2.	Female	10	10	10	10	10	10
3.	Births	180	180	180	180	180	180
(i)	Male	92	92	92	92	92	92
(ii)	Female	88	88	88	88	88	88
4.	Disposal	20	20	20	20	20	20
(i)	Male	12	12	12	12	12	12
(ii)	Female	8	8	8	8	8	8
5.	Sale due to replacement						
(i)	Male	–	–	1	–	1	–
(ii)	Female	–	–	10	–	10	–
6.	Sold pig pnit	80	160	160	160	160	240
7.	Unit of due piglet	80	80	80	80	80	80
8.	Age of due piglet	60 days	49 days	107 days	40 days	150 days	5 days
9.	Gross profit by sale of fatteners	512000	1024000	1024000	1024000	1024000	1408000
10.	Income by selling of manure	19200	31200	34200	32400	33900	32550
	Total	**531200**	**1055200**	**1058200**	**1056400**	**1057900**	**1440550**

F. Bank loan required

Sl.No.	Description	Amount in Rs
1.	Farmer's Contribution @ 25% of the capital investment	93400
2.	Working capital needed for an period of one year	366802
3.	Input required as loan (Capital investment – Farmer's replacement share + working capital for one year) (373600 – 93400 + 366742)	647002
4.	Interest rate (per annum)	10%
	Total Bank loan required	**647002**

Profitability (Gross profit = Income - Fixed cost - Variable cost - Interest)

H. Cash flow statement for unit of 10 sows

Sl.No.	Item description	I year	II year	III year	IV year	V year	VI year
1.	Gross income	62049	361427	368996	386796	404156	825383
2.	Loan outstanding	647002	647002	485252	323501	161751	–

Table 21.5 (*Contd...*)

3.	Loan installment	–	161751	161751	161751	161751	–
4.	Net surplus	62049	199677	207246	225046	242406	825383
5.	Net income/month	5171	16640	17270	18754	20200	68782

Table 21.6 Financial Scheme for Pig Unit for 30 Sows and 3 Boars

Sl.No.	Item description	Amount in Rs
A. Capital investment		
1.	Cost of the male Rs 2000 each	6000
2.	Cost of the female Rs 2000 each	60000
3.	Cost of housing.	
(i)	Covered area 4279 sq ft @ Rs 300/sq ft	1283700
(ii)	Open area 2912 sq ft @ Rs 100/sq ft	291200
4.	Cost of equipment	12000
5.	Miscellaneous	6000
	Total capital investment	**1658900**

B. Working cost (variable cost)

Sl.No.	Details	I year	II year	III year	IV year	V year	VI year
1.	Green fodder	25407	42855	43503	42999	38751	30159
2.	Concentrates						
(i)	Feed for two boar	22119	22119	22119	22119	22119	22119
(ii)	Feed for 5 gilts (during breeding)	54564	39204	34962	39204	34962	39204
(iii)	Feed for sow during pregnancy	146712	159438	163680	159438	165942	158418
(iv)	Feed for sow during lactation	58536	58536	58536	58536	58536	63414
(v)	Feed for piglets (0–30)	9180	9180	9180	9180	9180	9180
(vi)	Feed for grower (31–90)	175608	271212	255606	234,144	234144	234144
(vii)	Feed for Grower (91–150)	175608	351216	400968	412674	41602	351216
(viii)	Feed for finisher (151–240)	310272	698112	698112	698112	698112	802824
3.	Labour 1500/month	90000	90000	90000	90000	90000	90000
4.	Veterinary aid	14400	14400	14400	14400	14400	14400
	Total variable cost	**1082406**	**1756272**	**1791066**	**1780806**	**1781748**	**1815078**

C. Working cost (fixed cost)

Sl.No.	Item description	Amount in Rs
1.	Insurance on cost of sows @ 2.25%	1485
2.	Depreciation @ 10% on (a) cost of sows. (b) cost of equipment's and (c) cost of shed	165290
	Total (fixed cost)	**166775**

Table 21.6 (*Contd...*)

D. Income

Sl.No.	Item description	Amount in Rs
1.	By sale of live piglets at the rate of Rs 80/kg.	1536000
2.	By sale of manure 2 ton/year adult, 5q./year piglet @ Rs 300/ton	57600
	Total Income	**1593600**

E. Birth and death register

Sl.No.	Item description	I year	II year	III year	IV year	V year	VI year
1.	Male	3	3	3	3	3	3
2.	Female	30	30	30	30	30	30
3.	Births	540	540	540	540	540	540
(i)	Male	276	276	276	276	276	276
(ii)	Female	264	264	264	264	264	264
4.	Disposal	60	60	60	60	60	60
(i)	Male	36	36	36	36	36	36
(ii)	Female	24	24	24	24	24	24
5.	Sale due to replacement						
(i)	Male			3		3	
(ii)	Female			30		30	
6.	Sold Pig Unit	240	480	480	480	480	720
7.	Unit of due piglet	240	240	240	240	240	240
8.	Age of due piglet	60 (days)	49 (days)	107 (days)	40 (days)	150 (days)	5 (days)
9.	Gross profit by sale of fatteners	1536000	3072000	3072000	3072000	3072000	4224000
10.	Income by selling of manure	57600	93600	102600	97200	101700	97650
	Total	**1593600**	**3165600**	**3174600**	**3169200**	**3173700**	**4321650**

F. Bank loan required

Sl.No.	Description	Amount in Rs
1.	Farmer's contribution @ 25% of the capital investment	414725
2.	Working capital needed for an period of one year	1082406
3.	Input required as loan (Capital investment - Farmer's Share + working capital for one year) (1658900–414725+1082406)	2326581
4.	Interest rate (per annum)	10%
	Total bank loan required	**2326581**

G. Profitability (Gross profit = Income -Fixed cost -Variable cost-interest)

1. Gross Profit	Income	Fixed cost	Var.cost	Interest	Amount
I[st] year	1593600	166775	1082406	232658	111761
IInd year	3165600	166775	1756272	232658	1009895
IIIrd year	3174600	166775	1791066	174494	1042265
IVth year	3169200	166775	1780806	116329	1105290
V[th] year	3173700	166775	1781748	58165	1167012
VIth year	4321650	166775	1815078	–	2339797

Table 21.6 (*Contd...*)

2. Repayment of loan ($^1/_4$ annually) of Rs 2326401	–	–	581645
3. Time to pay back loan			4 year

H. Cash flow statement for unit of 30 sows

Item description	I year	II year	III year	IV year	V year	VI year
1. Gross income	111761	1009895	1042265	1105290	1167012	2339797
2. Loan outstanding	2326581	2326581	1744936	1163291	581645	–
4. Loan installment	–	581645	581645	581645	581645	–
5. Net surplus	111761	428250	460620	523645	585367	2339797
6. Net income/month	9313	35687	38385	43637	48781	194983

CHAPTER 22

INTEGRATED PIG PRODUCTION

22.1 Introduction

In India, pig keeping is an important activity, especially among the weaker section of the society. Though a considerable proportion of the rural people depend upon the income from pigs for their livelihood, the pig rearing has not been given due recognition until recently. The pig keeping system in the country is mainly subsistence oriented and heavily constrained by availability of limited resources. The revolution that has occurred in other livestock species, especially in dairy sector, has not even touched the pig, mainly due to lack of emphasis on pig production. The reasons are many folds including the religious taboo among people. Nevertheless, the preference for pork is increasing across the population; especially among youngsters and in some places the cost of pork is as high as the price of mutton/chicken. This clearly shows the potential that lies within the country to earn substantial income from pig enterprises on one hand and sustainable livelihood of rural and tribal people on the other.

To obtain good income from pig husbandry, the "*pig keeping or rearing*" need to be transformed into "*pig production*" enterprises. This transformation will be viable when the availability of inputs is adequately addressed. The programme of entrepreneurship has to be based on modern system of management which is financialy viable.

In this chapter, effective management of *in situ* resources, exploitation of complementary role of crop, fish and pig production and economics of integrated pig production models are discussed for boosting integrated pig production in the country.

22.2 Current Scenario of Pig Production System

Pig farming in India is primarily a small-scale unrecognized rural activity and is an integral part of diversified agriculture. The pig production systems, across the country, show wide variations in respect to the system of rearing, feeding and other management practices due to several factors including demographical, ecological, social characteristics etc. of the place. In general, the pig production system in Indogangetic plain areas, from south to north and from east to west, exhibit more or less similar characteristics. However, the same is not true in case of hill eco system *i.e.* the pig production system in northeastern hills is different from north and northwest hills. Similarly, the difference in the production system is clearly visible between the plain and hilly areas.

There have been two major trends where in small farrow have increased their pigs units and used bank loans to move to become pig enterprises using government schemes as the foundations.

North eastern States ahve invested in pig development. A large number of pig industrial units have been established by a major busness to take the import market share for imported products.

In plain areas, where the social discrimination of people is comparatively higher than the hilly area, pig keeping is mainly taken up by weaker (both in monitory and social hierarchy) section of the society and hence mostly scavenging system is followed. In this system, the pigs are allowed to roam around during daytime to find their food by scavenging. Only at night they return back home for bunking and get kitchen waste to eat. They are not provided with any health or other management care. This system aims at some output with zero input. Whereas in northeastern hilly region, pig keeping is socio-culturally intermingled with the life style of the people and at times the person's social value is determined by the number of pigs he owns. Hence, due importance is given to pig production and health management practices. Further, pig husbandry in tribal belt of northeastern region is an inseparable and integral part of agriculture, as most of the people, due to multifarious reasons, depends on pig for their economic support. Backyard system is most commonly practiced in this region; in this system of rearing, pigs are housed in temporary sheds constructed using locally available materials like wood, bamboo etc. The location of the pig house is usually in the backyard of the house of owner for easy operation. They allow their pig to stay in those houses for 6 to 10 months from farrow fattening purposes. The pigs are usually provided with kitchen waste mixed with vegetable waste and other leafy materials. Occasionally concentrate feed is provided to them. Only few farmers keep their pigs for breeding purpose.

Small scale unorganized or smallholder pig production, low output-low input system is supported mainly with locally available resources. It is not only the major means of converting low quality feed into high quality protein, but also the way of meeting out livelihood and household expenses. The farmers based on the local resources have evolved this system, in which pigs are mainly dependent on local vegetations, crop residues and kitchen waste. This system has been followed generation after generation and needs to be studied in detail. Smallholder pig production systems received insufficient attention in the past and have not been considered seriously because of the introduction of "exotic" systems based on high inputs, high technology and breeds of high genetic merit. Under the resource driven production system, local pigs which are more prolific than the exotic pigs, because of limited inputs availability are not able to complete. However, due to increasing demand for pork, replacement of local pigs with high performance exotic pigs and their crossbreds has started. It is well known that performance of the pigs depends upon the production practises. The smallholder resource poor pig production system is a form of integrated mode of food production. Under this system, especially in hilly areas, compost making from the bedding material is to an important element. Analysis of the practice revealed few important points. In areas at high altitude, the temperature goes down sharply during winter season, which may cause cold bites to the pigs, if reared on floor without bedding. The grass provided as bedding material reduces this problem. Further, the nutrient content of the grass is enriched with the nutrients from the pig dung and urine and thus a good source of nutrients for crops if managed *in situ*. Moreover, during winter season, it may be difficult to produce compost as the environmental temperature is low; rearing of pigs with grass as bedding materials may hasten the process of fermentation of the grass due to the heat generated by the pig during resting on the grass soiled with dung and urine. Under this system of pig rearing there is little smell in the pigsty with dried grass used as bedding material. Because of these and other undealt advantages, the integrated pig production offers a great scope to optimize the pork productivity at small holder level.

22.3 Need for Integrated Pig Production

The main factors that emphasis integrated pig production are (i) continuous and increasing erosion of natural resources, (ii) under usage/in efficient use of *in situ* resources, (iii) horizontal shrinking of agricultural land, (iv) increasing human and animal population etc. Internationally, importance is being given to "*integrated mode of food production*" for sustaining the productivity as well as improving the livelihood and nutritional security. This mode of food production is very pertinent to India as the crop as well as individual animal productivity is either stagnating or decreasing while the input resources are gradually decreasing over the time. The aftermath of the "green revolution" has been clearly visible in the soils of the several states where high intensive inputs in agriculture has been practiced. That is why the concept of "ever green revolution" using the natural resources, animal manure and other organic inputs has been evolved and being promoted nowadays. This is very much applicable to animal production too, especially for smallholder pig production system, as the success depends upon how best the inputs are being generated *in situ* to reduce the total cost of production.

Integrated pig production system is based on the concept that "*there is no waste*". This system addresses the rational utilization of the *in situ* resources effectively. The system also recycles the excreta of pig in the form of manure or compost for production of feed ingredients for pigs as well as food for human. The bio-resource inflow and out flow in a integrated pig production system is given in figure 22.1.

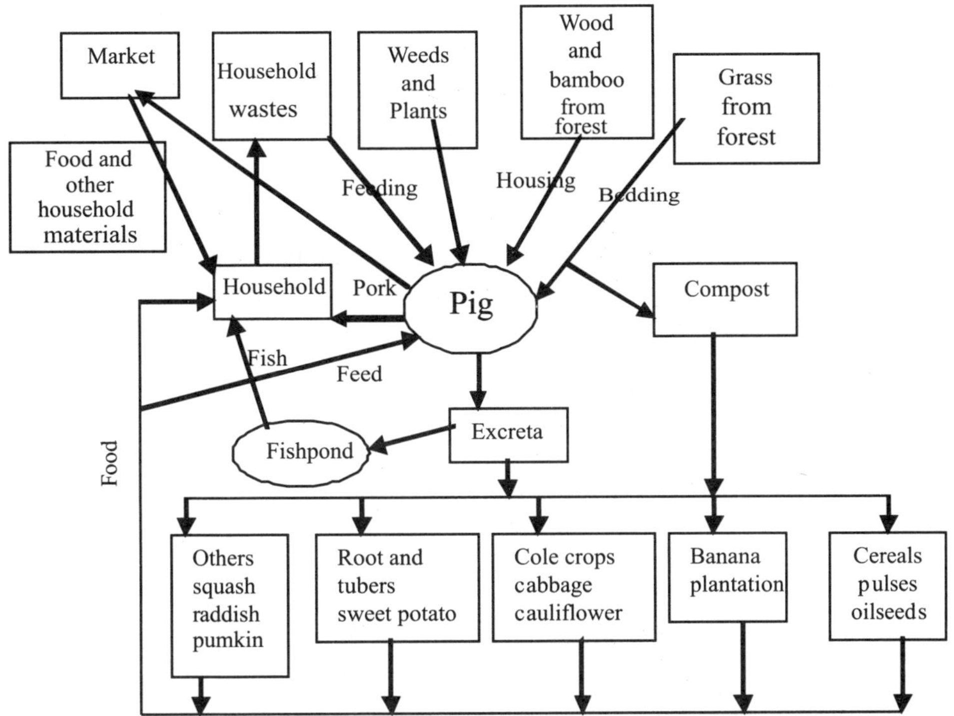

Fig. 19.1. Bio resource inflow and out flow in integrated pig production system

22.4 Integrated pig production system

Integrated food production system involves a variety of components like agriculture, horticulture, agroforestry, livestock, fish farming etc.and is a proven, environmentally sustainable and economicallyviabletechnology that encompasses rational utilization of available resources. For small farmers, these systems enable a means of diversifying the use of meager resources in the context of a rational means of reducing risk. Additionally, it also enables increased efficiency in the use of these resources in a manner that there will be livelihood and nutritional security.

Integrated agri-livestock and fish farming is an old practice consisting of the culture of fish associated with the husbandry of domesticated animals such as pigs, ducks, chicken, etc. and a variety of crops. In order to prevent serious environmental problems due to the excreta of livestock, there is a possibility of recycling these organic wastes, manures and farm effluents in fish ponds. The end product is an improved production of animal protein, particularly needed in developing countries. Through farming systems researches, new techniques are continuously being developed, particularly for small holding farmers. The major advantages of integrated pig production are transformation of scavenging to intensive system, encashing the complimentarily of crop-pig-fish and other components, environmentally and economically sustainable and green and clean pig production.

1. Integrated pig production models

There are several types of integrated pig production models depending upon the location, availability of resources, agricultural priority of the area, food habit etc; few general models are listed below.

Model 1

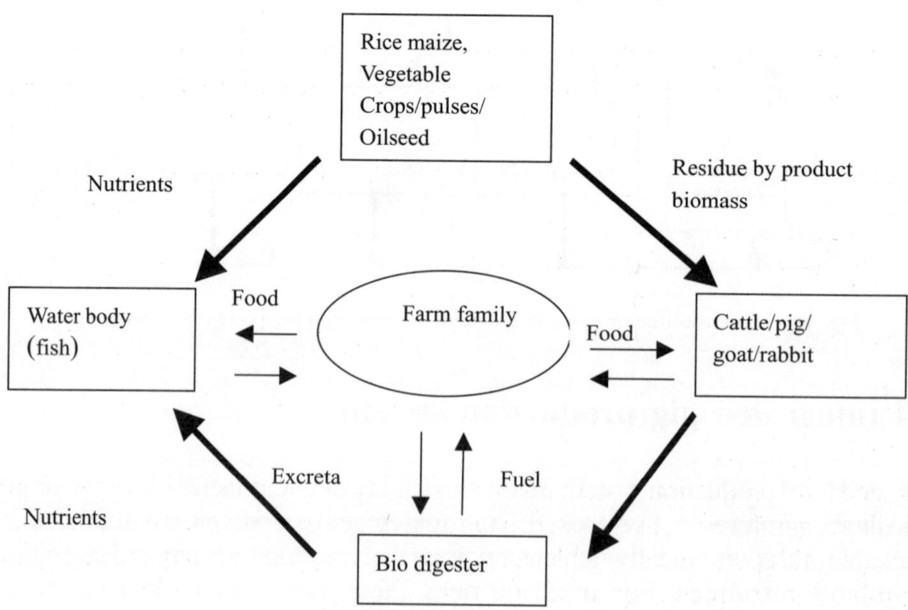

Model 2

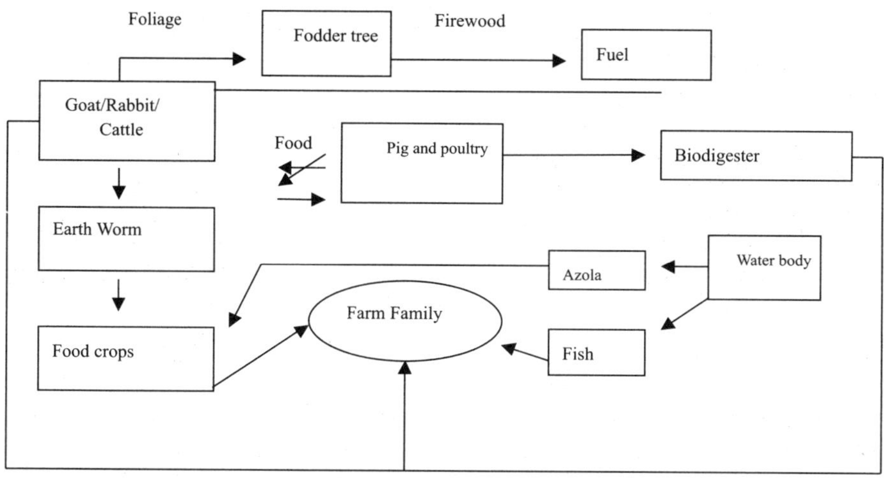

2. Integrated pig feed food crop production

This system involves raising of crops (parts of which is used as human food and the rest is used as animal feed) and pigs together and the energy from one component is used as input for the other. In general, pig feeding in smallholder production system mainly dependent on the local vegetations, agro-wastes, household and kitchen wastes. Since the inception of pig rearing, farmers fed their pigs with locally available plants suitable for feeding pigs, left out rice and kitchen wastes. Generally, the commonly used non–conventional pig feeds are *Spilanthus sp., Bidens biternata, Hibiscus sp, Mikenia scanden, Conyza auriculata, Polygonum chinensis*, Pumpkin, Bamboo shoots, etc. The leaves and stem of plants are cut into pieces and mixed with little wheat bran and rice, cooked together as slurry and then the prepared feed is offered to pigs twice daily in nearly equal quantity. In this feeding system, local pigs proved more prolific than exotic breeds. Of late, when the necessity of increasing pork production was felt, importing exotic pigs having high body weight gain and upgrading local pigs became inevitable. Though the germ plasm improvement has been undertaken, no significant steps have been taken up in improving the managemental practices especially feeding strategies, which is very important to get desirable results from the upgraded and exotic pigs. Hence, at present, the results are not up to the expected level with exotic pigs.

To supply more energy to the pigs, it is wise to cultivate tuber and root crops that requires less cost. The following are some of the food-feed crops that can very well be integrated with pig production.

a. Job's tear (*Coix lachryma– jobi*)

This is grown wildly and the grain yield range between 25–30q/ha. The grain contains 14–15% protein and 1.2–1.4% fiber. Full grain meal can be incorporated at 60% level in grower finisher ration.

b. Sweet potato (*Ipomea batata*)

Although the main nutritional importance of sweet potato is in the starch content of the root, it is also a source of important vitamins such as vitamin A, ascorbic acid, thiamine, riboflavin and niacin. The fresh sweet potato root contains 4.6% CP, 4.2 – 7.2 g lysine/100 g of protein and 14 MJ of digestible energy/kg dry matter. The vines of sweet potato contain 18.5% CP and 4–6 g lysine/100 g of protein. Sweet potato meal can also be prepared and fed to the pigs. Sweet potato meal is a good energy feed and can be used in compounded rations instead of cereal grains. It must be mixed with protein feeds, mineral supplements and vitamin supplements. The energy that pigs can get from sweet potato meal is similar to that they get from maize and hence sweet potato meal can completely replace maize in pig rations. However, tubers from some cultivars have high levels of trypsin inhibitor and hence they can be cooked to increase the digestibility and availability of protein. Therefore, it is safe to replace only 50% of the maize in pig ration with sweet potato meal. It has also been claimed that feeding sweet potato reduces the parasitic load, thus having an additional advantage in body weight gain of pigs. Sweet potato vines can be fed to pigs either in the fresh form or after drying. Pigs readily eat the vines. Vines can be dried and grind into a meal. By drying 100 kg of fresh vines, about 30 kg of dried vines can easily be obtained. Sweet potato vine meal can be used in compounded pig rations, but only at low levels. It should not be used more than 5% level in pig rations.

c. Yam (*Dioscorea* spp.)

There are several species of yam and the nutritional composition varies with species. The average CP content of tubers ranges from 8–8.5%. It contains high tryptophan content but deficient in lysine and other sulphur containing amino acids.

d. Taro or cocoyam (*Colocasia esculanta*)

The starch grains of this corn are very small, which make the digestibility of the tuber high. The level of crude protein (9%), although slightly higher than that in yam cassava or sweet potato, contains low amounts of amino acids like histidine, lysine, isoleucine, tryptophan and methionine.

e. Cassava (*Manihot esculanta* Grantz)

Cassava is one of the major sources of carbohydrates for humans as well as animals. Although the CP content of cassava root is 2 to 4% in dry matter, the true protein content is less than half of this amount, due to the fact that 50% of the nitrogen in roots is in the form of non-protein-nitrogen. Furthermore, the available protein is deficient in sulphur containing amino acids. But, the roots contain significant amount of vitamins particularly vitamin C, thiamine, riboflavin and niacin. The root contains 15–16% digestible energy and 2.2 g-lysine/100 g nitrogen.

f. Bananas and plantains

Although banana (*Musa cavendishii*) and plantains (*Musa paradisiaca*) are mainly used as human food, a considerable amount of reject fruit can be fed to pigs. The CP content of banana pseudo stem, leaf meal, plantain pseudo stems, plantain leaves, green banana and ripe bananas is 2.5%, 9–10%, 2–3%, 9–10%, 4–5% and 5–6% respectively. Both green and ripe bananas, however, are deficient in lysine and other sulphur containing amino acids. The digestible energy of green and ripe banana range from 13 to 14 MJ/kg dry matter. The green banana meal can be fed to the pigs up to 20% of the feed.

Besides the feeds mentioned above, an additional place is reserved for kitchen waste as they are frequently consumed by pigs in rural areas. These usually consist of cooked vegetable, rice waste, brewery waste and other feed wastes. As the kitchen waste is cooked, its digestibility is very high but they must be supplemented with concentrate feed, vitamins and minerals.

22.5 Crop-Pig-Fish Production

Culture of fish in paddy fields either as a secondary crop after paddy or along with paddy cultivation are being carried out with promising results in various parts of India, particularly in West Bengal and Orissa. The advantages associated with this type of farming system are increased fertilization of the rice plants through fish culture and increased production (rice and fish). The rice plants provide feed for fish including the pollen from the rice panicles. Pig can be reared on or adjacent to the pond bank and the manure can be used as fertilizer for paddy as well as for ponds. The synchronous system of pig-paddy-fish culture may be adopted for productive utilization of land and to supplement the income to farmers. Culturing fish in rice field has the characteristics of low cost quick effectiveness, better return and has been recognized as an additional source of food and income of farmers in rural areas. Pig-Rice-fish integration also helps in enrichment of soil organic matter and nutrients. Reduced nutrient loss in rice field due to introduction of fish and effective utilization of weeds, plankton, macro and micro aquatic animals, insects, bacteria and organic detritus through introduction of omnivorous fishes give additional benefit in terms of productivity. The fish seed can be stocked in paddy field @ 10000 fingerlings per hectare. Fish can also be reared by making trenches besides the rice field. The plot can be renovated by excavating canals, pools or trenches to retain water, which will provide shelter to fish. While two crops of paddy can be harvested in a year, the fish also would be ready for harvesting. In this kind of integrated system, fish production may be about 3000 kg/ha/year. In paddy fields with water harvesting structure, pig-rice-fish farming can very well be integrated (Figure 22.2). This will not only enhance the land productivity but also provides employment throughout the year.

The fish species which can thrive well in very shallow water, withstand turbidity of water, able to tolerate relatively high temperature and having faster growth rate should be selected for rice cum fish culture. Katla, rohu and mrigal are commonly used for culture with paddy in India. In addition to fishes, prawn can also be grown in conjunction with paddy.

Fig. 22.2 Integrated pig–paddy–fish culture at ICAR Mizoram

Pond banks/bunds provide a suitable place which can be economically used for raising fruit crops like banana, papaya and also some vegetables. The pond dikes can also be used for cultivation of fodder grass, maize, alfa-alfa etc., which are used as feed for animals as well as grass carp, thus cutting down the production cost. Other supplementary component may include biogas, vermicompost, enriched pond slurry etc. depending upon the need.

22.6 Pig-Fish Production

As compared to the progress achieved in the Eastern countries of Asia, little progress has been made in India on integrated pig-fish farming. This method is another classical Chinese integrated fish farming system widely practiced in its original geographic area. Small pigsties are constructed over the fish ponds (Figure 22.3), while the bigger ones are constructed on the dikes and pig manure is allowed to enter the pond directly or collected and fermented in suitable pits before applying in the ponds. Fresh pig manure is regarded as highly efficient for pond fertilization and fish can utilize directly the feed spilled by the pigs, which would other wise go as waste. In this system, supplementary fertilization and feeding are not required for fish culture. Pig manure is rich in phosphorus and nitrogen, which are highly essential to sustain a good stocking density of fish fingerlings per hr. The nutrient content of pig manure is about 0.6, 0.5 and 0.2 % N, P and K, respectively. On an average, 30 to 40 pigs are sufficient to fertilize one ha pond water area.

Fig. 22.3 Integrated pig–fish culture at farmer's field

Depending upon the need and prevailing local market, non-descript as well as exotic or improved varieties of pig are recommended for rearing. The pig– fish culture system is economically viable when good management practices are followed. Results of a study to assess the total biomass production based on the live body weight of pig and fishes in an integrated pig-fish production system revealed that the biomass production/unit area was higher in integrated production system compared to the non-integrated system. In both commercial and traditional feeding system, the total biomass production was significantly higher in pig-fish integration compared to other integration and non-integration. The economics of pig-fish systems was worked out for a period of two years and found that the system was economically viable with input-output ratio of 1:1.2, when the animals were reared under commercial feeding system. The input-output ratio was 1:1.5 under traditional production system.

By adopting the integrated animal-fish production system, it has been demonstrated that the production of total animal and fish biomass could be increased by 159%. Adoption of integrated animal-fish production system by farmers is recommended for increased economic benefit and sustainable animal protein production. Judicious integration of crop, pig and fish components may be suggested as the foundation block for a viable agricultural production system for smallholder pig production. However, more research work is needed for injecting an element of accuracy in component selection, integration and standardization of various methodologies/technologies involved therein.

CHAPTER 23

MEAT PRODUCTION AND MARKETING

23.1 Status of Meat Industry

At present the industry is based on 97 million buffaloes, 61 million sheep, 124 million goats and 13.5 million pigs and other animals (based on FAO production year book 2003). The meat eating population of the country is about 65 to 70%. India being a multiracial country, has developed a complex meat industry, with different religious beliefs and taboos prevalent in this country. This untapped animal wealth is a high resource to provide increased meat production so as to meet the enhanced requirement of animal protein for human consumption, vast employment opportunities, through export of meat and meat products.

According to Indian Council of Medical Research, about 24 grams of meat and fish for balance diet per capita per day is needed and taking meat by weight of about 30%, works out to 2.55 kg per annum per person. In view of this enhanced need and huge gap in actual production and requirement of meat, production is to be enhanced not only quantitatively but also qualitatively.

Importance of meat industry has to be understood in terms of variety of activities which have beneficial impact on (i) employment and rural incomes, number of persons engaged in livestock rearing, (ii) production of animal feed, as well as in ancillary industries, (iii) like leather, wool and other by-products etc. in addition to direct employment in meat industry, including export of these products.

The country is at present highly deficient in modern methods of slaughtering, storage, processing and preservation, meat inspection, quality control, and marketing, so much so that hygienic conditions in slaughter houses are not available resulting in poor quality meat which is a public health hazard. Many of these slaughter houses in the country lack even elementary facilities for hygienic production, handling and utilization of animal by-products. Since meat is a perishable commodity and its poor handling creates a public health and economic problem, there is no room for any complacency in hygienic production of meat.

Meat industry in this country has made some progress during last two decades. About 90 meat processing plants of different capacities are operating and for improving the quality of meat production in the country, different states in India with assistance of Government of India planned establishment of modern slaughter complexes at Kolkata, Goa, Bangalore, Madras, Hyderabad, Delhi. Eight modern composite meat plants are operating in the country at Aligarh (Uttar Pradesh), Alwar (Rajasthan), Borivili (Maharastra), Gannavaram (Andhra Pradesh), Koothattukulam (Kerala), Kharar (Punjab), Ranchi (Bihar), Haringhata (West Bengal) for attending to meat production and marketing of meat and meat products. Six Meat Corporation in different States *viz.* (1) The West Bengal Livestock Processing Development Corporation Ltd., Kolkata, (2) Bangalore Animal Food Corporation Ltd., Bangalore, (3) Tamilnadu Meat Corporation, Chennai, (4) Andhra Pradesh State Meat and Poultry Development Corporation Ltd., Hyderabad, (5) J and K State Sheep and Sheep Products Development Board, and (6) Uttar Pradesh Poultry Specialties are functioning.

23.2 Meat Trade and Export

At present 3 to 4% of domestic meat production is hardly exported. Efforts should be made to utilize additional high quality meat of the livestock species including pig towards exports so as to increase employment and income potential of people engaged in the meat trade.

A survey carried out by Indian Institute of Foreign Trade for Meat Production has shown that selected markets for meat produced in India is available in middle east countries like Saudi Arabia, Kuwait, Dubai, Abu Dhabi, Oman etc., Sterilized canned meat products have export potential in South East Asian countries, UK and East European countries. For carrying out meat and meat products export, slaughter houses, meat processing units and plants need to be modernized to ensure hygienic production of wholesome and quality meat, besides well organized meat inspection service so as to create confidence in purchase of foreign countries in respect of export of wholesome and hygienically produced meat.

Table 23.1 Meat Production in India (Figures in thousand tones)

Type of meat	1975	1985	1986	1987	2000	2003	2004	2005
Beef and buffalos meat	1504	304	324	329	2781	613	750	731
Sheep and goat meat	386	500	519	531	680	614	734	762
Pig meat	56	86	87	90	47	125	221	233
Poultry meat	101	161	175	138	527	1626	1705	1773
Total	2027	1167	1331	1208	4035	2978	3410	3499

Source: FAO Year Book 1975, 2000

Table 23.2 Countrywise share of pork production

Countries	Pig meat million tonnes	%
China	48117.790	48.27
USA	9392.000	9.42
Brazil	3110.000	3.12
Vietnam	2288.315	2.30
Germany	4499.991	4.51
Spain	310z0.718	3.11
Poland	1955.500	1.96
France	2257.000	2.26
Canada	1913.520	1.92
Mexico	1102.940	1.11
India	497.000	0.50

Meat production

Special features of pig slaughter and dressing

Pigs slaughtered for meat production are mostly in rural areas than meat produced in the slaughter houses or meat processing units or meat and meat product factories located in urban areas. It is, therefore, important to take due care in pre-slaughter handling. There is weight loss in pigs due to stress on account of crowding, overloading, and long time taken in transportation. Generally about 3 to 5 kg of live weight is lost in 24 hr in transit. There is also risk of mortality in transit, if proper care for providing adequate ventilation and timely feeding of pigs are not done. Stress in transit and handling at the time of loading and unloading also affects the quality of meat.

Care before slaughter

Pigs after arrival at slaughter house from pig markets, pig farm or production units are not immediately slaughtered. Pigs have to undergo stressful conditions during transit in loading, transport, off loading and some dehydration on long journeys.

They should be given rest and kept in lairages and this is essential for ensuring proper bleeding. Sufficient water and rest be provided to avoid any chance of deteriorations in quality of meat. Generally 18 to 30 hr rest should be provided in lairage and if they are retained for more than this duration, light food be provided in lairage, but pigs must be fasted for at least 12 hr before slaughter. Lairage should have proper shade, drainage, adequate supply of fresh water and connected to slaughtering chamber by long, narrow and slopping passage. Pre-slaughter fasting reduces burden of faeces and undigested food and depletes glycogen reserves on muscles to improve keeping quality of meat.

23.3 Marketing of Pigs and Meat

Marketing of pigs, pig meat and its products need special attention for profitability of pig enterprises.

Marketing of pigs

The pig farmers' main interest end as soon as finished pigs leave their farm premises. Efficient pig producers plan their breeding, feeding and management operations in a way that they are able to produce their hogs of weight and conformation desired by the market at the time they are in demand. Farmer must analyze the requirement of the swine enterprise and decide on its own marketing strategy to procure maximum price for their produce.

The meat type hog is getting popular due to change in demand of pork products. The law of supply and demand affects the fluctuation in prices of pigs.

There are many components of production, processing and marketing leading to the consumer. This inter-dependency controls the product finally offered to the consumer. As such, it is essential that consumer's requirements are relayed efficiently up through the chain of links to the pig producers. Increase demand has favourable effect on the price paid for pig meat and opens opportunities for expansion of the pig meat industry.

In developing countries there are two market segments whcih are operative. One which is local consumption in the village and tis vicinity. These animals are sold and purchased from local markets on the days fixed by the local authorities like town committee/panchayets etc. Here the sale is done through a system based on eye estimates, where the looks of the animal the price through a complex negotiating system. Second segmement is the one which supplies the upmarket and is based on European cuts and products. This system follows a procedure where higher pricess are paid based on exotic breeds and on the basis of carcass characters, as well as such items as ham, becon and other processing characters.

Factories prefer pigs live weight ranging between 80–90 kg and pay less or keep a discount per kg of live weight if the pig is over and above 90 kg less than 50 kg, as the output of meat is less in under weight pigs and there is more fat in over weight pigs. In foreign countries and now even in India there is demand of leaner cuts. As such the pig producer will have to produce as per market demand.

In India the price adapted for purchase of live pig is through producers who can be registered existing in the area adjoining the meat production unit. Either producers directly or through middle men (who may be one of the producers) obtain order from meat production unit and collect the required number of pigs from different pig producer of the area and transport them to the meat production unit in hired transports (truck) where they are paid on the basis of live weight taken before slaughter after resting these pigs for 24 hr. Transportation cost is also paid on the basis of number of pigs and distance covered. In such cases, often the pig producers also accompany the middlemen to be sure of their transaction. Bigger pig enterprises send their pigs in one lot to the meat unit in truck load and get their payment. Pigs purchased directly from market by the meat units are weighed and the price is paid on the basis of the weight.

In some meat units pigs are purchased on the basis of carcass weight and rates are fixed on that basis. Pigs are slaughtered in meat factory and after dressing, weight of the carcass is taken and prices are paid. In foreign countries carcasses are graded and price is paid on the basis of grades. When cuts are purchased, they are also purchased on the basis of quality of cut and weight.

Pig markets in India are hardly provided with any facilities of shed or water supply, where purchasers and sellers are assembled and transaction is finalized only in a market area where they are required to stay for a few days. Temporary shelter and watering facilities are provided but arrangement for feed has to be made by the producer. There are no regulatory practise controlling their markets, except that the owner charge same token money for each transaction.

In foreign countries like UK and others, the livestock markets and sale yards are regulated for which requirements are laid down regarding the construction of lairage and shed and cleaning and disinfection of these markets.

A number of factors are involved in selection of market *viz.* distance, transportation problem, dependability and grading process differed and methods of purchase like live weight and grading of carcasses and other conditions of contract prescribed by purchases including terms of payment. Every unit has to decide its own purchase policy, arrangements for collection of pigs and their transport, transit losses involved, stress and exhaustion during transport to long distances and shrinkage. Many of the poor pig producers find difficult to transport their animals to the factory. Efforts by meat factories to purchase pigs from collection/purchase centres located in rural areas around which sizable pig produces are available and then arrange transport to the factory after weighment, will greatly facilitate marketing. These centres may be equipped with shed/lairage for collection of live pigs and weighment facilities etc. when live pigs are produced.

23.3.1 Transportation and care during transport

In India, specially built transport trucks for pigs or for that matter for any livestock are not available and common goods trucks are used for transportation of pig loads. From by distances of about 15 to 20 km around meat production units, pigs are mostly driven on foot by road. Contractors and pig producers generally prefer transport through trucks for quicker transport and transaction. During transport by truck animals should be provided with sufficient bedding paddy straw which is commonly used as bedding in India so that they don't get bruised in transit. Animals should not be overloaded in the truck to avoid exhaustion and if possible properly restrained during transit to avoid injury. In case of long distances some light food on way should also be provided to them.

The pigs are also transported to long distances through rail, in railway vans especially to big metropolitan towns, like Kolkata and Mumbai from various places like Uttar Pradesh, Madhya Pradesh etc. as higher prices are available in such big towns. In such cases, bedding of paddy straws, watering arrangements and feed be provided and an attendant has to accompany them. They can be kept in small enclosures, temporarily provided in the railway vans.

Care in preparing animals for transportation to long distances either by truck or rail is required to be taken during loading and handling of pigs so as to reduce greatly the losses due to bruises, crippling and death of animals in transit. Following considerations be given in transport of pigs:

(i) Clean the truck or railway wagon before loading the pigs;
(ii) Pigs are not to be fed heavily before transportation;
(iii) Use paddy straw for bedding specially during winters;
(iv) Don't put too many or too few pigs in truck or railway wagon that is maintain optimal stocking density.
(v) In hot weather, generally transport them during night;
(vi) Handle the animals quickly and with care;
(vii) Check rail heads and other obstructions in the transport truck before loading and remove them to avoid possibilities of any injury to animals;

(viii) Wet or sprinkle animals and bedding during hot weather. The greater the distance, the greater the value of sprinkling and there is less danger of mortality from heat and less bruising from transportation, crowding and less shrinkage;

(ix) Heavy and light pigs depending on weight be kept separately;

(x) Animals should not be overloaded. Allow about 2:2 pigs of about 90 kg weight per running foot of length and about 1.8 pigs per running foot of length of truck or railway wagon floor space;

(xi) Side openings should be closed and cover over trucks should be provided specially during summer/rains to avoid heat stress;

(xii) The vehicle should be driven carefully and sudden stops should be avoided.

23.3.2 Disinfection and precautions in transport

It is always advisable to clean and disinfect rail wagon or truck, through which pigs are transported, especially those in which some diseased animals have been transported or in which any pig may have died during transit and may be suspected to be suffering from some diseases.

Every fitting in the railway wagon or truck should be thoroughly scrubbed, sweeped and washed with washing soda and water, then disinfected with any approved disinfectant. The scrapings and sweepings of the truck or railway wagon floor and the sides along with solid material be first removed from there and may be mixed with quick lime and can be destroyed by burning in case there is suspicion of infectious disease in animal during earlier transport. All head gear and halts used for securing animals on truck or rail wagon, after use be disinfected through immersion in the disinfectant. The carcass of dead animal should be buried in pit, about 4 to 5 ft deep below the surface of the earth and covered with sufficient quantity of quick line. The carcass can also be destroyed by exposure to a high temperature, chemical at a site nearest available to the premises. The disinfectant used is standard phenol of the dilution which can be one part of phenol to nineteen parts of water or any other disinfectant equivalent to it.

Marketing of meat products and care in their transportation

Pig meat must be processed, packaged and sent to market for disposal through retail outlets or to whole sales outlets for distribution, under not only hygienic condition but also at temperatures which may not affect the quality of meat. The wholesomeness of meat and the health of the consumer are to be fully safeguarded.

Pigs are converted into attractive variety of pork and processed products at low cost and with maximum appeal to modern consumer.

Market survey

For taking up sales in any area, the proposed market should first be surveyed. Market survey should be first carried out, assessing the sale of each variety of pork products, various brands of products popular in market, price at which they are sold, type of packaging, incentives provided to wholesalers/retailers along with discount and commission provided, potential demand of market, consumer/ retailer and wholesaler wise, additional facilities demanded and improvement in terms and condition of sales and the views regarding quality of product. Once market survey is completed then the forecast of demand of various products for different markets at which these can be sold and proposed planning of product sales and market potentiality and proposed marketing strategy be prepared in the form of a report which may enable the management of meat product unit to take decisions on organizing sales in a particular market.

Publicity and advertising

For organizing and improving sales of pig meat and its products, sales promotion and publicity campaign be organized, through different media *viz*. insertion about products in newspaper and magazine, through audio-visual media, slides in cinema halls or television and radio hoardings, small display boards and sampling to parties etc. Proper publicity campaign should be organized beforehand in new markets and also for any new product, once they are introduced in any market. Packaging and labeling of different products have to be attractive to draw attention of consumers. Incentive for retailer and wholesalers in the form of cold storage facilities and deep freezer at attractive terms, will help them for organization of exhibitions and cinema shows for promoting sales. Proper publicity campaign be organized and results of this publicity campaign be monitored as this expenditure will pay off with increased sales.

Marketing strategy

For formulating marketing strategy, markets have to be chosen after carrying out market survey and sales budget, indicating product wise sales for each market both in terms of quantity and value. In this, anticipated increase due to normal growth as well as due to publicity campaign indicated if existing market is being taken. Market strategy for existing markets indicating expansion of sales, product wise and for new markets in 2 to 3 years be forecasted. Careful monitoring of the market strategy by rough market performance reporting and periodical review of performance with marketing staff is a must. Production in the factory has to be planned on the basis of this sales budget so that proper inventory of finished products is maintained so that consumer's wholesaler's and retailers orders are supplied promptly. The inventory of fresh products have to be less as they have limited shelf life, being highly perishable in nature, but inventories of smoked, cooked and tinned products can be higher as they can be kept for longer periods.

Marketing intelligence

Regular market intelligence regarding various products, quantum of sales, prices of competitive products and consumer complaints, comments etc. have to be collected by marketing staff by regular visits to market for different brands of products. These market intelligence reports be properly recorded and periodically reviewed for making amendments in market strategy as well as for revision of sale target and pricing policy of the products.

There are various factors affecting demand of meat and its products:

(i) Price

It's a competitive market and reduction in price of products in comparison to other competitive products, will increase sales but it has to bear relation with one's own cost of production. The extent to which the demand will be increased for a unit of reduction in price is known as 'price elasticity of demand' which varies with each product and market.

(ii) Competitive meats and substitute products

The price of pork in relation to mutton/chicken or buffalo meat also affects demand of pork.

(iii) Income

As income of consumers in developing countries increase, the demands increase. The change in demand for a product for a unit change in income is called 'income elasticity'. This has to be kept in view in planning marketing policy.

(iv) Consumer taste

In this country there is intolerance to pig meat on grounds of religion of some persons and variety of other objections or superstition *i.e.* pig meat should not be consumed in summer months etc. Some associate pork with high fat content, being high in calories and is not considered desirable to be consumed by people suffering with heart conditions.

(v) Sales promotion

The sales promotion efforts also affect sales.

Precautions during transport, packing and dispatch

In case of fresh pork product and smoked or cooked products, care should be taken to provide ice packing in tin container, insulated with paddy straw and kept in wooden boxes for transport through rail or ordinary truck, as the boxes are roughly handled in transit. These products can also be packed in thermocole boxes in ice packing and should be packed in cool conditions, after taking out these products directly from cold storage. For nearby markets refrigerated vans are used for transportation. Of these products, the tinned products can be transported in wooden boxes which should be properly marked and labeled on the outside and should be properly secured. In case of fresh products there is difference in weight of products, get deteriorated in quality due to delay in transit or due to melting of ice. Due precaution in transport, especially of fresh meat products is essential. Sometimes the insulation material like paddy straw, soil the products due to tin packing being damaged in transit. So the tin container having meat products be properly secured and preferably transported through own truck so that there is least risk of damage to the packing. Being perishable product, delay in transit through railway may completely spoil the goods. Immediate action to get these goods released from railway is necessary to avoid loss due to spoilage.

Proper marketing of pigs, pork and pork products are the key consideration for running any profitable pig enterprise.

CHAPTER 24

BEHAVIOUR OF PIGS

24.1 Introduction

Animal behaviour is the overt and composite functioning of animals individually and collectively. Ethology, in fact, enables the experts to develop production systems that provide animal comfort yet efficiently utilize the animals that were domesticated to serve man. The following is a summary of behaviors exhibited by domestic swine.

24.2 Neonatal Behaviour

Within a few minutes of birth piglets can walk, see and hear (Precocial offspring). Certain physiological mechanisms, such as temperature regulation, are not mature at this time and temperature conservation by huddling is, therefore, a prominent feature of neonatal behaviour in the pig.

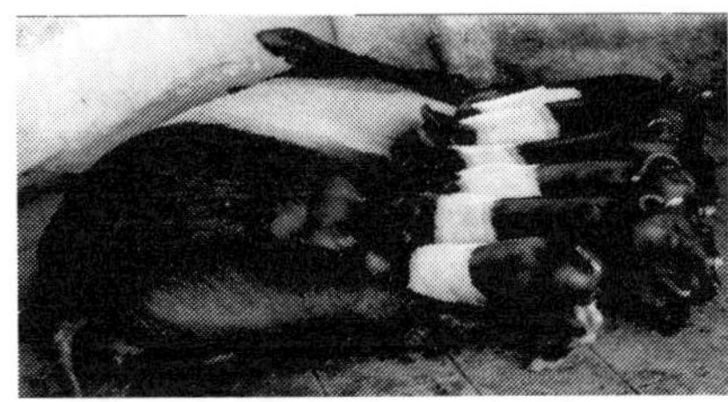

Fig. 24.1. Suckling behaviour in piglets

High piglet mortality is mostly due to wandering away from the litter. Piglet wandering is an early sign of inanition. Normal healthy and well-fed piglets stay close to their littermates and the sow's mammary region. Wandering piglets are very liable to become crushed, chilled or traumatized by other pigs.

The formation of the social organization within the litter, which takes the form of the 'teat order', is a notable behavioural phenomenon (McBridge, 1963). Piglets form a teat order after about four days of age and after this, each piglet goes straight to its own position (Jeppesen, 1981).

24.3 Feeding Behaviour

Eating habits of meals by pigs are (1) meals separated by long intervals; (2) meals consisting of clusters of eating bouts separated by shorter intervals, sometimes associated with drinking; (3) within each eating bout short intervals occur, as pigs constantly move in and out of the feeder (Morgan *et al.* 2000). Daily food intake increase linearly with time, but there is considerable individuality in the degree of order. Pigs make between 18.8 and 80.3 (mean 47.9) daily visits to the feeder. Intervals between visits could be described as too long-normal distributions. Within and between meal intervals are then estimated to be 4.2 and 93.9 min, respectively.

It has been observed that as the pigs grew the daily feed intake increased nearly threefold, while eating bout frequency fell from 14 to 7 per day; consequently both eating bout size and inter-bout interval increased. However, bout size was increased primarily by an increased rate of eating during bouts without any consistent increase in bout duration. Neither pre-meal nor post-meal intervals were correlated with meal size. Of the pigs' daily water intake, 75% was closely associated with eating bouts and over $^1/_3$ of this (25%) was preprandial. 64% of daily food intake and 68% of water intake was during the 12 hr light period. Nocturnal eating bouts were less frequent, but larger (Bigelow and Houpt, 1988).

Obese pigs consumed more feed per unit body weight, spent more time eating per day, and exhibited a slower rate of eating, compared to lean pigs. Diurnal distribution of feeding bouts also differed in the two pig strains with obese pigs exhibiting a tendency to equalize feeding activity over a 24 hr period.

Contra freeloading is the phenomenon that animals prefer to "work" for food even though "free" food is available nearby. The pigs are found to express contra freeloading when using a natural foraging task and the reinforcing effects of anticipation, which occurs during natural foraging, in the delays between searching and finding food, may contribute to the observed expression of contra freeloading in pigs (De Jong *et al.,* 2008).

24.4 Agonistic Behaviour

It is most severe among adults. Mixing adult pigs together, therefore, is an operation which must be carried out with care. If a strange sow is introduced to an established group of sows, the collective aggressive behaviour of the group directed at the stranger is likely to be so severe that physical injuries may result in death.

24.5 Behavioural Thermoregulation

The ability of newborn piglets to adapt to their environmental temperature is very limited as they lose body heat rapidly. They deal with this problem by the way of huddling. During huddling, they lie parallel to each other, often with head and tail ends alternating along the row, so that heat lost by the piglets is lessened to a great extent. Although huddling behaviour is shown in the litter early in life, it is retained by the groups of piglet into adult life, as means of conserving body heat.

Wallowing in mud, in addition to permitting heat to be conducted from the body, also causes a reduction in the body temperature by radiation. Actual heat stress is brought under control by pigs' voluntary wallowing activities (Fraser, 1980). A pig through wallowing in mud quickly acquires a thick coating of mud over its lateral and ventral surfaces and its limbs. After the pig has left the wallow this coating remains adherent and becomes dried out to form a protective insulation against sun rays.

24.6 Elimination Behaviour

Elimination occurs on the day of birth using postures characteristic of older animals. Piglets do not re-use the same site of defaecatiom, so the site of elimination can not be predicted. Pigs have a keen sense of territory and even in the most limited quarters, they reserve an area for sleeping accommodation and an area for excretion. This sleeping area is kept as clean and dry as possible.

24.7 Sexual Behaviour

Sows live in small family groups and the boars are not present. The sow takes the initiative in the search of a sexual partner and she may be attracted towards any boar. The sow begins her search at least one day before the onset of oestrus (Signoret, 1970). The regular surveillance of sow by boars has a synchronizing effect (Wood-Gush and Stolba, 1982).

Commonly observed signs of oestrous are:

1. Boar seeking; 2. Genital sniffing and nuzzling; 3. Mock fighting 4. Hyperactivity.

24.8 Parturient Behaviour

The sow remains in lateral recumbency for delivery. In the interval between births, sows sometimes change to the ventral position and occasionally stand up. Delivery is often heralded and accompanied by vigorous tail swishing. The average time taken to give birth of whole litter varies with litter size and other factors, but averages about 3 hr.

24.9 Nursing and Maternal Care

In the sow, nursing and suckling follow a strict pattern. The sow receives the piglet for suckling either lying or occasionally, standing. Each piglet then massages around its respective teat with rapid upward and downward movements of the snout, during which time the sow grunts at slow, regular intervals. This massage stage lasts for about a minute and ends when milk flow begins.

24.10 Cannibalism

It is wounding of specific individuals. Biting is the first step towards cannibalism. Cannibalism occurs in the form of tail, ear and vagina biting, as well as biting of joints. Factors responsible for cannibalism are overly large group, stock density too high, unregulated feeding, not enough feeding space for all pigs, parasite infestation, humidity too high, drafts, intensive unrest, boredom, etc.

24.11 Bar-Biting

Bar-biting in swine is a behavioural abnormality which generally appears in the breeding sows or boars kept single in the crate. The cause of this abnormal behaviour may be due to poor husbandry and boredom (Schunke, 1980).

Tail-biting

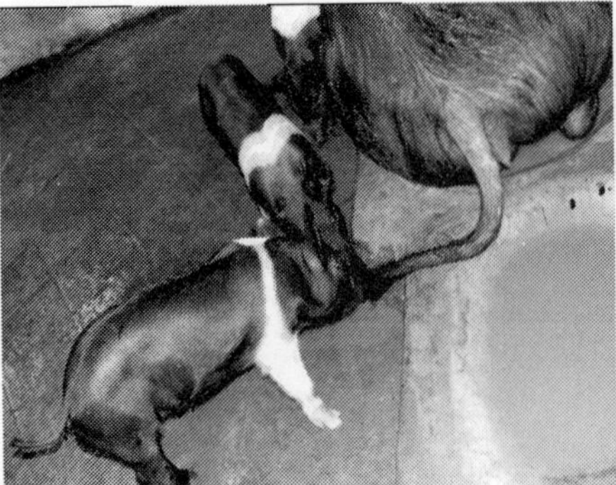

Bar-biting

Table 24.1 Commonly Encountered Behavioural Problems in Pigs

Behavioural problems	Possible cause (s)
Tail biting	Dietary deficiency of fibre, iodine, iron, calcium,salt and lysine; boredom, restlessness, poor ventilation, mange mites, endocrine disturbances, overcrowding, wet floor, slatted floor
Excessive aggression	Food competition, overcrowding, forced encounters
Pica	Boredom, dietary deficiency, excessive restraint, confinement, delayed feeding
Excessive eating	Boredom
Excessive huddling	Cold stress
Oral stereotype or bar biting	Boredom, lack of bedding material, increased stocking rate
Cannibalism	Calcium deficiency

Source: Fraser, A.F. (1980). Ethology of farm animals. Elsvier Science Pub. B.V., New York, World Animal Science, A5, pp-45 and Merck Veterinary Manual.

CHAPTER 25

ORGANIC PIG FARMING

25.1 Introduction

Organic farming, by definition, is a holistic approach which avoids or largely excludes the use of synthetically compounded fertilizers, pesticides, growth promoters and livestock feed additives. Livestock production is an important part of organic farming that aims at achieving a balanced relationship between the soil, plants and animals in a farming system. It is established that, the conventional livestock farming has been impressively successful in increasing the performance of farm animals and decreasing the production costs. However, the recent day's production intensification has pushed the issues of environmentally friendly production, animal health and welfare into the background, especially because these are cost and labour intensive.

Confronted with the effects of environmental degradation, as well as with the increasing consciousness on animal welfare, there is a search for alternative livestock production systems, allowing for preservation of the environment and with high standard of animal welfare without compromising food security and food safety. Organic livestock production could fulfill these goals and it includes all aspects of food, from soil management and environmental and social impact at primary production level through processing to sale and distribution of food. It is expected under organic livestock production system that organic milk, meat, poultry and egg products come from farms that have been inspected to verify that they meet rigorous standards which mandate the use of organic feed, prohibits the use of antibiotics and give animals' access to outdoor fresh air and sunlight.

25.2 Advantages of Organic Livestock Farming

Absence of residues

As the organic livestock farming restrict the indiscriminate use of antibiotics, the chance of emerging drug resistant bacteria and their transfer through the food chain is limited. Also, the products from organically raised animals will be free of growth hormone residues, which will otherwise have deleterious effects in the consumers. Further, the chance of getting the harmful effects of pesticides through the consumption of meat, milk and other livestock products could be minimized as such animals are raised only on certified organic feed which is produced without the use of synthetic fertilizers or pesticides for at least three years prior to the harvest.

Environment friendly

The specifications concerning the organic farming have to do with the denial of pesticides and nitrogenous fertilizers in feed and fodder production. There is need to restrict the number of farm animals per unit area and to avoid the use of bought in feed stuffs. It must also rely on the efficient nutrient circulation within the farm to maintain the soil fertility and high production.

Better animal welfare practises

Many questions have been raised on individual welfare of animals with respect to health care, as the organic livestock farming is prohibiting the conventionally used veterinary medicines (except in emergencies). There are several studies which indicate that the welfare of animals was better in organic herds as compared to conventional herds in terms of general health *viz.* production, body condition, hock lesion, chronic infection etc. The most common health problem on organic farms is parasitism. All the leading animal welfare organizations support organic farming. Royal Society for the Prevention of Cruelty to Animals (RSPCA) have stated "we hope that more people/consumers will become aware of the potential for organic farming as one means of alleviating the suffering of farm animals".

Better product quality

A clear comparison between organic and conventional products is difficult to establish due to the great variation within the production methods, ration fed and the type of animals used. Earlier studies suggest that the characteristics of product quality such as nutritional, hygienic, sensorial and technological factors do not differ much between the production methods. In some cases, organic food get better market, in some others conventionally produced food scores higher, which suggest that the product quality is primarily a function of farm management.

25.3 Indian Scenario

The area under organic farming in India is reported to be 2.5 million hectors, representing about 10% of world area. Despite the fact that, the farming system in India can be considered as organic by default, only less than 0.01% of the total food production in India (about 14000 tonnes out of total production of about 200 million tonnes) is considered as organically produced. India's share in global organic trade is only about 0.8%. Officially, less than 0.03% of India's land is considered to be under organic agriculture, which means that the farms which are basically following organic farming practises are not officially considered as organic.

The problems of developing countries like India are entirely different from those of developed countries and our basic problems are poverty, malnutrition and unemployment, so food security is the prime goal rather than food safety. In this situation, development of the organic farming sector itself is very difficult. Development of an organic livestock sector is more difficult. In India most of the animal husbandry practises are traditional with close resemblance to prescribed organic practises, but we failed significantly to convert our advantages into fruitful gains. Small land holding, low level of literacy, lack of information, inadequate production of feed and fodder, high cost of certification and absence of marketing facilities are some of the hindrances prevailing in India, in the way of conversion from traditional to organic farm culture.

25.4 Requirements for Organic Livestock Production

Production of organic livestock is founded upon a number of basic principles, which are embodied within the standards for organic production. In India, the National Standards for Organic Production developed by Ministry of Commerce and Industry, Government of India, provide guidelines for organic production. Some of those relevant to organic livestock production are discussed below to illustrate the concept.

Maintenance

All animals intended for final sale as organic, must be raised on an organic farm. There should be sufficient space for free movement, protection against adverse climatic conditions, adequate resting and lying area and ample access to fresh water and feed.

Conversion period

Livestock products may be certified organic only after the farm has been under conversion for at least 12 months and the required standards have been achieved. Length of conversion period can be extended at the discretion of the certification agency. All organic animals should be born and raised under organic holding. However, when organic livestock is not available, certification programme shall allow brought in conventional animals according to the following age limits: two day old chickens for meat production, 18 week old hens for egg production, piglets up to six weeks and after weaning, calves up to 4 weeks that have received colostrum and have been fed a mainly milk diet. Breeding stock may be brought in from conventional farms but maximum replacement rate will be 10%.

Reproduction

Breeds should be chosen which are adapted to local conditions. Reproduction techniques should be natural. Hormonal treatments are not allowed. Artificial insemination is allowed, but embryo transfer techniques are not allowed.

Feeding

The livestock should be fed 100% organically grown feed. More than 50% of the feed shall come from the farm unit itself or shall be produced within the region. However, in some cases 15–20% of total feed could be obtained from conventional farms. Synthetic growth promoters, feed preservative, artificial colouring agents and genetically engineered organisms are strictly prohibited in feed. However, vitamins, trace elements and supplements of natural origins may be used.

Treatment

An important objective of organic livestock husbandry is the avoidance of reliance up on routine and/or prophylactic use of conventional veterinary medicines. The use of conventional veterinary medicines are allowed when no other non-allopathic alternative is available and where these are used, the withholding period shall be twice the legally required period. Vaccinations shall be used only when diseases are known and are expected to be a problem.

Record keeping

Proper documentation is essential for tracing the sources of animals, sources of feed, amount of feed given, feed supplements, treatments and animal health etc.

25.5 Certification and Standards

Implementing organic standards require inspection and the end product of the inspection is certification. Certification ensures that organic products are produced, processed and packaged according to organic standards. Certification also ensures that consumers, producers and traders against fraudulent labeling of non-organic products. There are few international standards for organic production like the IFOAM Basic Standards (developed by International Federation of Organic Agriculture Movement), EU Regulation No-1804/1999 and Codex Alimentarius ALINORM 99/22A. India too has developed National Standards on Organic Production and Handling in the year 2000.

25.6 Areas to be Strengthened

- Organic standards should be modified according to regional agro-climatic conditions.
- Regional standards should be developed to bridge the gap between the National and International standards.
- A low cost certification agency must be established, that small farmers can afford.
- A strong domestic market should be developed, otherwise the benefits of producers cannot be protected, as international markets are always fluctuating.
- Organic farming needs research and development in order to apply the most modern knowledge and improve its performance.
- Training and extension should be provided to all categories of stakeholders.
- Government has to make legislation in order to ensure the regulatory framework, where all stakeholders can play a fair level ground.

25.7 Speciality of Organic Pig Farming

To fully understand how to raise pigs organically, we first need to understand how pigs behave in a natural environment. Feeding, housing and raising pigs organically depend on matching breeds to the environment and to the market, and then matching management to the needs of the animals.

In western countries pigs raised on organic farms, are on strictly approved farms who participate in an organic farming scheme that is subject to frequent random audits. Organic farmers do not use artificial chemical fertilizers, pesticides and GMOs. The use of antibiotics and other drugs is very restricted compared with conventional pig farms and only used under very controlled conditions.

All pigs are allowed to roam outdoors. They have access to dry bedding and shelter. Their feed cannot contain GMO's, meat, bonemeal, animal fat, antibiotics, growth promoters or other drugs and preservatives.

Small scale organic pig production based on low cost grass based systems will have the greatest chance of success. It has been observed by Bert Dening, Business Development Officer with Alberta Agriculture that using older breeds, feeding special local diets will develop unique specialty meat. Using modern breeds and the same diets as the modern pig industry will result in commodity meat and poor prices.

Modern pig breeds were developed to maximize production in large scale confinement operations. These pig breeds are not as suited to being raised out of doors as older breeds. Some of the older breeds were bred for specific purposes, such as grazing apple orchards. A little research may be needed before selecting an ideal breed. Although the gene pool in Canada is small, Bert suggests using older breeds such as Berkshire, Large Black, Tamworth, Hampshire and Lacombe. Older breeds can be too fat for consumer preference. A solution is to use older breeds of sows, and lean modern breed boars (such as Yorkshire, Landrace or Duroc) to combine desirable traits.

Pigs, like human and unlike cattle, have a single stomach and cannot digest cellulose well. Forages for pigs need to be leafy, with less stems and straw than a cow would enjoy. Older pigs can handle up to 70% leafy forages, but young pigs need more of the high quality grain and protein. Bert recommends rotational grazing on high quality pasture, supplemented with local grains and legumes. In winter, pasture can be replaced with young grass, hay or silage. In pork production, "you are what you eat" seems to apply. The flavour of the meat depends on the diet of the animal. This can be the key to niche marketing.

Organic management depends more on prevention of health issues than on cure. The key to healthy pigs is fresh air, good feed, and rotating the pigs through pastures so disease does not build up. Pasture rest and sunlight as a disinfectant is one of the best ways to control disease. Of course, starting with healthy, parasite free animals is also important. Reducing stress is also important to healthy animals. Letting pigs wean themselves, not crowding of animals, providing lots of bedding, reasonable shelter, clean water and good nutrition, all help to keep a healthy herd.

Rotating pastures quickly reduces the damage that pigs cause to hay land with their rooting, keeps fresh forage available, and also reduces disease. Pigs respond well to electric fences. For young pigs, a wire at 6 inches, and for larger animals, a wire at 12 inches is adequate. Pigs tend to chew through or dig under other fence types.

In summer, pigs need a mud hole or sprinkler to keep cool. They can't sweat, so this is good for more than piggy morale. In winter, some shelter is required. Tarp covered straw bale shelters with lots of fresh air and dry straw can be ideal.

In some organic farms, six weeks after farrowing, the sows and their piglets are placed together in family group. A few days later, boars join the group in turn so that the sows become pregnant again. This enables late weaning at three months old whist the sows can still produce two litters of piglets per year.

Pigs can have more than two litters per year of 8 to 12 piglets. A sow prefers to go off on her, build a nest and give birth away from the herd. She will need plenty of clean bedding, and in the winter, well insulated structures (or heat). The sow and piglets will return to the herd after a week or two.

Pigs can be very prolific. A single sow can produce 20 piglets per year. These can be ready for market at about 120 kg in 7 months. This 2400 kg live weight of pig would make 1000–1200 kg of meat per year depending on the amount of boneless cuts. Alternately, those 7 month old pigs can be bred, and produce their own litters before they are a year old. With such potential, a sound plan for butchering, processing and marketing is important.

The options for marketing are varied, but best prices are likely to go to producers who develop speciality local products, that reflect special breeds, special diets and special processing. Pigs can be a good fit in an organic operation, but success will depend on avoiding the commodity trap and marketing into niche markets.

REFERENCES

Abdulali, H. (1962). Wild pigs in the Andamans. *J. Bombay Nat. Hist. Soc.* **59**: 280–283.

Adamstone, P.B., Krider, J.D., James, M.F. and Blomquist, C.A. (1949). Response of swine to vitamin E-deficient rations. Ann. N.Y. Acad. Sci. **52**:260–268.

Agricultural research council (1981). The nutrient requirements of pigs: Technical Review. Rev. ed. Slough, England. Commonwealth Agricultural Bureaux. xxii, 307 pp.

Aherne, F.X. and Kirwood, R.N. (1985). Nutrition and sow prolfificacy. *J. Reprod. Fertil. Suppl.* **33**: 169–183.

Albayrak, I. and Inci, S. (2007). The Karyotype of the Wild Boar Sus scrofa Linnaeus, 1755 in Turkey (Mammalia: Artiodactyla). *Turk. J. Zool.* **31** (1): 65–68.

Alcantara, P.F., Hanson, L.E. and Smith, J.D. (1980). Sodium requirements, balance and tissue composition of growing pigs. *J. Anim. Sci.* **50**: 1092–1101.

Andresen, E. (1957). Investigations on blood groups of the pigs. Nord. Vetmed. **9:** 274–284.

Anderson, G.C. and Hogan, A.C. (1950b). Requirements of the pig for vitamin B12. *J. Nutr.* **40**: 243–250.

Andersen, E., Hogaard, N., Birth Jylling, Larsen, B., Moller, F., Moustgarrd, J. and Neimann-Sorensen, A. (1959). Blood and serum, group investigations on cattle, pig and dog in Denmark, Prod. 6th int. blood Grp. Cong. 24–39.

Andresen, E. (1962). Blood groups in pigs. Ann. N.Y. Acad. Sci. **97**: 205–225.

Andresen, E and Baker. L.N. (1964). The C blood group system in pigs and the detection and estimation of linkage between the C and J systems. Genetics **49**: 379–386.

Andresen, E. (1966). Blood groups of the J system in pigs. Association with variants of serum amylase. Science **153**: 1660–1661.

Andresen, E. and Wroblewski, A. (1961). The G and H blood group systems of pig. Acta. Vet. Scand. **2**: 267–280.

Andersson, L. Haley, C.S. Ellegren, H. (1994). Genetic mapping of quantitative trait loci for growth and fatness in pigs. Science. **263**(5154):1771–1774.

Arthur, John, R. (1994). The biochemical functions of selenium: Relationships to thyroid metabolism and antioxidant systems. Pp. 11–20 in Rowett Research Institute Annual Report for 1993, Rowett Research Institute, Bucksburn, Aberdeen, UK.

Arthur, S.R., Kornegay, E.T., Thomas, H.R.,Veit, H.P., Notter, D.R. and Barczewski, R.A. (1983a). Restricted energy intake and elevated calcium and phosphorus intake for gilts during growth. III. Characterization of feet and limbs and soundness scores of sows during three parities. J. Anim. Sci. **56**:876–886.

Arthur, S.R., Kornegay, E.T., Thomas, H.R., Veit, H.P., Notter, D.R., Webb, K.E.Jr. and Baker, J.L. (1983b). Restricted energy intake and elevated calcium and phosphorus intake for gilts during growth. IV. Characterization of metacarpal, metatarsal, femur, humerus and turbinate bones of sows during three parities. *J. Anim. Sci.* **57**: 1200–1214.

Baker, D.H. (1997). Ideal amino acid profiles for swine and poultry and their applications in feed formulation. Biokyowa Technical Review—9. Chesterfield, MO: Nutri-Quest, Inc.

Baker, D.H. and Allee, G. (1970). Effect of dietary carbohydrate on assessment of the leucine need for maintenance of adult swine. *J. Nutr.* **100**: 277–280.

Baker, D.H. and Chung, T.K. (1992). Ideal protein for swine and poultry. BioKyowa Technical Review–4. Chesterfield, MO: Nutri-Quest, Inc.

Baker, D.H., Becker, D.E., Norton, H.W., Jensen, A.H. and Harmon, B.G. (1996a). Quantitative evaluation of the threonine, isoleucine, valine and phenylalanine needs of adult swine for maintenance. *J. Nutr.* **88**:391–396.

Baker, D.H., Becker, D.E., Norton, H.W., Jensen, A.H. and Harmon, B.G. (1996b). Quantitative evaluation of the tryptophan, methionine and lysine needs of adult swine for maintenance. *J. Nutr.* **89**:441–447.

Baker, D.H., Hahn, J.D., Chung, T.K. and Han, Y. (1993). Nutrition and Growth: The application of an ideal protein for swine growth. Pp. 133–139 in Growth of the Pig. Wallingford, U.K.: CAB International.

Barnhart, C.E., Catron, D.V. , Ashton, G.C.and Quinn, L.Y. (1957). Effects of dietary pantothenic acid levels on the weanling pig. *J. Anim. Sci.* **16**: 396–403.

Bartley, J.C., Reber, E.F., Yusken, J.W. and Norton, H.W. (1961). Magnesium balance study in pigs three to five weeks of age. *J. Anim. Sci.* **20**: 137–141.

Beck, J., Knorr, C. Habermann, F., Fries, R. and Brenig, B. (2002). Assignment of the beta-glucuronidase (GUSB) gene to porcine chromosome SSC3p16—>p14 by FISH and confirmation by hybrid panel analyses. Cyt. Gen. Res. **97**: 277G.

Beeson, W.M., Andrews, F.N., Witz, H.L. and Perry, T.W. (1947). The effect of thyroprotein and thiouracil on the growth and fattening of swine. *J. Anim. Sci.* **6**: 482. (Abstr.)

Bengtsson, G., Hakkarainen, J., Jonsson, L., Lannek, N. and Lindberg, P. (1978a). Requirement for selenium (as selenite) and vitamin E (as alphatocopherol) in weaned pigs. I. The effect of varying alpha-tocopherol levels in a selenium deficient diet on the development of the VESD syndrome. *J. Anim. Sci.* **47**: 143–152.

Bengtsson, G., Hakkarainen, J., Jonsson, L., Lannek, N. and Lindberg, P. (1978b). Requirement for selenium (as selenite) and vitamin E (as alphatocopherol) in weaned pigs. II. The effect of varying selenium levels in a vitamin E-deficient diet on the development of the VESD syndrome. *J. Anim. Sci.* **46**: 153–160.

Bertram, M.J., Stahly, T.S. and Ewan, R. C. (1994). Impact of lean growth genotype and dietary phosphorus regimen on rate and efficiency of growth and carcass characteristics of pigs. *J. Anim. Sci.* **72** (Suppl. 2):68 (Abstr.).

Bethke, R., Burroughs, M.W., Wilder, O.H.M., Edgington, B.H. and Robison, W.L. (1946). The comparative efficiency of vitamin D from irradiated yeast and cod liver oil for growing pigs, with observations on their vitamin D requirements. Ohio Agricultural Experiment Station Bulletin 667:1–29. Wooster, Ohio: Ohio Agricultural Experiment Station

Beyer, M.W., Jentsch, L., Hoffmann, R., Schiemann, and M. Klein. (1994). Untersuchungen zum energie and stickstoffumsatz von graviden and laktierend saun sowie von saugferkeln 4, Mitteilung Chemische Zusammensetzung and energiegehalt der Konzeptionsprodukte, der reproduktiven organe und der lebendmassezunahmmen order abnahmen bei graviden und laktierenden Sauen. Arch. Anim. Nut. **46**: 7–35

Bigelow, J.A. and Houpt, T.R. (1988).Feeding and drinking patterns in young pigs. Physiol Behav. **43**: 99–109.

Black, J.L. and Carr, J.R. (1993). A symposium stocking density and pig performance. In: Manipulating pigproducstionIV (E.S.Batterham, ed). Australasian Pig Science Association, Werribee, Victoria, Australia, p. 84.

Black, J.L. (2000). Swine production past, present and future. In: XXXVIII Reuniao Annual da SBZ. Vicosa MG de Julho.

Bokonyi, S. (1974). History of Domestic Mammals in Central and EasternEurope. Akademiai Kiado, Budapest.

Bonnette, E.D., Kornegay, E.T., Lindemann, M.D. and Hammerberg, C. (1990). Humoral and cell-mediated immune response and performance of weaned pigs fed four supplemental vitamin E levels and housed at two nursery temperatures. *J. Anim. Sci.* **68**: 1337–1345.

Bornemann-Kolatzki, Knorr, Peters, Harlizius, Brenig. (2002). A total genome scan for porcine hernia inguinalis and scrotalis. Abstract presented at the ISAG 2002 meeting, Gottingen, Germany.

Bosma, A.A. (1976). Chromosomal olymorphism and G banding patterns in the wild boar (Sus scrofa L.) from the Netherlands. Genetica **46**: 391–399.

Bosma, A.A. de Haan, N.A. and MacDonald, A.A. (1991). The current status of cytogenetics of the Suidae; and review. Bongo, Berlin **18**: 258–272.

Brady, P. S., Ku, P. K., Ullrey, D.E. and Miller, E.R. (1978). Evaluation of an amino acid-iron chelate hematinic for the baby pig. *J. Anim. Sci.* **47**:1135–1140.

Braidman, I.P. and D.C. Anderson. (1985). Extra-endocrine functions of vitamin D. Clin. Endocrinol. **23**:445–460.

Braude, R., Chamberlain, A. G., Kotarbinska, M. and Mitchell, K.G. (1962). The metabolism of iron in piglets given labeled iron either orally or by injection. *Br. J. Nutr.* **16**: 427–449.

Braude, R., Foot, A.S., Henry, K.M., Kon, S.K., Thompson, S.Y. and Mead, T.H. (1941). Vitamin A studies with rats and pigs. *Biochem. J.* **35**: 693–707.

Braude, R. and Cotchin, E. (1949). Thiourea and methylthiouracil as supplements in rations of fattening pigs. *Br. J. Nutr.* **3**:171–186.

Brief, S. and Chew, B.P. (1985). Effects of vitamin A and â-carotene on reproductive performance in gilts. *J. Anim. Sci.* **60**: 998–1004.

Brooks, C.C., Nakamura, R. M. and Miyahara, A.Y. (1973). Effect of menadione and other factors on sugar-induced heart lesions and hemorrhagic syndrome in the pig. *J. Anim. Sci.* **37**: 1344–1350.

Brown, R.G., Buchanan-Smith, J. G. and Sharma,V. D. (1975). Ascorbic acid metabolism in swine. The effects of frequency of feeding and level of supplementary ascorbic acid on swine fed various energy levels. *Can. J. Anim. Sci.* **55**: 353–358.

Calabotta, D.F., Kornegay, E.T., Thomas, H.R., Knight, J.W., Notter, D.R. and Veit, H.P. (1982). Restricted energy intake and elevated calcium and phosphorus intake for gilts during growth. I. Feedlot performance and foot and leg measurements and scores during growth. *J. Anim. Sci.* **54**: 565–575.

Carson, T.L. (1986). Toxic chemicals, plants, metals and mycotoxins. Pp. 688–701 in Diseases of Swine, 6th Ed., A. D. Leman, B. Straw, R. D. Glock. W. L. Mengeling, R. H. C. Penny, and E. Scholl, eds. Ames: Iowa State University Press.

Carter, S.D. and Cromwell, G.L. (1998a). Influence of porcine somatotropin on the phosphorus requirement of finishing pigs. I. Performance and bone characteristics. *J. Anim. Sci.* **76**:584–595.

Carter, S.D. and Cromwell, G.L. (1998b). Influence of porcine somatotropin on the phosphorus requirement of finishing pigs. II. Carcass characteristics, tissue accretion rates, and chemical composition of the ham. *J. Anim. Sci.* **76**: 596–605.

Cartwright, G.E., Tatting, B., Robinson, J., Fellows, N.M., Gunn, F.D. and Wintrobe, M.M. (1951). Hematologic manifestations of vitamin B12 deficiency in swine. Blood **6**: 867–891.

Catron, D.V., Richardson, D., Underkofler, L. A., Maddock, H.M. and Friedland,W.C. (1952). Vitamin B12 requirement of weanling pigs. II. Performance on low level of vitamin B12 and requirements for optimum growth. *J. Nutr.* **47**: 461–468.

Charles, A.B (1985). Factors affecting the growth of sheep and goats in AfricaSmall ruminants in African agriculture-Les petits ruminants dans l'agriculture africaine - Proceedings of a conference held at ILCA, Addis Ababa, Ethiopia

Chavez, E. R. (1985). Nutritional significance of selenium supplementation in a semipurified diet fed during gestation and lactation to first-litter gilts and their piglets. *Can. J. Anim. Sci.* **64**: 497–506.

Cheng Peilieu, (1984). Livestock breeds of China. FAO Animal Production and Health paper No. **46**. FAO: Rome.

Chew, B.P., Rasmussen, H., Pubols, M.H. and Preston, R.L.(1982). Effects of vitamin A and â-carotene on plasma progesterone and uterine protein secretions in gilts. Theriogenology **18**: 643–654.

Chung, A.S., Hoekstra, W.C. and Grummer, R.H. (1976). Supplemental cobalt or nickel for zinc-deficient G.F. pigs. *J. Anim. Sci.* **42**: 1352 (Abstr.).

Cline, J.H., Mahan, D.C. and Moxon, A.L. (1974). Progeny effects of supplemental vitamin E in sow diets. *J. Anim. Sci.* **39**: 974. (Abstr.).

Close, W.H. (1999). A review pig research and development: What needs to be done? Who should pay and who do the work? In: Manipulating Pig production VII (p>D.Cranwell, edn.) Australasian Pig Science Association, Werribee, Victoria, Australia, pp 1–4.

Coffey, M.T., Stainer, R.W., Funderburker, D.W. and McCampbell, H.C. (1982). Effect of heat increment and level of dietary energy and environmental temperature on the performance of growing finishing swine. *J. Anni. Sci.* **54**: 95–105.

Cole, D.J.A. (1982). Nutrition and reproduction. Pp 603–619 in control of pig reproduction, D.J.A. Cole and G.J. Foxcroft, eds. London: Butterworth.

Combs, G. E., Berry, T.H. Wallace, H.D. and Crum, Jr. R.C. (1966). Levels and sources of vitamin D for pigs fed diets containing varying quantities of calcium. *J. Anim. Sci.* **25**: 827-830.

Combs, N.R. and Miller, E.R. (1985). Determination of potassium availability in K2CO3, KHCO3, corn and soybean meal for the young pig. *J. Anim. Sci.* **60**: 715–719.

Combs, N.R., Miller, E.R. and Ku, P.K.(1985). Development of an assay to determine the bioavailability of potassium in feedstuffs for the young pig. *J. Anim. Sci.* **60**: 709–714.

Combs, N.R., Kornegay, E.T., Lindemann, M.D. and Notter, D.R. (1991a). Calcium and phosphorus requirement of swine from weaning to market: I. Development of response curves for performance. *J. Anim. Sci.* **69**: 673–681.

Combs, N.R., Kornegay, E.T., Lindemann, M.D., Notter,D.R., Wilson, J.W. and Mason, J.P. (1991b). Calcium and phosphorus requirement of swine from weaning to market: II. Development of response curves for bone criteria and comparison of bending and shear bone testing. *J. Anim. Sci.* **69**: 682–693.

Cox, J.L., Becker, D.E. and Jensen, A.H. (1966). Electrocardiographic evaluation of potassium deficiency in young swine. *J. Anim. Sci.* **25**: 203–206.

Crenshaw, T.D., Peo, E.R.Jr., Lewis, A.J., Moser, B.D. and Olson, D. (1981). Influence of age, sex and calcium and phosphorus levels on the mechanical properties of various bones in swine. *J. Anim. Sci.* **52**: 1319–1329.

Cromwell, G.L. (1991). Antimicrobial agents. Pp. 297–314 in Swine Nutrition, Miller, E.R. Ullrey, D.E. and Lewis, A. J. eds. Stoneham, MA: Butterworth-Heinemann.

Cromwell, G. L., Pierce, J., Sauber, T.E., Rice, D. W., Etrl, D. S. and Raboy, V. (1998). Bioavailability of phosphorus in low-phytic acid corn for growing pigs. *J. Anim. Sci.* **76** (Suppl. 2): (Midwestern Section Abstr. #115).

Cromwell, G.L., Stahly,T.S. and Monegue, H.J. (1981a). Effects of sodium and chloride on performance of pigs *J. Anim. Sci.* **53** (Suppl. 1): 237 (Abstr.).

Cromwell, G.L., Stahly, T.S. and Overfield, J.R. (1979). Effect of dietary phosphorus level on growth, carcass and bone traits of boars. *J. Anim. Sci.* **49** (Suppl. 1): 101 (Abstr.).

Cromwell, G.L., Hays, V.W., Chaney, C.H. and Overfield, J.R. (1970). Effects of dietary phosphorus and calcium level on performance, bone mineralization and carcass characteristics of swine. *J. Anim. Sci.* **30**: 519–525.

Cunha, T.J., Lindley, D.C. and Ensminger, M.E. (1946). Biotin deficiency syndrome in pigs fed desiccated egg white. *J. Anim. Sci.* **5**: 219–225.

Cunha, T.J., Colby, R.W., Bustad, L.K., and Bone, J.F. (1948). The need for and interrelationship of folic acid, anti-pernicious anemia liver extract, and biotin in the pig. *J. Nutr.* **36**: 215–229.

Cunnane, S.C. (1984). Essential fatty-acid/mineral interactions with reference to the pig. Pp. 167–183 in fats in animalnutrition, J. Wiseman, ed. London; Butterworth

Czaker, R. and Mayr, B. (1982). Comparative studies on the polymorphism of nucleolus organizer regions (NORs) in four breeds of domestic pigs (Sus scrofa domestica L.), with special emphasis on the development of breeds. Z. Zool. Syst. Evolut. **20**: 233–241.

Darwin, C. (1868). The Variation of Animals and Plants under Domestica- Ruvinsky and M. F. Rothschild. CAB International, Oxon, UK.

Davey, R.J. and Stevenson, J.W. (1963). Pantothenic acid requirement of swine for reproduction. *J. Anim. Sci.* **22**: 9–13.

De Jonge, A.M., Tilly, F.H, Baars, S.L, and Spruijt, B.M. (2008). On the rewarding nature of appetitive feeding behaviour in pigs (Sus scrofa): Do domesticated pigs contrafreeload?.Applied Animal Behaviour Science, **114**: 359–372.

Devilat, J. and Skoknic, A. (1971). Feeding high levels of rapeseed meal to pregnant gilts. Can. *J. Anim. Sci.* **51**: 715–719.

Dickerson, G.E. (1973). Inbreeding and heterosis in animals. In Proceedings Animal Breeding and Genetics Symposium in Honor of Dr. J.L. Lush, p. 54. American Society of Animal Science, Champaign, Illinois.

Dinkalage, H. and Major, F. (1968). Eaegu (=E8) a new allele in the E blood group system of pig. Vox Sang. **14**: 315–317.

Dinkalage, H., Sachahmirzadi, H., Hradecky, J.H. and Hojny, J.H. (1969). Eedghi (=E9), a new allele in the E blod group system of pig. Vox. Sang. **17**: 129–133.

Doige, C.E., Owen, B. D. and Mills, J.H.L. (1975). Influence of calcium and phosphorus on growth and skeletal development of growing swine. *Can. J. Anim. Sci.* **55**: 147–164.

Done, J.T. and Wijeratne, W.V.S. (1972). Genetic disease in pigs. In: Pig Production, (ed. D.J.A. Cole), pp. 53–67. Butterworths: London.

Dourmand, J.Y., Etienne, M., Prunier, A. and Noblet, J. (1994). The effect of energy and protein intake of sows ontheir longevity; A review livest. Prod. Sci. **40**: 87–97.

Ducos A., Pinton, A., Berland H.M., Seguela A., Brun-Baronnat, C., Bonnet N. and Darre, R. (2002). Controle chromsomique des populations porcines en France: bilan de 5 annees d' activite, in: Journees Rech. Porcine France, 5–7 February 2002, Vol. 34, Institut technique du porc, Paris, pp. 269–275.

Ducsay, C.A., Buhi, W.C., Bazer, F.W., Roberts, R. M. and Combs, C.E. (1984). Role of uteroferrin in placental iron transport: Effect of maternal iron treatment on fetal iron and uteroferrin content and neonatal hemoglobin. *J. Anim. Sci.* **59**: 1303–1308.

Edwards, R.L. and Omtvedt, I.T. (1971) Genetic analysis of Swine Control Population, Estimation of Population Paramet. *J.Anim. Sci.* **32**: 185

Edwards, R.L., Omtvedt, I.T., Tuman, E.J., Stephens, D.F. and Mahoney, G.W.A. (1968). Reproductive performance of gilts following heat stress prior to breeding and in early gestation. *J. Anim. Sci.* **27**: 1634–7.

Edwards, S.A. and Turner, S.P. (1999). Housing and management of pigs in large social groups. In: Proc. Pig Vet. Soc., Market Bosworth, U.K.

Eeckhout, W., De Paepe, M., Warnants, N. and Bekaert, H. (1995). An estimation of the minimal P requirements for growing-finishing pigs, as influenced by the Ca level of the diet. Anim. Feed Sci. Tech. **52**: 29–40.

Egbunike, G.N. and Dede, T.I. (1980). The influence of short term exposure to tropical sunlight on boar seminal characteristics. *Int. J. Biometor.* **24** (2): 129–135.

Ehrllich, P. and Morgenroth, J. (1900). Ueber Haemolysine. Klin. Wschr. **37**: 453–458.

Ellis, N.R. and Madsen, L.L. (1944). The thiamine requirements of pigs as related to the fat content of the diet. *J. Nutr.* **27**: 253–292.

Ellis, R.P. and Vorhies, M.M. (1976). Effect of supplemental dietary vitamin E on the serologic response of swine to an Esherichia coli bacterin. *J. Am. Vet. Med. Assoc.* **168**: 231–232.

Emmert, J.L., Garrow, T.A. and Baker, D.H. (1996). Development of an experimental diet for determining bioavailable choline concentration and its application in studies with soybean lecithin. *J. Anim. Sci.* **74**: 2738–2744.

Enser, M. 1984. The chemistry, biochemistry and nutritional importance of animal fats. Pp. 23–51 in Fats in Animal Nutrition, J. Wiseman, ed. London: Butterworth.

Epstein, H. (1969). Domestic animals of China. 1st ed. Published by the common wealth agricultural bureaux. Farnham Royal, Bucks, England.

Epstein, H. (1971). The origin of the domestic animals of Africa, vol.2. Africana publishing Corporation: New York, NY.

Erencin, Z. (1977). Av Hayvanlari ve Av. Ankara Universitesi Veteriner Fakultesi yayinlari, Ankara.

Esch, M.W. Easter, R. A. and Bahr, J. M. (1981). Effect of riboflavin deficiency on estrous cyclicity in pigs. Biol. Reprod. **25**: 659–665.

Eusebio, A.N. (1980). Animal genetic resources in the Philippines. In proceeding of SABRAO workshop on animal genetic resources in Asia and Oceania, held at Tsukub science city, Japan. Pp 313–338.

Ewan, R.C., Wastell, M.E., Bicknell, E.J. and Speer, V.C. (1969). Performance and deficiency symptoms of young pigs fed diets low in vitamin E and selenium. *J. Anim. Sci*. **29**: 912–915.

Feed Additive Compendium. (1998). Minneapolis, Minn.: Miller Publishing Co.

Feldman et al. (2000). Schalm's Veterinary hematology, Lippinkott Williams and Wilkins, USA

Fidge, N.H., Smith, F. R. and Goodman, D. S. (1969). Vitamin A and carotenoids. The enzymatic conversion of â-carotene into retinal in hog intestinal mucosa. *Biochem. J*. **114**: 689–694.

Follis, R.H., Miller, M.H., Wintrobe, M.M. and Stein, H.J. (1943). Development of myocardial necrosis and absence of nerve degeneration in thiamine deficiency in pigs. *Ani. J. Pathol*. **19**: 341–357.

Forbes, R.M. and Haines, W.T. (1952). The riboflavin requirement of the baby pig. *J. Nutr.* **47**: 411–424.

Frank, G.R., Bahr, J.M. and Easter, R.A. (1984). Riboflavin requirement of gestating swine. *J. Anim. Sci*. **59**: 1567–1572.

Frape, D.L., Speer, V.C., Hays, V.W. and Catron, D.V. (1959). The vitamin A requirement of the young pig. *J. Nutr*. **68**: 173–187.

Fraser, A.F. (1980). Farm Animal Behaviour. 2nd Edition, Bailliere Tindall, London, pp-291.

Friend, D.W. and Wolynetz, M.S. (1981). Self-selection of salt by gilts during pregnancy and lactation. *Can. J. Anim. Sci*. **61**: 429–438.

Fritschen, R.D., Grace, O.D. and Peo, E.R.Jr. (1971). Bleeding pig disease. Nebr. Swine Rep. EC71 **219**: 22–23.

Froseth, J.A., Honeyfield, D.C. and Barke, R.J. (1982a). Dietary sodium and chloride levels for young pigs. *J. Anim. Sci*. **55** (Suppl. 1):271 (Abstr.).

Fuller, M.F., McWilliam, R., Wang, T.C. and Giles, L.R. (1989). The optimum dietary amino acid pattern for growing pigs. 2. Requirements for maintenance and for tissue protein accretion. Br. *J. Nutr*. **62**: 255–267.

Glienke, L. R. and Ewan, R.C. (1977). Selenium deficiency in the young pig. *J. Anim. Sci.* **45**: 1334–1340.

Goodman, D. S. (1979). Vitamin A and retinoids: Recent advances. Fed. Proc. **38**: 2501–2503.

Goodman, D. S. (1980). Vitamin A metabolism. Fed. Proc. **39**: 2716–2722.

Goodwin, R.F.W. and Coombs, R.R.A. (1956). The blood groups of pig. IV. The A-antigen antibody system and haemolytic disease in newborn piglets. *J. Comp. Pathol*. **66**: 317–331.

Graetzer, M.A., Hesselholt, M., Moustgaard J. and Thymann, M. (1965). Studies on protein polymorphism in pigs, horses and cattle. Proc. 9th European Anim. Blood Group Con. Prague, 1964. pp. 279–288.

Grandhi, R. R. and Strain, J. H. (1980). Effect of biotin supplementation on reproductive performance and foot lesions in swine. *Can. J. Anim. Sci*. **60**: 961–969.

Grofekd, R., Sieg, B., Struckmann, C., Frenzel, A., Maxwell, W.M. and Rath, D. (2008). Institute of Farm animal genetics, Friedrich-Loeffler-institut, Federal Research Institute for animal Health (FLI), Hoeltystrasse 10, 31535 Neustadt, Germany Theriogenology.

Groce, A.W., Miller, E.R., Ullrey, D.E., Ku, P.K., Keahey, K.K. and Ellis, D.J. (1973a). Selenium requirements in corn-soy diets for growing-finishing swine. *J. Anim. Sci*. **37**: 948–956.

Groce, A.W., Miller, E.R., Hitchcock, J.P., Ullrey, D.E. and Magee, W.T. (1973b). Selenium balance in the pig as affected by selenium source and vitamin E. *J. Anim. Sci*. **37**:942–947.

Groce, A.W., Miller, E.R., Keahey, K.K., Ullrey, D.E. and Ellis, D.J. (1971). Selenium supplementation of practical diets for growing-finishing swine. *J. Anim. Sci.* **32**: 905–911.

Guilbert, H.R., Miller, R.F. and Hughes, E.H. (1937). The minimum vitamin A and carotene requirements of cattle, sheep, and swine. *J. Nutr.* **13**: 543–564.

Gustavsson, I. (1988). Standard karyotype of the domestic pig. Hereditas **109**: 151–157.

Gustavsson, I., Hageltorn, M., Zech, L. and Reiland, S. (1973). Identification of the chromosomes in a centric fusion/fission polymorphic system of the pig (Sus scrofa L.). Hereditas **75**: 153–155.

Haag, J. and Nizza, P. (1969). Le karyotype du pore normal. Ann. Genet. **12**: 242–246.

Hahn, J.D. and D.H. Baker. (1995). Optimum ratio of lysine to threonine, tryptophan, and sulphur amino acids for finishing swine. *J. Anim. Sci.* **73**: 482–489.

Hakkarainen, J., Lindberg, P., Bengtsson, G., Jonsson, G. and Lannek, N. (1978). Requirement for selenium (as selenite) and vitamin E (as alphatocopherol) in weaned pigs. III. The effect on the development of the VESD syndrome of varying selenium levels with a low- tocopherol diet. *J. Anim. Sci.* **46**: 1001–1008.

Hall, D.D., Cromwell, G.L. and Stahly, T.S. (1986). The vitamin K requirement of the growing pig. *J. Anim. Sci.* **63** (Suppl. 1): 268 (Abstr.).

Hall, D.D., Cromwell, G.L. and Stahly, T.S. (1991). Effects of dietary calcium, phosphorus, calcium: phosphorus ratio and vitamin K on performance, bone strength and blood clotting status of pigs. *J. Anim. Sci.* **69**: 646–655.

Hancock, J.E., Peo, E.R.Jr., Lewis, A.J., Crenshaw, J.D. and Moser, B.D. (1986). Vitamin D toxicity in young pigs. *J. Anim. Sci.* **63** (Suppl. 1): 268 (Abstr.).

Hansen, B.C., Lewis, A.J. and Peo, E.R.Jr. (1987). Bone traits of growing boars, barrows and gilts fed diffeıent levels of dietary protein and available phosphorus. *J. Anim. Sci.* **65**(Suppl. 1): 126 (Abstr.).

Hansen-Melander, E. and Melander, Y. (1974). The karyotype of the pig. Hereditas **77**: 149–158.

Hanson, L.E. and Hathaway, I.L. (1948). The fertility of boars fed a vitamin E deficient ration. *J. Anim. Sci.* **7**: 528 (Abstr.).

Harmon, B.G., Miller, E.R., Hoefer, J.A., Ullrey, D.E. and Luecke, R.W. (1963). Relationship of specific nutrient deficiencies to antibody production in swine. II. Pantothenic acid, pyridoxine or riboflavin. *J. Nutr.* **79**: 263–268.

Hart, E.B. and Steenbock, H. (1918). Hairless pigs: The cause and remedy. Wis. Agric. Exp. Stn. Bull. **297**: 1–11.

Hart, E.B., Elvehjem, C.A., Steenbock, P., Kemmerer, A.R., Bohstedt, G. and Fargo, J.M. (1930). A study of the anemia of young pigs and its prevention. *J. Nutr.* **2**: 277–294.

Hays, V.W. (1976). Phosphorus in Swine Nutrition. West Des Moines, Iowa: National Feed Ingredients Association.

Hays, V.W., Ashton, G.C., Liu, C.H., Speer, V.C. and Catron, D.V. (1957). Studies on the utilization of urea by growing swine. *J. Anim. Sci.* **16**: 44–54.

Heinemann, W.W., Ensminger, M.E., Cunha, T.J. and McCulloch, E.C. (1946). The relation of the amount of thiamine in the ration of the hog to the thiamine and riboflavin content of the tissue. *J. Nutr.* **31**: 107–125.

Heitman, H.Jr. and Hughes, E.H. (1949). The effects of air temperature and relative humidity on the physiological well beingof swine. *J. Anim. Sci.* **8**: 171–81.

Henderson, C.R. (1975). Rapid method for computing the inverse of a relationship matrix. *J. Dairy Sci.* **58**: 1727.

Hendricks, D.G., Miller, E.R., Ullrey, D.E., Struthers, R.D., Baltzer, B.V., Hoefer, J.A. and Luecke, R.W. (1967). â-Carotene vs. retinyl acetate for the baby pig and the effect upon ergocalciferol requirement. *J. Nutr.* **93**: 37–43.

Hentges, J.F.Jr., Grummer, R.H., Phillips, P.H. and Bohstedt, G. (1952a). The minimum requirement of young pigs for a purified source of carotene. *J. Anim. Sci.* **11**: 266–272.

Hesselholt, M. (1969). Studies on blood and serum types in Icelandic horses. Acta vet. Scand. **7**: 206–225.

Hetzer, H.O. (1948). Inheritance of coat colour in swine. VII Results of landrace by Hampshire crosses. *J. Hered.*, **39**: 123–128.

Heusner, A.A. (1982). Energy metabolism and body size. 1. is the mass exponent of Kleiber's equation a statistical artificat? Respir. Physiol. **48**: 1–12.

Hickman, D.S., Mahan, D.C. and Cline, J.H. (1983). Dietary calcium and phosphorus for developing boars. *J. Anim. Sci.* **56**: 431–437.

Hill, G.M., Miller, E.R. and Stowe, H.D. (1983b). Effect of dietary zinc levels on health and productivity of gilts and sows through two parities. *J. Anim. Sci.* **57**: 114–122.

Hill, G.M., Miller, E.R., Whetter, P.A. and Ullrey, D.E. (1983c). Concentrations of minerals in tissues of pigs from dams fed different levels of dietary zinc. *J. Anim. Sci.* **57**: 130–138

Hill, G.M., Ku, P.K., Miller, E.R., Ullrey, D.E., Losty, T.A. and O'Dell, B.L. (1983a). A copper deficiency in neonatal pigs induced by a high zinc maternal diet. *J. Nutr.* **113**: 867–872.

Hitchcock, J.P., Ku, P.K. and Miller, E.R. (1974). Factors influencing iron utilization by the baby pig. Pp. 598–600 in Trace Element Metabolism in Animals, Volume a, W.G. Hoekstra, J.W. Suttie, H.E. Ganther, and W. Mertz, eds. Baltimore: University Park Press.

Hjarde, W., Neimann-Sorensen, A., Palludan, B. and Sorensen, P.H. (1961). Investigations concerning vitamin A requirement, utilization and deficiency symptoms in pigs. Acta Agric. Scand. **11**: 13–53.

Hoekstra, W.G. (1970). The complexity of dietary factors affecting zinc nutrition and metabolism in chicks and swine. Pp. 347–353 in Trace Element Metabolism in Animals, C.F. Mills, ed. Edinburgh: E. and S. Livingstone.

Hoekstra, W.G., Faltin, E.C., Lin, C.W., Roberts, H.F. and Grummer, R.H. (1967). Zinc deficiency in reproducing gilts fed a diet high in calcium and its effect on tissue zinc and blood serum alkaline phosphatase. *J. Anim. Sci.* **26**: 1348–1357.

Hoekstra, W.G., Lewis, P.K., Phillips, P.H. and Grummer, R.H. (1956). The relationship of parakeratosis, supplemental calcium and zinc to the zinc content of certain body components of swine. *J. Anim. Sci.* **15**: 752–764.

Hojny, J. and Hala, K. (1965). A contribution to the study of the blood group system A in pigs. Proc. 9th Eur. Anim. Blood Grp. Biochem. Polymorph., Prague, 1964, pp. 155–161. pp. 151–158.

Hojny, J. and Hradecky, J. (1972). Ei-Em, the 4th closed E subsystem of blood groups in pigs. Anim. Blood group biochem. Genetic. **3**: 51–57.

Hojny, J., Gavalier, M., Hradecky, J. and Linhart, J. (1966). New blood facators in ;pigs. Proc. 10th Eur. Con. Anim. Blood group biochem. Polymorph., Paris, pp. 151–158.

Holmes, C.W. (1971). Growth and backfat dept of pigs kept at a high temperature. Anim. Prod. **13**: 521–7.

Honeyfield, D.C. and Froseth, J.A. (1985). Effects of dietary sodium and chloride on growth, efficiency of feed utilization, plasma electrolytes and plasma basic amino acids in young pigs. *J. Nutr.* **115**: 1366–1371.

Honeyfield, D.C., Froseth, J.A. and Barke, R.J. (1985). Dietary sodium and chloride levels for growing-finishing pigs. *J. Anim. Sci.* **60**: 691–698.

Horst, R.L., Napoli, J.L. and Littledike, E.T. (1982). Discrimination in the metabolism of orally dosed ergocalciferol and cholecalciferol by the pig, rat, and chick. *Biochem. J.* **204**:185–189.

Hsu, J. and Benirschke, K. (1967). An atlas of mammalian chromosomes vol. 1. Folio 39. Springer-Verlag, New York.

Hughes, E.H. (1940a). The minimum requirement of riboflavin for the growing pig. *J. Nutr.* **20**: 233–238.

Hughes, E.H. (1940b). The minimum requirement of thiamine for the growing pig. *J. Nutr.* **20**: 239–241.

Hughes, E.H. and Ittner, N.R. (1942). The minimum requirement of pantothenic acid for the growing pig. *J. Anim. Sci.* **1**: 116–119.

Hughes, E.H. and Ittner, N. R. (1942). The potassium requirement of growing pigs. *J. Agric. Res.* **64**: 189–192.

Hus, S. (1967). Av Hayvanlari ve Avcilik. Kutulmus Matbaasi, Istanbul.

Imlah , P. (1964). A study of blood groups in pigs. Proc 9th European Animal Blood Group Conference. PP 109–122.

Imlah, P. (1964). Inherited variants in serum ceruloplasmins of the pig. Nature, London **203**: 658–659.

Info Paks: Republic of South Africa, Department of Agriculture. www.nda.agric.za/publications

Jensen, A. H., Terrill, S. W. and Becker, D. E. 1961. Response of the young pig to levels of dietary potassium. *J. Anim. Sci.* **20**: 464–467.

Jeppesen, L.E. (1981). An artificial sow to investigate the behaviour of suckling piglets. Appl. Anim. Ethol., **7**: 359–367.

Johanson, I. and Rondel, J. (1968). Genetic and animal breeding O.W.H. Freeman and company, Sanfancsiso.

Johnson, B.C. and James, M.F. 1948. Choline deficiency in the baby pig. *J. Nutr.* **36**: 339–344.

Jollans, J.L. (1959). A preliminary report onthe indigenous pigs of ashanti, Ghans. *J. West African Sci. Asso.* **5**: 133–145.

Jones, G.F. (1998). Genetic aspects of domestication, common breeds and their origin, pp. 17–50 in *The Genetics of the Pig*, edited by A. Ruvinsky and M.F. Rothschild. CAB International, Oxon, UK.

Jongbloed, A.W. (1987). Phosphorus in the Feeding of Pigs: Effect of Diet on the Absorption and Retention of Phosphorus by Growing Pigs. Instituut voor Veevoedingsanderzoek. Lelystad. XVI, 343 pp.

Jorge, L.A. and Ana C.V.S. (1993). Rev. Brasil. Genet. 16, 3, 653–659.

Kalkus, J.W. (1920). A study of goiter and associated conditions in domestic animals. Wash. *Agric. Exp. Stn. Bull.* **156**: 1–48.

Kanadkhedkar, H.L., Nehete, S.B., Suryawanshi, A.R. and Umrikar, U.D. (2006). The Journal of Bombay Veterinary college year 2006, Col. 14, Issue: 1–2.

Kernkamp, H. C.H. and Ferrin, E.F. (1953). Parakeratosis in swine. *J. Anim. Vet. Med. Assoc.* **123**: 217–220.

Kesel, G. A., Knight, J.W., Kornegay, E.T., Veit, J.P. and Notter, D.R. (1983). Restricted energy and elevated calcium and phosphorus intake for boars during growth. 1. Feedlot performance and bone characteristics. *J. Anim. Sci.* **57**: 82–98.

King, J.W.B. (1991). Pig breeds of the world: their distribution and adaption. In: Genetic Resources of Pigs, Sheep and Goats, (ed. K. Maijala), pp 51–62. Elsevier: Amsterdam.

Kloster, G.F., Larsen, B. and Neilson, P.B. (1970). Carbonic anhydrasee polymorphism in cattle and seine. Acta Vet. Scand. **11**: 318–321.

Knorr, Peters, Henne, Woerner, Harlizius, Brenig (2002). A total genome scan and analysis of candidate genes for scrotal hernia in pigs. Abstract presented at the ISAG 2002 meeting, Gottingen, Germany.

Koh. F.K. (1952). Swine production in Taiwan. In bulletin no. 17 economic research lab. bank of Taiwan. pp 50–56.

Koch, M. E., Mahan, D. C. and Corley, J. R. (1984). An evaluation of various biological characteristics in assessing low phosphorus intake in weanling swine. *J. Anim. Sci.* **59**: 1546–1556.

Kormann, A.W. and Weiser, H. (1984). Protective functions of fat-soluble vitamins. Pp. 201–222 in Proc. 37th Nottingham Feed Manufacturer's Conference, Nottingham, England. London: Butterworth.

Kornegay, E.T. (1985). Calcium and Phosphorus in Animal Nutrition. Pp. 1–106 in Calcium and Phosphorus in Animal Nutrition. West Des Moines, Iowa: National Feed Ingredients Association.

Kornegay, E.T. and Thomas, H.R. (1981). Phosphorus in swine. II. Influence of dietary calcium and phosphorus levels and growth rate on serum minerals, soundness scores and bone development in barrows, gilts and boars. *J. Anim. Sci.* **52**: 1049–1059.

Kornegay, E.T., Diggs, B.G., Hale, O.M., Handlin, D.L., Hitchcock, J.P. and Barczwski, R.A. (1984). Reproductive performance of sows fed elevated calcium and phosphorus levels during growth and development. A cooperative study. Report S-145 of the Committee on Nutritional Systems for Swine to Increase Reproductive Efficiency. *J. Anim. Sci.* **59** (Suppl. 1): 253 (Abstr.).

Kornegay, E.T., Miller, E.R., Ullrey, D.E., Vincent, B.H. and Hoefer, J.A. (1965). Influence of dietary urea on performance, antibody production and hematology of growing swine. *J. Anim. Sci.* **24**: 951–954.

Kornegay, E.T., Lindemann, M.D. and Bartlett, H.S. (1991). The influence of sodium supplementation of two phosphorus sources on performance and bone mineralization of growing-finishing swine evaluated at two geographical locations. *Can. J. Anim. Sci.* **71**: 537–547.

Kösters, W.W. and Kirchgessner, M. (1976a). Gewichtenwicklung und Futterverwertung Friiheutwijhnter Ferkel bei untersehiedlicher Vitamin B6-Versorgung. (Growth rate and feed efficiency of early-weaned piglets with varying vitamin B6 supply.) Z. Tierphysiol. Tierernahr. Futtermittelkd. **37**: 235–246.

Kösters, W.W. and Kirchgessner, M. (1976b). Zur Veranderugn des Futterverzeha Friiheutwohnter Ferkel bei Untersehiedlicher Vitamin BeVersorgung. (Change in feed intake of early-weaned piglets in response to different vitamin B6 supply.) Z. Tierphysiol. Tierernahr. Futtermittelkd. **37**: 247–254.

Krider, J.L., Albright, J.L., Plumlee, M.P., Conrad, J.H., Sinclair, C.L., Underwood, L., Jones, R.G and Harrington, R.B. (1975). Magnesium supplementation, space and docking effects on swine performance and behavior. *J. Anim. Sci.* **40**: 1027–1033.

Krider, J.L., Terrill, S.W. and VanPoucke, R.F. (1949). Response of weanling pigs to various levels of riboflavin. *J. Anim. Sci.* **8**: 121–125.

Kristjansson, F.K. (1960). Genetic control of two blood serum proteins in swine. *Can J. Genet. Cytol.* **2**: 295–300.

Kristjansson, F.K. (1961). Genetic control of three haptoglobins in pigs Genetics **46**: 907–910.

Kristjansson, F.K. (1966). Fractionation of serum albumin and genetic control of two albumin fractions in pigs. Genetics **52**: 627.

Kristjansson, F.K. and Cipera J.D. (1963). The effect of sialidase on pig transferrins. Can. J. Biochem. Physiol. **41**: 2523–2527.

Kristjansson, F.K. (1961). Genetic control of three heptaglobins in pigs. Genetics **46**: 907–910.

Ku, P.K., Ullrey, D.E. and Miller, E.R. (1970). Zinc deficiency and tissue nucleic acid and protein concentration. Pp. 158–164 in Trace Element Metabolism in Animals, C.F. Mills, ed. Edinburgh: E. and S. Livingstone.

Ku, P.K., Ely, W.T., Groce, A.W. and Ullrey, D.E. (1973). Natural dietary selenium, Á-tocopherol and effect on tissue selenium. *J. Anim. Sci.* **37**: 501–505.

Kuhajek, E.J. and Andelfinger, G.F. (1970). A new source of iodine for salt blocks. *J. Anim. Sci.* **31**: 51–58.

Kumerloeve, H. (1978). Tukiyenin memeli hayvanlari. Istanbul Universitiesi Orman Fakultesi Dergisi, Seri. B. **28**: 178–204.

Kyriazakis, I., Emmans, G.C. and McDaniel, R. (1993). Whole body amino acid composition of the growing pig. *J. Sci. Food Agric.* **62**: 29–33.

Leach, R.M. Jr. and Muenster, A.M. (1962). Studies on the role of manganese in bone formation. 1. Effect upon the mucopolysaccharide content of chick bone. *J. Nutr.* **78**: 51–56.

Lehrer, W.P.Jr., Wiese, A.C. and Moore, P.R. (1952). Biotin deficiency in suckling pigs. *J. Nutr.* **47**: 203–212.

Lehrer, W.P.Jr. and Wiese, A.C. (1952). Riboflavin deficiency in baby pigs. *J. Anim. Sci.* **11**: 244–250.

Lewis, A.J., Cromwell, G.L. and Pettigrew, J.E. (1991). Effects of supplemental biotin during gestation and lactation on reproductive performance of sows: A cooperative study. *J. Anim. Sci.* **69**: 207–214.

Lindemann, M.D. and Kornegay, E.T. (1986). Folic acid additions to weanling pig diets. *J. Anim. Sci.* **63**(Suppl. 1): 35 (Abstr.).

Lindemann, M.D. and Kornegay, E.T. (1989). Folic acid supplementation to diets of gestating-lactating swine over multiple parities. *J. Anim. Sci.* **67**: 459–464.

Lindley, D.C. and Cunha, T.J. (1946). Nutritional significance of inositol and biotin for the pig. *J. Nutr.* **32**: 47–59.

Liptrap, D.O., Miller, E.R.,Ullrey, D.E., Whitenack, D.L., Schoepke, B.L. and Luecke, R.W. (1970). Sex influence on the zinc requirement of developing swine. *J. Anim. Sci.* **30**: 736–741.

Livingston, A.L., Nelson, J.W. and Kohler, G.O. (1968). Stability of alpha-tocopherol during alfalfa dehydration and storage. *J. Agric. Food Chem.* **16**: 492–495.

Louca, A. and Robinson, O.W. (1967). Components of variance and covariance in purebred and crossbred swine. *J. Anim. Sci.* **26**: 267-273.

Lowry, K.R., Mahan, D.C. and Corley, J.R. (1985a). Effect of dietary calcium on selenium retention in postweaning swine. *J. Anim. Sci.* **60**: 1429–1437.

Lowry, K.R., Mahan, D.C. and Corley, J.R. (1985b). Effect of dietary phosphorus on selenium retention in postweaning swine. *J. Anim. Sci.* **60**: 1438–1446.

Luecke, R. W., Hoefer, J. A. and Thorpe, F. Jr. (1952). The relationship of protein to pantothenic acid and vitamin B12 in the growing pig. *J. Anim. Sci.* **11**: 238–243.

Luecke, R.W., Hoefer, J.A. and Thorpe, F.Jr. (1953). The supplementary effects of calcium pantothenate and aureomycin in a low-protein ration for weanling pigs. *J. Anim. Sci.* **12**: 605–610.

Luecke, R.W., Hoefer, J.A., Brammell, W.S. and Schmidt, D.A. (1957). Calcium and zinc in parakeratosis of swine. *J. Anim. Sci.* **16**: 3–11.

Luecke, R.W., McMillen, W.N. and Thorpe, F.Jr. (1950). Further studies of pantothenic acid deficiency in weanling pigs. *J. Anim. Sci.* **9**: 78–82.

Luecke, R.W., McMillen, W.N., Thorpe, F.Jr., and Tull, C. (1948). Further studies on the relationship of nicotinic acid, tryptophane and protein in the nutrition of the pig. *J. Nutr.* **36**: 417–424.

Lylian Rodríguez and Preston, T.R. (1997). Local feed resources and indigenous breeds: fundamental issues in integrated farming systems. Livestock Research for Rural Development (9) **2**: 92–99.

Lynch, P.B., Hall, G.E., Hill, L.D., Hatfield, E.E. and Jensen, A.H. (1975). Chemically preserved high-moisture corns in diets for growing-finishing swine. *J. Anim. Sci.* **40**: 1063–1069.

Maak, S., Jaesert S., Neumann, K. and von Lengerken, G. (2003). Characterization of the porcine CDKN3 gene as a potential candidate for congenital splay leg in piglets. Genet Sel Evol. **35**: 157–65.

Macchi, E., Tarantola, M., Perrone, A., Paradiso, M.C. and Ponzio, G. (1995). Cytogenetic variability in the wild boar (Sus scrofa scrofa) in Piedmont (Italy): Preliminary data. *Ibex J. of Mountain Ecology* **3**: 17–18.

MaConnel, J., Fecheimer, N. and Gilmore, L.O. (1963). Somatic chromosomes of the domestic pig. *J. Anim. Sci.* **22**: 374–379.

Madsen, A., Mortensen, H. P., Hjarde, W., Leebeck, E. and Leth, T. (1973). Vitamin E in barley treated with propionic acid with special reference to the feeding of bacon pigs. Acta Agric. Scand. Suppl. **19**: 169–173.

Mahan, D.C. (1982). Dietary calcium and phosphorus levels for weanling swine. *J. Anim. Sci.* **54**: 559–564.

Mahan, D. (1991). Assessment of the influence of dietary vitamin E on sows and offspring in three parities: Reproductive performance, tissue tocopherol, and effects on progeny. *J. Anim. Sci.* **69**: 2904–2917.

Mahan, D.C. (1994). Effects of dietary vitamin E on sow reproductive performance over a five-parity period. *J. Anim. Sci.* **72**: 2870–2879.

Mahan, D., Lepine, A.J. and Dabrowski, K. (1994). Efficacy of magnesium-L-ascorbyl-2-phosphate as a vitamin C source for weanling and growing-finishing swine. *J. Anim. Sci.* **72**: 2354–2361.

Mahan, D.C., Moxon, A.L. and Hubbard, M. (1977). Efficacy of inorganic selenium supplementation to sow diets on resulting carry-over to their progency. *J. Anim. Sci.* **45**: 738–746.

Mahan, D.C. and Moxon, A.L. (1978a). Effect of adding inorganic or organic selenium sources to the diets of young swine. *J. Anim. Sci.* **47**: 456–466.

Mahan, D.C. and Moxon, A.L. (1978b). Effect of increasing the level of inorganic selenium supplementation in the postweaning diets of swine. *J. Anim. Sci.* **46**: 384–390.

Mahan, D.C. and Moxon, A.L. (1984). Effect of inorganic selenium supplementation on selenosis in postweaning swine. *J. Anim. Sci.* **58**: 1216–1221.

Mahan, D.C. and Parrett, N.A. (1996). Evaluating the efficacy of seleniumenriched yeast and sodium selenite on tissue selenium retention and serum glutathione peroxidase activity in grower and finisher diets. *J. Anim. Sci.* **74**: 2967–2974.

Mahan, D.C. and Magee, P.L. (1991). Efficacy of dietary sodium selenite and calcium selenite provided in the diet at approved, marginally toxic, and toxic levels to growing swine. *J. Anim. Sci.* **69**: 4722–4725.

Mahan, D.C. and Kim, Y.Y. (1996). Effect of inorganic selenium at two dietary levels on reproductive performance and tissue selenium concentrations in first parity gilts and their progeny. *J. Anim. Sci.* **74**: 2711–2718.

Mahan, D.C., Newton, E.A. and Cera, K.R. (1996a). Effect of supplemental sodium chloride, sodium phosphate, or hydrochloric acid in starter pig diets containing dried whey. *J. Anim. Sci.* **74**: 1217–1222.

Mahan, D.C., Weaver, E.M. and Russell, L.E. (1996b). Improved postweaning pig performance responses by adding NaCl or HCl to diets containing animal plasma. *J. Anim. Sci.* **74** (Suppl. 1): 58 (Abstr).

Mahan, D.C., Jones, J.E., Cline, J.H., Cross, R.F. ,Teague, H.S. and Crifo, A.P.Jr. (1973). Efficacy of selenium and vitamin E injections in the prevention of white muscle disease in young swine. *J. Anim. Sci.* **36**: 1104–1108.

Mahan, D.C., Ekstrom, K.E. and Fetter, A.W. (1980). Effect of dietary protein, calcium and phosphorus for swine from 7 to 20 kilograms body weight. *J. Anim. Sci.* **50**: 309–314.

Mahan, D.C., Pickett, R.A., Perry, T.W., Curtin, T.M. ,Featherston, W.R. and Beeson, W.M. (1966). Influence of various nutritional factors and physical form of feed on esophagogastric ulcers in swine. *J. Anim. Sci.* **25**:1019–1023.

Malm, A., Pond, W.G., Walker, E.F.Jr., Homan, M., Aydin, A and Kirtland, D. (1976). Effect of polyunsaturated fatty acids and vitamin E level of the sow gestation diet on reproductive performance and on level of alpha-tocopherol in colostrum, milk and dam and progeny blood serum. *J. Anim. Sci.* **42**: 393–399.

Manners, M.J. and McCrea, M.R. (1964). Estimates of the mineral requirements of 2-day weaned piglets derived from data on mineral retention by sow-reared piglets. Ann. Zootechnol. **13**:29–38.

Mathur, K.K. (1967). Nicobar Islands. National Book Trust, New Delhi.

Matte, J.J., Girard, C.L. and Brisson, G.J. (1984a). Folic acid and reproductive performance of sows. *J. Anim. Sci.* **59**: 1020–1025.

Matte, J.J., Girard, C.L. and Brisson, G.J. (1984b). Serum folates during the reproductive cycle of sows. *J. Anim. Sci.* **59**: 158–163.

Matte, J.J., Girard, C.L. and Brisson, G.J. (1992). The role of folic acid in the nutrition of gestating and lactating primaparous sows. Livestock Prod. Sci. **32**: 131–148.

Maxson, P.F. and Mahan, D.C. (1983). Dietary calcium and phosphorus levels for growing swine from 18 to 57 kilograms body weight. *J. Anim. Sci.* **56**: 1124–1134.

Mayer, J.J. and Brisbin, I.L. (1991). Wild pigs inthe United States. Their history, comparative morphology, and current status. The university of Georgia press, Athens and London.

Mayo, R.H., Plurnlee, M.P. and Beeson, W.M. (1959). Magnesium requirement of the pig. *J. Anim. Sci.* **18**: 264–273.

McBridge, G. (1963). The 'teat order' and communication in young pigs. Anim. Behav. **11**: 53–56.

McDonaled, P., Edwards, R.A. and Green Lalgh, J.F.D. (1988). Animal Nutrition, 4th edn. Longman, London.

McFee, A.F., Banner, M.W. and Rary, J.M. (1966). Variation in chromosome number among European wild pigs. Cytogenetics **5**: 75–81.

McGlone, J.J. (2001). Farm animal welfare in the context of other society issues: toward sustainable systems. Livst. Prod. Sci. **72**: 75–81.

McLaren, D.G. (1990). Potential of Chinese pig breeds to improve pork production efficiency in the USA. Animal Breed. Abstr., **58**: (5), 347–369.

Meade, R.J., Hanson, L.E., Hanke, H.E., Miller, K.P., Rust, J.W., Grant, R.S. and Tumbleson. M.E. (1969). B-vitamin supplementation of conventional diets for growing swine. Minnesota Agricultural Experiment Station Technical Bulletin 263. St. Paul: University of Minnesota Press.

Mellink, C.H.M., Bosma, A.A., Haan, N.A. and Wiegant, J. (1991). Distribution of rRNA genes in breeds of domestic pig studied by non-radioactive in situ hybridization and selective silver-staining. Genet. Sel. Evol. **23** (Suppl. 1): 169–172.

Meyer, J.H., Grummer, R.H., Phillips, P.H. and Bohstedt, G. (1950). Sodium, chlorine, and potassium requirements of growing pigs. *J. Anim. Sci.* **9**: 300–306.

Meyer, J.N. and Verhorst, D. (1973). The evidence of erythrocyte acid phposphatase by starch-gel electrophoresis. Anim. Blood Group Biochem. Genetic **4**: 129–131.

Meyer, W.R., Mahan, D.C. and Moxon, A.L. (1981). Value of dietary selenium and vitamin E for weanling swine as measured by performance and tissue selenium and glutathione peroxidase activities. *J. Anim. Sci.* **52**: 302–311.

Michel, R.L., Whitehair, C.K. and Keahey, K.K. (1969). Dietary hepatic necrosis associated with selenium-vitamin E deficiency in swine. *J. Anim. Vet. Med. Assoc.* **155**: 50–59.

Miller, C.O. and Ellis, N.R. (1951). The riboflavin requirement of growing swine. *J. Anim. Sci.* **10**: 807–812.

Miller, C.O., Ellis, N.R., Stevenson, J.W. and Davey, R. (1953). The riboflavin requirement of swine for reproduction *J. Nutr.* **51**: 163–170.

Miller, E.R. (1980). Bioavailability of minerals. P. 144 in Proc. Minnesota Nutrition Conference. St. Paul: University of Minnesota Press.

Miller, E.R., Schmidt, D.A., Hoefer, J.A. and Luecke, R.W. (1955). The thiamine requirement of the baby pig. *J. Nutr.* **56**: 423–430.

Miller, E.R., Schmidt, D.A., Hoefer, J.A. and Luecke, R.W. (1957). The pyridoxine requirement of the baby pig. *J. Nutr.* **62**: 407–419.

Miller, E.R., Ullrey, D.E., Zutaut, C.L., Baltzer, B.V., Schmidt, D.A., Vincent, B.H., Hoefer, J.A. and Luecke, R.W. (1964). Vitamin D2 requirement of the baby pig. *J. Nutr.* **83**: 140–148.

Miller, E.R., Ullrey, D.E., Zutaut, C.L., Hoefer, J.A. and Luecke, R.W. (1965). Comparisons of casein and soy proteins upon mineral balance and vitamin D2 requirement of the baby pig. *J. Nutr.* **85**: 347–353.

Miller, E.R., Ullrey, D.E., Zutaut, C.L., Hoefer, J.A. and Luecke, R.W. (1965c). Mineral balance studies with the baby pig: Effects of dietary magnesium level upon calcium, phosphorus, and magnesium balance. *J. Nutr.* **86**: 209–212.

Miller, E.R., Ullrey, D.E., Zutaut, C.L., Hoefer, J.A. and Luecke, R.W. (1965d). Mineral balance studies with the baby pig: Effects of dietary vitamin D2 level upon calcium, phosphorus, and magnesium balance. *J. Nutr.* **85**: 255–258.

Miller, E.R., Liptrap, D.O. and Ullrey, D.E. (1970). Sex influence on zinc requirement of swine. Pp. 377–379 in Trace Element Metabolism in Animals, C.F. Mills, ed. Edinburgh: E. and S. Livingstone.

Miller, E.R. ,Waxler, G.L., Ku, P.K., Ullrey, D.E. and Whitehair, C.K. (1982). Iron requirements of baby pigs reared in germ-free or conventional environments on a condensed milk diet. *J. Anim. Sci.* **54**: 106–115.

Miller, E.R., Stowe, H.D., Ku, P.K. and Hill, G.M. (1979). Copper and zinc in swine nutrition. P. 109 in National Feed Ingredients Association Literature Review on Copper and Zinc in Animal Nutrition. West Des Moines, Iowa: National Feed Ingredients Association.

Miller, E.R., Luecke, R.W., Ullrey, D.E., Baltzer, B.V., Bradley, B.L. and Hoefer, J.A. (1968). Biochemical, skeletal and allometric changes due to zinc deficiency in the baby pig. *J. Nutr.* **95**: 278–286.

Mitchell, H.H., Johnson, B.C., Hamilton, T.S. and Haines, W.T. (1950). The riboflavin requirement of the growing pig at two environmental temperatures. *J. Nutr.* **41**: 317–337.

Mohr, E. (1960). Wild schweine. A. Ziemsen Verlag-Witenberg Lutherstadt, Leipzig.

Morgan, C.A., Beilsen, B.L., Lawrence, A.B. and Mendl, M.T. (1998). Describing the social environment and it's effect on food intake and growth. In: A Quantitative Bology of the pig (I. Kyriazakis, edn.) CABI Publishing, Wallingford, Oxon, UK, pp. 99–125.

Morgan, C.A., Emmans, G.C., Tolkamp, B.J., Kyriazakis, I. (2000).Analysis of the feeding behavior of pigs using different models. Physiol Behav. **68**: 395–403.

Moser, B.D. and Lewis, A.J. (1980). Adding fat to sow diets. Feedstuffs **52**: 36–37.

Mraz, F.R., Johnson, A.M. and Patrick, H. (1958). Metabolism of cesium and potassium in swine as indicated by cesium-134 and potassium-42. *J. Nutr.* **64**: 541–548.

Muhrer, M.E., Cooper, R.G., Cornell, C.N. and Thomas, R.D. (1970). Diet related hemorrhagic syndrome in swine. *J. Anim. Sci.* **31**: 1025 (Abstr.).

Mukherjee, T.K. (1980). Animal genetic resources in Malaysia. In proceeding of SABRO workshop on animal genetic resources in Asia and Oceania, held at Tsukub science city, Japan. pp 201–311.

Muramoto, J., Mahino, S. Ishikawa, T. and Kanagawa, H. (1965). On the chromosomes of the wild boar and the boar pig hybrids. Proc. Japanese Acad. **41**: 236/0239.

Mursaloglu, B. (1964). Turkiye nin azalan memelileri hakkinda. Turk Biologi Dergisi **14**: 65–70.

Na Puket, S.R. (1980). Animal genetic respurces in Thailand. In proceeding of SABRAO workshop on animal genetic resources in Asia and Oceania, held at Tsukub science city, Japan. pp 313–338.

NAHMS. (2001). Part I: Reference of swine health and management in the United States, 2000, National Animal Health Monitoring System. N338.0801. Fort Collins, CO.

National Research Council. (1980). Mineral Tolerance of Domestic Animals. Washington, DC: National Academy Press. 577 pp.

National Research Council. (1988). Nutrient Requirements of Swine. Ninth Edition. Washington, D.C.: National Academy Press. 93 pp.

National Research Council. (1997). The Role of Chromium in Animal Nutrition. Washington, DC: National Academy Press. 80 pp.

Nelson, E.C., Dehority, B.A., Teague, H.S., Grifo, A.P.Jr. and Sanger, V.L. (1964). Effect of vitamin A and vitamin A acid on cerebrospinal fluid pressure and blood and liver vitamin A concentration in the pig. *J. Nutr.* **82**: 263–268.

Nelson, E.C., Dehority, B.A., Teague, H.S., Sanger, V.L. and Pounden, W.D. (1962). Effect of vitamin A on some biochemical and physiological changes in swine. *J. Nutr.* **76**: 325–332.

Neumann, A.L. and Johnson, B.C. (1950). Crystalline vitamin B12 in the nutrition of the baby pig. *J. Nutr.* **40**:403–414.

Neumann, A.L., Thiersch, J.B., Krider, J.L., James, M.F. and Johnson, B.C. (1950). Requirement of the baby pig for vitamin B12 fed as a concentrate. *J. Anim. Sci.* **9**: 83–89.

Neumann, A.L., Krider, J.L., James, M.R. and Johnson, B.C. (1949). The choline requirement of the baby pig. *J. Nutr. 38*: 195–214.

Newland, H.W., McMillen, W.N. and Reincke, E.P. (1952). Temperature adaptation in the baby pig. *J. Anim. Sci.* **11**: 118–33.

Nielsen, F.H. (1984). Ultratrace elements in nutrition. Annu. Rev. Nutr. **4**:21–41.

Nielsen, H.E. Danielsen, V. Simesen, M.G. Gissel-Nielsen, C. Hjarde, W. Leth, T. and Basse, A. (1979). Selenium and vitamin E deficiency in pigs. I. Influence on growth and reproduction. Acta Vet. Scand. **20**: 276–288.

Nimmo, R.D., Peo, E.R.Jr., Moser, B.D and Lewis, A.J. (1981b). Effect of level of dietary calcium-phosphorus during growth and gestation on performance, blood, and bone parameters of swine. *J. Anim. Sci.* **52**: 1330–1342.

Nimmo, R.D., Peo, E.R.Jr., Crenshaw, J.D., Moser, B.D. and Lewis, A.J. (1981a). Effects of level of dietary calcium-phosphorus during growth and gestation on calcium-phosphorus balance and reproductive performance of first-litter sows. *J. Anim. Sci.* **52**: 1343–1349.

Noblet, J. and Etienne, M. (1989). Estimation of sow milk nutrient output. *J. Anim. Sci.* **67**: 3352–3359.

Noblet, J., Fortune, H., Shi, X.S. and Dubois, S. (1994). Prediction of net energy value of feeds for growig pigs. *J. Anim. Sci.* **72**: 344–354.

Noblet, J. ,Dourmad, J.Y. and Etienne, M. (1990). Energy utilization in pregnant and lactating sows. Modeling of energy requirements. *J. Anim. Sci.* **68**: 562–572.

Noblet, J., Dourmad, J.Y., Le Dividich, J. and Dubois, S. (1989b). Effect of ambient temperature and addition of straw or alfa-alfa in the diet on energy metabolism of pregnant sows. Livestock Prod. Sci. **21**: 309–324.

Noblet, J., Shi, X.S. and Dobois, S. (1994). Effect of body weight on net energy value of feeds for growing pigs. *J. Anim. Sci.* **72**: 645–657.

Nockels, C.F. (1979). Protective effects of supplemental vitamin E against infection. Fed. Proc. **38**: 2134–2138.

Nordskog, A.W. ,Comstock, R.E and Winters, L.M. (1944). Hereditary and Environmental Factors Affecting Growth Rate in Swine. *J. Anim Sci.* 1944. **3**: 257–272.

Nuoranne, P.J., Raunio, R.P., Saukko, P. and Karppanen, H. (1980). Metabolic effects of a low-magnesium diet in pigs. *Br. J. Nutr.* **44**: 53–60.

Oliver, W.L.R. (1984). Introduced and feral pigs. In feral mammals problems and potentials. Proc. of the workshop on feral mammals at the 3rd Int. Theriol. Cong. Helsinki, 1982; IUCN, Gland: 87–126.

Oishi, T. and Abe, T. (1970). Studies on blood groups of pigs VI usefulness of blood groups and serum protein types for parentages test Jap. *J. Zootech. Sci.* **41**: 501–506.

Oishi, T. and Tomita, T. (1976). Blood groups and serum protein polymorphisms in the Pitman-Moore and Ohmini strains of miniature pigs. Anim. Blood groups biochem. Genet. **7**: 27–37.

Oishi, T., Esaki, K. and Tomita, T. (1980). Genetic relationship among Gottingen miniatuare, European and East Asian pigs investigated from blood groups and biochemical polymorphism Jap. J. Zootech. Sci. **51**: 226–228.

Okonkwo, A.C., Ku, P.K., Miller, E.R., Keahey, K.K. and Ullrey, D.E. (1979). Copper requirement of baby pigs fed purified diets. *J. Nutr.* **109**: 939–948.

Okumura, N., Ishiguro, N., Nakano, M. and Hirai, K. (1996). Geographic population structure and sequence divergence in the mitochondrial DNA control region of the Japanese wild boar (Sus scrofa leucomystax), with reference to those of domestic pigs. Biochem. Genet. **34**: 179–189.

Omtvedt, I.T., Nelson, R.E., Edwards, R.L., Stephens, D.F. and Turman, E.J. (1971). Influence of heat stress during early, mid and late pregnancy of gilts. *J. Anim. Sci.* **32**: 312–317.

Osborne, J.C. and Davis, J.W. (1968). Increased susceptibility to bacterial endotoxin of pigs with iron deficiency anemia. *J. Anim. Vet. Med. Assoc.* **152**: 1630–1632.

Osweiler, G.D. (1970). Porcine hemorrhagic disease. No. AS3531 in Proc. Pork Producers Day. Ames, IA. Iowa State Univ. Press.

Palm, B.W. ,Meade, R.J. and Melliere, A.L. (1968). Pantothenic acid requirement of young swine. *J. Anim. Sci.* **27**: 1596–1601.

Paszek, A.A., Flickinger, G.H., Fontanesi, L., Beattie, G.A. (1998). Evaluating evolutionary divergence with microsatellites. *J. Mol. Evol.* **46**: 121–126.

Peo, E.R.Jr. (1976). Calcium in Swine Nutrition. West Des Moines, IA: National Feed Ingredient Association. 65 pp.

Peo, E.R.Jr. (1991). Calcium, phosphorus, and vitamin D in swine nutrition. Pp. 165–182 in Swine Nutrition, E.R. Miller, D.E. Ullrey, and A.J. Lewis, eds. Stoneham, ME: Butterworth-Heinemann Publishing.

Peo, E.R.Jr., Libal, G.W., Wehrbein, G.F., Cunningham, P.J. and Vipperman, P.E.Jr. (1969). Effect of dietary increments of calcium and phosphorus on G-F swine. *J. Anim. Sci.* **29**: 141(Abstr.).

Peplowski, M.A., Mahan, D.C., Murray, F.A., Moxon, A.L., Cantor, A. H. and Ekstrom, K. E. (1980). Effect of dietary and injectable vitamin E and selenium in weanling swine antigenically challenged with sheep red blood cell. *J. Anim. Sci.* **51**: 344–351.

Perez, R. (1997). Feeding pigs in the tropics. FAO animal production and health paper 132, FAO. Rome.

Pettigrew, J.E. (1993). Amino acid nutrition of gestating and lactating sows. BioKyowa Technical Review–5. Chesterfield, MO: Nutri-Quest.

Pettigrew, J.E.Jr and Moser, R.L. (1991). Fat in swine nutrition. Pp. 133–146 in swine nutrition, E.R. Miller, D.E. Ullrey, and A.J. Lewis, eds. Stoneham, U.K. Butterworth Heinemann.

Phillips, R.W., Johnson, R.G. and Moyer, R.T. (1945). The livestock of China. Pub. 2249 (Far East Series 9). Dept. State Washington.

Piatkowski, T.L., Mahan, D.C., Cantor, A.H., Moxon, A.L., Cline, J.H. and Grifo, A.P.Jr. (1979). Selenium and vitamin E in semipurified diets for gravid and nongravid gilts *J. Anim. Sci.* **48**: 1357–1365.

Pickett, R.A., Plumlee, M.P., Smith, W.H. and Beeson, W.M. (1960). Oral iron requirement of the early-weaned pig. *J. Anim. Sci.* **19**: 1284. (Abstr.).

Plumlee, M.P., Thrasher, D.M., Beeson, W.M., Andrews, F.N. and Parker, H.E. (1956). The effects of a manganese deficiency upon the growth, development and reproduction of swine. *J. Anim. Sci.* **15**: 352–368.

Pond, W.G. and Jones, J.R. (1964). Effect of level of zinc in high-calcium diets on pigs from weaning through one reproductive cycle and on subsequent growth of their offspring. *J. Anim. Sci.* **23**: 1057–1060.

Pond, W.G., Kwong, E. and Loosli, J.K. (1960). Effect of level of dietary fat, pantothenic acid, and protein on performance of growing-fattening swine. *J. Anim. Sci.* **19**: 1115–1122.

Pond, W.G., Lowrey, R.S., Maner, J.H. and Loosli, J.K. (1961). Parenteral iron administration to sows during gestation or lactation. *J. Anim. Sci.* **20**: 747–750.

Porter, V. (1993). Pigs A handbook to the breeds of the world. Helm information: Mountfield.

Prasad, A.S., Oberleas, D., Miller, E.R. and Luecke, R.W. (1971). Biochemical effects of zinc deficiency: Changes in activities of zinc-dependent enzymes and ribonucleic acid and deoxyribonucleic acid content of tissues. *J. Lab. Clin. Med.* **77**: 144–152.

Prasad, A.S., Oberleas, D., Wolf, P., Horwitz, J.P., Miller, E.R. and Luecke, R.W. (1969). Changes in trace elements and enzyme activities in tissues of zinc-deficient pigs. *Anim. J. Clin. Nutr.* **22**: 628–637.

Qian, H., Kornegay, E.T and Conner, D.E.Jr. (1996). Adverse effects of wide calcium:phosphorus ratios on supplemental phytase efficacy for weanling pigs fed two dietary phosphorus levels. *J. Anim. Sci.* **74**: 1288–1297.

Quintanilla, R., Milan, D. and Bidanel, J.P. (2002). A further look at quantitative trait loci affecting growth and fatness in across between Meishan and Large White pig populations. Genet Sel Evol. **34**: 193–210.

Ramisz, A., Balicka-Laurans, A. and Ramisz, G. (1993). The influence of selenium on production, reproduction and health in pigs. Advances Agri. Sci. **2**: 67.

Rary, M.J Vernon, G.H. and Matschke, G.H. (1968). The cytogenetics of swine inthe Tellico wildlife management area, Tennessee. *The J. of Heredity* **59**: 201–204.

Rasmusen B.A. (1965). Eecf (E) 6th allele at the E blood group locus in Yorkshire pigs voz. Sang. **10**: 242–245.

Reinhart, G. A and Mahan, D. C. (1986). Effect of various calcium: phosphorus ratios at low and high dietary phosphorus for starter, grower and finisher swine. *J. Anim. Sci.* **63**: 457–466.

Richardson, D., Catron, D.V., Underkofler, L.A., Maddock, H.M. and Friedland, W.C. (1951). Vitamin B12 requirement of male weanling pigs. *J. Nutr.* **44**: 371–381.

Rickes, E.L., Brink, N.G., Koniuszy, F.R., Wood, T.R. and Folkers, K. (1948). Vitamin B12, cobalt complex. Science 108:134.

Rosyara, U.R., Maxson-stein, K.L., Glover, K.D., Stein, J.M. and Gonzalez-hernandez, J.L. (2007). "Family-based mapping of FHB resistance QTLs in hexapl,oid wheat", Proceedings of National Fusarium head blight forum.

Roth-Maier, D.A and Kirchgessner, M. (1977). Utersuchungen zum optimalen Pantothensaurebedarf von Mastschweinen. (Studies on the optimal pantothenic acid requirement of market pigs.) Z. Tierphysiol. Tierernahr. Futtermittelkd. **38**: 121–131.

Rothschild, M.F., Zhi-liang, H. and Jiang, Z. (2007). Advances in QTL Mapping in Pigs. *Int. J. Biol Sci* 2007; **3**: 192–197.

Rotruck, J.T., Pope, A.B., Canther, A.L., Swanson, H.E., Hafeman, D.C. and Hoekstra, W.G. (1973). Selenium: Biochemical role as a component of glutathione peroxidase. Science **179**: 588–590.

Russett, J.C., Krider, J.L., Cline, T.R. and Underwood, L.B. (1979b). Choline requirement of young swine. *J. Anim. Sci.* **48**: 1366–1373.

Ruvinsky and M.F. Rothschild. CAB International, Oxon, UK.

Ruvinsky, A and M.F. Rothschild, (1998). Systematics and evolution of the pig, pp. 1–16 in *The Genetics of the Pig*, edited by A.

Rydberg, M.E., Self, H.L., Kowalczyk, T. and Grummer, R.H. (1959). The effect of prepartum intramuscular iron treatment of dams on litter hemoglobin levels. *J. Anim. Sci.* **18**: 415–419.

Sahoo, N.R. (2009). Evaluation and characterization of indigenous pigs. Annual report, NRC on Pog, Assam.

Saison, R. (1967). A new reagebt abtu-ke, in K blood group system of pigs. Vox. Sang. **12**: 286–292.

Schendel, H.E. and Johnson, B.C. (1962). Vitamin K deficiency in the baby pig. *J. Nutr.* **76**: 124–130.

Schunke, B. (1980). Verhaltensanomalien bei Zuchtsauen in Kastenstand. Diss. Med. Cet. Munchem.

Seerley, R.W., Charles, O.W., McCampbell, H.C. and Bertch, S.P. (1976). Efficacy of menadione dimethylpyrimidinol bisulfite as a source of vitamin K in swine diets. *J. Anim. Sci.* **42**: 599–607.

Seerley, R.W. (1984). The use of fat in sow diets. Pp. 333–352 in fats in animal nutrition, J. Wiseman, ed. London; Butterworth.

Seerley, R.W. and Ewan, R.C. (1983). An overview of energy utilization in swine nutrition. *J. Anim. Sci.* **57** (Suppl. 2) 300–314.

Serres, H. (1992). Manual of pig production in the tropics. CAAB International: Wallingford.

Sewell, R.F., Price, D.G. and Thomas, M.C. (1962). Pantothenic acid requirement of the pig as influenced by dietary fat. Fed. Proc. **21**: 468.

Sewell, R.F., Nugara, D., Hill, R.L. and Knapp, W.A. (1964). Vitamin B-6 requirement of early-weaned pigs. *J. Anim. Sci.* **23**: 694–699.

Sharma,B.D. and Sharma, N. (2000). Food packaging of meat, dairy and poultry proudcts. Indian Veterinary Reseach Institute, Izatnagar, India.

Signoret, J.P. (1970). Reproductive behaviour of pigs. *J. reprod. Fertile. Suppl.* **11**: 105–117.

Sihombing, D.T.H., Cromwell, G.L. and Hays, V. W. (1974). Effects of protein source, goitrogens and iodine level on performance and thyroid status of pigs. *J. Anim. Sci.* **39**: 1106–1112.

Slatter, E.E. (1955). Mild iodine deficiency and losses of newborn pigs. *J. Anim. Vet. Med. Assoc.* **127**: 149–152.

Smith, C., Jensen, E., Baker, L.W. and Cox, D.F. (1968). Quantitative studies on blood group and serum protein systems in pigs. *J. anim. Sci.* **27**: 4.

Stahly, T.S. (1984). Use of fats in diets for growing pigs. Pp. 313–331 in fat in animal nutrition, J. Wiseman, ed. London, Butterworth.

Stant, E.C., Martin, T.C. and Kassler, W.V. (1969). Potassium content of the porcine body and carcass at 23, 46, 68 and 91 kilograms live weight. *J. Anim. Sci.* **29**: 547–556.

Steinbach, J. (1972a). Bioclimatic influencesonsexual activity inboars. In: Proceedings of the 7th InternationalCongress on Animal Reproduction and Artificial Insemination, Munich. Summaries, P. 427.

Steinbach, J. (1972b). The oestral cycle of gilts in a tropical environment. In: Proceedings of the 7th International Congress on Animal Reproduction andArtificialinsemination,Munich. Summaries pp. 425–426.

Steiner, H.M. and Vauk, G. (1966). Saugetiere aus dem Beysehir gebiet (vil. Konya, kleinasien). Zoologischer Anzeiger **176**: 97–102.

Stigler, J., Distl, O., Kruff, B. and Kraeusslich, H. (1991). Segregation analysis of hereditary defects in pigs.Zuechtungskunde 63: 294-305.

Stothers, S.C., Schmidt, D.A., Johnston, R.L., Hoefer, J.A. and Luecke, R.W. (1955). The pantothenic acid requirement of the baby pig. *J. Nutr.* **57**: 47–54.

Suomi, K. and Alaviuhkola, T. (1992). Responses to organic and inorganic selenium in the performance and blood selenium content of growing pigs. Ag. Sci. Finland 1: 211.

Suttie, J.W. 1980. The metabolic role of vitamin K. Fed. Proc. **39**: 2730–2735.

Suttie, J.W. and Jackson, C.M. (1977). Prothrombin structure, activation and biosynthesis. Physiol. Rev. **57**: 1–70.

Svajgr, A. J., Peo, E.R.Jr. and Vipperman, P.E.Jr. (1969). Effects of dietary levels of manganese and magnesium on performance of growing-finishing swine raised in confinement and on pasture. *J. Anim. Sci.* **29**: 439–443.

Tagliaro, C.H., Frenco, M.H.L.P and Meincke, W. (1993). Biochemical polymorphisms and genetic relationship among landrace large white and duroc pigs from Sourthern Brazil. Rev. Brazil. Genetic **16**: 671–678.

Tanaka, K., Oishi, T., Kurosawa, Y. and Suzuki, S. (1983). Genetic relationship among several pig populations in East Asia analysed by blood groups and serum protein polymorphisms. Anim. Blood groups biochem. Gent. **14**: 191–200.

Tanake, K. and Masangkey, J.S. (1978). Body conformation, coat colour variations blood group and serum protein polymorphism of native pigs in the Philippines. Rep. Soc. Res. Native livestock **8**: 63–70.

Terrill, S.W., Ammerman, C.B., Walker, D.E., Edwards, R.M., Norton, H.W. and Becker, D.E. (1955). Riboflavin studies with pigs. *J. Anim. Sci.* **14**: 593–603.

Tess, M.H., Dickerson, G.E., Nienabar, J.Y., Yen, J.Y. and Farrell, C.L. (1984). Energy costs of protein and fat deposition in pigs fed and libitum. *J. Anim. Sci.* **58**: 111–122.

Theuer, R.C. and Hoekstra, W.C. (1966). Oxidation of 14C-labeled carbohydrate, fat and amino acid substrates by zinc-deficient rats. *J. Nutr.* **89**: 448–454.

Thomas, H.R. and Kornegay, E.T. (1981). Phosphorus in swine. I. Influence of dietary calcium and phosphorus levels and growth rate on feedlot performance of barrows, gilts and boars. *J. Anim. Sci.* **52**: 1041–1048.

Thornton, K. (1988). Outdoor Pig Production. Farming Press, Ipswich, U.K.

Thurley, D.C. Gilbert, F.R. Done, J.T. (1967). Congenital splayleg of piglets: Myofibrillar hypoplasia. The Vet. Rec. **80**: 302–304.

Tiege, J., Jr. 1977. The generalized Shwartzman reaction induced by a single injection of endotoxin in pigs fed a vitamin E-deficient commercial diet. Acta Vet. Scand. **18**: 140–142.

Tikhonov, V.N. and Troshina, A.I. (1974). Identification of chromosomes and their aberrations in karyotypes of subspecies fo Sus scrofa L. by differential staining. doklady Akademi Nauk, SSR **214**: 932–935.

Tomkins, E.C., Heidenreich, C.J. and Stob, M. (1967). Effect of post-breeding thermalstress on embryonic mortality in swine. *J. Anim. Sci.* **26**: 377–80.

Tucker, H.F. and Salmon, W.D. (1955). Parakeratosis or zinc deficiency disease in the pig. Proc. Soc. Exp. Biol. Med. **88**: 613–616.

Tuncok, S. (1935). Yaban Domuzlari ve Avcilik. Koyhocasi Basimevi, Ankara.

Turan, N. (1984). Turkiye nin av ve yaban hayvanlari (memeliler). Ongun Matbaacilik Sanayi, Ankara.

Ullrey, D.E. (1974). The selenium deficiency problem in animal agriculture. Pp. 275–293 in Trace Element Metabolism in Animals, Volume 2, W. C. Hoekstra, J. W. Suttie, H.E. Ganther, and W. Mertz, eds. Baltimore: University Park Press.

Underwood, E.J. (1977). Trace Elements in Human and Animal Nutrition, Fourth edition. New York: Academic Press.

Van Etten, C., Ellis, N.R. and Madsen, L.L. (1940). Studies on the thiamine requirement of young swine. *J. Nutr.* **20**: 607–624.

Van Kempen, G.J.M., van der Kerk, P. and Crimbergen, A.H.M. (1976). The influence of the phosphorus and calcium content of feeds on growth, feed conversion and slaughter quality and on the chemical, mechanical and histological parameters on the bone tissue of pigs. *Neth. J. Agric. Sci.* **24**: 120–139.

Van Vleet, J.F., Meyer, K.B. and Olander, H.J. (1973). Control of selenium-vitamin E deficiency in growing swine by parenteral administration of selenium-vitamin E preparations to baby pigs or to pregnant sows and their baby pigs. *J. Am. Vet. Med. Assoc.* **163**: 452–456.

Venn, J.A.J., McCance, R.A. and Widdowson, E.M. (1947). Iron metabolism in piglet anemia. J. Comp. Pathol. Ther. **57**: 314–325.

Vipperman, P.E.Jr., Peo, E.R.Jr. and Cunningham, P.J. (1974). Effect of dietary calcium and phosphorus level upon calcium, phosphorus and nitrogen balance in swine. *J. Anim. Sci.* **38**: 758–765.

Van Zeveren, A., Bouquet, Y. Van de Weghe A. and Coppieters, W. (1990). A genetic blood marker study on 4 pig breeds. I. Estimation and comparison of within breed variation. *J. Anim. Breed. Genet.* **107**: 104–112.

Verstegen, M.W.A., Verhagen, and J.M.F. den Hartog, L.A. (1987). Energy requirements of pigs during pregnancy: A review. Livest. Prod. Sci. **16**: 75–89.

Verstegen, M.W.A., Close, W.H., Start, I.B. and Mount, L.R. (1973). The effect of environmental temperature and plane of nutrition on heat loss, energy, relation and deposition of protein and fat in groups of growing pigs. *Brit. J. Nut.* **30**: 21–35.

Vijalainen, P. and Rimaila-Parnanen, E. (1978). A case of chromosomal polymorphism n an inbred Yorkshirepig. Herditas **88**: 276–279.

Vischnevskaya, S.A. and Vsevodolov, E.B. (1986). Individual variability and inheritance of the number and size of the nucleolar organizer regions of the domestic pig chromosomes. Genetika **2**: 690–698.

Vogt, D.W. and Ellersieck, M.R. (1990). Heritability of susceptibility to scrotal herniation in swine. Am *J. Vet Res.* **51**: 1501-3.

Wahlstrom, R.C. and Stolte, S. (1958). The effect of supplemental vitamin D in rations for pigs fed in the absence of direct sunlight. *J. Anim. Sci.* **17**: 699–705.

Wald, G. (1968). Molecular basis of visual excitement science. **162**: 230–239.

Walsh, B. and Lynch, M. (2000). Selection under inbreeding, University of Arizona.

Ward, P.S. (1978). The splayleg syndrome in newborn pigs: A review Part I and II. Vet. Bull. **48**: 279–295, 381–399.

Watanabe, T.Y., Hayashi, J., Kimura, Y., Yasuda, N. and Saitou, T. (1986). Pig mitochondrial DNA: polymorphism, restriction map orientation, and sequence data. Biochem. Genet. **24**: 385–396

Webb, N.G., Penny, R.H.C. and Johnston, A.M. (1984). The effect of a dietary supplement of biotin on pig hoof horn strength and hardness. Vet. Rec. **114**: 185–189.

Weeden, T.L., Nelssen, J.L., Goodband, R.D., Hansen, J.A., Fitzner, G.E., Fiesen, K.G. and Laurin, J.L. (1993b). Effects of porcine somatotropin and dietary phosphorus on growth performance and bone properties of gilts. *J. Anim. Sci.* **71**: 2674–2682.

Weeden, T.L., Nelssen, J.L., Goodband, R.D., Hansen, J.A., Fiesen, K.G. and Richert, B.T. (1993a). The interrelationship of porcine somatotropin administration and dietary phosphorus on growth performance and bone properties in developing gilts. *J. Anim. Sci.* **71**: 2683–2692.

Wehrbein, G.F., Vipperman, P.E.Jr., Peo, E.R.Jr. and Cunningham, P.J. (1970). Diammonium citrate and diammonium phosphate as sources of dietary nitrogen for growing-finishing swine. *J. Anim. Sci.* **31**: 327–332.

Weller, J.I. (2001). QTL analysis in Animals. CABI publishing London.

Wenk, C. Pfirter, H.P. and Bickel, H. (1980). Energetic aspects of feed conversion in growing pigs. Livest. Prod. Sci. **7**:483–495.

Werf, J van der. (2009). http://www-personal.une.edu.au/~jvanderw/07_Interval_mapping_of_QTL.PDF.

Werf, J van der and Marshall, K. (2005).Applications of Gene-Based Technologies for Improving Animal Production and Health in Developing Countries. In Combining Gene-Based Methods and Reproductive Technologies to Enhance Genetic Improvement of Livestock in Developing Countries.(Eds, Harinder P.S. Makkar and Gerrit J. Viljoen). Springer Netherlands.

Whitaker, R. (1988). Endangered Andamans. Environmental services group; WWF India and M.B.A India, Department of environment 50 pp.

Wiese, A.C., Lehrer, W.P.Jr., Moore, P.R., Pahnish, O.F. and Hartwell, W.V. (1951). Pantothenic acid deficiency in baby pigs. *J. Anim. Sci.* **10**: 80–87.

Wilde, R.O. de, and Jourquin, J. (1992). Estimation of digestible phosphorus requirements in growing-finishing pigs by carcass analysis. *J. Anim. Phys. Anim. Nutr.* **68**: 218.

Wilkinson, J.E., Bell, M.C., Bacon, J.A. and Melton, C.C. (1977b). Effects of supplemental selenium on swine. II. Growing-finishing. *J. Anim. Sci.* **44**: 229–233.

Wilkinson, J.E., Bell, M.C., Bacon, J.A. and Masincupp, F.B. (1977a). Effects of supplemental selenium on swine. I. Gestation and lactation. *J. Anim. Sci.* **44**: 224–228.

Williams, I. H., Close, W.H. and Cole, D.J.A. (1985). Strategies for sow nutrition: Predicting the response of pregnant animals to protein and energy intake. Pp. 133–147 in Recent Advances in Animal Nutrition, W. Haresign, and D.J.A. Cole, eds. London: Butterworth.

Wintrobe, M.M., Miller, M.H., Follis, R.H.Jr., Stein, H.J., Mushatt, C and Humphreys, S. (1942). Sensory neuron degeneration in pigs. IV. Protection afforded by calcium pantothenate and pyridoxine. *J. Nutr.* **24**: 345–366.

Wintrobe, M.M., Alcayaga, R., Humphreys, S. and Follis, R.H.Jr. (1943a). Electrocardiographic changes associated with thiamine deficiency in pigs. Bull. Johns Hopkins Hosp. **73**: 169.

Wintrobe, M.M., Follis, R.H.Jr., Alcayaga, R., Paulson, M. and Humphreys, S. (1943b). Pantothenic acid deficiency in swine with particular reference to the effects on growth and on the alimentary tract. Bull. Johns Hopkins Hosp. **73**: 313.

Wintrobe, M.M., Buschke, W., Follis, R.H. Jr. and Humphreys, S. (1944). Riboflavin deficiency in swine with special reference to the occurrence of cataracts. Bull. Johns Hopkins Hosp. 75: 102–110.

Wood-Gush, D.G.M. and Stolba, A. (1982). Etology and welfare of farm livestock. Abstr. 20th Congr. Int. Assoc. Appl. Psychol. Edinburgh, July 1982, pp-76.

Wuryastuti, H., Stowe, H.D., Bull, R.W and Miller, E.R. (1993). Effects of vitamin E and selenium on immune responses of peripheral blood, colostrum, and milk leukocytes of sows. *J. Anim. Sci.* **71**: 2464–2472.

Yamamoto, F. and Yamamoto, M. (2001). Molecular genetic basis of porcine histo-blood group AO system. **97**: 3308–3310.

Yen, J.T., Jensen, A.H. and Baker, D.H. (1976). Assessment of the concentration of biologically available vitamin B6, in corn and soybean meal. *J. Anim. Sci.* **42**: 866–870.

Yen, J.T and Pond, W.G. (1981). Effect of dietary vitamin C addition on performance, plasma vitamin C and hematic iron status in weanling pigs. *J. Anim. Sci.* **53**: 1292–1296.

Young, L.G., Lun, A., Pos, J., Forshaw, R.P and Edmeades, D. (1975). Vitamin E stability in corn and mixed feed. *J. Anim. Sci.* **40**: 495–499.

Young, L.G., Lumsden, J.H., Lun, A. ,Claxton, J and Edmeades, D.E. (1976). Influence of dietary levels of vitamin E and selenium on tissue and blood parameters in pigs. *Can. J. Comp. Med.* **40**: 92–97.

Young, L.G., Miller, R.B., Edmeades, D.E., Lun, A., Smith, G.C. and King, G.J. (1977). Selenium and vitamin E supplementation of high-moisture corn diets for swine reproduction. *J. Anim. Sci.* **45**: 1051–1060.

Young, L.G., Miller, R.B., Edmeades, D.E., Lun, A., Smith, G.C. and King, G.J. (1978). Influence of method of corn storage and vitamin E and selenium supplementation on pig survival and reproduction. *J. Animssss. Sci.* **47**: 639–647.

Young, S.S.Y. (1961). A further examination of the relative efficiency of three methods of selection for genetic gains under less restricted conditions. Genet. Res. **2**: 106.

Zhiliang, H. and James, M. Reecy. (2007). Animal QTLdb: Beyond a Repository - A Public Platform for QTL Comparisons and Integration with Diverse Types of Structural Genomic Information. Mammalian Genome, Volume **18**: 1–4 (2007).

Zhiliang H, Eric Ryan Fritz and James, M. Reecy (2007). AnimalQTLdb: a livestock QTL database tool set for positional QTL information mining and beyond. Nucleic Acids Research, 2007, 35 (Database issue):D604–D609; doi: 10.1093/nar/gk l946.

Zhiliang, H., Svetlana Dracheva, Wonhee Jang, Donna Maglott, John Bastiaansen, Max F., Rothschild and James, M. Reecy (2005). A QTL resource and comparison tool for pigs: PigQTLDB. Mammalian Genome. Volume **16** (10): 792–800.

Zimmerman, D.R. (1980). Iron in swine nutrition. In National Feed Ingredient Association Literature Review on Iron in Animal and Poultry Nutrition. Des Moines, Iowa: National Feed Ingredient Association.

Zimmerman, D.R. (1986). Role of subtherapeutic antimicrobials in pig production. *J. Anim. Sci.* **62** (Suppl. 3): 6.

Zivkovic, S., Jovanociv, V., Isakovic, I. and Milosevic. M. (1971). Chromosome complement of the European wild pig (Sus scrofa L.). Experientia **27**: 224–226.

Zhiliang H, [illegible] (2009) [illegible] QTL [illegible] database for [illegible] QTL information [illegible] doi:10.1016/[illegible]

Zhang H, [illegible] (200[illegible]) [illegible] QTL [illegible] and comparative [illegible] Animal Science [illegible] 16(10): 792–808.

Zimmerman, D.R. (1980) Iron in swine [illegible] In National Feed Ingredient Association, [illegible] in Animal and Poultry Nutrition. Des Moines, Iowa: National Feed Ingredient Association.

[illegible] (1982) [illegible] Suppl. [illegible]

Zukowski S., [illegible] and [illegible] M. (1977) [illegible] composition of the [illegible] 27: [illegible]

INDEX

A

B

C

D

E

F

G

H

I

J

K

L

M

N

O

P

Q

R

S

T

U

V

W

X

Y

Z